WILDLIFE TOXICITY ASSESSMENTS FOR CHEMICALS OF MILITARY CONCERN

WILDLIFE TOXICITY ASSESSMENTS FOR CHEMICALS OF MILITARY CONCERN

Edited by

MARC A. WILLIAMS

GUNDA REDDY

MICHAEL J. QUINN Jr.

MARK S. JOHNSON

ELSEVIER

AMSTERDAM • BOSTON • HEIDELBERG • LONDON • NEW YORK • OXFORD
PARIS • SAN DIEGO • SAN FRANCISCO • SYDNEY • TOKYO

Elsevier
Radarweg 29, PO Box 211, 1000 AE Amsterdam, Netherlands
225 Wyman Street, Waltham, MA 02451, USA
The Boulevard, Langford Lane, Kidlington, Oxford OX5 1GB, UK

First edition 2015

Library of Congress Cataloging-in-Publication Data
A catalog record for this book is available from the Library of Congress

British Library Cataloguing in Publication Data
A catalogue record for this book is available from the British Library

ISBN: 978-0-12-800020-5

For information on all Elsevier publications
visit our web site at **store.elsevier.com**

This book has been manufactured using Print On Demand technology. Each copy is produced to order and is limited to black ink. The online version of this book will show color figures where appropriate.

DISCLAIMER

This book, and all book chapters herein, were prepared for educational use. The views and assertions expressed in this volume are the private views of authors and do not necessarily reflect, nor should they be construed as reflecting the views, and official policy of the Department of Defense, the Department of the Army, the U.S. Army Medical Department or the U.S. Federal Government.

Further, Use of trademarked name(s) does not imply endorsement by the U.S. Army but is intended only to assist in identification of a specific product.

CONTENTS

EDITOR BIOGRAPHIES

Mark S. Johnson, PhD, DABT, *is the Director of the Toxicology Portfolio, U.S. Army Institute of Public Health, Aberdeen Proving Ground, MD.* Dr. Johnson has worked extensively in the evaluation of the toxicity of military unique compounds and development and evaluation of sensitive indicators of stress (e.g., immunotoxicity) for use in field applications and toxicity testing. He has extensive experience in risk assessment and has developed and tested new methods in improving exposure/effects relationships. Current work involves the development of a phased approach to the gathering toxicity data for new compounds under development (ASTM E-2552). Other work has included the development of a process to derive toxicity reference values for terrestrial wildlife. He has authored over 50 peer-reviewed publications, book chapters, and technical reports. He has been a member of SETAC since 1997 and a member of SOT since 2009. Within SETAC, Dr. Johnson is a Steering Group Member of the Wildlife Toxicology World Advisory Group, Vice Chair of Ecological Risk Assessment World Advisory Group, and a member of the Science Committee for SETAC North America. He is also Chair of the Tri-Service Toxicology Consortium (TSTC); Steering Committee Chair of the Joint Army-Navy-NASA-Air Force (JANNAF) Propulsion Committee, Subcommittee on Safety and Environmental Protection; and the past Chair of the Terrestrial Toxicity Subcommittee of the Biological Fate and Effects Committee of the American Society for Testing and Materials (ASTM). He is the current Vice President of the American Board of Toxicology (ABT).

Michael J. Quinn, Jr., PhD, *is the Program Manager of the Health Effects Research Program – Toxicology Portfolio, U.S. Army Institute of Public Health, Aberdeen Proving Ground, MD*. His early education at Fairleigh Dickinson University in Madison, NJ focused on Marine Biology, where he studied ichthyological and marine invertebrate behavior and physiology. Upon graduation, he completed an internship at the U.S. Geological Survey where he developed a Master's thesis at Shippensburg University, PA, on the effects of polychlorinated biphenyls on the development of falcon plumage. Dr. Quinn received a PhD in Animal Science at the University of Maryland, College Park, MD where he studied the effects of estrogen- and androgen-active endocrine disruptors on the functional development of the reproductive and immune systems in Japanese quail. In addition to contributing to many toxicology studies at the PHC with explosives and propellants in a variety of taxa (e.g., mammals, birds, reptiles, and amphibians), Dr. Quinn's primary passion is method development. His most recent research has centered on developing an avian two-generation toxicity test for the U.S. Environmental Protection Agency's Endocrine Disruptor Screening Program's Tier 2 battery of tests.

Gunda Reddy, PhD, DABT, *is a Senior Toxicologist with the Toxicology Portfolio, U.S. Army Institute of Public Health, Aberdeen Proving Ground, MD*. He received his PhD from Osmania University, Hyderabad, India. After postdoctoral work, Dr. Reddy joined the U.S. Army Biomedical Research and Development Laboratory at Fort Detrick, MD as a Pharmacologist, and then the U.S. Army's Aberdeen Proving Ground, where he focuses on toxicity of munitions compounds and chemicals, toxic metabolites, and degradation products of chemical warfare agents using *in vivo* and *in vitro* model systems. He conducts in-house research, and manages extramural research projects with universities, government agencies, and private institutions. He reviews toxicity data and performs risk assessments of Army-related chemicals. Dr. Reddy has published over 100 peer-reviewed publications and book chapters and has presented his work at national and international meetings. Dr. Reddy serves as an Associate Editor of *Drug and Chemical Toxicology Journal* and as a regular editor with the editorial boards of *International Journal of Toxicology*, *Journal of Toxicology and Environmental Health*, and *Toxicology Mechanisms and Methods*. He is recognized by the Army as a Master Consultant in his specialized field. Dr. Reddy is a member of the Society of Toxicology, the American College of Toxicology, the Association of Government Toxicologists, the Association of Scientists of Indian Origin in America, and the Indian Science Congress.

Marc A. Williams, PhD, FAAAAI, *is a Biologist with the Toxicology Portfolio – Health Effects Research Program, U.S. Army Institute of Public Health, Aberdeen Proving Ground, MD.* Dr. Williams was previously a Research Biologist with the U.S. Environmental Protection Agency (U.S. EPA), with research interests in allergic asthma, and exposure to respirable ambient airborne particulate matter, and nanoparticulate pollutants that he showed were both pro-inflammatory and pro-oxidative instigators of allergic immunity. Prior to joining the U.S. EPA, Dr. Williams was an Assistant Professor of Medicine and Environmental Medicine at the University of Rochester School of Medicine and Dentistry, NY and an Instructor of Medicine at the Johns Hopkins University School of Medicine where he led research programs in human immunology, immunotoxicology, and disease pathways. Dr. Williams holds a PhD in Hematological Oncology and Immunobiology from Queen Mary – University of London, UK, where his group was one of the first to combine successfully adoptive dendritic and T cell-mediated immunotherapy in the treatment of a hematological malignancy. Dr. Williams did postdoctoral training in infectious disease and HIV-1 immunology at the University of California San Diego, La Jolla, CA, and in autoimmunity and immunological gene therapy at Johns Hopkins University, Baltimore, MD. He is a Fellow of the American Academy of Allergy, Asthma, and Immunology, and a nominated member of the Faculty of 1000 (Immunology Section) post-publication review panel with emphasis on allergy and immunotoxicology, and the Sigma Xi scientific honor society. Dr. Williams has authored over 100 peer-reviewed articles including books, book chapters, and technical reports, and is an associate editor of the *Journal of Innate Immunity*, and *Stem Cells and Development*. He is also a regular editor of *Toxicology Letters, Biomarker Insights*, and several others. His primary interests are in chemical risk assessment analysis, inhalation toxicology of nanoparticles, health outcomes of complex chemical mixture exposures, adverse outcome pathways, and immunotoxicological mechanisms of disease.

AUTHOR BIOGRAPHIES

Valerie H. Adams, MS, PhD, *is a Biologist in the Toxicology Portfolio – Health Effects Research Program, U.S. Army Institute of Public Health, Aberdeen Proving Ground, MD.* As an undergraduate at California State University Fresno she majored in Biology, minored in Chemistry, and worked as an intern at an agricultural research facility. During her master's program at University of Guam, she studied a predation-induced plant defense response, participated in coral reef field studies, and completed coursework in aquatic toxicology and environmental contaminants. Dr. Adams received a PhD in Cellular and Structural Biology from the University of Texas Health Sciences Center, San Antonio, TX where she studied the immune response to chlamydial infection using a mouse model. Dr. Adams completed postdoctoral research at Colorado State University in the field of chromatin biology and protein biochemistry. A prior recipient of NIH predoctoral and postdoctoral training grants, Dr. Adams has published over 30 peer-reviewed articles, abstracts, and technical reports. She is a member of the Society of Toxicology and has previously served as chair of the Triservice Environmental Risk Assessment Workgroup. She volunteers annually for the U.S. Army e-Cybermission as a science project judge and was selected as a 2014 National Junior Science and Humanities Symposium Judge. In her current position, Dr. Adams is the principle investigator for *in vivo* and *in vitro* toxicity studies and explores novel approaches for *in vitro* toxicity assessment.

Christine Ann Arenal, MS, *is a Biologist and Ecological Risk Assessor with CH2M HILL, Inc.* She earned her BA in Biology at Occidental College in Los Angeles, CA, where she was an instructor for the Classroom at Sea program aboard the college's research vessel R/V Vantuna. Following this teaching program, Ms. Arenal spent time tracking and studying grizzly bears with the Interagency Grizzly Bear Study Team, in Billings, MT. Ms. Arenal received an MS in Zoology at Southern Illinois University in Carbondale, where she conducted field studies to evaluate the reproductive and behavioral effects of polychlorinated biphenyls and metals on avian species. She also studied liver enzyme activity in nesting birds exposed to metals. As a postgraduate fellow at the Oak Ridge National Laboratory (ORNL), TN, Ms. Arenal developed models of wildlife risk assessment, and methods for interspecies extrapolation of avian and mammalian toxicity data. She also performed risk analyses for the endangered gray bat, and completed a regulatory analysis of the Marine Mammal Protection Act of 1972 as it applies to underwater explosions in military sea range activities. Immediately following this fellowship in 2000, she joined CH2M HILL and currently specializes in ecological risk assessment, wildlife toxicology, ecology, and wildlife biology. As a project scientist, Ms. Arenal has continued to develop methods and tools for ecological risk assessment, including avian and mammalian toxicity reference values for a wide range of chemicals such as petroleum hydrocarbons and military relevant chemicals. She also developed and parameterized an exposure model for incorporation into the Army Risk Assessment Modeling System (ARAMS). Ms. Arenal has given over 20 oral or poster presentations at national and regional professional conferences, authored or coauthored 10 peer-reviewed publications, and has been a member of the Society of Environmental Toxicology and Chemistry (SETAC) since 1995. She currently lives and works in Sacramento, CA, with her husband and four children.

Desmond I. Bannon, PhD, DABT, *is a Toxicologist in the Toxicology Portfolio – Health Effects Research Program, U.S. Army Institute of Public Health, Aberdeen Proving Ground, MD.* Dr. Bannon previously spent 10 years at the Kennedy Krieger Institute Lead Poisoning Program in Baltimore, MD. In addition to his PhD from the Department of Environmental Health Sciences at Johns Hopkins University School of Public Health (2002), Dr. Bannon has been a Diplomate of the American Board of Toxicology since 2006. With a focus on toxicology and human health, his interests include metal toxicology (tungsten alloys and lead) bioinformatics/genomics, mechanisms of toxicology, and toxicology studies in support of risk assessment. He recently led a U.S. Army Public Health Command effort to propose new blood lead guidelines for lead exposure and medical management in the Department of Defense. His range of journal publications includes, among others, Biomarkers, Clinical Chemistry, Chemical Research in Toxicology, Environmental Health Perspectives, Environmental Science and Toxicology, Toxicology and Applied Pharmacology, and Nature Communications, indicating broad interests and strong interdisciplinary collaborations. Dr. Bannon currently serves as the Contracting Officer's Representative for an ongoing Department of Defense contract with the U.S. National Academy of Sciences in Washington, DC.

Adam T. Deck, BS, *is an Environmental Scientist and Health Risk Assessor with the Environmental in the Health Risk Management Portfolio – Health Risk Assessment Program, U.S. Army Institute for Public Health (AIPH).* Throughout his tenure at the AIPH, his work has focused on ecological and human health risk assessments in both garrison and deployed settings. Mr. Deck currently serves on the AIPH Institutional Animal Care and Use Committee and the administrative team for the chairperson of the Joint Environmental Surveillance Working Group. Mr. Deck's primary passion is conducting field studies and serving as the primary investigator and team lead for the Comprehensive Reproductive Assessment for Wild Rodents at Contaminated Army Properties. Mr. Deck holds a BS in Biology from the University of Mary Washington with a focus in ecology and zoology.

William S. Eck, PhD, *is a Biologist (Chemist/Toxicologist) in the Toxicology Portfolio – Health Effects Research Program, U.S. Army Institute of Public Health, Aberdeen Proving Ground, MD.* His principle efforts at the U.S. Army Institute of Public Health have been directed toward support of the U.S. Army's Ordnance Environmental Program as Team Leader for preparation of Toxicology Assessments. Dr. Eck is also responsible for QSAR modeling of newly developed explosive, propellant, and pyrotechnic compounds, and is a Subject Matter Expert on perchlorates. Dr. Eck has also served as a Study Director and regularly participates in the general toxicology effort within the Directorate. Prior to joining the Directorate of Toxicology in 2008, Dr. Eck was a university professor, Senior Intelligence Analyst with the Defense Intelligence Agency, and a serving Army officer in the Medical Service and Chemical Corps. Dr. Eck is the primary or coauthor of numerous publications relating to compounds of military interest. He received his PhD in Chemistry (Biochemistry) from the University of Maryland, College Park, MD and a BS in Chemistry, from the College of William and Mary, Williamsburg, VA.

Rachel M. Hebert, BS, *is a Biological Science Technician with the Oak Ridge Institute for Science and Education, positioned with the Toxicology Portfolio – Health Effects Research Program, U.S. Army Institute of Public Health, Aberdeen Proving Ground, MD.* Ms. Hebert received a BS in Biology, with a concentration in Organismal Biology and Ecology and a minor in Geographic Information Systems from Towson University in Towson, MD. While at Towson University, she studied the reproductive biology and behavior of Mountain Bluebirds in the Bighorn Mountains of Wyoming. Ms. Hebert also conducted the first systematic mammal species survey at Eden Mill Park in Pylesville, MD as her departmental honors thesis and spent two years studying plant–mammal interactions in wetlands using exclosures.

Allison M. Jackovitz, BS, *is a Biologist with the Toxicology Portfolio – Toxicity Evaluation Program, U.S. Army Institute of Public Health, Aberdeen Proving Ground, MD.* Upon graduating from Pennsylvania State University, Ms. Jackovitz arrived at the U.S. Army U.S. Army Institute of Public Health to work under Dr. Michael Quinn, Jr., in the development of an avian two-generation toxicity test for the U.S. Environmental Protection Agency's Endocrine Disruptor Screening Program's Tier 2 battery of tests. Ms. Jackovitz attends the University of Maryland School of Medicine and is working on a PhD in Toxicology. Her most recent research focuses on the endocrine disrupting potential of insensitive munitions.

Glenn J. Leach, PhD, DABT, *is a former Senior Toxicologist and Program Manager in the Toxicology Portfolio, U.S. Army Institute of Public Health, Aberdeen Proving Ground, MD.* He retired in 2011 and returned to work part-time as a consultant in the same organization. Dr. Leach received a PhD in Biological Sciences from the University of Delaware and had postdoctoral training at the Louisiana State University Medical Center. He became a Diplomate of the American Board of Toxicology in 1982 and is a member of the Society of Toxicology. He served for over 10 years on the Acute Exposure Guidelines Committee. His interests include evaluating the toxicity of munitions compounds and risk assessment. Dr. Leach has authored or coauthored more than 50 peer-reviewed papers, book chapters, and technical reports.

Sang Ho Lee, PhD, DVM, *is an Officer in the U.S. Army Veterinary Corp., U.S. Army Institute of Public Health, Aberdeen Proving Ground, MD, currently serving as the Executive Officer in the Toxicology portfolio at the U.S. Army Public Health Command.* His work contributes to the collaborative data sharing in the field of toxicology, cooperation between toxicology and veterinary medicine, and issuing toxicity clearances for chemicals and platforms of interest to the U.S. Army. Dr. Lee has previously worked in the field of medical diagnostics of chemical warfare agents while serving as the Branch Chief of the Medical Diagnostic and Chemical Branch at the U.S. Army Medical Research Institute of Chemical Defense. He received his PhD in Chemistry at the University of Maryland. He also has Doctorate degree in Veterinary Medicine from the Konkuk University, Korea.

Emily May Lent, PhD, *is a Toxicologist with the Toxicology Portfolio – Toxicity Evaluation Program, U.S. Army Institute of Public Health, Aberdeen Proving Ground, MD.* Previously, Dr. Lent was a Research Assistant at the New England Aquarium Edgerton Research Lab where her research interests included effects of habitat degradation on species assemblages, by-catch survivability, and effects of nutrient loading on eelgrass beds. Dr. Lent received a Master's degree in Natural Resources – Wildlife Management from the University of New Hampshire where she studied the effects of multiple stressors on phenotypic plasticity in larval anurans. Dr. Lent holds a PhD in Natural Resources and Environmental Studies from the University of New Hampshire. As a U.S. Environmental Protection Agency (EPA) Science to Achieve Results (STAR) Fellow, Dr. Lent studied the effects of endocrine-disrupting contaminants on gonadal histology and reproductive steroidogenesis in anurans. Prior to joining the U.S. Army Institute of Public Health, Dr. Lent was a National Research Council (NRC) Postdoctoral Research Associate with the U.S. Army Edgewood Chemical and Biological Center where she conducted investigations on the effects of nerve agents on gene expression patterns. Dr. Lent has authored over 20 peer-reviewed publications and technical reports. She received the Federal Executive Board Award for Excellence in Federal Career in 2007. Current work includes investigations on the reproductive toxicity and genotoxicity of militarily relevant compounds. Her primary interests are in endocrine disruption, reproductive toxicology, and genotoxicology.

Matthew J. McAtee, BA, *is the Section Chief, Strategic Risk Assessment in the Environmental Health Risk Assessment Program, U.S. Army Institute of Public Health, Aberdeen Proving Ground, MD.* He provides 19 years of subject matter expertise in environmental health risk assessment. He is responsible for directing a staff of risk assessment subject matter experts from several science and engineering disciplines. Their mission is to support Army Public Health by identifying and executing initiatives to develop methods, guidance, and information to improve the value of Army health risk assessments, with special emphasis on chemical and microbiological hazards. Mr. McAtee and his section are currently focused on designing risk assessment methods, information databases, and military exposure guidelines (MEGs) for chemical and biological hazards. He is personally interested in expanding collaboration opportunities across risk assessment disciplines, supporting risk managers' decision-making processes, training junior risk analysts, facilitating proper risk assessment design and risk characterization, and integrating risk assessment concepts throughout the Army Public Health enterprise and its collaboration with chemical, biological, radiologic, and nuclear defense (CBRND) planning and response operations. He has authored numerous technical reports, technical guidance documents, and peer-reviewed publications. His risk assessment experience also includes the development of health risk assessments for hazardous waste sites, military training ranges, deployment locations, and combustion activities, including facilities designed to destroy the US chemical-warfare agent stockpile. He developed the ecological risk assessment methodology for Army combustion facility permitting decisions before guidance was available from the Environmental Protection Agency. He codeveloped the Army's first process to derive toxicity reference values for terrestrial wildlife. Mr. McAtee holds a BA in Biology from the University of Colorado, Boulder and a Risk Sciences and Public Policy Certificate and training in Epidemiology at the Johns Hopkins Bloomberg School of Public Health.

Wilfred C. McCain, PhD, *is a Toxicologist with the Toxicology Portfolio – Health Effects Research Program, U.S. Army Institute of Public Health, Aberdeen Proving Ground, MD, and works on a variety of toxicological issues affecting global military operations.* He is involved in all phases of toxicology including risk characterization, risk assessment, and risk management. He represents the Army at public forums and meetings with various regulatory agencies. Dr. McCain's primary function is to conduct toxicological research for regulatory and safety purposes under GLP guidelines. Dr. McCain has a strong background in general and neurotoxicology with extensive experience in physiological, pharmacological, and statistical evaluation of cardiovascular, neuromuscular, and pulmonary function both *in vivo* and *in vitro*. He has more than 30 years of technical experience in academic, industrial, and government laboratories including supervisory and management positions.

Concurrent with his present position, Dr. McCain is adjunct associate professor of veterinary bioscience at the VA MD Regional College of Veterinary Medicine, VA Tech, where he has both teaching and advisory duties. He also teaches toxicology in the Public Health graduate program at the Uniformed Services University of the Health Sciences.

Emily N. Reinke, PhD, *is a Biologist with the Toxicology Portfolio – Health Effects Research Program, U.S. Army Institute of Public Health, Aberdeen Proving Ground, MD, where she evaluates the toxicity of Army-relevant compounds.* She holds a PhD in Pathology from the University of Illinois at Chicago and an MS in Animal Sciences from Washington State University. During her PhD studies, Dr. Reinke received the Science, Mathematics, and Research for Transformation (SMART) Scholarship from the Department of Defense. She is also a postdoctoral member of the Society of Toxicology. Dr. Reinke has authored or coauthored more than six peer-reviewed journal articles or book chapters, and several toxicological technical reports. Her primary interests are in the development and adoption of alternative methods of assessing toxicology and in the mechanisms of carcinogenesis and cancer treatment efficacy.

Stephen W. Rice, BS, *is an Intern with the Toxicology Portfolio – Health Effects Research Program, U.S. Army Institute of Public Health, Aberdeen Proving Ground, MD.* Mr. Rice graduated from Salisbury University in 2014 with a BS in Biology and a minor in Chemistry. During his time at Salisbury University and the U.S. Army Institute of Public Health, he has dedicated much of his time preparing for a career in medicine by volunteering and shadowing physicians at Peninsula Regional Medical Center and Kirk U.S. Army Health Clinic respectively. Working with the U.S. Army Institute of Public Health, he focuses on analyzing dose-response data and the derivation of threshold reference values using the benchmark dose approach.

Christopher Salice, PhD, *is the current Director of the Environmental Science and Studies Program at Towson University, MD.* His research interests have previously focused on understanding the ecological and evolutionary consequences of anthropogenic and natural stressors. Examples of recent research projects include studies to better understand long-term exposures to stressors (multigeneration); transgenerational effects of stress, ecology, and pesticide resistance in mosquitoes; and the ecotoxicology and risk of perfluorinated compounds to aquatic species. Dr. Salice received a BS in

Environmental Science with an emphasis in ecology from Drexel University, and a PhD in environmental toxicology from the University of Maryland, Baltimore. He conducted his doctoral research at the University of Maryland's Chesapeake Biological Laboratory. Previously, Dr. Salice joined the U.S. Army Center for Health Promotion and Preventive Medicine where he conducted laboratory studies and developed tools for ecological risk assessment. Dr. Salice was then appointed to conduct research with the U.S. Environmental Protection Agency as an ecological risk assessor. From 2008 to 2014, Dr. Salice was an Assistant Professor at Texas Tech University before being appointed as the Director of the Environmental Science and Studies Program at Towson University.

Bradley E. Sample, PhD, *is an Ecotoxicologist/Ecological Risk Assessor with Ecological Risk, Inc., with over 20 years of experience designing and leading large-scale risk assessments for both private and public-sector clients.* His work includes evaluation of exposure and effects to wildlife, aquatic, sediment, and soil biota from radionuclides, metals, munitions, chlorinated organics, and petroleum compounds. He is the author of more than 100 peer-reviewed publications and presentations. Dr. Sample served on the steering committee and developed the wildlife exposure model for the USEPA's Ecological Soil Screening Levels (Eco-SSLs), and developed the wildlife modeling component for the U.S. Army's Adaptive Risk Assessment Modeling System (ARAMS). He has twice served on peer-review committees for the ecological risk assessment and bioaccumulation modeling components of the USEPA's Hazardous Waste Rule. In 2004, he served on an independent peer-review committee evaluating the EPA's ecological risk assessment for the General Electric Housatonic River site. Also in 2004, he served as a peer-review committee member for the development of USEPA's national metals risk assessment framework. In 2008, Dr. Sample served on the USEPA Science Advisory Panel reviewing the Office of Pesticide Programs Preliminary Notice of Intent to Cancel the pesticide Carbofuran. Dr. Sample is coauthor of a book on ecological risk assessment at contaminated sites, and has served as Editor for Hazard/Risk Assessment for the Society of Environmental Toxicology and Chemistry (SETAC) since 2006.

Joseph P. Sullivan, PhD, *is a Certified Wildlife Biologist and owner of Ardea Consulting, Woodland, CA. Dr. Sullivan established Ardea Consulting in 1997 in Morrisville, PA.* Before that, Dr. Sullivan conducted field studies to evaluate the effects of pesticides on wildlife, primarily birds, and worked for American Cyanamid Company as an avian toxicologist. Dr. Sullivan holds a PhD in Fisheries and Wildlife Sciences from Virginia Tech University and an MS in Biology/Ecology from Utah State University. He is a nominated member of Beta Beta Beta Biology Honor Society, Gamma Sigma Delta Agricultural Honor Society, and Sigma Xi Scientific Honor Society. He has authored or coauthored 14 peer-reviewed articles or book chapters and 20 technical reports along with many company confidential reports. His primary interests are in ecological risk assessment, mostly of organics such as pesticides and wildlife toxicology. He also conducts monitoring of special-status wildlife on contaminated sites.

Lawrence R. Williams, PhD, *is a Biologist with the Toxicology Portfolio – Health Effects Research Program, U.S. Army Institute of Public Health, Aberdeen Proving Ground, MD.* Dr. Williams did his undergraduate and graduate work in chemistry and anatomy, respectively, at the University of Iowa and served 2 years as a Lieutenant in the Army Chemical Corp. Most of Dr. Williams's career was spent in the pharmaceutical industry focused on the discovery of therapeutics for neurotraumatic injuries, ischemic stroke, and neurodegenerative diseases. He has over 90 peer-reviewed articles, book chapters, and technical reports, and holds a patent for the use of Glial Cell-line Derived Neurotrophic Factor for the treatment of Alzheimer's disease. After brief periods of employment with Battelle at the MRICD in Edgewood, MD, and the NIAAA in Rockville, MD, Dr. Williams joined the USAIPH in 2008. There he has focused on the subacute toxicity of novel energetics, and discovered the mechanism of action of seizures induced by RDX overexposure.

David Ziolkowski, MSc, *is a Wildlife Biologist at the U.S. Geological Survey's Patuxent Wildlife Research Center, Laurel, MD.* He has accrued over 20 years of professional experience in the United States and abroad, ranging from studies of avian hormones and behavior, toxicology, taxonomy and systematics, and biogeography. In his current position, he serves as the program biologist for the multinational North American Breeding Bird Survey, where the focus of his recent research interests is the analysis and modeling of long-term population data to assess the impacts of anthropogenic and climatic change on the health of avian populations. Prior to joining the USGS, Mr. Ziolkowski developed toxicology assessments and conducted terrestrial exposure and bioaccumulation studies using avian models as a postgraduate fellow in the Health Effects Research Program, U.S. Army Center for Health Promotion and Preventive Medicine, Aberdeen Proving Ground, MD. He received an MSc degree in Biology (Evolutionary Ecology) from the University of Missouri, Saint Louis, MO and a BS degree in Biology, from Towson University, MD.

ACKNOWLEDGEMENTS

We thank the Army Environmental Command, Installation Restoration Program; Strategic Environmental Research Program (SERDP); and, the Army Environmental Quality Program for supporting the research, data and review of this effort.

PART I

Introduction

CHAPTER 1

Characterizing Potential for Toxicity: Estimating Risks to Wildlife

Mark S. Johnson, Matthew J. McAtee

Contents

INTRODUCTION

Adverse effects to wildlife from unintended exposure to substances have been described in various sources and have provided the basis for many regulations that are intended to protect valued species. Additionally, many animal species, because of their greater exposure potential to substances in the environment, have been proposed as sentinels of adverse events as early warning signals for humans that may be exposed [1]. In the United States, there are regulations that protect migratory birds, threatened and endangered species, and eagles from "harm," which may include adverse effects from chemical exposure in the environment. The Resource Conservation and Recovery Act and the Comprehensive Environmental Response, Compensation, and Liability Act (i.e., "Superfund") were established, and require addressing adverse effects to the environment as well as to humans. Additionally, the Federal Insecticide, Fungicide, and Rodenticide Act (FIFRA) also requires an environmental evaluation of risk for applications of new pesticides to determine whether nontarget species could be at risk from exposure. These regulatory requirements have provided the initiative to develop the means to assess health risks to valued wildlife species. An important part of wildlife risk assessment for any given substance is the derivation of a toxicity benchmark.

Wildlife Toxicity Assessments for Chemicals of Military Concern
http://dx.doi.org/10.1016/B978-0-12-800020-5.00001-6

The military utilizes and controls relatively large tracts of land, some of which are hundreds of square kilometers in size. Often land and personnel uses are highly restricted and controlled, resulting in highly valuable habitats for many wildlife species. In fact, the Nature Conservancy and the U.S. Department of Defense have been partnering for more than 30 years in managing the natural resources, long range sustainability of lands, and the more than 330 threatened or endangered species on military installations while simultaneously ensuring the success of the military mission [2]. The program known as the Readiness and Environmental Protection Initiative (REPI) has resulted in partnerships in over 24 states, protecting sensitive and valuable habitats while protecting the military's ability to test and train at some of the nation's most important military installations. Although these efforts are habitat-focused, clearly the value of these areas and the species that use them are of national interest.

Some activities that have occurred on military lands since before World War I have resulted in releases of substances in the environment that are, or may be, harmful to wildlife. On military properties, many of these substances represent those that are relatively military-unique, such as explosives, propellants, and pyrotechnics (collectively known as energetics). Their release often was a result of [1] earlier production and use practices before regulation that modified such practices (e.g., ammunition production during World War II), or [2] intended use where either incomplete combustion has occurred (e.g., detonation ranges), or where otherwise allowed by law (e.g., military facilities with RCRA permits). Other releases have occurred through training or use, or during war through theater operations. Characterization of risks at these locations where releases have occurred or to where substances have been transported is important to natural resource personnel, regulators, and society as in many cases, these substances are unique and data are scarce. A subset of this characterization includes those to valued wildlife species. In some cases, these releases were not of military unique substances, but of substances of military relevance. Examples include specific metals, pesticides, or organic compounds. Textbooks have been written to help provide central sources for this and other information [3,4].

A primary challenge in characterizing risk to wildlife from environmental exposure requires the interpretation and translation of toxicity information into useful risk assessment tools. Federal entities and jurisdictions collect information and do this for human health risk assessment. However, to date, there is no single updated source for toxicity values for wildlife.

HISTORY OF WILDLIFE TOXICITY REFERENCE VALUES

Most evaluations of risk from exposure to man-made substances (xenobiotics) require the determination of the threshold for toxicity—to describe the point where exposure exceeds the organisms' capability to adapt and illicit no adverse health effects. This dose-response function is the fundamental principal of toxicology—that the dose makes the poison. Ascertaining the exposure point at which adverse effects occur is the keystone in determining level of exposure that could be used in risk assessment applications. In principal, this seems a straightforward task; however, there are many confounding aspects of this exercise and determining this threshold is often a complex problem.

Estimating safe levels of exposure at or near the suspected threshold of toxicity is and has been an active area for human health risk assessment. Various methods are used. Examples include procedural "cook book" methods where points of departures (PODs) are determined from the dose-response relationship from a key or seminal study that are then modified with uncertainty factors that result in a toxicity benchmark. Some use a weight-of-evidence approach where professional judgment is used and described in determining the toxicity benchmark. Each endeavor to characterize the threshold for adverse responses. Examples include the U.S. Environmental Protection Agency (U.S. EPA) Integrated Risk Information System (IRIS) Reference Dose and the American Conference for Governmental Industrial Hygienist (ACGIH®) Threshold Limit Values, respectively. Toxicity benchmarks developed for wildlife and described herein are referred to as Toxicity Reference Values (TRVs).

Toxicity benchmark derivation for wildlife was initially patterned after human health risk assessment methods [5]. Here we define wildlife as wild terrestrial or riparian vertebrates that predominantly reside in terrestrial or riparian environments.

Most controlled toxicity data used for human health assessment are from animal data or "terrestrial mammal" data, which are largely collected from laboratory rodent animal models. To be comprehensive in the protection of humans, an incredible number of measurements or toxicity endpoint data are gleaned – from histopathology of numerous organs and organ systems to blood cell differentials, plasma clinical chemistry values, feed consumption, and body mass changes. Data are typically reviewed, and the toxicity of the most sensitive endpoint is modeled and evaluated for POD determination.

Most methods used for wildlife and humans have employed the collective use of no observed adverse effect levels (NOAELs) and/or lowest observed adverse effect levels (LOAELs) used in determining PODs or used together from many studies and species to estimate or characterize the variation in toxicity between species. NOAELs and LOAELs, or NOEC/LOECs where "C" represents a concentration in media, such as air, water, or soil; and the former "L" is typically reserved for oral exposures or doses in mg of substance/kg body mass each day or a daily oral exposure estimate, in fact, represent treatment levels from controlled studies that are statistically different in the parameter measured from those organisms that are not exposed or the controls. They are statistical artefacts of study design and have been criticized for providing little information on the dose response relationship between exposure and effect [6–12]. Additionally, care must be taken in their interpretation as statistical significance within the confines of the study and the biological relevance of the endpoint may not be the same.

The NOAEL/LOAEL approach, though used for years in human health risk assessment, has been criticized for many years. The U.S. EPA Office of Research and Development Integrated Risk Information System have now moved to the exclusive use of the benchmark dose (BMD) method in determining the POD. However, the challenge of developing chemical-specific TRVs that are applicable to many species remains.

The primary challenge in TRV derivation is endpoint selection and extrapolation to the species of interest. In the human health example, only one species (humans) is considered, though it is recognized that there is considerable range in response between humans; young and old, ones with existing conditions, genetic profiles, and so forth.[i] However, although there are data intended for human health concerns that can be relevant to wildlife species (e.g., rodent data can be reasonably extrapolated to other wild rodents such as voles and deer mice), the relevance of extrapolating rodent data to other species of wildlife is questionable. The variation in response that is expected from the controlled rodent data that are largely from isogenetic strains to outbred wild white-tailed deer in which the risk assessor is interested must be considered relative to the physiological differences between the species and the nature of the toxic endpoint of effect.

Early efforts in TRVs focused on NOAELs/LOAELs that were limited to endpoints expected to be most directly related to effects that would cause changes in populations [13]. Initially, they were patterned on the same

[i] It is important to note that intra- and interspecies variability is an issue for wildlife.

measures typically measured in aquatic organisms: mortality, growth, and reproduction [5]. However, these efforts have been criticized in that other, more sensitive effects may indirectly influence population growth and sustainability. For example, exposures to substances that cause lethargy have the potential to be directly related to an increased mortality rate from predators that would reduce fitness. Therefore, the method used here takes a conservative approach and uses the most sensitive endpoint determined as adverse, not adaptive, to derivate the TRV.

The USEPA employed a transparent system of literature review, quality assurance, and TRV derivation based on the evaluation of NOAELs and LOAELs [14]. The Eco Soil Screening Level (EcoSSLs) effort developed TRVs for 10 metals, based on NOAEL/LOAEL spread for growth, mortality, and reproduction, though other endpoint data were also plotted for comparison purposes. Typically, the highest NOAEL for mortality, growth, or reproduction was used above the lowest LOAEL. Acute data were not included and this effort has not been revised recently.

RECENT ADVANCES IN TOXICITY REFERENCE VALUE DERIVATION

Recent efforts in TRV derivation have become more sophisticated. Species sensitivity distributions (SSDs) have been used by many to consider variation in response between species [15]. However, they require discreet data; continuous data such as growth and histopathological data are difficult to plot. Additionally, the perceived variation attributed to species differences may be a reflection of differences in study designs instead [16]. Attempts to keep SSD models within the endpoints assumed to be ecologically relevant (i.e., mortality, growth, and reproduction) often ignore other endpoints of potential significance.

Many have suggested a best fit straight-line extrapolation of the dose response curve to be an improvement over the NOAEL/LOAEL approach [9,16,17]. These values are typically referred to as Effective or Ecological Dose of Concentrations at a specific level of population response (e.g., EC_{20}); however, their predominant use has been typically relegated to soil invertebrates [14].

The application of the benchmark dose method used in human health benchmark derivation has been used as a preferred method for extrapolating laboratory data for TRVs and is the method used in this present effort [17]. As described, the benchmark dose method can be used with data of many types to include discreet (incidence), continuous, and even mixed (e.g., categorical regression). The model allows the user to define the population level of interest. It is important to note that estimates from the 1 to 5% or the 95 to 99% ranges are highly variable and confidence intervals are large as data are rarely available for the tails of the dose response distribution. The mean modeled 10% level of a sublethal effect is usually referred to as the benchmark dose (BMD) and the lower 95% confidence limit on the mean 10% response is referred to as the benchmark dose limit (BMDL). Typically, the BMDL has been used for screening purposes for ecological risk assessment purposes and for determining the point of departure for estimating noncancer values for human health risk assessment. When a less protective, more predictive value is needed, the risk assessor can move up the dose response curve and choose a higher level, or frequency of response, for a population effect or a lower level of confidence (e.g., choosing a 15 or 20% sublethal effect to a population or lower the confidence bound on the mean rather than using the lower 95% confidence bound). The 50% level of confidence is known as the BMD and is the point on the best fit modelled curve at the determined population level of effect.

Toxicity data developed for the species of interest, on the substance of interest, should be used in the derivation of the TRV. However, rarely are data available for wildlife species. Therefore, data collected on a substance within a vertebrate class (i.e., mammals, birds, reptiles, and amphibians) are used collectively to make a toxicity assessment and derive a benchmark. Typically this level of comparison can provide data that are sufficient for this purpose. Challenges remain in evaluating these data since the variation in toxicological response may be due to a host of factors. Few studies use the same methods to be able to determine if differences are due to species or methodological differences. Differences between feeding and gavage studies and vehicle choices can influence absorption, hence toxicity. There are few methods described for wildlife toxicity testing and fewer published intended for TRV derivation.

Ideally, many dose response curves could be used at one time for a given endpoint using data from many species to obtain a more accurate representation of cross-species variation. However, currently there are no models developed for this purpose. Therefore, the current method is constrained by selecting the most relevant, high quality data set to model the effect determined to be representative of the class representing the species of interest. This requires professional judgment of the entire data set and a weight of evidence determination.

Recent reviews of wildlife risk assessments suggest that the TRV can greatly influence the risk calculation more than the exposure estimation [18]. It has been suggested that the BMD method can be applied to data of species of interest, not all data within a class, to assist in deriving a predictive value for remedial decision making [16]. This and moving from the BMDL to the BMD may provide flexibility and reduce the uncertainty sufficiently to assist in making rational remedial decisions—many of which have other habitat impacts on wildlife populations. Clearly a balance between predictive adverse effects and necessary habitat alterations must be considered. Other methods are being considered. These enable the calculation of a best-fit dose response curve to a collection of curves for a specific endpoint using data for various species. This approach permits a more robust evaluation when considering species variability; however, care is exercised to ensure the variation in response is due to differences between species, and not differences in study design.

IMPORTANT RESEARCH NEEDS

A lack of adequate toxicity data is often a limiting factor in deriving safe levels of exposure. This statement is accurate for public health concerns and for ensuring wildlife populations are protected. The latter is often confounded by the sheer number of species to be assessed. Animal data used for human health risk assessment is clearly inadequate for many reasons; the predominant one is the lack of such data for many substances used in commercial applications. New methods have been developed (and are under development) to help streamline the toxicology process and provide more useful low-level data applicable to many human exposures (e.g., Tox21) [19], but these will require further development.

Data developed for human health purposes using laboratory rodents can be useful in estimating effects on nonhuman mammals as well. Naturally, those rodent data are most applicable to other rodents and likely less useful in characterizing effects to Carnivores or Mustelids. Variation in levels of exposure that causes adverse effects can be attributed to species differences, but also differences in study design. For example, data from gavage and feeding studies can vary enormously attributed to differences in toxicokinetic differences between a single bolus and a longer-sustained daily exposure, respectively. Standardized methods in exposure and endpoint measurement are needed to compare differences between species accurately.

Still, it is impractical to test all species of concern for each potential environmental xenobiotic. Evaluations need to be conducted that compare species sensitivities to help determine which species would provide results that could best be extrapolated to others. This may be most useful where specific classes of compounds are concerned. Greater refinement, development, and use of *in silico* (e.g., Quantitative Structural Activity Relationship; QSAR) models may also be incorporated. Further, additional statistical tools that enable the characterization of multiple dose-response curves are needed to determine the BMD for several species simultaneously, which may evaluate all variables in the results at once. Although many terrestrial vertebrates share or show homology between biological systems that enable life, differences in how these functions are performed due to various selective pressures through history create the basis for selective toxicity at their respective targets. Finding the "canary" sentinel equivalent is likely to remain elusive for the vast array of environmental substances that have selective targets of toxicity.

REFERENCES

[1] National Research Council (U.S.) Committee on Animals as Monitors of Environmental Hazards. Animals as sentinels of environmental health hazards. Washington (DC): National Academy Press; 1991.

[2] Nature Conservancy. The military and nature: our partnership with the department of defense, [Internet] 2014 [cited 2014 Sept 14]. Available from, http://www.nature.org/newsfeatures/specialfeatures/partnership-with-the-department-of-defense.xml.

[3] Kendall RJ, Smith PN. Perchlorate ecotoxicology. Pensacola, FL: SETAC Press; 2006.

[4] Sunahara GI, Lotufo G, Kuperman RG, Hawari J. Ecotoxicology of explosives. Boca Raton, FL: CRC Press; 2009.

[5] Suter GW, Sample BE. Oak Ridge National Laboratory, Lockheed Martin Energy Research Corp. Guide to the ORNL ecotoxicological screening benchmarks: background, development, and application. Oak Ridge, TN: Oak Ridge National Laboratory; 1998, Pub. No. 4793.

[6] Chapman PM, Cardwell RS, Chapman PF. A warning: NOECs are inappropriate for regulatory use. Environ Toxicol Chem 1996;15(2):77–9.

[7] Warne M St J, van Dam R. NOEC and LOEC data should no longer be generated or used. Australas J Ecotoxicol 2008;14(1):1–5.

[8] Laskowski R. Some good reasons to ban the use of NOEC, LOEC and related concepts in ecotoxicology. Oikos 1995;73(1):140–4.

[9] Kodell RL. Replace the NOAEL and LOAEL with $BMDL_{01}$ and $BMDL_{10}$. Environ Ecol Stat 2009;16(1):3–12.

[10] Yanagawa T, Kikuchi Y. Statistical issues on the determination of the no-observed-adverse-effect levels in toxicology. Environmetrics-Chichester 2001;12(4):319–26.

[11] Leisenring W, Ryan L. Statistical properties of the NOAEL. Reg Regul Toxicol Pharmacol 1992 Apr;15(2):161–71.

[12] Bokkers BG, Slob W. A comparison of ratio distributions based on the NOAEL and benchmark approach for subchronic-to-chronic extrapolation. Toxicol Sci 2005;85(2):1033–40.

[13] U.S. Environmental Response Team. Ecological risk assessment guidance for Superfund: process for designing and conducting ecological risk assessments. Edison, NJ: U.S. Environmental Protection Agency, Environmental Response Team; 1997, Interim Final. EPA 540-R-97-006.
[14] U.S. Environmental Protection Agency. Guidance for developing ecological soils screening levels. Washington DC: Office of Solid Waste and Emergency Response; 2005.
[15] Awkerman JA, Raimondo S, Barron MG. Development of species sensitivity distributions for wildlife using interspecies toxicity correlation models. Environ Sci Technol 2008 May 1;42(9):3447–52.
[16] Mayfield DB, Johnson MS, Burris JA, Fairbrother A. Furthering the derivation of predictive wildlife toxicity reference values for use in soil cleanup decisions. Integr Environ Assess Manag 2014 Jul;10(3):358–71.
[17] U.S. Army Center for Health Promotion and Preventive Medicine (USACHPPM). Standard practice for wildlife toxicity reference values. Aberdeen Proving Ground, MD: U.S. Army Center for Health Promotion and Preventive Medicine; 2000, Technical Guide No. 254.
[18] Sample BE, Fairbrother A, Kaiser A, Law S, Adams B. Sensitivity of ecological soil-screening levels for metals to exposure model parameterization and toxicity reference values. Environ Toxicol Chem 2014 Jun 18; http://dx.doi.org/10.1002/etc.2675 [Epub ahead of print].
[19] National Research Council (U.S.). Committee on Toxicity Testing and Assessment of Environmental Agents. Toxicity testing in the 21st century: a vision and a strategy. Washington, DC: National Academies Press; 2007.

PART II

Methodological Derivations

CHAPTER 2

Methods for Derivation of Wildlife Toxicity Values for Use in Ecological Risk Assessments

Adam T. Deck, Mark S. Johnson

Contents

INTRODUCTION

Understanding the potential for chemicals in the environment to cause adverse effects to wildlife requires knowledge of exposure and toxicology. The potential for exposure can be measured directly (e.g., through a chemical analysis of crop or gastrointestinal contents [1] or modeled from environmental media (e.g., soil and/or sediment chemical concentrations). In the latter case, food web models can be used to determine oral uptake [2]. However, toxicity data are typically used to determine a "safe" exposure level, most often assuming exposure occurs predominantly through oral exposure routes, although inhalation and dermal pathways can also be modeled. In this paradigm, exposure is considered the numerator, and a safe level of exposure based upon toxicity data is considered the denominator. The potential for adverse effects exists when exposure estimates (numerator) exceed the safe level of exposure (denominator).

Controlled toxicity data are typically used to understand the levels at which chemicals could be considered safe. Such data are from studies conducted in a controlled laboratory environment where the only difference

Wildlife Toxicity Assessments for Chemicals of Military Concern
http://dx.doi.org/10.1016/B978-0-12-800020-5.00002-8

between groups of test animals is the concentration of the chemical to which the animals are exposed. These studies can be expensive and laboratory animal wildlife models are limited. Regardless, the data from these studies can be used to establish a dose-response curve from which a threshold for adverse events occurs and a safe level can be identified. This level is often referred to as a Toxicity Reference Value (TRV).

Optimally, toxicity data would be available for chemicals of environmental concern and the wildlife species of interest; this is rarely the case. Typically, toxicity data collected for another purpose are used to develop a TRV. These data are from studies that use different designs, a factor that also contributes to the variability in results. Although TRVs are designed to apply to classes of vertebrates (e.g., mammals, birds, reptiles, and amphibians), chemical-specific data are often scarce, and TRV developers use whatever data are available to determine the threshold for adverse effects that would be useful for an entire class of vertebrates. An essential element of this process is to collect all potentially relevant information and to document the procedure. Once obtained, this information is documented in the form of a chemical-specific toxicity profile. Pertinent studies are gleaned and highlighted, professional judgment is used to ascertain the toxicity endpoints of relevance, and TRVs are then derived logically and transparently. Herein, we describe the procedure used to develop the TRVs provided in this book.

DATA COLLECTION AND LITERATURE SEARCH

The first step in creating the toxicity profile is to search the available scientific literature and collect data that will be used to develop the toxicity profile. Background information on manufacturing, environmental fate and transport, field studies, and any other data that provide context for the need to develop TRVs for wildlife are also helpful. When searching for study data, it is recommended that all available and pertinent databases be searched to ensure the toxicity profile represents a complete and comprehensive summary of all relevant material and data. The literature search should be conducted based on the vertebrate class (e.g., mammals, birds, and reptiles) and cross-referenced with the specific chemical that is being detailed. It is also important to document the parameters and results of the search. Important information includes databases and dates searched, keywords used, number of database accessions, and the abstracts and papers retrieved. Primary sources of information are preferred.

IDENTIFYING RELEVANT STUDIES

After the literature search has been completed and the pertinent studies gathered, they are analyzed to determine their applicability for use in developing the TRV. Seven criteria are considered in determining the relevance of the studies: endpoint, exposure duration, effect level, exposure pathway, validity of the study design, quality of the study, and corroboration of study results with other data.

While human risk assessments are concerned with protecting the health of individual human lives, ecological risk assessments are concerned with sustaining populations of species [2]. Because of this, a study must have endpoints that can conceivably affect the sustainment of the population in order to be included as relevant. For example, studies that find endpoints of uncertain toxicological significance (e.g., enzyme induction, adaptive responses, changes in clinical chemistry parameters within the normal ranges, etc.) are often not included in TRV development, but they may be used for corroboration of the relevant adverse effect. Studies that clearly identify adverse effects such as those that result in mortality; impaired reproduction, development, or growth; changes in behavior related to reproduction, feeding, and predator avoidance; and resistance to disease are considered because they are most likely to affect the sustainment of the population [2]. Additionally, endpoints that are adverse on an individual level are assumed to conceivably result in an adverse population level impact through changes in somatic repair, which could result in higher energetic demands in a food-limited environment in which reproduction could also be indirectly affected.

The exposure duration, defined as the length of time to which the study subjects were exposed to the test chemical, must be indicated for each study. For the purposes of developing a TRV, chronic exposures are the most relevant because representative species in each of the vertebrate classes could be resident and, as such, could conceivably be exposed for sustained periods. Data for chronic exposures are few, however, and there are certain circumstances under which shorter exposures are more relevant to field populations. Lengthy, sustained exposure periods typically result in lower exposure levels that often result in toxicity at much lower daily exposure rates than acute, short-term exposures. Thus, a conservative approach is used in preferring studies of chronic duration. For the purposes of developing TRVs, chronic exposures are those equal to or greater than 10% of the lifespan of the test species. While chronic exposures are preferred, subchronic exposures (less than 10% of the lifespan of the test species but greater than 14 days), subacute exposures (repetitive exposure of up to 14 days), and acute exposures (a single exposure event) may be relevant in providing information on mechanism, etiology, or species differences [3].

Studies are considered relevant when both no adverse effects and sublethal adverse effects are reported. Statistical comparisons are used to define effect levels that include the no observed adverse effect levels (NOAELs), the lowest observed adverse effect levels (LOAELs), and the doses, which produce health effects in 50% of the exposed population (ED_{50}). Studies with no and low adverse effect are desirable because they capture the threshold for adverse effects, which is useful for developing TRVs for screening purposes.

For each study, the exposure pathway (i.e., the route by which a chemical is introduced into a receptor) should be considered along with the exposure pathway for the receptor, and the two should match. The exposure pathway of a wild animal that eats contaminated wild vegetation is more similar to that of a study animal that eats contaminated food than to a study animal that has been dosed by gavage. However, the characterization of dose in feeding studies can be problematic. In these studies, exposure is quantified by evenly coating the feed material with the test chemical and weighing the feed before and after its exposure to the test animal. The ability of animals to select feed particles that are less toxic than others has occurred [4], resulting in an increase in the chemical concentration of uneaten feed over time. Moreover, animals typically spill feed or change the mass of the test feed by urinating or defecating in feed troughs, adding variability to the exposure estimate [5].

Typically, test animals exposed to toxic compounds in feed resist feeding and frequently show a marked reduction in feed consumption and body mass. This often changes with continued exposure, during which feed consumption is resumed, and body mass increases to normal levels. However, the wide fluctuations in body mass contribute to the variation in the dose estimate because dose is calculated on a body mass basis: milligrams/chemical/kilogram body mass/day (mg chemical/kg body mass/day). This variation is rarely captured in dose estimates. Still, when these factors are best controlled for, feeding studies best emulate environmental conditions in many cases where exposures occur throughout the day, and that metabolism, disposition, and excretion may be able to prevent toxic systemic levels from occurring. Although quantification of exposure remains a challenge, compounds administered in the test animals' drinking water are less prone to some of these drawbacks.

Gavage studies, that is, those in which a precise amount of chemical is directly administered in the gut of each individual test animal specific to its body weight and treatment, can provide conservative, less variable responses. Such studies are conservative in that each animal is given one daily dose each day, a bolus that can result in a greater probability of eliciting adverse effects by overwhelming the ability of the liver to metabolize the test compound, or overwhelming the ability for excretion via the urine or feces. Clearly, there are drawbacks and benefits associated with each.

The validity and quality of the study needs to meet minimum requirements for inclusion. The study must be statistically and biologically relevant. Thus, the chemical being investigated should occur in such a form as to best replicate the bioavailability of the test compound under natural conditions, and the dose of the chemical should be derived with limited inherent variability. Sufficient information must be provided to allow the study to be repeated, and the study must corroborate results from similar studies [3].

After each study has been considered and evaluated as relevant, the final step prior to writing the toxicity profile is to determine whether the collection of relative studies provides adequate data with which to derive the TRV. Typically, studies that contain data from three species within a class, data for species from at least two orders, and a minimum of two chronic LOAELs and one chronic NOAEL are required [3]. However, for many substances, these criteria were not satisfied. In such cases, weight-of-evidence techniques have been employed to use the best science in developing a value that would be relevant for screening purposes. Where insufficient data exist, uncertainty factors spanning an order of magnitude have been applied. Lastly, each value has been qualified with a level of confidence based on the abundance and concordance of the data and the biological systems that are expected to be perturbed to cause an adverse effect.

TOXICITY PROFILE PREPARATION

With the relevant studies in hand, the toxicity profile can be prepared. In addition to providing a summary of the review of the literature on the toxicology of a given compound, the toxicity profile includes the logic on the relevance of the studies in deriving a TRV. Specifics include the dates on which the search was conducted, which sources were reviewed, and the search strategy used. The summarized toxicity data are organized by taxonomical class, route of exposure, and exposure duration, respectively. When sufficient, the quantitative data are depicted in a scatter diagram and a table indicating the most relevant data, to include the study endpoint, the study species, and the chemical's effect levels.

DERIVATION OF TOXICITY REFERENCE VALUES

Data from the completed toxicity profile are used to develop two TRVs for each class of vertebrate and for each exposure pathway: low-value (useful for screening using conservative criteria) and high-value (uses mean values or the lowest dose at which adverse effects occur). By creating two TRVs, risk assessors have a range of toxicity reference values upon which risk estimates in an ecological risk assessment can be developed.

The methods recommended for deriving TRVs include the benchmark dose approach, the NOAEL/LOAEL approach, and the approximation approach [3]. Where possible, the benchmark dose was the preferred method; it fits a dose-response curve to the most relevant data and endpoint and considers the variation in the measurement of the endpoint. This variation is either captured in the low-value TRV, based on the lower 95% confidence interval on a 10% effect level (i.e., a dose about which there is 95% confidence that 10% or fewer animals are expected to develop an adverse effect) or captured using the 95% confidence level of one standard deviation. The latter is used when normal ranges of values do not exist for a particular test species. The high-value TRV is the dose that either corresponds to a 10% population effect level or represents the threshold for adverse effects that is on the best fit dose-response curve. The benchmark dose is preferred because it uses all the data from a specific endpoint from a study that was considered relevant and plots the best fit dose-response function that can be the means of departure for the derivation of other values (e.g., for remedial decision making) [6]. However, there are cases where dose-responses are multimodal (e.g., essential nutrients, hormetic responses, etc.), where a dose-response curve cannot be mathematically fitted with a level of confidence; in such cases, the NOAEL/LOAEL and/or the approximation approach can be used.

Historically, many entities and stakeholders have used the NOAEL/LOAEL approach to develop TRVs. This approach follows ecological risk assessment guidance provided by the U.S. Environmental Protection Agency (U.S. EPA) [2] and is used when sufficient data are not available to create a dose-response curve for a benchmark dose. To derive the high-value TRV, the lowest LOAEL of a toxic endpoint that could conceivably result in a change in population density is chosen from all the

LOAELs in the relevant studies. To derive the low-value TRV, the highest NOAEL that matches the endpoint of the high-value TRV is chosen. If a NOAEL that matches the endpoint of the high-value TRV is unavailable, the highest NOAEL of a different relevant endpoint is chosen. When choosing the NOAELs and LOAELs for this approach, chronic exposures are considered and are used unless there are no other data available or the mechanism/mode of action suggests that data from exposures of shorter duration are adequate (i.e., longer exposures are unlikely to produce effects that are more pronounced). Since reproduction includes stressors that could be manifested through changes in reproduction indirectly, these studies are considered equivalent to chronic exposures unless contraindicated.

In cases where there is an absence of available toxicological data, the approximation approach can be used. In this approach, the highest-quality single study having a population sustainment endpoint is selected. Uncertainty factors are applied to that study's dose to approximate a chronic NOAEL and a chronic LOAEL. The uncertainty factors used vary based on differences in the exposure duration, the endpoint, and interspecies variation between the dose in the chosen study and the TRV that is generated. If the study with the best quality presents a subchronic LOAEL, a three-fold uncertainty factor is applied to approximate it to a chronic NOAEL: one uncertainty factor to approximate from subchronic to chronic, another uncertainty factor to approximate from a LOAEL-based value to a NOAEL-based value, and a final uncertainty factor to account for interspecies variation. The specific uncertainty factors that are recommended for use are found in the U.S. Army Center for Health Promotion and Preventive Medicine (USACHPPM) Technical Guide (TG) 254 [3]. The resulting approximated chronic NOAEL and chronic LOAEL values are then used as low- and high-value TRVs, respectively. If the data have previously been used to fit a dose-response function via the benchmark dose approach, uncertainty factors were applied to the data where the database was lacking.

ASSIGNING A CONFIDENCE LEVEL

All TRV derivations are not alike. There is always uncertainty both in the wildlife toxicity data and the derived TRV that is generated, regardless of which approach was used. Therefore, each TRV is assigned a level of confidence to provide the risk assessor with an indication of the TRV's robustness. These qualifiers are based on the concordance of the data among species, the robustness of the database, and the number of species represented. Acknowledging the confidence in the generated TRVs aids risk assessors in understanding the uncertainty that the TRVs introduce into risk estimates. The confidence levels are rated by professional judgment and are ranked as low, medium, or high.

CREATION OF THE TOXICITY REFERENCE VALUE REPORT

The final step of the process is to summarize the relevant information and provide the logic for selecting the toxic endpoint of importance and the studies that best describe the adverse effects from exposure. The TRV report provides the method used, the number and magnitude of uncertainty factors used, and the logic for the exclusion of any other pertinent information, if necessary. The values are based on the state of the current knowledge and best professional judgment available. The user of these values is advised to exercise caution in their application as they are not designed to apply to every situation for every species within a class. They are intended to be used as a screening guideline that, with other site-specific data, can be used to help make decisions or rank relative hazards and/or risks. Users do so at their own risk. The TRVs are not intended to be used as clean-up values in themselves. Current practices for making site-specific clean-up decisions can be found elsewhere [7].

REFERENCES

[1] Mellott RS, Woods PE. An improved ligature technique for dietary sampling in nestling birds. J Field Ornithol 1993;64:205–10.
[2] U.S. EPA. Ecological risk assessment guidance for superfund: process for designing and conducting ecological risk assessments. Interim final. Washington (DC): U.S. Environmental Protection Agency; 1997.
[3] USACHPPM. Standard practice for wildlife toxicity reference values. Aberdeen Proving Ground (MD): U.S. Army Center for Health Promotion and Preventive Medicine; 2000, Technical Guide No. 254.
[4] Gogal Jr RM. Influence of feed exposure with TNT derivatives on quail (Northern bobwhite, *Colinus virginianus*) immunity as measured by National Toxicology Program immunity endpoints. 2000. Avian Toxicology Study No.990503, 981124G Final Report.
[5] Fairbrother A, Barron MC, Johnson MS. Method in environmental toxicology. In: Hayes AW, Kruger CL, editors. Hayes' principles and methods of toxicology. 6th ed. Boca Raton (FL): CRC Press; 2014.
[6] USEPA. Risk assessment guidance for superfund. Volume 1, Human health evaluation manual (part A). Interim final. Washington (DC): U.S. Environmental Protection Agency; 1989.
[7] Mayfield DB, Johnson MS, Burris JA, Fairbrother A. Furthering the derivation of predictive wildlife toxicity reference values for use in soil cleanup decisions. Integr Environ Assess Manag 2013;10(3):358–71.

PART III

Wildlife Toxicity Assessments of Explosives

CHAPTER 3

Wildlife Toxicity Assessment for 2,4,6-Trinitrotoluene (TNT)

Mark S. Johnson, Gunda Reddy

Contents

INTRODUCTION

Trinitrotoluene (TNT, 2, 4, 6-trinitrotoluene) is a nitroaromatic compound that is used extensively in military high explosives, and was largely used during World Wars I and II. Trinitrotoluene is one of the most important military bursting-charge explosives in use today. It is widely used in the

Wildlife Toxicity Assessments for Chemicals of Military Concern
http://dx.doi.org/10.1016/B978-0-12-800020-5.00003-X

manufacture of shells, bombs, grenades, demolition explosives, and propellant compounds, and from 1969 to 1971, approximately 45 million pounds of TNT were produced per month [1,2]. Crude TNT contains approximately 5% of the meta-isomers but impurities vary according to the final product. Wastewaters that contain TNT, and its isomers or other intermediate products, are formed during production and from the washings. Wastewaters, collectively referred to as "red water," often contain a high content of solids [2]. "Pink water" arises from red-water discharge and consists of dinitrotolune (DNT) and other contaminants that are generated from load and packing plant effluent and shell washout operations. Pink water often contains other breakdown products and impurities (e.g., DNT isomers, hexhydro-1,3,5-trinitro-1,3,5-triazine (RDX), and other additives). Pink water will take on a brick color under natural or basic conditions on exposure of wastewater to sunlight [3].

Environmental Fate and Transport

The distribution of TNT at many U.S. military sites is substantial. At least 17 Army installations have reported soil concentrations ranging from 0.08 to 64,000 micrograms per gram (μg/g) [4]. The detectable concentrations from five installations had samples in which surface soils exceeded 10 milligrams (mg) TNT/g soil dry weight [5].

Table 3.1 provides a summary of the physical and chemical properties of TNT. Discharge of large aqueous effluents of rinse pink water has

Table 3.1 Summary of Physical-Chemical Properties of 2, 4, 6-Trinitrotoluene

CAS No.	118-96-7
Molecular weight	227.13
Color	Yellow-white
State	Monoclinic needles
Melting point	80.1 °C
Boiling point	240 °C
Odor	Odorless
Solubility	130 mg/L in water at 20 °C; soluble in acetone, benzene, alcohol, and ether
Partition coefficients	
Log K_{OW}	1.60; 2.2 (measured), 2.7 (estimated)
K_{OC}	300 (estimated), 1,100 (measured)
Vapor pressure (at 20 °C)	1.99E-04 mm Hg
Henry's Law constant (at 20 °C)	4.57E-07 atm m^3/mole
Conversion factors	1 ppm = 9.28 mg/m^3 1 mg/m^3 = 0.108 ppm

Source: ATSDR [6].

historically been an important route of TNT contamination of surface and ground water, and surface soils [5,6]. Some reported wastewater effluents ranged from 61 to 210 pounds/day [3]. Due to its relatively low vapor pressure and low water solubility, TNT does not actively partition from surface waters to the atmosphere [6]. By comparing river waters and distilled water, studies have shown a direct relationship between increased pH and organic matter content with the rate of TNT photolysis [7].

Soil contamination by TNT results from spills, disposal of solid waste, open incineration, detonation of explosives, or leaching from poorly engineered impoundments [8]. Retrieval and subsequent destruction of unexploded ordnance (UXO), including open burning/open detonation (OB/OD) areas results in soil contamination with TNT. Based primarily upon the physical and chemical properties of TNT (i.e., octanol-water partition coefficient (K_{OW}) and water solubility), TNT is not expected to bioaccumulate or biomagnify in terrestrial systems [9], and data collected to date do not support it [10].

Soil biotransformation of TNT can be significant, and is readily reduced under anaerobic microbial conditions, and primarily through successive reduction of the nitro groups. Microbial transformation of TNT leads to many reduction products, including 2-amino and 4-amino dinitrotoluene and azoxydimers [6,8,9], though some oxidation products have been identified [11]. Biological transformation by bacterial and fungal species occurs slowly in the environment, with slightly higher rates seen in the presence of other carbon sources. Following reduction, the amines irreversibly adhere to the organic (humic) fractions of the soil [10].

Although biological degradation may not extend to cleavage of the TNT ring, successive reductions of each nitro group to an amine is followed by oxidative deamination to a phenol that releases ammonia or nitrite. Numerous transformation products have been identified in pink water, and the predominant ones include 1,3,5-trinitrobenzene, 4,6-dinitroanthranil, 2,4,6-trinitrobezaldehyde, 2,4,6-trinitrobenzonitrile, in addition to several azo and azoxy derivatives formed by the coupling of nitroso and hydroxylamine products [12,13]. The biotransformation of TNT under aerobic and anaerobic conditions forms several amino DNTs to 2,4,6-triaminotoluenes (TAT). Under abiotic conditions, many possible routes of TNT transformation are possible and can form several degradation products under photolytic conditions in the presence of Fe, TiO_2, and under alkaline conditions [14]. Plant uptake studies have shown concentrations of TNT and that of its reduction products in the roots and stems [15,16]. These degradation products have also been found present in water, soil, and sediment [17].

SUMMARY OF MAMMALIAN TOXICITY

Mammalian Oral Toxicity: Acute/Subacute

The acute oral toxicity of TNT in mammals including some wildlife species has been reported. For example, the oral median lethal dose (LD_{50}) value of 660 mg/kg was determined for both male and female mice; for male and female rats, it was 1320 and 795 mg/kg, respectively [18]. These animals developed seizures (grand mal), followed by mild convulsions 1 to 2 hours after exposure. All deaths occurred within 24 hours following exposure, and other signs of exposure included red-colored urine and lethargy [18]. Fourteen days following exposure, animals that survived convulsions remained alive. Additionally, cyanosis was evident 12 hours following administration of 10 mg/kg TNT in dogs. Severe incoordination and tremors followed dosing [19]. Most species, including mice, rats, and dogs, showed signs of ataxia after dosing with TNT [18,19].

Cats injected intraperitoneally with 100 to 150 mg/kg TNT died within 5.5 hours [20]. Injections of 40 mg/kg caused convulsions, hind limb paralysis, decreased body temperature, and enhanced saliva secretion. Methemoglobin was also present in the blood of cats. Daily subcutaneous injections of 50 mg/kg TNT caused cats to die within 4 to 9 days [21]. These studies showed signs of splenic congestion, hepatic fat accumulation (steatosis), and Kupffer cell hemosiderosis in treated animals.

Oral LD_{50} estimates for cotton rats (*Sigmodon hispidus*) exposed to TNT in corn oil were 607 and 767 mg TNT/kg for males and females, respectively [22]. Animals also exhibited an increased respiratory rate within 90 minutes of dosing. Orange-colored urine and urinary bladder distension were observed in all animals at necropsy. No other abnormal observations were reported.

Seven-day gavage exposures representing an LD_{50} for male (75.9, 151.8, and 303.5 mg TNT/kg body weight per day (bw/day)) and female (96, 192, and 384 mg TNT/kg bw/day) cotton rats, respectively, was conducted using corn oil [22]. Histopathology of major organs, hematology, assays of hepatic metabolizing enzymes, and serological clinical chemistry were evaluated. Increased splenic weights were seen in the 192 and 384 mg/kg/day treatment groups (females), and liver weights were increased in the 152 and 303.5 mg/kg/day treatment groups (males). Both high-dose groups showed hematological endpoints that were consistent with erythrolytic anemia. Following exposure of rats to the lowest dose of TNT (75.9 mg/kg), the spleen contained hemosiderin laden macrophages. Subtle testicular lesions were also noted in the two high-dose groups.

Oral exposures of TNT to the white-footed mouse (*Peromyscus leucopus*; 10/group/sex) occurred at 0, 0.042, 0.083, 0.165, and 0.330% TNT in the feed for 14 days [23]. The results suggested hemolysis was evident in the 0.330% TNT diet treatment group of both genders. These treatments were calculated by the authors as being equivalent to 66, 145, 275, and 602 mg TNT/kg bw/day for males and 70, 142, 283, and 550 mg/kg/day for females for the 0.042, 0.083, 0.165, and 0.330% doses of TNT, respectively. Moreover, only males in the 0.165 and 0.330% treatment groups had suppressed splenic phagocyte hydrogen peroxide production, and a reported reduction in phagocytosis for males in all TNT exposures [23].

Mammalian Oral Toxicity: Subchronic

Subchronic oral exposure of rats, mice, and dogs to TNT has produced consistent hematologic effects [24–27]. Exposures of 13 weeks were sufficient to provoke anemia (i.e., reduced red blood cell numbers, reduced hemoglobin, and hematocrit) in all species studied. Increases in immature red blood cells (reticulocytes), reductions in blood hemoglobin, hematocrit, and corpuscle volumes were evident after only 15 days of exposure in dogs that were administered TNT in gelatin capsules of doses ranging from 5 to 33 mg/kg [19].

TNT exposure might result in direct hemolysis within circulating blood, leading to an increase in spleen weight. Dilley *et al.* [18] reported similar findings in beagles including pathological assessment of the spleen that suggested hemolytic anemia. Other important effects included increased liver weight (including hepatocytomegaly), intestinal inflammation (and mucoid stools), enlarged kidneys, and splenic congestion in mice, rats, and dogs [25–29]. Most animals in the highest dose group of all species displayed differential splenic hemosiderosis [25]. Rats and dogs had dose-related increases in serum cholesterol and lower iron and serum glutamic-pyruvic transaminase (SGPT) levels following the 13-week exposure. By contrast, mice were more resistant to treatment [25]. Increased serum cholesterol was consistent with doses in rats and dogs [25,28]. Additionally, other endpoints consistent with anemia included decreased erythrocytes, hemoglobin and hematocrit values, and occasional bone marrow hyperplasia.

Testicular atrophy was most pronounced in rats [25] and consisted of dose-dependent degeneration of the germinal epithelium lining the seminiferous tubules, and hyperplasia of the interstitial Leydig cells (i.e., in the high-dose group, 300 mg/kg/day (28)). The no observable adverse effect levels (NOAELs) were 0.20, 1.42, and 7.76 mg/kg/day for dogs, rats, and mice, respectively, which suggested that dogs were the most sensitive to repetitive oral exposures to TNT [25]. However, the effects appeared to be completely reversible (up to a 4-week exposure) following a 4-week recovery period [25].

Results of a 90-day feeding study using white-footed mice (*Peromyscus leucopus*) suggested that Nearctic mice are more resistant than Palearctic (Old-World: *Mus*) species. McCain [30] exposed 100 male and female *P. leucopus* to TNT concentrations of 660, 1,320, and 2,640 parts per million (ppm) in animal feed. The calculated doses were approximately 165, 330, and 660 mg/kg/day, respectively. The highest concentration used in this study (2,640 ppm or 660 mg/kg/day) was equivalent to an LD_{50} of 660 mg/kg in mice [25]. None of the animals died during this study. Reductions in the initial animal weights were consistent with reduced palatability, although all groups gained weight over time. The adverse physiological changes (e.g., organ weight, chromaturia, and hemosiderin, etc.) were associated only with exposures to 1,320 (165 mg/kg/day) and 2,640 ppm (660 mg/kg/day) [30]. A NOAEL of 660 ppm (33 mg/kg/day) was reported for white-footed mice for a 90-day oral dietary exposure to TNT.

Mammalian Oral Toxicity: Chronic

The hematological effects of chronic exposure to TNT were consistent with those of subchronic exposures. Two studies using Fisher 344 rats [29] and beagle dogs [26] reported dose-dependent indicators suggesting hemolytic anemia (e.g., reduced hemoglobin, hematocrit, and erythrocyte counts, and increased levels of reticulocytes). These effects varied from controls at doses 8.0 mg/kg or greater (i.e., 32 mg/kg/day for dogs [26]) and for all TNT treatments for rats (i.e., 0.4, 2.0, 10.0, and 50.0 mg/kg/day [29]). Exposures for the rat study were continued for 106 weeks and 26 weeks in dogs. Compensatory responses to anemia were minimal in rats (e.g., erythrocytic macrocytosis and reticulocytosis [29]). Methemoglobinemia was apparent in both studies in animals of the higher-dose groups. Reduction in body weight was apparent in rats following exposure to 10 mg/kg/day or greater, or 8.0 mg/kg/day or greater for dogs [26,29]. Dose-related hepatomegaly (and increased kidney weights) was evident in rats receiving over 2.0 mg/kg/day, and this was only evident in the high-dose group for dogs. Splenomegaly was evident in the higher-dose groups of rats and dogs.

Hemosiderosis in Kupfer cells was seen in various dogs at most dose levels [26]. Gross and tissue morphological examinations in the high dose groups supported evidence of renal injury in rats [29]. Increased pigment deposition occurred in the kidneys, and there was evidence of bone marrow fibrosis in rats exposed to 2.0 mg/kg/day or more oral daily doses of TNT [29]. Observations of enteritis of the small intestine were related to TNT treatment in dogs [26]. Urinary bladder carcinomas were evident in some rats (2 male and 4 female rats) that were exposed for 106 weeks [29]. An NOAEL of 0.4 mg/kg/day was determined for chronically exposed rats [29], and none was found for dogs [26].

Studies Relevant for Mammalian TRV Development for Ingestion Exposures

Primary target organs for TNT include the nervous system (primarily from acute effects), testicular toxicity and the peripheral blood (Table 3.2 and Figure 3.1). Since TNT causes erythrolysis, the primary blood conditioning organs might be affected (e.g., liver, spleen, and kidney). Testicular toxicity occurs at levels higher than that which causes indicators of anemia. These conditions were found in *Peromyscus* [30], beagle dogs [25,26], rats [29], and laboratory mice (*Mus*; [25]). Several studies were identified that were well designed, and appropriate for the development of Toxicity Reference Values (TRVs) for mammals.

The work of Dilley *et al.* [25], Levine *et al.* [26–28], and Furedi *et al.* [29] were particularly valuable since they included chronic, subchronic, and acute exposures of mammals, and they used several species as had been identified above. Two orders (e.g., Carnivora and Rodentia), and three families (e.g., Canidae, Cricetidae, and Muridae) of *Mammalia* were identified. Two wildlife species including cotton rats and white-footed mice were also evaluated. Effects from exposure were consistent, yet slightly variable in their magnitude of effect. Each study identified several NOAELs and low observable adverse effect levels (LOAELs) for various endpoints, and the investigations were inclusive for other potential organ systems. For these reasons, this review was considered sufficient to derive class-specific TRVs for TNT.

With few exceptions, data from acute studies using injection (e.g., intraperitoneal (IP), intravenous (IV)) methods were considered irrelevant exposure pathways and not used for comparative (TRV derivation) purposes. All other reports that evaluated TNT from exposure in animal feed or from gavage exposures had sufficient and consistent quality reporting of the methods and were included in this evaluation.

Animal models of dermal studies were not found; however, information suggesting the importance of human dermal exposure has been reported

Table 3.2 Summary of Relevant Mammalian Data for TRV Derivation

Test Organism	Test Duration	Test Results: NOAEL (mg/kg-bw/day)	Test Results: LOAEL (mg/kg-bw/day)	Test Results: Effects Observed at the LOAEL	Study
Mouse *Peromyscus leucopus*	90 days	165	330	Increased kidney, liver, spleen weights; presence of hemosiderin in spleen; chromaturia; increased extramedullary hematopoiesis in spleen	[30]
Mouse *Peromyscus leucopus*	14 days	142	550 (f)	Indicators of erythrolytic anemia (increased spleen weight, histopathology); decreased intracellular hydrogen peroxide of splenic phagocytes); phagocytosis results of uncertain biological significance	[23]
Cotton rat *Sigmodon hispidus*	7 days	76 (m)	152 (m)	Erythrolytic anemia; changes in spleen and liver pathology, hematology; changes in hepatic glutathione S-transerfase for females, not males of uncertain biological significance; male dose protective of female dose for all other endpoints	[22]
Rat (Sprague Dawley)	13 weeks	1.4	160	Anemia and leukocytosis	
		34.7	160	Increased cholesterol, decreased body weight (10–20%), increased spleen weight, hemosiderosis, lymphocytosis; testicular atrophy	[25]
		7	34.7	Decreased food consumption	
Mouse (Swiss-Webster)	13 weeks	35.7	193	Decreased hematocrit/RBC, liver necrosis	
Dog (Beagle)	13 weeks	2	20	Mucoid stools (red), diarrhea, anemia, increased liver weight, bilirubin, and cholesterol; lethargy	

Rat (Fisher 344)	13 weeks	5	25 (m)	Anemia, increased serum cholesterol	[28]
		25	125	Lipofuscin-like pigment in renal cortex, splenic enlargement with congestion, slight lethargy and ataxia; reduced food intake and body weight; atrophic seminiferous tubules, degenerated germinal epithelium	
Dog (Beagle)	6 months	2 (m)	8 (m)	Anemia, methemoglobinemia, increased platelets, slight ataxia; chromaturia	[26]
		8 (f)	32 (f)		
		2 (m)	8 (m)	Decrease in body weight (16.4%; females at 32)	
Rat (Fisher 344)	13 weeks	5	125	Increased spleen weight with diffuse congestion	[27]
Rat (Fisher 344)	24 months	10 (f)	50 (f)	Bone marrow fibrosis	[29]
		2 (♀)	10 (♀)	Increased cholesterol, enlarged liver; 14% decrease in body weight gain; splenic congestion, extramedullary hematopoiesis	
Mouse (B6C3F1)	24 months	10 (♀)	70 (♀)	Mild anemia, increased liver weight, reduced serum globulin levels; 10–15% decrease in body weight gain; enlarged spleen and lymph nodes	[29]

f=female; m=male.

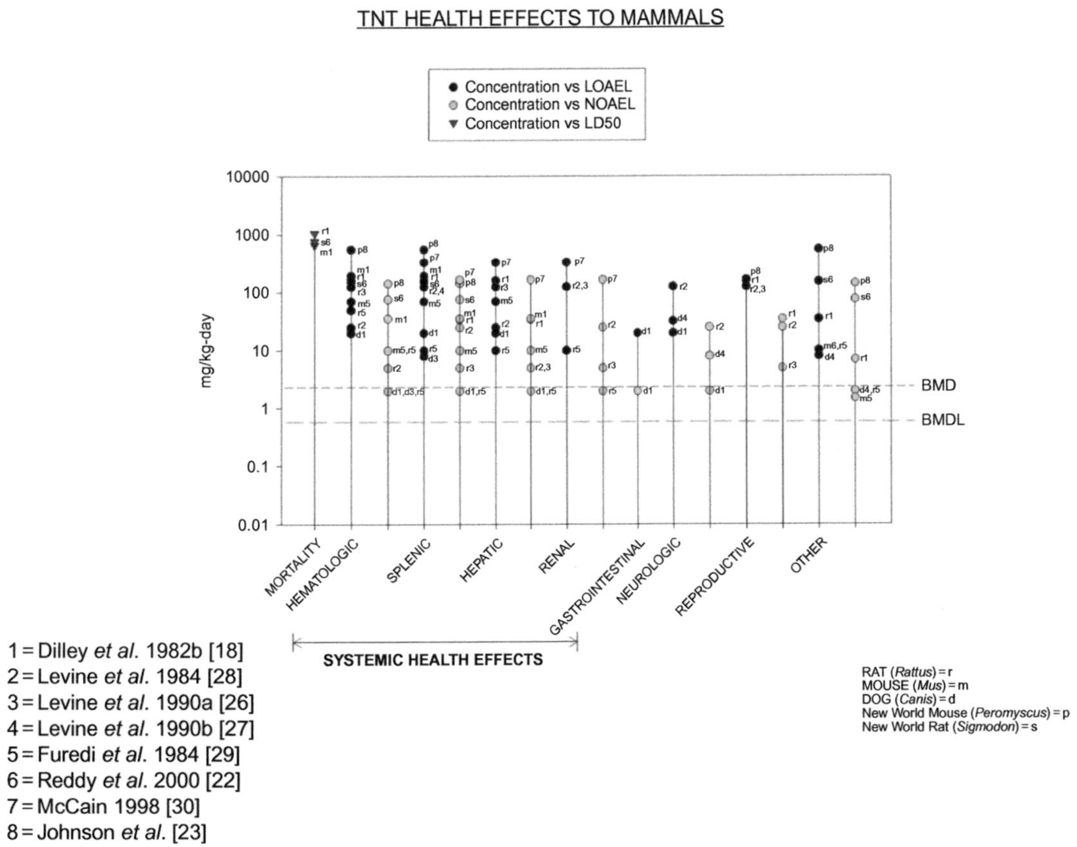

Figure 3.1 TNT health effects to mammals. Letters by symbols represent test species, and numbers represent study.

[31,32]. In addition, studies of the potential of TNT to penetrate mammalian skin *in vitro* from a soil matrix demonstrated that dermal exposures to TNT in soil may add to the total systemic dose [33–35].

SUMMARY OF AVIAN TOXICITY

Avian Oral Toxicity: Acute

The toxicity of TNT in avian species is limited to acute and subchronic studies. The approximate lethal dose (ALD) for TNT was reported for the Northern Bobwhite (*Colinus virginianus*) [36]. Both male and female birds were gavaged with single oral doses of ranging from 263 to 4,508 mg TNT/kg bw in corn oil in three-phased trials and observed for 14 days. Male birds exposed to 1,335 and 2,003 mg/kg died within 6 days in the first trial; no females died from single oral exposure of up to 2,003 mg/kg. The second phase involved fasted birds where three male and three females were exposed to either 2,003, 3,005, or 4,508 mg/kg. Here, both sexes at the high dose and a female exposed to 2,003 mg/kg died within 4 days. The third phase consisted of four fasted birds of both sexes exposed to 2,003 or 3,005 mg/kg. Here, all died with the exception of a single female at 2,003 mg/kg that showed severe ataxia. With the exception of one female, all birds that were exposed to 3,005 mg/kg died within 5 days. Therefore, the authors report the ALD to be 2,003 mg/kg [36].

Avian Oral Toxicity: Subchronic

Johnson *et al.* [37] studied the oral toxicity of TNT in the common pigeon (*Columba livia)* by gavage exposures of 0, 20, 70, 120, and 200 mg/kg/day in corn oil for 60 days. Signs of toxicity included weight loss, neuromuscular (ataxia, tremors) effects, and traces of red feces (chromaturia). Mortality rates of 8/12 were observed at a dose of 200 mg/kg/day in males and females, and mortality rates of 2/12 were observed at a dose of 120 mg/kg/day in females. The changes in hematological parameters, liver, kidney, and ovary weights were all dose related. Histopathology data of male birds exposed to a dose of 200 mg/kg/day showed moderate to severe hypospermatogenesis in the testes and mild to moderated hemosiderosis in the spleen of female birds. No other changes were observed that could be attributable to TNT exposure. There were also no significant differences in food consumption and changes in body weight of pigeons following dosing at 200 mg/kg/day. The differences in the body weights of the liver and kidney, and body weights to organ weight ratios were present at all doses. Hematological and clinical parameters showed treatment-related effects beginning at the 70 mg/kg/day level. The NOAEL of 20 mg/kg/day for a subchronic (60 days) oral exposure in pigeons was also reported in this study.

In another subchronic study, 50 adult male and female Northern Bobwhite (*Colinus virginianus*) were provided TNT in animal feed at concentrations of 3.3, 1.56, 0.863, and 0.16 g TNT/kg feed for a 90-day exposure [38]. Based on the feed consumed, the calculated doses to males and females consuming the experimental diets were approximately 178, 97.5, 48.0, and 7.0 mg/kg/day. Feed consumption was similar between genders. Initially, 4/10 birds died from exposure to 3.3 g/kg TNT, but none of the birds died thereafter during the study. Histopathology and sensitive indicators of immune function were evaluated. The effects included a dose-dependent and nonsignificant dampening of the total red blood cell count, packed cell volume, total plasma protein, blood prolymphocytes, blood lymphocytes, an increase in late apoptotic or necrotic blood leukocytes, and minor hemosiderosis of the liver. The authors noted that significant erythrolytic anemia was not the major target of toxicity in quail, which was most likely due to the refractory nature of the avian hematological and vascular system. No adverse histopathology was associated with any animal exposed to the 0.16 g/kg feed (7.0 mg/kg/day) treatment group. Homogeneity of TNT within the feed was variable, which may have contributed to the variation in response. A NOAEL of 7.0 mg/kg/day TNT in the animal feed given to the Bobwhite was determined from the 90-day exposure. The LOAEL of 178 mg/kg/day was reported as the incidence of mortality occurred at this dose. No data are available for chronic exposures in birds.

Studies Relevant for Avian TRV Development for Ingestion Exposures

The 90-day subchronic TNT oral exposure study conducted in the Northern Bobwhite evaluated many toxicological parameters that could be relevant in the development of TRVs for avian species [38]. The 90-day results suggest that birds are much less sensitive to the hemolytic mechanisms found in mammals, yet there is evidence of a mild erythrolytic effect. Given the refractory nature of the avian hematopoietic system and the magnitude of these observations, the biological significance of these findings is uncertain. Consistent with the mammalian data are the initial central nervous system (CNS)-related effects of exposure where individuals exhibited ataxia and neuromuscular effects, which were observed prior to death of the quail in the high dose group 178 mg/kg/day (3.30 g/kg feed). Therefore, authors suggested an NOAEL of 7.0 mg/kg/day based upon the lack of adverse pathological and immunotoxicological observations for any individual in the low dose group (Table 3.3). A LOAEL was identified as 178 mg/kg/day (3.30 g/kg feed) based on the four deaths that occurred in the high dose group, and that possible adverse histopathology was associated with some individuals in the group. The other study reported oral dosing of TNT in corn oil to pigeons [37] was conducted more precisely than the earlier feeding study using Northern Bobwhite [38]. In this study, effects were more definitive and followed a dose response relationship. Here, the NOAEL and the LOAEL for the pigeon study was determined to be 20 and 70 mg/kg/day, respectively.

Inhalation or dermal toxicity data were not available for birds.

SUMMARY OF AMPHIBIAN TOXICITY

There are limited toxicity data on TNT available for amphibians. These studies include acute and subchronic (28 days) studies in the salamander, and chronic (90 days) studies conducted in bullfrog tadpole larvae.

Amphibian Toxicity

The embryonic toxicity of TNT, 2ADNT, and 4-amino-2,6-dinitrotoluene (4ADNT) was studied in a standardized assay using *Xenopus laevis* embryos (Frog Embryo Teratogenesis Assay-*Xenopus*, FETAX). At 96 hours, the LC_{50} values of TNT (16.7 μM), 2ADNT (166 μM), and 4ADNT (115 μM) showed that TNT was more toxic than its isomers [39]. At 96 hours, the median effective concentration (EC_{50}) value of TNT was 9.78 μM, a level that induced tadpole malformations and teratogenic effects. These nitroaromatic compounds may show lethal or teratogenic effects on aquatic animals when exposed to high levels of TNT.

The acute oral toxicity of 2,4,6-TNT, 2,4-DNT, and 2,6-DNT was tested on adult male bullfrogs (*Lithobates catesbeiana*) via oral gavage [40]. The acute oral LD_{50} value for TNT was 1,060 mg/kg and 1,098 mg/kg for 2.4-DNT and 2,6-DNT. All compounds produced similar toxic symptoms (e.g., change of skin color, body weight, development of seizures, and changes in respiratory rhythm). Liver and kidney necrosis and lung cyanosis was also observed at 14 days into the observation period [40]. Chronic

Table 3.3 Summary of Relevant Avian Data for TRV Derivation

Test Organism	Test Duration	Test Results: NOAEL (mg/kg/day)	Test Results: LOAEL (mg/kg/day)	Test Results: Effects Observed at the LOAEL	Study
Common Pigeon (*Columba livia*)	60 days		20	8/12 mortality in both sexes at 200 mg/kg, 12/12 in females at 120 mg/kg, weight loss, neuromuscular effects, changes in hematological parameters, liver, kidney and ovary weight increase, mild to moderate hemosiderosis in spleen and liver (200 mg/kg), reduced Hb, RBC at 20 mg/kg	[37]
Northern Bobwhite (*Colinus virginianus*)	Lowest lethal dose detected (LD_{LOW}) 2003 mg/kg	2003		Male dies during the determination of the approximate lethal dose at 2003 mg/kg; female did not (one female only)	[36]
Northern Bobwhite (*Colinus virginianus*)	90 days	7	178	4/10 initial deaths in high dose group (3300 mg/kg); dose-dependent nonsignificant decreasing trend in: total red blood cell counts, packed cell volume, total plasma protein, blood prolymphocytes, blood lymphocytes, an increase in late apoptotic/necrotic blood leukocytic cells, and slight hemosiderosis in the liver	[38]

toxicity studies of 2,4,6-TNT, 2,4-DNT, and 2,6-DNT on bullfrog tadpoles were determined by exposing them to contaminated water that was dose-dependently supplemented with TNT (0.125, 0.25 1.0, 2.0, and 4.0 mg/L) for 90 days [41]. Exposures to 0.125 mg/L nominal (0.12 mg/L measured) TNT reduced both the survival time and growth (e.g., increase in body weight and snout vent length (40)). In general, the survival time of the tadpole was low at high concentrations of TNT. The percent survival of tadpoles that were exposed to TNT at 0, 0.125, 0.25, 0.5, 1.0, and 4.0 mg/L were 77, 22, 22, 21, 22, and 0%, respectively. These results also show that effluents containing as low as 0.125 mg/L TNT negatively affect the early stages of life in bullfrog eggs following TNT exposure. In this study, TNT exposure showed increased rates of mortality to a peak of about 77% at the lowest dose of TNT.

Amphibian Microcosm Study

Tiger salamanders (*Ambystoma tigrinum*) were exposed to TNT in a soil matrix and were fed earthworms that were exposed to TNT in soil using a microcosm design for 14 days [42]. Previous dermal exposures to TNT were considerable as compared oral exposures in *Ambystomid* salamanders [35]. The TNT concentrations in the soil decreased with time, and ranged from 280 μg/g soil at the beginning of the study to 59 μg/g at the conclusion. At this time, the primary reduction products of TNT increased (39 and 62 μg/g at the beginning of the study to 58 and 78 μg/g of 2ADNT and 4ADNT at the conclusion of the study, respectively). Concentrations of TNT in earthworms ranged from 0.25 to 0.62 μg/g, and from 2.1 to 2.6 μg/g of the primary reduction products mentioned previously. Immune function, histopathology, weight changes, and blood parameters were investigated. No adverse health effects were observed, and the animals gained weight during exposure to TNT in soil.

The subchronic toxicity studies of TNT in soil was conducted using red-backed salamanders (*Plethodon cinereus*) that were exposed to soils containing TNT at concentrations of 0, 325, 750, 1,500, or 3,000 mg/kg dry weight for 28 days [43]. These studies were similar in design to those previously conducted for 14 days using *Amystoma tigrinum* [42], though food items were not exposed and exposures were primarily dermal. The levels of TNT in the soils during the 28 days period were determined in the beginning, midpoint, and conclusion of the exposures. The mean concentration in the soils at the end of the exposures were, 0.64, 41.7, 373, 1,200, 2,667 mg/kg as loss of TNT was measured, which was presumably degraded to its reduced products (e.g., 2ADNT and 4ADNT) during exposure. The levels of reductive breakdown products at the end of the exposures were 0.03, 20, 28, 27, 26; and 0.03, 28, 31, 29; and 26 mg/kg of 2ADNT and 4ADNT, respectively. The survival was reduced in salamanders that were exposed to 1500 and 3000 mg/kg by 10 and 55%, respectively. The reduction in food consumption, body weight, and hematological effects (e.g., a reduction in hemoglobin and splenic congestion) in the 750 mg/kg/soil (373 mg/kg) dose group was observed. No effects were observed at the lower dose of 325 mg/kg (41.7 mg/kg). The concentrations in parentheses represent mean concentrations determined at the end of the experiments, which were considered exposure concentrations in the determination of effect levels for salamanders. The NOAEL of 41.5 mg/kg and LOAEL of 373 mg/kg were suggested by the authors based on the results of this study.

Relevance for Amphibian TRV Development

Some of the studies described herein used a microcosm design that considered all pathways of exposure (oral, soil, and dermal) and potential variation in feeding regimes [42]. Since soil concentrations of TNT were monitored, these data were used to derive a NOAEL for terrestrial salamanders. A soil concentration of 59 mg/kg reflected all exposure pathways; however, these studies were limited to a 14-day exposure [42]. Since adverse effects were not observed in the study, a LOAEL was not available.

Bazar *et al.* [43] exposed another species using a subchronic study of TNT in a soil matrix, which suggested a NOAEL of 41.7 mg/kg, and a LOAEL of 373 mg/kg from a 28-day exposure. The toxic effects included body weight reduction and systemic toxic effects. Reduced red blood cell count was observed, and adverse effects on both the liver and spleen at LOAEL levels. The levels of TNT present in the soil were determined at the beginning, middle, and conclusion of the study. The average values of three determinations derived the exposure dose. The reduction in hemoglobin in salamanders exposed to a soil concentration of 373 mg/kg soil for 28 days is shown in Table 3.4. The benchmark dose (BMD) analysis using this reduction in hemoglobin content provided a BMD of 358 mg/kg and a BMDL of 231 mg/kg. No other amphibian toxicity was available for dermal and inhalation exposures. These data are consistent with those of Johnson *et al.* [41] where no adverse effects were reported up to 59 mg TNT/kg soil.

Table 3.4 Summary of Relevant Amphibian Data for TRV Derivation

Test Organism	Test Duration	Test Results: NOAEL (mg/kg/day)	Test Results: LOAEL (mg/kg/day)	Test Results: Effects Observed at the LOAEL	Study
South African Clawed frog (*Xenopus leavis*), FETAX Assay	96 hours	LC_{50} (16.7 µM)		Mortality at 9.78 µM; Induced malformation	[39]
Bullfrog (*Lithobates catesbeiana*)	14 days	LD_{50} (1060 mg/kg bw)	—	Changes in respiratory rhythm decrease in motor activities, salvation, and muscle tone changes, changes in skin color and ocular signs	[40]
Bullfrog tadpoles (*Rana catesbeiana*)	90 days	—	0.125 mg/L	Mortality about 20%, decrease in body weight and snout vent lengths from lowest exposure concentration	[41]
Tiger salamanders (*Ambystoma tigrium*)	14-day soil matrix	59 mg/kg soil		No adverse health effects were observed during 14-day exposure in soil matrix	[42]
Red-backed salamanders (*Plethodon cinereus*	Acute	The median lethal dose (2,481 mg/kg)	NA	Two salamanders died during the first week in 1,500 mg/kg and two others from 3000 mg/kg. Early signs of toxicity observed lethargy, and unresponsive to touch	[43]
Red-backed salamanders (*Plethodon cinereus*)	28-day soil matrix	41.7 mg/kg soil	373 mg/kg soil	Mortality occurred in first week, reduction in food consumption, body weight, hematological effects reduction in hemoglobin, and spleen congestion	[43]

mg/kg = milligrams per kilogram; bw = body weight.

SUMMARY OF REPTILIAN TOXICITY

Reptilian Toxicity: Acute and Subacute

The oral toxicity data of TNT for reptiles is limited to studies that reported acute, subacute, and subchronic toxicity in the Western Fence Lizard (*Sceloporus occidentalis*) [44]. The oral median LD_{50} values for TNT in corn oil were 1.038 and 1.579 g/kg bw for male and female lizards, respectively. Clinical signs of toxicity included anorexia, weight loss, abdominal enlargement, lethargy, tremors, and chromaturia. Lizards showed signs of recovery at 72 hours after dosing. In the 14-day subacute study, male lizards were dosed with TNT in corn oil at 0, 33, 66, 132, 263, 525, and 1,050 mg/kg each day. Mortality started to occur at a dose of approximately 5% (66 mg/kg), 67% (132 mg/kg), and 93% in high-dose treated reptiles during the 14-day study. Clinical signs of toxicity such as enlarged abdomen, tremors, seizures, and chromaturia were similar to those that were observed in the acute toxicity study. No dose-related effects were found for any of the hematological or clinical parameters measured [44].

Reptilian Toxicity: Subchronic

In a subchronic (60-day) study, male lizards were orally dosed with TNT in corn oil at 0, 3, 15, 25, 35, and 45 mg/kg/day. Animals in all dose groups showed survival rates above 90%. The decrease in food consumption and body weight was observed in lizards dosed with 35 and 45 mg/kg/day, and alterations in hematological, clinical chemistry parameters, and histopathological changes in the kidney, liver, spleen, and testis were observed in lizards dosed at 25 and 45 mg/kg/day. The benchmark dose at the lower 95% confidence limit (BMDL) was calculated using the dose-response relationship and the treatment-related decrease in hemoglobin (Hb) that was observed in male fence lizards. The LOAEL of 25 mg/kg/day and the NOAEL of 15 mg/kg/day was based on these Hb levels [44] (Table 3.5). No data were found on chronic toxicity in reptiles.

RECOMMENDED TOXICITY REFERENCE VALUES

Toxicity Reference Values for Mammals

Based on the information from five species, and as described earlier (mammalian oral toxicity), the dog is the most sensitive mammal from oral exposures to TNT. The lowest LOAEL is 8.0 mg/kg/day, where Levine *et al.* [26] reported evidence of blood effects and decreased weight gain in dogs

Table 3.5 Summary of Relevant Reptilian Data for TRV Derivation

Test Organism	Test Duration	Test Results			Study
		NOAEL (mg/kg/day)	LOAEL (mg/kg/day)	Effects Observed at the LOAEL	
Fence lizard *Sceloporus occidentalis*	acute	LD_{50} (1,038 mg/kg bw males) (1,579 mg/kg bw females)		Abdominal enlargement, excitability, tremors, chromaturia, and orange-red discoloration	[44]
Fence lizard *Sceloporus occidentalis*	14 days	—	Mortality occurred starting about 5% from 333 mg /kg to 93% at high dose, 1050 mg/kg	Threshold motility between 66 and 132 mg/kg. Adverse effects inconsistent to acute toxicity. No dose- related effects in hematological parameters	[44]
Fence lizard *Sceloporus occidentalis*	60 days	15 mg/kg	25 mg/kg	Survival rate was about 90%; decreased food consumption and body weight; altered hematological and clinical chemistry; histopathological changes in kidney, liver, and spleen; treatment- related decrease in hemoglobin	[44]

mg/kg = milligrams per kilogram; bw = body weight.

receiving 8.0 mg/kg/day. The NOAEL within the same endpoints and species was the dose of 2.0 mg/kg/day reported by the same authors. Since decreased gain of weight (an indicator of reduced growth and/or energy efficiency) and anemia have the potential to adversely affect future fitness, these endpoints are considered ecologically relevant.

In addition, this and other studies satisfy the minimum data set requirements to derive TRVs. Thus, uncertainty factors were not used to derive TRVs. The data were appropriate for a benchmark dose derivation. A BMD of 1.7 mg/kg/day was calculated from the model fit of the mean response at one standard deviation using the hemoglobin data. A lower-bound on the BMDL was calculated as 0.9 mg/kg/day from the lower 95% confidence interval (CI) of the modeled curve (Figure 3.2). These values were selected as the class-specific TRVs (see Table 3.6). Since these studies were well calibrated, data are available from five species of two taxonomic orders, and the results are consistent with those of others, this TRV is given a high degree of confidence.

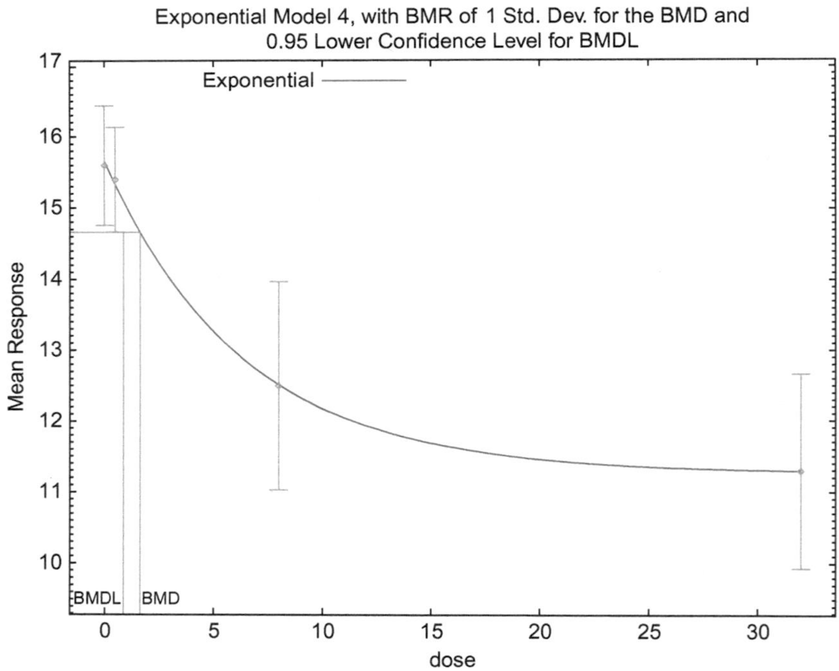

Figure 3.2 Benchmark dose fit of hemoglobin data from chronic oral TNT exposures from the dog using the exponential model [26].

Table 3.6 Selected Ingestion TRVs of TNT for the Class Mammalia

TRV	Dose	Confidence
BMD	1.7 mg/kg-bw/day	High
BMDL	0.9 mg/kg-bw/day	High

Source: [26].

TRVs for Ingestion Exposures for Mammalian Foraging Guilds

TRVs specific to particular guild associations (e.g., small herbivorous mammals) have not yet been derived. However, since the dog is the most sensitive mammal tested, the class-specific TRVs shown in Table 3.6 are considered as being protective of noncarnivorous mammals. TRVs that are more specific may be developed when considering the data provided in Table 3.2. No TRVs were derived for inhalation or dermal exposures of TNT at this time.

Toxicity Reference Values for Birds

The subchronic studies in the Northern Bobwhite and Common Pigeon [37] were well conducted using two different methods of oral administration to two species representing two orders within the Class Aves. The authors of both studies suggested that indicators of anemia were not considered to be biologically relevant as adverse effects, since the mean values were within normal ranges for each species of review. However, those data provided a sensitive indicator of sublethal effect slightly protective of other adverse findings (e.g., kidney, ovarian, and liver mass changes [37]). Therefore, the authors used changes in hemoglobin levels in male and female Pigeons to model the benchmark dose from the 60-day dosing study in Pigeons [36]. They modeled male and female hemoglobin levels and RBC counts using the suite of BMDS continuous models. They found the female Hb data gave acceptable estimates of the BMD and BMDL, and based on the graphical data plots and the various statistical tests performed, the most reliable estimates were provided by the Exp2, Exp4, and linear models using one standard deviation to establish the BMDL. Since the BMD estimates from these models were very similar, they averaged the three estimates. Using this approach, the BMD was 89.9 mg/kg/day, and the BMDL was 61.4 mg/kg/day, which resulted in a TRV_{high} and TRV_{low} of 90 and 60 mg/kg/day, respectively (Figure 3.3, Table 3.7). These estimates are roughly consistent with the NOAELs and LOAELs described by the author for both studies. A medium level of confidence was associated with these

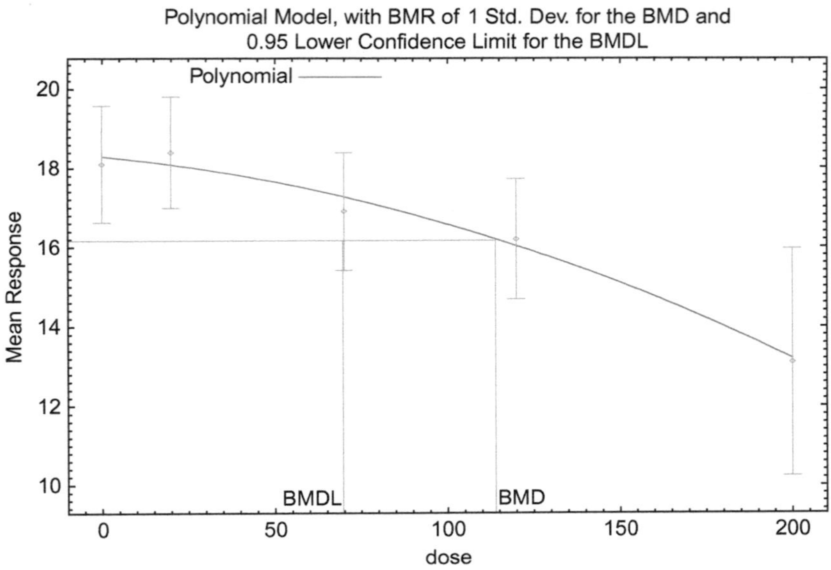

Figure 3.3 Polynomial model example of female hemoglobin changes relative to dose in the Common Pigeon (*Columba livia*).

Table 3.7 Selected Ingestion TRVs of TNT for the Class Aves

TRV	Dose	Confidence Level
BMD	90 mg/kg-bw/day	Medium
BMDL	60 mg/kg-bw/day	Medium

Source: [37].

estimates as both studies were considered of sufficient quality and were comprehensive in the endpoints investigated.

TRVs for Amphibians

Given the average life span of *Ambystomid* salamanders (>10 years), and that the exposures were relatively brief, they were thus classified as subacute exposures (14-day) [42]. In addition, since dermal exposures to TNT were reported as considerable, an oral TRV would be inappropriate. However, since this study used a media exposure regime, a media-based value for soil was derived. In this study, the NOAEL (14-day) of TNT in soil (59 μg/g) was determined. The NOAEL of 41.7 mg/kg and LOAEL of 373 mg/kg were determined for hematological effects in red-backed salamanders following soil exposures of TNT at various concentrations for 28 days [43].

This microcosm study provides amphibian data that were useful in deriving TRVs for terrestrial amphibians and results were consistent with the data from the Johnson *et al.* [42] study using an Ambystomid species. Data are available for two species of different families within the same order (Caudata). The data were limited for aquatic exposures of TNT to *L. catesbeiana*; therefore, a TRV for aquatic amphibians was not derived.

Using the data from Bazar *et al.* [43], a BMD analysis was performed on the hemoglobin content, which showed a dose-dependent decrease. The soil matrix concentrations were determined at the beginning, middle, and conclusion of the study. The average TNT concentration at the conclusion of the study was used to provide data for BMD calculation, since it decreased with time and represented a conservative estimate for derivation of BMDL. The concentrations were 41.7, 37.3, 1,200, and 2,667 mg/kg of dry soil. The model predicted a BMD of 358 and a BMDL of 231 mg/kg of dry soil considering one standard deviation from the predicted BMD for hematological effects [43]. An uncertainty factor (UF) of 10 was used to extrapolate subacute/subchronic to chronic exposures to derive a BMDL and BMD of 23 and 36, respectively (Table 3.8). These values have low to medium level of confidence, as data are limited in duration and species diversity.

TRVs for Reptiles

Only one study provided acute, 14-day and 60-day oral toxicity data for TNT dosing using reptiles [44]. No other data were found for reptiles (Table 3.9). The hematological effects that were observed in lizards were considered relevant to derive a BMD and a BMDL. A decrease in

Table 3.8 Selected Ingestion TRVs for Terrestrial Amphibians

TRV	Dose	Confidence Level
BMD	36 mg/kg/day	Low
BMDL	23 mg/kg/day	Low

Sources: [42,43].

Table 3.9 Selected Soil TRVs of TNT for the Class Reptilia

TRV	Dose	Confidence Level
BMD	19 mg/kg	Low
BMDL	14 mg/kg	Low

Source: [44].

hemoglobin content that was proportional to increased TNT exposure followed a dose response relationship in this study [44].

The hemoglobin and mean corpuscular hemoglobin concentration (MCHC) exhibited the greatest change at a dose of 25 mg/kg/day. The measurements were approximately 20% lower than were found in the control group. These hematological effects have the potential for physiological and ecological consequences in wild populations. The decreases in hematological parameters were also reported for TNT exposure in other animals, which suggested that these hematological effects are sensitive and a physiologically relevant target organ for TNT toxicity in the Western Fence Lizard.

A BMD analysis was performed on the lizard's hematological data from a 60-day study. The initial trails could not fit a dose-response curve to the data. We determined the variance in the two high-dose groups were skewing the data plots and resulting in an inadequate model fit of the dose response function. Data from the two highest doses were dropped and the data were fitted with an appropriate curve. Examination of these new plots showed that estimated control value was far from the actual controls so we specified the control value to adapt the curves through the control mean. Once these adjustments were made a BMD of 19.4 mg/kg/day and BMDL of 13.5 mg/kg/day was derived. There are no other toxicity studies on TNT reported for reptiles. The BMD of 19 mg/kg/day and BMDL of 14 mg/kg/day were therefore derived to be protective of reptiles exposed to oral TNT concentrations (Figure 3.3). A low level of confidence was given since no data are available for any other species of reptile.

REFERENCES

[1] U.S. Department of the Army. Technical manual. Military explosives. Washington (DC): Headquarters, U.S. Materiel Command; September 1984, with Change 4; 25 September 1990. TM 9-1300-214.

[2] Ryon MG, Pal BC, Talmage SS, Ross RH. Database assessment of the health and environmental effects of munition production waste products. Final report. Oak Ridge (TN): Oak Ridge National Laboratory; 1984, Contract No. PO83PP3802-AD ORNL-6018.

[3] Rosenblatt DH, Small MJ, Barkley JJ. Munitions production products of potential concern as waterborne pollutants Phase I. Edgewood Arsenal, (MD): U.S. Army Medical Environmental Engineering Research Unit; 1973, Report No. 73-07.

[4] Hovatter PS, Talmage SS, Opresko DM, Ross RH. Ecotoxicity of nitroaromatics to aquatic and terrestrial species at Army Superfund sites. In: Doane TR, Hinman ML, editors. Environmental toxicology and risk assessment: modeling and risk assessment, Sixth Vol. West Conshohocken (PA): American Society for Testing and Materials; 1997. p. 117–29.

[5] Walsh ME, Jenkins TF. Identification of TNT transformation products in soil. U.S. Army Corps of Engineers, Cold Regions Research & Engineering Laboratory; 1992. Prepared for: U.S. Army Waterways Experiment Station, Vicksburg, Mississippi. Special Report 92-16; ADA 255308. Available at, www.dtic.mil/cgi-bin/GetTRDoc?AD=ADA255308.
[6] Agency of Toxic Substances and Disease Registry (ATSDR). Toxicological profile for 2,4,6-trinitrotoluene. Atlanta (GA): U.S. Department of Health & Human Services, Public Health Service; 1995.
[7] Spanggord RJ, Mill T, Chou TW, Mabey WR, Smith JH, Lee S. Environmental fate studies on certain munition wastewater constituents, final report, Phase II-Laboratory studies. Menlo Park (CA): SRI International; 1980, ADA0099256.
[8] Burrows EP, Rosenblatt DH, Mitchell WR, Parme DL. Organic explosives and related compounds: environmental and health considerations. Fort Detrick, Frederick (MD): U. S. Army Biomedical Research and Development Laboratory; 1989, Technical Report No.: ADA 210554.
[9] Hazardous Substance Data Bank (HSDB). TNT [Internet]. Bethesda (MD): National Library of Medicine (US), National Institutes of Health [cited 2009]. Available from: http://toxnet.njm.nih.gov/.
[10] Johnson MS, Salice CJ, Sample BE, Robidoux P-Y. Bioconcentration, bioaccumulation, and biomagnification of nitroaromatic and nitramine explosives in terrestrial systems. In: Sunhara GI, Lutufo G, Kuperman RG, Hawari J, editors. Ecotoxicology of explosives. Boca Raton (FL): CRC Press; 2009. p. 227–52.
[11] Won WD, Heckly RJ, Glover DJ, Hoffsommer JC. Metabolic disposition of 2,4,6-trinitrotoluene. Appl Microbiol 1974;27(3):513–6.
[12] Jerger DE, Simon PB, Weitzel RL, Schenk JE. Aquatic field surveys at Iowa, Radford, and Joliet Army Ammunition plants. Volume III. Microbial investigations, Iowa and Joliet Army Ammunition plants. Washington (DC): U.S. Army Medical Research and Development Command; 1976, Accession Number: ADA036778.
[13] Spanggord RJ, Mill T, Chou TW, Mabey WR, Smith JH, Lee S. Environmental fate studies on certain munition wastewater constituents, final report, Phase II-Laboratory studies. Menlo Park (CA): SRI International; 1980, ADA0099256.
[14] Monteil-Rivera F, Halasz A, Groom C, Zhao J-S, Thibuottot S, Ampleman G, et al. Fate and transport of explosives in the environment: a chemist's review. In: Sunahara HI, Lotufo G, Kuperman RG, Hawari J, editors. Ecotoxicology of explosives. Boca Raton (FL): CRC Press; 2009.
[15] Harvey SD, Fellows RJ, Cataldo DA, Bean MR. Analysis of 2,4,6-trinitrotolune and its transformation products in soils and plant by high-performance liquid chromatography. J Chromato 1990;518:361–74.
[16] Pennington JC. Plant uptake of 2,4,6-trinitrotoluene, 4-amino-2,6-dinitrotoluene and 2-amino-4-,6-dinitotoluene using 14c-labeled and unlabeled compounds. Vicksburg (MS): Department of the Army Waterways Experiment Station, Corps of Engineers; 1988, Prepared for: U.S. Army Biomedical Research and Development Laboratory, Fort Detrick, MD. Technical Report No.: EL-88-10. Accession Number: AD-A203690.
[17] Talmage SS, Opresko DM, Maxwell CJ, Welsh CJ, Cretella FM, Reno PH, et al. Nitroaromatic munition compounds: environmental effects and screening values. Rev Environ Contam Toxicol 1999;161:1–156.
[18] Dilley JV, Tyson CA, Spanggord RJ, Sasmore DP, Newell GW, Dacre JC. Short-term oral toxicity of a 2,4,6-trinitrotoluene and hexahydro-1,3,5-trinitro-1,3,5-triazine mixture in mice, rats, and dogs. J Toxicol Environ Health 1982;9(4):587–610.
[19] Voegtlin C, Hooper CW, Johnson JM. Trinitrotoluene poisoning – its nature, diagnosis, and prevention. J Ind Hygiene 1921;3:239–54.

[20] Bredow M, Jung F. Methemoglobin formation. XXV. Comparative toxicity of some aromatic nitro derivatives. Arch Exp Pathol Pharmakol 1942;200:335.
[21] Lillie RD. Notes on the pathology of experimental trinitrotoluene poisoning. Publ Health Rep (1896–1970) 1943;58:436–8, U.S. Public Health Service.
[22] Reddy G, Chandra S, Lish J, Qualls C. Toxicity of 2,4,6-trinitrotoluene (TNT) in hispid cotton rats (*Sigmodon hispidus*): hematological, biochemical, and pathological effects. Int J Toxicol 2000;19(3):1–9.
[23] Johnson MS, Ferguson JW, Holladay SD. Immune effects of oral 2,4,6-trinitrotoluene (TNT) exposure to the white-footed mouse, Peromyscus leucopus. Int J Toxicol 2000;19(1):5–11.
[24] von Oettingen WF, Donahue DD, Synder RK, Horecker BL, Monoco RA, Lawton AH, et al. Experimental studies on the toxicity and potential dangers of trinitrotoluene (TNT). Washington (DC): U.S. Public Health Service; 1944, Public Health Bulletin No. 285.
[25] Dilley JV, Tyson CA, Spanggord RJ, Sasmore DP, Newell GW, Dacre JC. Short-term oral toxicity of 2, 4, 6-trinitrotoluene in mice, rats, and dogs. J Toxicol Environ Health 1982;9(4):565–85.
[26] Levine BS, Rust JH, Barkley JJ, Furedi EM, Lish PM. Six-month oral toxicity study of trinitrotoluene in beagle dogs. Toxicology 1990;63(2):233–44.
[27] Levine BS, Furedi EM, Gordon DE, Barkley JJ, Lish PM. Toxic interactions of the munitions compounds TNT and RDX in F344 rats. Fundam Appl Toxicol 1990; 15(2):373–80.
[28] Levine BS, Furedi EM, Gordon DE, Lish PM, Barkley JJ. Subchronic toxicity of trinitrotoluene in Fischer 344 rats. Toxicology 1984;32(3):253–65.
[29] Furedi EM, Levine BS, Gordon DE, Rac VS, Lish PM. Determination of the chronic mammalian toxicological effects of TNT (Twenty-four month chronic toxicity/carcinogenicity study of trinitrotoluene (TNT) in the Fischer 344 rat). Final report Phase III, vol. I. Chicago (IL): IIT Research Institute; 1984, Prepared for: U.S. Army Medical Research and Development Command, Fort Detrick, Frederick, Maryland. Accession Number: ADA168637.
[30] McCain WC. Fourteen-day range finding and ninety-day feeding study of 2,4,6-trinitrotoluene in the white-footed mouse (*Peromyscus leucopus*). Aberdeen Proving Ground (MD): U.S. Army Center for Health Promotion and Preventive Medicine; 1998, Toxicological Study No. 2340-38-95-6-1.
[31] Hathaway JA. Trinitrotoluene: a review of reported dose-related effects providing documentation for a workplace standard. J Occup Med 1977;19(5):341–5.
[32] Woollen BH, Hall MG, Craig R, Steel GT. Trinitrotoluene: assessment of occupational absorption during the manufacture of explosives. Br J Ind Med 1986;43(7):465–73.
[33] Reifenrath WG. Assessment of skin penetration of environmental contaminants in air and bioremediated soil utilizing the pig skin model: percutaneous absorption of carbon-14 labeled trinitrotoluene from air and soil. Novato (CA): Reifenrath Consulting and Research; 1994, Prepared for: U.S. Army Medical Research, Development, Acquisition and Logistics Command, Fort Detrick, Frederick, Maryland. Contract No. DAMD17-93-C-3167. Accession Number: ADA290333.
[34] Reifenrath WG, Kammen HO, Palmer WG, Major MM, Leach JG. Percutaneous absorption of explosives and related carbon compounds: an empirical model of bioavailabilty of organic nitro compounds from soil. Toxicol Appl Pharmacol 2002; 182(2):160–8.
[35] Johnson MS, Franke LS, Lee RB, Holladay SD. Bioaccumulation of 2,4,6-trinitrotoluene and polychlorinated biphenyls through two routes of exposure in a terrestrial amphibian: Is the dermal route significant? Environ Toxicol Chem 1999; 18(5):873–8.

[36] Gogal Jr. RM. Influence of feed exposure with TNT derivatives on quail (Northern bobwhite, *Colinus virginianus*) immunity as measured by the National Toxicology Program Immune Endpoints. Washington (DC): National Toxicology Program; 2000, Avian toxicology study No.990503, 981124G Final Report (Contract Report) to the U.S. Army Center for Health Promotion and Preventive Medicine.
[37] Johnson MS, Michie MW, Bazar MA, Salice CJ, Gogal RM. Responses of oral 2,4,6-trinitrotolune (TNT) exposure to the common pigeon (*Columba livia*): a phylogenic and methodological comparison. Int J Toxicol 2005;24(4):221–9.
[38] Gogal Jr. RM, Johnson MJ, Larsen CT, Prater MR, Duncan RB, Ward DL, et al. Influence of dietary exposure to 2,4,6-trinitrotoluene exposure in the northern bobwhite (*Colinus virginianus*). Environ Toxicol Chem 2002;21(1):81–6.
[39] Saka M. Developmental toxicity of p, p'-dichlorodiphenyltrichloroethane, 2,4,6-trinitrotoluene, their metabolites, and benzo[a] pyrene in Xenopus laevis embryos. Environ Toxicol Chem 2004;23(4):1065–73.
[40] Paden NE, Smith EE, Kendall RJ. Acute toxicity of 2,4,6-trinitrotoluene, 2,4-dinitrotoluene, 2,6-dinitrotoluene in the adult bullfrog (*Lithobates catesbeiana*). Bull Environ Contamin Toxicol 2008;80(6):487–91.
[41] Paden NE, Smith EE, Maul JD, Kendall RJ. Effects of chronic 2,4,6-trinitrotoluene, 2,4-dinitrotoluene, and 2,6-dinitrotoluene exposure on developing bullfrog (*Rana catesbeiana*) tadpoles. Ecotoxicol Environ Saf 2011;74(4):924–8.
[42] Johnson MS, Holladay SD, Lippenholz KS, Jenkins JL, McCain WC. Effects of 2,4,6-trinitrotoluene in a holistic environmental exposure regime no a terrestrial salamander: *Ambystoma tigrinum*. Toxicol Pathol 2000;28(2):334–41.
[43] Bazar MA, Quinn MJ, Mozzachio K, Johnson MS. Toxicological response of red-backed salamanders (*Plethodon cinereus*) to subchronic soil exposure of 2,4,6-trinitrotoluene. Environ Toxicol Chem 2008;27(6):1393–8.
[44] McFarland CA, Quinn Jr. MJ, Bazar MA, Remick AK, Talent LG, Johnson MS. Toxicity of oral exposure to 2,4,6-trinitrotoluene in western fence lizard (*Sceloporus occidentalis*). Environ Toxicol Chem 2008;27(5):1102–11.

CHAPTER 4

Wildlife Toxicity Assessment for 1,3,5-Trinitrohexahydro-1,3,5-Triazine (RDX)

Desmond I. Bannon, Lawrence R. Williams

Contents

INTRODUCTION

The explosive 1,3,5-Trinitrohexahydro-1,3,5-Triazine (CAS No. 121-82-4) has several common synonyms, including 1,3,5-triaza-1,3,5-trinitrocyclohexane, hexahydro-1,3,5-trinitro-1,3,5-triazine, cyclotrimethylenenitramine, hexogen, and cyclonite, but is most frequently known as RDX (Royal Demolition Explosive) [1]. RDX has found widespread application in detonators, grenades, bombs, and a variety of other military ordnance. Structurally, the compound is a trinitro-substituted triazine with the empirical formula, $C_3H_6N_6O_6$. Due to the presence of unexploded ordinance, or manufacturing sites due to spills and releases, RDX can contaminate military sites, and this poses potential toxicity to wildlife and other ecological receptors.

Wildlife Toxicity Assessments for Chemicals of Military Concern
http://dx.doi.org/10.1016/B978-0-12-800020-5.00004-1

This Wildlife Toxicity Assessment (WTA) summarizes current knowledge of the likely harmful impacts of RDX on wildlife based on published toxicity studies and, where applicable, uses available studies to derive Toxicity Reference Values (TRVs) for RDX. The TRVs are intended to serve as protective exposure standards for wildlife ranging in the vicinity of contaminated sites. The basis for this WTA is the U.S. Army Center for Health Promotion and Preventive Medicine (USACHPPM) Technical Guide (TG) 254, the *Standard Practice for Wildlife Toxicity Reference Values* [2].

TOXICITY PROFILE

Environmental Fate and Transport

Military grade RDX (containing about 10% high melting explosive (HMX) by weight) has been a widely used explosive mixture since the early years of World War II, when it began to either replace or supplement trinitrotoluene (TNT) as the primary ingredient in shells, bombs, and detonators. RDX may occur in many commercial and military activities including manufacturing of munitions, waste discharge during manufacture (up to 12 micrograms per milliliter (μg/mL) in process wastewater during the manufacture of RDX [3], testing and training at military facilities, demilitarization programs, and open burning/open detonation [4]. These activities represent the primary release mechanisms of RDX in the environment, and have resulted in extensive RDX contamination of soil, surface water, and groundwater at many military installations [4,5]. Although the sole manufacturing site in the United States is the Holston Army Ammunition Plant (AAP) in Kingsport, Tennessee, a pattern of manufacturing and assembling practices has resulted in RDX release to the environment. RDX also accumulates at other military sites on training and testing ranges, and due to its weak adsorption to soil, RDX migrates into groundwater [3,4,6]. Talmage *et al.* [7] reported that concentrations of up to 30 milligrams per liter (mg/L) RDX were detected in groundwater at Milan AAP, while surface water impoundments and associated sediments at this facility also displayed concentrations in the parts per million (ppm) ranges. Talmage *et al.* [7] list soil concentrations of up to 13,900 mg RDX per kilogram (kg) for this and other military sites. Table 4.1 lists the physicochemical properties of RDX that are relevant to its environmental fate and transport.

In the environment, RDX can break down to mono-, di-, and trinitroso derivatives (MNX, DNX, and TNX) and the ring cleavage products

Table 4.1 Summary of Physical-Chemical Properties of RDX

CAS No.	121-82-4
Molecular weight	222.26
Color	White
Physical state	Crystalline solid
Melting point	205 – 206 °C
Boiling point	Decomposes
Odor	No data
Solubility water	38.4 mg/L; slightly soluble in methanol, ether, ethyl acetate, glacial acetic acid
Partition coefficients:	
Log K_{ow}	0.87
Log K_{oc}	0.84 – 2.2
Vapor pressure at 20 °C	1.0×10^{-9}, 4.0×10^{-9} mm Hg
Henry's Law constant	1.2×10^{-5} atm.m3/mole
Conversion factors	1 ppm = 9.1 mg/m^3 1 mg/m^3 = 0.11 ppm

Sources: ATSDR [1], Talmage *et al.* [7], HSDB [8].

methylenedinitramine (MEDINA) and 4-nitro-2,4-diazabutanal (NDAB) [3,4,6,9]. Published toxicological studies on TNX, MNX, and DNX are uncommon compared with RDX. Available toxicity data on breakdown products are primarily studies of potential carcinogenicity of TNX. Where appropriate, these breakdown products and/or metabolites of RDX will be discussed when relevant to wildlife toxicity.

RDX has an estimated vapor pressure of 1–4 × 10^{-9} millimeters of mercury (mmHg) at 25 °C, a low value implying that partitioning to air is unlikely. Furthermore, the compound is soluble only to a limited extent in a number of common organic solvents and in water (38.4 mg/L at 20 – 25 °C). However, despite its limited solubility, the compound has been detected in both surface water and groundwater (see Talmage *et al.* [7] for review). Hovatter *et al.* [10] and Talmage *et al.* [7] also present RDX soil concentration data from other studies for a number of AAPs, depots, and arsenals.

RDX can absorb ultraviolet light strongly at wavelengths between 240 and 250 nanometers (nm). In addition, biodegradation of RDX has been demonstrated under anaerobic conditions in the presence of a number of microbial isolates and mixed cultures, with total degradation in 5 days or less. Thus, when RDX was incubated in an anaerobic test system containing sewage sludge and mixed cultures in nutrient broth, the disappearance of RDX was accompanied by the formation of a range of metabolites including hexahydro-1-nitroso-3,5-dinitro-1,3,5-triazine (MNX), hexahydro-1,3-dinitroso-5-nitro-1,3,5-triazine (DNX), hexahydro-1,3,5-trinitroso-1,3,5-triazine (TNX), hydrazine, 1,1-dimethyl-hydrazine, 1,2-dimethyl-hydrazine, formaldehyde, and methanol [3,9]. Biodegradation of RDX is considered a key factor in determining its fate, rather than abiotic factors [11].

The relative oral bioavailability of RDX in contaminated soil has been examined in rats by Crouse *et al.* [12]. Two soils containing aged RDX were sieved to less than 600 micron and measured 2,300 μg/g and 527 μg/g RDX. Rats were orally dosed with either contaminated soil or an equivalent amount of neat RDX and the area under the curve over time used to compare the aged soil with the neat RDX. The relative bioavailability was 20 to 30% lower for the first soil and 15 to 38% lower for the second, indicating a potential for bioavailability adjustments in clean-up of RDX contaminated lands. However, unlike lead (to which bioavailability adjustments have been made), RDX does not have established models of bioavailability nor allowable limits in soil.

Studies have shown that plants are able to absorb RDX from soil and to a lesser degree from irrigation water. Radio-labeled RDX accumulated in bush bean grown on RDX amended soil with the highest concentration of RDX in the seeds followed, in order of decreasing concentration, by leaves, stems, roots, and pods [13]. Concentration ratios were on the order of 20 to 60% for seeds and leaves, which suggests an efficient uptake mechanism and high plant mobility. Analysis of bush bean grown in RDX amended hydroponic solution showed that approximately 23 and 50% of the radiolabel present in the root and leaves, respectively, was parent RDX after a 7-day exposure [14]. The efficiency of RDX absorption varies with species and is inversely proportional to organic matter content of the soil. Studies on uptake of RDX from spiked irrigation water showed a lower uptake of RDX by tomato, bush bean, corn, soybean, alfalfa, lettuce, and radish [15]. Concentrations of RDX in the plants were less than that of the irrigation water. Thus, research to date indicates that plant uptake of RDX is highest from RDX contaminated soils and, importantly, that RDX in plants can be a potential exposure route for herbivorous terrestrial wildlife.

Metabolism of RDX in mammalian systems has been studied by several investigators. Bhushan *et al.* [16], used rabbit liver cytochrome P450 *in vitro*. The disappearance of RDX was simultaneously accompanied by the release of nitrite and formaldehyde. In addition, two novel metabolites, 4-nitro-2, 4-diazabutenal (NDAB) and MEDINA, were produced under aerobic or anaerobic conditions, respectively. The rate of RDX biotransformation was about three-fold higher under anaerobic than aerobic conditions. This was thought to be because the Fe^{2+} site of the prosthetic group of the heme-containing P450 also binds oxygen (O_2) so that O_2 competes with the substrate (RDX). This provides a mechanistic explanation for the inhibition of RDX metabolism under aerobic conditions. An interesting paper by Jackson *et al.* [17; see also 18] examined the biochemical properties of RDX using XPIA, an unusual bacterial P450, which was used to degrade RDX. XPIA has been found only in *Rhodococcus* bacteria isolated from RDX contaminated land indicating that XPIA may have evolved under selective pressure. Purified XPIA metabolized RDX under aerobic and anaerobic conditions. Under both conditions, nitrite and formaldehyde were metabolic products, but the distinct compounds, MEDINA and NDAB, were produced by denitration under anaerobic and aerobic conditions, respectively. These breakdown products, though unstable and short-lived, may have significance for long-term exposure studies. *In vivo* metabolism of RDX has been shown in both deer mice (*Peromyscus maniculatus*) [19] and a laboratory mouse strain (B6C3F1) [20], with nitroso-derivatives found at low concentrations (relative to RDX) in the stomach and brain and other organs, but none of these studies measured the ring cleavage products NDAB or MEDINA. The significance of these metabolites for acute or chronic toxicity is not clear, and there has not yet been a comprehensive examination of the mammalian metabolism of RDX that includes all known metabolites.

Mammalian Toxicity

Mammalian Oral Toxicity: Acute

The acute neurological effects of RDX at high doses converge into frank tonic-clonic seizure, with repeated seizure being lethal, possibly due to respiratory failure. Many studies have observed these seizures [21,22], though there has been little work carried out on their neurological or mechanistic basis. These seizures can confound subchronic studies since they may cause animals to become moribund, though they are often transient. They have been observed in humans [23–25], pigs [26,27], dogs [22], rodents [28–30], and birds [31–33]; the no observed adverse effect level (NOAEL) is consistent across taxonomic groups. *In vitro* studies suggested that RDX may affect calcium signaling in neurons [34], but *in vivo*, RDX-induced seizures were not blocked by verapramil, a potent calcium channel blocker [35]. In 2011, Williams *et al.* [36] discovered the mechanism of the convulsive effect of RDX to be due to the affinity and binding of RDX to the so-called convulsant site on the $GABA_A$ receptor in the brain, the major receptor for inhibitory neuronal input. Inhibition of this receptor by RDX blocks the $GABA_A$ chloride channel, disabling inhibitory input to excitatory neurons, and enables exacerbated excitatory activity resulting in seizure, and in most nonhuman species, death. Using single doses of RDX without seizure, Bannon *et al.* used microarrays to examine the temporal expression of genes in rat liver and brain over 48 hours, at which time the dose had been eliminated. There were dramatic changes in gene expression at the peak RDX concentration of 3.5 hours indicating an acute phase response to the internal dose of RDX. In addition, RDX has been reported to induce changes in expression deoxyribonucleic acid (DNA) and micro-RNA either directly or as a result of RDX-induced seizures [37–39].

Dilley *et al.* [40] conducted a number of toxicological tests on TNT, RDX, and a mixture of the two compounds styled LAP (load, assembly, and pack wastewater), employing mixtures of each compound in corn oil that were administered to the test animals by gavage. Mortality and clinical parameters were observed on all survivors for 14 days prior to termination. The acute oral LD_{50} for RDX in male Sprague Dawley rats was 71 mg/kg/day compared to values of 1,320 and 574 for TNT and LAP, respectively. Dilley *et al.* [40] reported respective acute oral LD_{50} values of less than 75 and 86 mg/kg/day for RDX in male and female Swiss-Webster mice, compared to 660 mg/kg/day for TNT in either sex of mouse and 947 and 1,131 mg/kg/day for LAP in male and female mice, respectively.

Cholakis *et al.* [41] also provided data on the acute oral lethality of RDX in experimental animals. This group determined LD_{50} values in male and female F344 rats (10/sex/group), and B6C3F1 mice (5/sex/group) that were given single doses of the compound by gavage in either 1% aqueous methylcellulose (rats) or a mixture of 1% methylcellulose and 1% polysorbate 80 (mice). Broadly consistent with the findings of Dilley *et al.* [40], Cholakis *et al.* [41] derived a combined (male and female) acute oral LD_{50} of 118.1 mg RDX/kg/day in F344 rats and 80.3 mg/kg/day in mice. Overall, the acute lethality data on RDX of Dilley *et al.* [40] and Cholakis *et al.* [41] have yielded lower values than the LD_{50} of 200 mg/kg/day obtained for the compound in the earlier studies of von Oettingen *et al.* [42].

In addition to acute oral lethality, single dose experiments with RDX have been used to determine the toxicokinetic behavior of the compound in experimental animals. For example, Schneider *et al.* [26] administered 100 mg RDX/kg by gavage to Sprague Dawley rats (n = 70, sex not specified) and Pittman-Moore miniature swine (n = 10 female) and monitored the partitioning of the compound between feces, urine, and the major organs and tissues. Only a small amount of RDX (less than 3%) was recovered in the feces of the rats, suggesting that the bulk of the material had been transported across the gastrointestinal absorption barrier. When 50 mg/kg ^{14}C-labeled RDX was administered to the rats, most of the radioactivity was found in the liver and urine after 24 hours, with further partitioning to other parts of the body during the next three days. Overall, 43% of the radioactivity was expired as $^{14}CO_2$ indicating that a significant amount of metabolism was taking place. Mammalian health effects of RDX are summarized in Figure 4.1.

Inferential support for the concept of the liver as a major catabolic site for RDX is provided by French *et al.* [43] who, in a meeting abstract, reported profound ultrastructural changes in the liver of rats (strain, sex, number

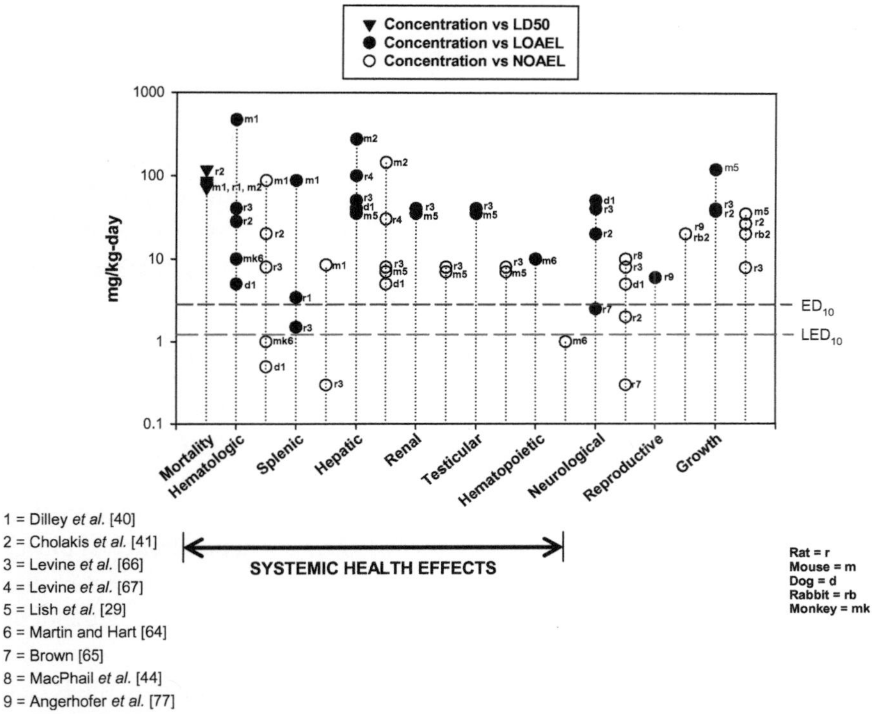

Figure 4.1 RDX health effects to mammals. Letters by symbols represent test species, and numbers represent study.

unstated) as a result of oral administration of a single dose of 100 mg RDX/kg. Among other membrane perturbations, the smooth endoplasmic reticulum was highly proliferated after 48 hours, possibly indicating the potential induction of the mixed function oxidase system. By contrast, ultrastructural changes to the kidney due to RDX were minor and inconsistent.

While the liver appears likely to be the primary site of RDX catabolism, the compound itself or another pharmacologically active metabolite has the capacity to induce neurotoxicological responses in male and female Sprague Dawley rats. Thus, in the acute section of a multiphase study, MacPhail *et al.* [44] administered single gavage doses of up to 50 mg RDX/kg in 2% carboxymethylcellulose and observed overall decreases in such responses as startle-response amplitude, startle-response latency, figure-8 maze motor activity, conditioned flavor aversions, and schedule-controlled responses.

Fellows *et al.* [45] used the prairie vole (*Microtus orchogaster*) to show that plants incorporated ^{14}C-labeled RDX wherein more than 75% of the radiolabelled dose was found in the fecal nonabsorbed bulk and 95% of all label was recovered. A study by Bannon [27] using juvenile Yorkshire female pigs showed that animals given 10 to 15 mg/kg pure RDX in a capsule rapidly absorbed the compound, reaching a peak at 3 hours. At the time of peak absorption, the highest concentrations of RDX were found in cerebral cortex and hippocampus, followed by heart, kidney, lungs, muscle, and liver. Interestingly, those pigs given a one-time oral dose of only 15 mg/kg RDX had seizures within 5 minutes, indicating that juvenile swine could be more sensitive to the neurological effects of RDX, since it takes greater than twice that concentration to induce seizure in adult rats.

An interesting approach using the U.S. Environmental Protection Agency's up-and-down procedure examined the age-dependent acute oral toxicity of RDX and its metabolites, MNX, DNX, and TNX [46]. The median lethal doses for 21-, 50-, and 200-day-old deer mice (*Peromyscus maniculatus*) for RDX were 136, 319, and 158 mg/kg; for MNX were 181, 575, and 542 mg/kg; and for TNX were 338, 338, and 999 mg/kg, respectively. Other factors on which the MLD might depend, such as purity of compound, absorption (age dependent), and precision of the technique, were not discussed. Nonetheless, there seems to be an age-dependent decrease in MLD with increasing age.

An extensive literature search was conducted on the toxicity of the RDX breakdown products: TNX, MNX, and DNX. Although numerous toxicity studies have been conducted for RDX, toxicity information on TXN,

MNX, and DNX were exceedingly rare. In fact, no toxicity studies were available for DNX. Two acute studies were found for TNX, and only one for MNX. A chronic study of TNX also appeared to have been conducted in association with the acute test, but no survival information or other reproductive/developmental information was reported [47]. Several papers discussed the carcinogenicity of TNX. However, all carcinogenicity studies focused on the measurement endpoints commonly performed in carcinogenicity tests such as histological examination and hepatocarcinogenic effects. In one of the studies, mitosis count increased in the presence of dimethylsulphoxide (DMSO) alone. Dimethylsulphoxide is often used as a solvent carrier of TNX and such finding indicates that DMSO may have contributed to the positive carcinogenicity test result for TNX. Because of the possible confounding factor due to DMSO, it is probable that TNX is not a carcinogen, as other researchers have observed [47–51].

Druckrey *et al.* [47] reported the oral LD_{50} for TNX to be 160 mg/kg in rats, and stated that no liver damage was observed. Smith *et al.* [46] reported an oral LD_{50} for TNX in deer mice at 12-, 50-, and 200-day-old mice to be 338, 338, and 999 mg/kg, respectively. Smith *et al.* [46] also determined the LD_{50} for MNX in deer mice at the same ages to be 181, 575, and 542 mg/kg, respectively. Meyer *et al.* [52] found the LD_{50} for MNX in female Sprague Dawley rats to be 187 mg/kg. Central nervous system (CNS) toxicity was observed at this exposure level prior to death as forelimb clonic seizures that progressed to generalized clonic and tonic seizures. The highest nonlethal dose was therefore determined to be 94 mg/kg. Meyer *et al.* [52] also reported an Embryonic Day $(ED)_{50}$ for neurotoxicity to be 57 mg/kg and showed splenic hemosiderosis and decreased blood hematocrit and hemoglobin concentration to occur at a threshold of 94 mg/kg in 14-day survivors. No acute toxicity data were located for DNX.

Mammalian Oral Toxicity: Subacute

Ferguson and McCain [53] conducted a 14-day subacute study on the oral toxicity of RDX to the white-footed mouse, *Peromyscus leucopus*. Ten male and 10 female mice were in each of five RDX exposure groups. RDX was mixed with feed in concentrations of 0.00, 0.05, 0.10, 0.20, 0.40, and 0.80 mg RDX/g feed, which corresponded to oral doses of 0, 8, 16, 31, and 59 mg RDX/kg body weight (bw)/day in males and 0, 8, 15, 32, and 68 mg RDX/kg bw/day in females. Exposure continued until day

14, at which point mice were euthanized by carbon dioxide asphyxiation. Data on feed consumption, body weight, organ weight, organ-to-body weight ratio, and organ-to-brain weight ratios were collected and statistically analyzed. Blood samples were obtained and used for hematological and clinical chemistry analyses; however, the analyses could not be conducted so the data were unavailable. After examining tissues for gross pathological lesions, the liver, kidney, spleen, brain, thymus, and testes were collected and weighed from half of the animals in each group and submitted for histological examination. The same tissues, minus the spleen, thymus, and brain from the remaining animals, were frozen and analyzed for biochemical parameters.

Results indicated very little compound-induced toxicity. In part, the authors attributed the lack of anticipated toxicity of RDX to *P. leucopus* to a higher metabolic rate and faster food transit time, which may increase the resiliency of this species, as compared to other *Mus* species. Mice exposed to RDX did show increased ovary, ovary-to-brain and ovary-to-body weights for the groups fed 0.05, 0.20, and 0.40 mg RDX/g feed. The effect, however, was not considered biologically significant since there was not a dose-response and the findings were unsupported by histological analyses. Similarly, liver weights were increased for females from the 0.05 and 0.20 exposure groups. Again, the finding was not considered biologically significant or compound related. Females in the two high dose groups showed an increase in spleen weight and spleen-to-brain weight ratios. Histopathological analyses did not reveal any treatment-related effects although the significant weight changes in the high dose group suggest possible RDX-induced toxicity. A potential NOAEL based on the weight change in the spleen is 16 mg RDX/kg bw/day while a potential lowest observed adverse effect level (LOAEL) is 31 mg RDX/kg bw/day [53].

Smith *et al.* [54] exposed pairs of deer mice (*Peromyscus maniculatus*) to 1, 10, or 100 μg/L TNX *ad libitum* in drinking water through production of three litters. Mice from the first two litters were exposed to the TNX water until postnatal day (PND) 45, and individuals from the last litter were exposed until PND 21. Although reproductive success remained unaltered by TNX, offspring mortality occurred from day of birth through PND 4. Unfortunately, the statistics used in this study do not allow the determination of a LOAEL for mortality. The LOAEL for decreases in body weight from birth until approximately weaning age, however, was determined to be 100 μg/L. TNX also accumulated in the liver tissue in levels approximately equal to exposure level, and a decrease in kidney weights relative to brain weights in offspring exposed to 100 μg/L was observed. No effects of TNX exposure in adults were described. No subchronic toxicity data on MNX and DNX were located.

A range of bioinformatic approaches have been used to investigate the impact of RDX toxicity in a range of species including rats [55], quail [31], fish [56], and earthworms [57–61] with a view to elucidating mechanisms of toxicity and developing predictive models for enhanced monitoring. While these exploratory studies reveal a consistent range of responses to RDX across species, the range of differentially expressed genes or enriched pathways still represents a somewhat complex descriptive toxicology, with the results highly dependent on the choice of statistical methods. Additionally, with such approaches, there is no estimate of uncertainty with respect to pathways or interactions, so that it remains difficult to envision a future application to wildlife toxicity assessments. Significantly, the use of this approach with RDX was insufficiently focused to identify a clear mechanism of seizures, and even though genes in the brain of several species were shown to be overrepresented in various neurological processes microarray analysis lacked the specificity to isolate $GABA_A$ inhibition as the underlying cause of seizures [39,62]. The reasons for this remain unclear.

Mammalian Toxicity: Subchronic

Mammalian species that have been used as models for testing the subchronic toxicological impact of RDX include beagle dogs, cyanomolgus (rhesus) macaques, Sprague Dawley and F344 rats, and B6C3F1 and Swiss-Webster mice. Litton Bionetics [63] exposed 3 beagles/sex/group to 0.1, 1, or 10 mg RDX/kg/day as a dietary additive for 90 days. Urinalysis was carried out after 4 weeks, 8 weeks, and at term, along with clinical chemistry and hematological determinations in blood samples collected at the same intervals. All survivors were subjected to a gross necropsy at term, organ weights were recorded, and histopathological comparisons of the brain, thyroid, lungs, heart, liver, spleen, kidney, adrenals, stomach, small intestine, and bone marrow were made between the high-dose and control groups. However, no abnormal findings in any measured parameter were noted at the doses chosen for the study. Similarly, in another study by Litton Bionetics [64], the same exposure duration and dose levels (by gavage) as for the dogs showed the subchronic effects of RDX to be comparatively benign in rhesus monkeys. The appearance of increased numbers of degenerate or necrotic megakaryocytes in sections of bone marrow from some high-dose monkeys led to a dose of 1.0 mg/kg/day as a NOAEL for this study. However, elevated numbers of megakaryocytes appeared in one of the three control animals examined for this feature, thereby suggesting the effect may not be compound-related.

Brown [65] reported a study on rats (number and strain unstated) in which RDX was administered in the diet at doses of 0, 0.3, 2.5, 6.5, 15, 50, or 100 mg/kg/day for 12 weeks. Increased levels of RDX in the blood in response to all doses except the lowest were associated with increases in the specific activities of brain monoamine oxidase and cholinesterase and in the capacity of excised brain tissue to take up oxygen; however, these changes were not related to overt neurotoxicity. Since these effects were negligible at the lowest dose, 0.3 mg/kg/day was chosen as a subchronic NOAEL for RDX.

The toxicological studies on RDX reported by Cholakis *et al.* [41] featured 90-day studies in F344 rats and B6C3F1 mice in which 10 animals/sex/group were exposed to RDX in feed at doses of 0, 10, 14, 20, 28, and 40 mg/kg/day. In a supplemental study, additional mice were exposed to 0, 40, 60, and 80 mg/kg/day for 2 weeks and then to 0, 320, 160, and 80 mg/kg/day, respectively, for the final 11 weeks of the investigation. A suite of toxicological endpoints were monitored, including clinical signs, body weights and food consumption, clinical chemistry and hematological parameters, gross pathology, and histopathology.

In the rats, there was a reduction in body weight gain in the high-dose males concomitant with a reduction in food consumption. In addition, sporadic though possibly compound-related hematological changes were noted, including a reduction in hemoglobin and hematocrit in high-dose males and males receiving 28 mg/kg/day after 30 and 60 days. Reticulocytes and platelets were increased in high-dose males after 90 days. There were few if any changes in clinical chemistry parameters, gross pathology, or histopathology in the rats receiving RDX, findings that, taken together, suggest a NOAEL of 20 mg/kg/day based on the hematological changes.

The absence of any compound-related toxicological consequences of the same doses of RDX in exposed mice led to a supplemental study in which a number of sporadic responses were observed. For example, a number of clinical signs were evident across the groups, with marked hyperactivity among the males. Four of 10 high-dose males and 2 out of 12 high-dose females died during week 11 of the study. Perhaps the most consistent treatment-related changes were observed at gross necropsy where dose-dependent and statistically significant increases in absolute and relative liver weights were observed in both sexes of mice. These changes appeared to be associated with the onset of hepatocellular vacuolization and other histopathological liver lesions, supporting the designation of a NOAEL at the time-weighted average mid-dose of 145 mg/kg/day.

Levine *et al.* [66] conducted a study similar to the 90-day study in F344 rats reported by Cholakis *et al.* [41], but with dose levels extending to 600 mg/kg/day. At 600 mg/kg/day, most of the subjects developed tremors and convulsions followed by death. Less severe toxicological responses were evident at the lower dose levels, including a concomitant reduction in body weight gain and food consumption in males receiving 100 mg/kg/day. Among the compound-related clinical chemistry changes was a dose-dependent reduction in plasma triglycerides that was statistically significant at 30 mg/kg/day and above. Increased relative liver weight in females receiving 100 mg/kg/day justified the choice of 30 mg/kg/day as a subchronic NOAEL for RDX in this strain of rat.

The report of acute neurological effects of RDX in male Sprague Dawley rats had a subchronic component in which animals were gavaged for 30 days with 0, 1, 3, or 10 mg RDX/kg/day in 2% aqueous carboxymethylcellulose [44]. Neurotoxicological tests were carried out before the onset of dosing and then on days 16 and 31. However, no significant effects of RDX were observed at any of the dose levels.

Dilley *et al.* [40] investigated the subchronic oral toxicity of a 1.6:1 mixture of TNT and RDX (LAP) in dogs, rats, and mice. The subchronic toxicity of TNT, but not RDX was evaluated in the study as well. Generally, the authors concluded that the results suggested that TNT dominated the toxicity of the LAP mixture. Similar to the studies of Dilley *et al.* [40], on LAP, Levine *et al.* [67] reported a 90-day dietary study in 10 F344 rats/sex/group in which the toxicological effects of mixtures of TNT and RDX (composition B) were evaluated. In this study, the authors concluded that many of the toxicological effects of each explosive individually were actually antagonized by the presence of the other compound.

In a subchronic oral study by Crouse *et al.* [68] young adult male and female Fischer 344 rats were gavaged with doses of RDX from 0, 8, 10, 12, and 15 mg/kg daily for 90 days. RDX was administered in a suspension of RDX/methylcellulose/Tween 80. Treatment related alterations in the weights of various organs were noted, including brain and testes, but no histopathological findings were reported for the highest dose group and there were no immunological findings. As expected, neurotoxic effects were noted, ranging from arousal and increased salivation to tremors and convulsions. Dose-related preterm deaths occurred in both sexes at doses greater than 4 mg/kg and in almost all cases, were preceded by convulsions. The lethality-based NOAEL was reported as 4 mg/kg/day. Interestingly, a functional observation battery of tests using hand-held observations, open arena observations, motor activity, and home cage observations revealed no patterns or biological meaning from the measured data, even in the face of neurological effects. This study reaffirms the "steep dose response" for RDX, where no seizures were observed in the 4 mg/kg dose group but lethality occurred at the 8 mg/kg dose.

No chronic toxicity data for MNX and DNX were located. Druckrey [47] examined TNX carcinogenicity as part of an investigation to determine the oral LD_{50}. The author noted that despite the relative high toxicity, TNX was not carcinogenic to the test species even after a chronic exposure of 546 days; no survivorship or other toxicity information was provided [47].

Mammalian Oral Toxicity: Chronic

The first study to examine the chronic toxicity of RDX in experimental animals was that of Hart [69], who administered the compound as a dietary supplement to 100 Sprague Dawley rats/sex/group for 104 weeks. The stated RDX levels of 0, 1.0, 3.1, and 10 mg/kg have been interpreted by the IRIS compilers [70] and other reviewers [7] as referring to doses in mg/kg (body weight)/day, though ambiguities in the study report suggest possibly that the above values might refer to the concentrations of RDX in feed. If this were the case, the actual dose levels would have been at least an order of magnitude lower than those normally assumed for this study, and possibly explain why, out of a full suite of clinical chemistry, hematology, urinalysis, gross pathology, and histopathological examinations, few if any compound-related changes were observed. However, as it stands, the data point to a NOAEL of 10 mg/kg/day for RDX, the highest dose tested.

Levine *et al.* [28] reported on the chronic toxicity of RDX in 75 F344 rats/sex/group exposed to the compound in feed in amounts equivalent to doses of 0, 0.3, 1.5, 8, or 40 mg/kg/day for a total of 2 years. Clinical signs were observed twice daily and food consumption and body weights were monitored weekly up to test week 14, after which they were monitored biweekly. Ophthalmic examinations were carried out on subjects during weeks 2, 25, 51, 76, and 103. Blood samples were taken at weeks 13, 26,

52, 78, and 104 for clinical chemistry and hematological determinations. Interim sacrifices of 10 rats/sex/group were carried out at weeks 27 and 52. At these points and at term, animals were subjected to a gross pathological examination. Samples of a wide range of organs and tissues were preserved by chemical fixation. Tissues from animals in control and high-dose groups were examined histopathologically, along with sections of brain, gonads, heart, liver, kidney, spleen, and spinal cord from all dosed groups.

Most rats receiving 40 mg RDX/kg/day died during the treatment period, many displaying profound clinical signs such as tremors, convulsions, hyperactivity, and discolored/opaque eyes. Body weight gain was also reduced in this and the intermediate-dose group, a change potentially associated with reduced food consumption. High-dosed rats had reduced red blood cell (RBC) counts, hemoglobin concentration, and hematocrit, while the platelet count was increased in intermediate-dose males; however, these hematological parameters fell within normal ranges [71]. There were some fluctuations in clinical chemistry parameters, including relative decreases in plasma cholesterol and triglycerides and in the activity of serum glutamate-pyruvate transaminase. High-dose females displayed an increased incidence of cataracts at week 78 and week 104. Organ weight changes were noted; in particular, an increase in the relative weights of liver and kidneys in both sexes of high-dose rats and a reduction in testis weights of high-dose males. Also, observations indicated toxic effects in the spleen as early as 6 months into the study. After 2 years, the appearance of a hemosiderin-like pigment in the spleen was evident in all dose groups from 1.5 mg/kg/day and up. This finding points to a NOAEL of 0.3 mg/kg/day, a value that was used as such by the IRIS compilers to derive a human health reference dose of 3×10^{-3} mg/kg/day [72].

A similar study to that described above was conducted by the same researchers on B6C3F1 mice in an experiment in which 85 animals/sex/group were exposed via diet to RDX at concentrations approximating doses of 0, 1.5, 7, 35, and 100 mg/kg/day [29]. The high-dose level, 175 mg/kg/day, had been lowered during the course of the experiment due to high mortality. Reduced body weight gain was noted in both sexes of high-dose mice, although food consumption was comparatively unaffected. Hematological parameters showed little change, although hematocrit and hemoglobin concentrations were reduced in high-dose females at an interim time point. Hypercholesterolemic and hypertriglyceridemic effects of RDX were observed, the former parameter displaying marked dose-response. A number of gross pathological and histopathological effects of RDX were evident in the mice, including increased relative liver and kidney weights in high- and intermediate-dose animals. Histopathological changes at the 2-year time point included degeneration of the testes in high- and intermediate-dose males, suggesting a NOAEL of 7 mg/kg/day for this response. Other important histopathological effects of RDX included a dose-dependent increase in the incidence of hepatocellular adenomas and carcinomas in the liver of females. This finding was used by the USEPA to classify RDX as a possible carcinogen in the IRIS database. Parker *et al.* [73] recently reevaluated this finding showing that the incidence of cancer was within the reported incidence of spontaneous hepatocellular neoplasms for female mice.

Mammalian Oral Toxicity: Other

With one exception, RDX has been negative in *in vitro* tests for mutagenicity and genotoxicity [74]; Pan *et al.* [75] reported that RDX was mutagenic in a non-standard 97a strain of *S. typhimurium*. RDX was not carcinogenic in Fisher rats after two years of dosing [28], but was reported to be carcinogenic in female B6C3F1 mice, but not male mice [29]. Sweeney *et al.* [74] presented strong arguments for a nonlinear, threshold approach to the assessment of mammalian cancer risk, and not a mutagenic or genotoxic mechanism of action for RDX, indicating that RDX is not likely to be carcinogenic.

Schneider *et al.* [76] followed their acute studies on the toxicokinetics of RDX in Sprague Dawley rats with subchronic studies in which the compound was administered either in drinking water or by gavage at 20 mg/kg/day for up to 90 days. Some animals were also exposed via drinking water to saturated unlabeled or ^{14}C-labeled RDX. The results pointed consistently to the relative inability of the compound to accumulate in the plasma or tissues. Overwhelmingly, the compound was released to the urine or as $^{14}CO_2$, with lesser amounts in the feces and carcass.

Angerhofer *et al.* [77] investigated the teratological potential of RDX in pregnant Sprague Dawley rats. In a pilot study, six pregnant rats/group were given 0, 10, 20, 40, 80, or 120 mg/kg by gavage in gum acacia on gestation days (GD) 6–15, and the parameters measured at GD 20 included the numbers of viable fetuses, nonviable fetuses, resorptions, implantations, and corpora lutea. Fetal parameters included weight, size, sex, and the incidence of external malformations and visceral abnormalities. The lowest dose inducing maternal toxicity in the pilot study (20 mg/kg/day) was chosen as the highest dose in the main part of the study. In the main study, 25 pregnant rats/group were given 0, 2, 6, or 20 mg/kg by gavage in gum acacia on GDs 6–15. Thirty-one percent of females receiving 20 mg RDX/kg died in the main study. For the survivors, there were few changes in reproductive parameters compared to controls and no compound-related anomalies among the teratological findings. The authors suggested a dose of 2 mg/kg/day as a LOAEL for the reductions in fetal size that were evident at the lowest dose tested. Inspection of the statistical results suggested that the original analyses may have been suspect. Statistical reanalysis of the data indicated that fetal size was significantly affected only at the highest dose, 20 mg/kg/day. Hence, the revised LOAEL is 20 mg/kg/day, and the NOAEL is 6 mg/kg/day.

Reproductive toxicity and teratological studies have also been conducted by Cholakis *et al.* [41], who administered 0.2, 2, or 20 mg RDX/kg/day by gavage to pregnant female F344 rats between GDs 6–19 and to New Zealand white rabbits between GDs 7–29. At sacrifice, the uteri were examined for live fetuses and resorptions, while the fetuses themselves were examined for skeletal abnormalities and visceral perturbations. Food consumption was reduced in high-dose rats through the first three days of dosing, though with subsequent recovery. In addition, this group displayed a reduction in body weight, marked neurological signs, and 24 percent (6/25) lethality. However, no changes in reproductive parameters were noted; there were no soft tissue or skeletal anomalies due to RDX exposure. Dosing pregnant New Zealand white rabbits at the same levels resulted in few changes in reproductive parameters but a catalogue of teratological responses that were essentially sporadic and therefore of uncertain significance. These responses included spina bifida, misshapen cranium, meningocele, misshapen and enlarged eye bulges, abdominal wall defects, gastroschisis, appendicular reduction anomalies and "tail problems."

Cholakis *et al.* [41] also reported a two-generational reproductive study in which male and female CD rats were fed diets adjusted to nominal daily doses of 0, 5, 16, or 50 mg RDX/kg for 13 weeks. F_0 adults were then mated within the groups with 26 of the resulting F_1 progeny maintained on the same diets for another 13 weeks. After a further round of mating, the F_2 progeny were necropsied and processed for histopathological examination.

There was a reduction in body weight gain in all generations of high-dose rats, which may have been related to a concomitant depletion in food consumption. Mortality reached 18% in high-dose rats of the F_0 generation with 17 and 52% stillbirths in the F_1 and F_2 high-dose progeny, respectively. Reductions in the number of fertile high-dose male and female rats were observed during the F_0 mating, although these differences were not statistically significant. Notwithstanding these changes, there were no specific reproductive or developmental changes due to treatment in this experiment, since feeding 16 mg/kg/day produced no apparent effects.

No toxicity studies were located for DNX. However, a number of studies investigated the carcinogenicity potential of TNX. Danz *et al.* [78] exposed Wister rats by single oral gavage of 50 mg/kg TNX and in sunflower oil measured the effect of TNX on the mitosis count/m^2 in the adrenal cortex as an indicator of potential carcinogenicity. An increased mitosis count is generally considered a promotion of cellular growth by the chemical stimulant, and would thus be categorized as a carcinogen. Based on the result of their study (i.e., no significant increase in mitosis count/m^2 in the adrenal cortex), Danz *et al.* [78] categorized TNX as noncarcinogen. However, in other studies where DMSO was used as the carrier solvent, TNX was categorized as a carcinogen because high mitosis count and hepatocarcinogenicity were observed in both Wistar and Sprague Dawley rats [48,50]. However, the authors noted that the carrier, DMSO, might have contributed to the increase in mitotic count was well as the hepatocarcinogenic effect from chronic exposure. Using DMSO as the carrier solvent to compare with a sunflower oil solvent, it was revealed that with DMSO, mitosis count increased significantly by as much as 3.4 times via intraperitoneal (i.p.) injection, and 5.7 times from the control via oral gavage. In contrast, mitosis count was not statistically different from that of the control when sunflower oil alone was used as the carrier solvent [79].

Urban, Amlacher, and Danz [79] noted that it was unclear whether DMSO was a cocarcinogen with TNX, or whether DMSO served as a facilitator in the permeation of TNX through the cell membrane. Danz *et al.* [50] acknowledged DMSO may have contributed to the increase in mitotic count, as well as the long-term hepatocarcinogenic effect, but categorized TNX as a carcinogen. Amlacher and Rudolph [51] used a single i.p. injection in CBA mice at approximately 15 to 30% of the LD_{50} (24–48 mg/kg). The carcinogenicity of TNX was evaluated by using the autoradiograph of nuclei of the renal tubular epithelial and liver epithelium. Although one part of DMSO was used against two parts of triethylenglycol, the result was not significantly different from the control. Therefore, TNX was considered a noncarcinogen by the authors. Table 4.2 summarizes the result of mammalian oral toxicity values.

To further assess the potential carcinogenicity of RDX's breakdown products, Pan *et al.* [75] used TNX and MNX in the *Salmonella tryphimurium* reverse mutation assay (Ames assay) with strains TA97a, TA98, TA100, and

Table 4.2 Summary of Relevant Mammalian Data for TRV Derivation

Test Organism	Test Duration	Test Results: NOAEL (mg/kg/day)	Test Results: LOAEL (mg/kg/day)	Test Results: Effects Observed at the LOAEL	Study
Rat (m,f) Fischer 344	90-day	4 mg/kg	8 mg/kg	Lethality	[68]
Rat (f) (Sprague Dawley)	GD 6–15	NA	2	Comparative reductions in fetal size	[77]
Rat (f) F344)	GD 6–19	2	20	Neurological signs/lethality	[41]
Rabbit (f) (NZ white)	GD 7–29	20	NA	Reproductive/Developmental toxicity	
Dog (beagle)	90-day	10	NA	NA	[63]
Monkey (rhesus)	90-day	1	10	Elevated megakaryocyte count	[64]
Rat (strain unstated)	12-week	0.3	2.5	Increased brain monoamine oxidase and cholinesterase activity	[65]
Rat (F344)	90-day	30	100	Increased liver weight	[66]
Rat (m) (Sprague Dawley)	30-day	10	NA	Neurological testing	[44]
Rat (F344)	90-day	26.4	37.7	Reduced hemoglobin and hematocrit; reduced body weight	[41]
Mice (B6C3F1)	90-day	145 (TWA)	277 (TWA)	Lethality and neurological signs, enlarged liver and hepatocellular lesions	
Rat (F344)	90-day	NA	5.0/29.8*	Reduced body weight gain in males	[67]
Rat (Sprague Dawley)	104-week	10	NA	NA	[69]
Rat (F344)	104-week	8	39.8	Decreased body weight	[28]
Mouse (B6C3F1)	104-week	7	35	Atrophy of the testis in males, increases in relative and absolute kidney and liver weights, decreased body weight	[29]
		NA**	NA**	Hepatocellular adenoma and carcinoma in females	

Key:
GD = gestation day RBC = red blood cell.
NA = not applicable TWA = time-weighted average.
*Doses are those of TNT/RDX mixed in various proportions.
**Identifying a NOAEL for tumorigenic responses may be unsafe, as per existing USEPA understandings on the identification of a subthreshold dose for a carcinogenic effect.

TA102. Although TNX and MNX were not found to be cytotoxic at doses up to 878.1 μg/plate and 1118.9 μg/plate, respectively, both were observed to induce moderate mutagenesis in some of the strains. TNX induced mutagenesis in strains TA97a at 22.7 μg/plate and TA100 at 1,200 μg/plate, and MNX induced mutagenesis in only strain TA97a at 21.7 μg/plate. This study used a higher S9 fraction (9%) than standard Ames assay procedures (4%) that previously showed no mutagenic effects of TNX and MNX (unpublished data reported in [75]). The high S9 with an additional preincubation step that was necessary in facilitating the activation of TNX and MNX caused the authors to suggest that these nitrosamines may require intensive metabolic activation to manifest carcinogenic effects.

Studies Relevant to Mammalian TRV Development: RDX Ingestion Exposures

The range of animal models in which responses to acute, subacute, and subchronic RDX administration have been monitored includes beagle dogs, cyanomolgus (rhesus) monkeys, Sprague Dawley and F344 rats, Swiss-Webster and B6C3F1 mice, miniature swine, and New Zealand white rabbits.

There is a striking contrast between the acute lethality of RDX in experimental animals and those of other explosive/energetic compounds such as TNT and HMX. For example, acute oral LD_{50} values for the latter compounds may be found in the 500–1,000 mg/kg/day range, suggesting low-to-moderate lethality, whereas the LD_{50} for RDX is in the 50–200 mg/kg/day range, with a median value closer to 100 mg/kg/day. This suggests that RDX has a higher acute toxicity than other explosive compounds. However, if RDX is characterized by comparatively high acute toxicity, the precise targets for these toxic effects remain to be fully identified. Toxicokinetic evidence indicates that the compound is readily absorbed at the gastrointestinal brush border but has a transitory existence in the body with rapid breakdown into a range of metabolic products including single carbon compounds occurring in the liver in some animals. The importance of the liver in response to RDX is underscored by the histological changes that take place when a receptor is challenged with the compound. Perturbations of clinical chemistry parameters potentially related to liver function, such as plasma lipid levels and enzyme activities such as serum glutamate-pyruvate transaminase, lend further weight to the concept that the liver is one of the primary sites of RDX toxicity.

Liver effects are also evident in a number of subchronic and chronic studies on RDX, the responses manifesting in dose-dependent increases in organ to body weight ratios and in changes to the cellular architecture revealed histologically. In the 24-month dietary study of RDX in B6C3F1 mice, histopathological evidence of compound-related hepatocellular adenoma and carcinoma formation was obtained in females. However, no effects were seen in male mice or either sex rats suggesting the response is not generally associated with rodent exposure to RDX. The ecological relevance of RDX-induced liver toxicity is questionable.

Other reasonably consistent responses that have been elicited in experimental animals exposed to RDX include changes in the levels of some hematological parameters associated with anemia and changes to the size and histopathology of the spleen. Although increased pigmentation of the spleen was used as the basis for a NOAEL of 0.30 mg/kg/day [28], associated hematological parameters, although significantly different than controls for the high dose group, fell within normal ranges [71]. This indicates that the increased pigmentation of the spleen was not associated with any hematological changes that would cause functional impairment. Given the lack of biological significance in this effect, increased pigmentation of the spleen is of questionable relevance.

As outlined in USACHPPM TG 254 [2], TRVs are derived from toxicological effects likely to be ecologically relevant. Decreased growth is regarded as an ecologically relevant parameter and was common to two studies on chronic ingestion of RDX, one on F344 rats [28] and one on B6C3F1 mice [29], and two studies on subchronic ingestion of RDX, one on F344 rats [41] and one on Swiss Webster mice [40]. These data suggest that reduced growth is a consistent feature of RDX-exposed rodents. From an ecological perspective, reduced growth and/or associated reductions in food consumption can affect the ecological performance of individuals by causing alterations in energy allocation patterns that could ultimately result in altered reproductive performance [80,81]. All three studies showing reduced growth in RDX-exposed rodents were well designed and well executed and can be considered high quality. For derivation of the TRV, the data on chronic toxicity in F344 rats [28] is most appropriate as these data meet the requirements of CHPPM TG 254 [2] and as such, require no uncertainty factors. Moreover, these data are protective of the data on B6C3F1 mice [29].

Mammalian Inhalation Toxicity

No inhalation studies conducted using mammals were found.

Mammalian Dermal Toxicity

No dermal studies conducted using mammals were found.

Summary of Avian Toxicology

Avian Oral Toxicity: Acute

One study was located on the toxicity of RDX to an avian species. Gogal [33,82] studied the acute, subacute, and subchronic toxicity of RDX to the Northern Bobwhite (*Colinus virginianus*). For the acute study, the objective was to determine the approximate lethal dose (ALD). RDX was administered to birds orally in a water vehicle. One male and one female per group were dosed with one of the following: 125, 187, 280, 420, 630, 945, 1,417, or 2,125 mg RDX/kg. Birds were observed for 14 days after administration of RDX and on day 14 surviving birds were weighed, bled, euthanized by electrocution and necropsied. The ALD values 14 days after the exposure were 280 mg/kg for male and 187 mg/kg for female Northern Bobwhite.

Avian Oral Toxicity: Subacute

Groups of six male and six female birds were exposed to RDX in the feed at concentrations of 0, 83, 125, 187, 280, and 420 ppm RDX for 14 days [33,82]. Daily doses of RDX were calculated to be 10.8, 13.4, 22.3, and 26.3 mg RDX/kg bw, respectively. Feed consumption, body weight, spleen weight/body weight ratio, liver weight/body weight ratio, and egg production were measured. Hematological analyses included whole blood cellularity, packed cell volume (PCV), total protein, and mean corpuscular volume (MCV). Histological analyses were done on liver, kidney, spleen, brain, spinal cord, intestine, heart, lung, pancreas, and gonad tissues.

Results [33,82] showed that there was a significant, linear decrease in feed consumed with increasing levels of dietary RDX and a concomitant decrease in body weight with increasing levels of RDX in the diet. The ratios of spleen weight/body weight in females and liver weight/body weight in both sexes were also significantly affected by dose and generally decreased with increasing RDX. Hematological effects of RDX exposure included an increase in PCV in females, a decline in total plasma protein in females, an increase in heterophils, and an increase in the heterophil/lymphocyte ratio in blood. Egg production showed a significant, linear decrease with increasing RDX for both week one and week two. The authors report a NOAEL of 8.7 mg RDX/kg/day and a LOAEL of 10.6 mg RDX/kg/day based on a dose-related decrease in body weight and egg production.

A more recent 14-day study by Quinn *et al.* [31] showed that RDX has a NOAEL and LOAEL of 3 and 8 mg/kg, respectively, based on mortality. Serum globulin and total leukocyte increases in the two highest dose groups were also observed.

In a study by Johnson *et al.* [32], *Colianus virginianus* (Northern Bobwhite) quail were dosed with either RDX or DNT in corn oil over the course of 14 days. Brain concentrations of RDX remained consistently high, at about 20 μg/g, at doses of 20, 80, 125, and 180 mg/kg indicating that saturation had occurred at even the low dose. This brain burden (20 μg/g) translates to about 90 μM RDX in brain, exceeding by almost 10-fold the average concentration of the abundant CNS neurotransmitter glutamate (~10 μM).

Avian Oral Toxicity: Subchronic

Five groups of 10 male and female Northern Bobwhite were provided with 0, 125, 187, 280, or 420 ppm RDX in the feed for 90 days [33,82]. The calculated daily oral doses were reported to be 0, 10.8, 13.4, 22.3, and 26.3 mg/kg for the 0, 125, 187, 280, and 420 ppm, respectively. Feed was weighed and replaced on a weekly basis. Parameters measured included those mentioned in the 14-day study including five-part leukocyte differentials, lymphocyte mitogen-induced proliferation, and leukocyte apoptosis/necrosis assays. Histological analyses were as earlier, with the addition of bone marrow. Changes in egg production were also evaluated. Although the same doses used in the subacute were identical to those in the subchronic study, no significant effects of RDX were seen after exposure for 90 days.

These data suggest that Northern Bobwhite develop a tolerance from exposure to RDX in the feed. However, although no significant effects were seen, there were dose-dependent trends apparent for several parameters, including a decrease in feed consumption, decrease in total protein, decrease in PCV, and decrease in egg production. No severe effects were noted. Since no significant effects of RDX were seen after 90 days of exposure, a LOAEL was not reported. Another study using 1-year old Northern Bobwhite administered an aqueous suspension of RDX 125 mg/kg bw to quail (n = 8) and examined the concentrations of RDX in whole blood and feces on days 1, 3, and 7 [83]. Concentrations of RDX were relatively high at day 1, compared to HMX and TNT, but were cleared from the blood by day 7. The microgram amounts of RDX in the feces were relatively low compared with the mg amounts dosed. Taken together these data show a rapid absorption and elimination of RDX, consistent with other studies in mammals.

Avian Oral Toxicity: Chronic

No data are available.

Avian Oral Toxicity: Other

No data are available.

Studies Relevant for Avian TRV Development: RDX Ingestion Exposures

Only one study was located on the effects of RDX on an avian species. Gogal *et al.* [33] investigated acute, subacute, and subchronic effects of orally administered RDX in Northern Bobwhite. In the 14-day study, there were significant effects of RDX on both body weight and egg production. In the 90-day study, the same doses of RDX were used as in the 14-day study; however, no significant effects of RDX were seen although there were dose-dependent decreases in body weight and total egg production. These data suggest that Northern Bobwhite develop a tolerance to prolonged dietary exposure to RDX. Although data from long-term exposures (i.e., subchronic and chronic) are preferred, in this case the subacute data on egg production is especially relevant. A rationale is provided below.

Birds are highly agile animals and thus often experience the environment in patchy distributions. Under these realistic exposure scenarios, birds are most likely to experience short-term exposures on the order of days as opposed to weeks. Therefore, a 14-day exposure to RDX may be more ecologically relevant than longer exposure scenarios. Moreover, these data are protective of longer exposure scenarios tested to date. Although these changes in egg production and other parameters (e.g., body weight gain) may be due to the reduction in consumed feed, food avoidance may also be an ecologically relevant parameter. Since the primary endpoint chosen is a reproductive one, under USACHPPM TG 254 and consistent with Sample *et al.* [84], data on quail egg production following exposure to RDX for 14 days are considered equivalent to a long-term study since the exposure occurred during a sensitive life cycle stage. Hence, the avian TRV for RDX was derived from the 14-day oral exposure in Northern Bobwhite [33,82]. A summary of the relevant avian data for TRV is shown in Table 4.3.

Avian Inhalation Toxicity

No data are available.

Table 4.3 Summary of Relevant Avian Data for TRV Derivation

Test Organism	Test Duration	Test Results: NOAEL mg/kg/day	Test Results: LOAEL mg/kg/day	Test Results: Effects at LOAEL	Study
Northern Bobwhite (*Colinus virginianus*)	ALD	NA	NA	187 mg/kg for female 280 mg/kg for male	[33,82]
	14 day	8.7	10.6	Decreased body weight in males and females and decreased egg production.	
	90 day	26.3	NA	No statistically significant effects; however, there were several dose-related trends: decreased egg production, feed consumption, total plasma protein, and packed cell volume.	

ALD = approximate lethal dose.
NA = not applicable.

Avian Dermal Toxicity

No data are available.

Amphibian Toxicology

To assess subchronic dermal exposure in amphibians, a 28-day study used a microcosm design to expose *Plethodon cinereus,* or red-backed salamander [85], to a range of RDX soil concentrations from 0, 10, 100, 1000, and 5,000 ppm per dry weight of soil. This highest dose group showed frank neuromuscular toxicity and weight loss, though histopathology showed no obvious treatment-related effects.

Reptilian Toxicology

McFarland *et al.* [86] carried out acute and subchronic oral studies to assess the toxicity of RDX on the Western Fence Lizard (*Sceloporus occidentalis*). Median lethal doses were 72 and 88 mg/kg for males and females respectively, with a significant decrease in survival at 20 mg/kg. A 60-day study at doses of 0, 1, 2.5, 5, 8, and 11 mg/kg resulted in lethargy and anorexia;

survival was decreased at 8 and 11 mg/kg, while reduced grown rate and food consumption occurred at 5 mg/kg. In animals where seizures were observed the measured RDX concentration in brain was at least 18 μg/g. Based on survival, growth rate, and food intake a LOAEL of 5 mg/kg and a NOAEL of 2.5 mg/kg were suggested.

RECOMMENDED TOXICITY REFERENCE VALUES

Toxicity Reference Values for Mammals

TRVs for Ingestion Exposures for the Class Mammalia

Decreased body weight was reported for both F344 rats [28] and B6C3F1 mice [29] after 2 years of oral dosing with RDX. Decreased body weight, an indication of a lower growth rate or a decrement in energy allocation, was used to determine the TRV because this endpoint may be ecologically relevant through effects on fitness. For example, indicated alterations in energy allocation patterns may impair reproductive function and/or schedules [80,81]. In addition, sustaining a smaller body size for longer time periods may increase risk of predation. Both chronic studies [28,29] indicated decreased growth in rats and mice fed RDX, and hence the effect may be a consistent feature of RDX exposure. For TRV determination, data on female F344 rats was used because these data were protective of males and exhibited a clear dose-response relationship [28]. In addition, the TRV based on the F344 rat data was protective of B6C3F1 mice. Growth, as indicated by body size, also meets the minimum data requirements of the Standard Practice, Section 2.2 [2], and therefore no uncertainty factors were required in the derivation of the TRV. The TRV was derived using the benchmark dose (BMD) approach [2] and the values presented in Table 4.4. This TRV is given a medium confidence rating since there were only two chronic studies and no wildlife toxicity data were available.

TRVs for Inhalation Exposures for the Class Mammalia

Not available at this time.

Table 4.4 Selected Ingestion TRVs for the Class Mammalia

TRV	Dose	Confidence Level
LED_{10}	1.19 mg/kg/day	Medium
ED_{10}	2.73 mg/kg/day	Medium

Source: Levine *et al.* [28].

TRVs for Dermal Exposures for the Class Mammalia

Not available at this time.

Toxicity Reference Values for Birds

TRVs for Ingestion Exposures for the Class Aves

The ecologically relevant parameter for RDX toxicity in birds was decreased fecundity (i.e., egg production) reported for Northern Bobwhite exposed to dietary concentrations of RDX for 14 days [33,82]. For this endpoint, the effect was significant and dose dependent and the study was of high quality. Decreased egg production was used to determine the TRV because it is an ecologically relevant parameter indicative of impaired reproductive performance, which can have direct impacts on population dynamics, particularly for this species.

Exposure to RDX in this study occurred during a sensitive life cycle stage, and therefore can be considered equivalent in value to a chronic exposure evaluation. Given the data quality, the dose-dependent nature of the effect, and the ecological relevance of effect, the BMD approach was used. The TRVs derived using the BMD approach [2] are presented in Table 4.5. It should be noted that although there was not a significant effect of RDX on egg production in quail for the 90-day exposure, there was a trend; egg production decreased with increasing concentrations of RDX. The BMD approach was applied to these data as well [2]. A benchmark dose (BMD or ED_{10}) of 8.14 mg/kg/day was calculated from the model fit of the mean response at the 10% response level. A lower bound on the benchmark dose (BMDL or LED_{10}) was calculated to be 3.65 mg/kg/day from the lower 95% confidence interval (CI) of the modeled curve. Comparison of BMDs for the 14-day and 90-day studies indicated that TRVs derived from the 14-day study are protective of TRVs derived from the 90-day study. Since data from only one study was located, the TRVs presented have been assigned a confidence rating of "low."

Table 4.5 Selected Ingestion TRVs for the Class Aves

TRV	Dose	Confidence Level
LED_{10}	3.65 mg/kg/day	Low
ED_{10}	8.14 mg/kg/day	Low

Sources: Gogal Jr., *et al.* [33], Gogal Jr., *et al.* [82].

Toxicity Reference Values for Amphibians

Not available at this time.

Toxicity Reference Values for Reptiles

The ecologically relevant parameter for RDX in reptiles was mean survival, food consumption, and change in body weight, which occurred at the same thresholds [86]. At 5 mg/kg RDX and above animals showed statistically significant decreased survival over 60 days while at 2.5 mg/kg there was no difference from controls or the lower dose of 1 mg/kg. Changes in body weight are consistent with potentially disruption to ability to breed and/or survival from predation. Reduced cricket consumption was a confounder in the reduction of body weight. A LOAEL of 5 mg/kg and a NOAEL of 2.5 mg/kg was established based on this data. Since there is data from only one study, these TRVs are given a low confidence rating.

IMPORTANT RESEARCH NEEDS

The limited availability of data on the toxicity of RDX to wildlife species precludes the development of a high-confidence TRV. Hence, more studies on the toxicity of RDX to wildlife species are needed. In particular, long-term toxicity studies on mammals and additional studies on nonmammalian wildlife such as birds, reptiles, and amphibians are particularly warranted. More information regarding the toxicity of RDX to wildlife would likely allow the derivation of a high confidence TRV. The metabolism of RDX, especially in warm-blood animals, is still not completely understood, and there are less than a handful of papers dealing with this subject. Metabolism studies may help to distinguish the neurotoxic effects of RDX from other systematic effects, especially for chronic exposures.

REFERENCES

[1] ATSDR. Toxicological profile for otto fuels II. Atlanta (GA): U.S. Department of Health and Human Services; 1995.
[2] USACHPPM. Standard practice for wildlife toxicity reference values. Aberdeen Proving Ground (MD): U.S. Army Center for Health Promotion and Preventive Medicine; 2000, Technical Guide No. 254.
[3] McCormick NG, Cornell JH, Kaplan AM. Biodegradation of hexahydro-1,3,5-trinitro-1,3,5-triazine. Appl Environ Microbiol 1981;42(5):817–23 [as cited by Talmage et al., 1999].

[4] Sheremata TW, Halasz A, Paquet L, Thiboutot S, Ampleman G, Hawari J. The fate of cyclic nitramine explosive RDX in natural soil. Environ Sci Technol 2001;35 (6):1037–40.
[5] Beller HR, Tiemeier K. Use of liquid chromatograpy/tandem mass spectrometry to detect distinctive indicators of in situ RDX transformation in contaminated groundwater. Environ Sci Technol 2002;36(9):2060–6.
[6] Paquet L, Monteil-Rivera F, Hatzinger PB, Fuller ME, Hawari J. Analysis of the key intermediates of RDX (hexahydro-1,3,5-trinitro-1,3,5-triazine) in groundwater: occurrence, stability and preservation. J Environ Monit 2011;13(8):2304–11.
[7] Talmage SS, Opresko DM, Maxwell CJ, Welsh CJ, Cretella FM, Reno PH, et al. Nitroaromatic munition compounds: environmental effects and screening values. Rev Environ Contam Toxicol 1999;161:1–156.
[8] Hazardous Substance Data Bank (HSDB). Propylene [Internet]. Bethesda (MD): National Library of Medicine (US), National Institute of Health [cited 2013]. Available from: http://toxnet.njm.nih.gov/.
[9] McCormick NG, Cornell JH, Kaplan AM. The anaerobic biotransformation of RDX, HMX, and their acetylated derivatives. Natick (MA): U.S. Army Natick Research and Development Center; 1984, Prepared for: U.S. Army Toxic and Hazardous Materials Agency, Aberdeen Proving Ground, MD. AD A149464, p. 817-23 (as cited by Talmage et al., 1999).
[10] Hovatter PS, Talmage SS, Opresko DM, Ross RH. Ecotoxicity of nitroaromatics to aquatic and terrestrial species at army Superfund sites. In: Dwyer FJ, Doane TR, Hinman ML, editors. Environmental toxicology and risk assessment: modeling and risk assessment. West Conshohocken (PA): American Society for Testing and Materials; 1997.
[11] Crocker FH, Thompson KT, Szecsody JE, Fredrickson HL. Biotic and abiotic degradation of CL-20 and RDX in soils. J Environ Qual 2005;34(6):2208–16.
[12] Crouse LC, Michie MW, Major MA, Leach GJ, Reddy G. Oral bioavailability of cyclotrimethylenetrinitramine (RDX) from contaminated site soils in rats. Int J Toxicol 2008;27(4):317–22.
[13] Cataldo DA, Harvey SD, Fellows RJ. An evaluation of the environmental fate and behavior of munitions materiel (TNT, RDX) in soil and plant systems. Environmental fate and behavior of RDX. Richland (WA): Pacific Northwest Laboratory; 1990, Prepared for the U.S. Army Biomedical Research and Development Laboratory, Fort Detrick, Fredrick MD. PNL-7529.
[14] Harvey SD, Fellows RJ, Cataldo DA, Bean RM. Fate of the explosive hexahydro-1,3,5-trinitro-1,3,5-triazine (RDX) in soil and bioaccumulation in bush bean hydroponic plants. Environ Toxicol Chem 1991;10(7):845–55.
[15] U.S. Army Environmental Center. Plant uptake of RDX and TNT utilizing site specific criteria for the Cornhusker Army Ammunition Plant (CAAP). Nebraska; 1996, Technical Report No 560786M8AA.
[16] Bhushan B, Trott S, Spain JC, Halasz A, Paquet L, Hawari J. Biotransformation of hexahydro-1,3,5-trinitro-1,3,5-triazine (RDX) by a rabbit liver cytochrome P450: insight into the mechanism of RDX biodegradation by Rhodococcus sp. strain DN22. Appl Environ Microbiol 2003;69(3):1347–51.
[17] Jackson RG, Rylott EL, Fournier D, Hawari J, Bruce NC. Exploring the biochemical properties and remediation applications of the unusual explosive-degrading P450 system XplA/B. Proc Natl Acad Sci U S A 2007;104(43):16822–7.
[18] Fuller ME, McClay K, Hawari J, Paquet L, Malone TE, Fox BG, et al. Transformation of RDX and other energetic compounds by xenobiotic reductases XenA and XenB. Appl Microbiol Biotechnol 2009;84(3):535–44.
[19] Pan X, Zhang B, Smith JN, Francisco MS, Anderson TA, Cobb GP. N-Nitroso compounds produced in deer mouse (Peromyscus maniculatus) GI tracts following

hexahydro-1,3,5-trinitro-1,3,5-triazine (RDX) exposure. Chemosphere 2007;67 (6):1164–70.
[20] Pan X, Ochoa KM, Francisco MJ, Cox SB, Dixon K, Anderson TA, et al. Absorption, distribution, and biotransformation of hexahydro-1,3,5-trinitro-1,3,5-triazine in B6C3F1 mice (Mus musculus). Environ Toxicol Chem 2013;32(6):1295–303.
[21] Schneider NR, Bradley SL, Andersen ME. The distribution and metabolism of cylotrimethylenetrinitramine (RDX) in the rat after subchronic administration. Toxicol Appl Pharmacol 1978;46(1):163–71.
[22] Berry AP, Arbuckle JB, Nicol J. Cyclonite poisoning in a dog. Vet Rec 1983;113(19):449.
[23] Stone WJ, Paletta TL, Heiman EM, Bruce JI, Knepshield JH. Toxic effects following ingestion of C-4 plastic explosive. Arch Intern Med 1969;124(6):726–30.
[24] Kucukardali Y, Acar HV, Ozkan S, Nalbant S, Yazgan Y, Atasoyu EM, et al. Accidental oral poisoning caused by RDX (cyclonite): a report of 5 cases. J Intensive Care Med 2003;18(1):42–6.
[25] Kasuske L, Schofer JM, Hasegawa K. Two marines with generalized seizure activity. J Emerg Nurs 2009;35(6):542–3.
[26] Schneider NR, Bradley SL, Andersen ME. Toxicology of cyclotrimethylenetrinitramine: distribution and metabolism in the rat and the miniature swine. Toxicol Appl Pharmacol 1977;39(3):531–41.
[27] Bannon DI. Biomarkers of RDX (trinitro hexahydro 1,3,5-trinitro-1,3, 5 triazone) exposure in breath of swine. Aberdeen Proving Ground (MD): U.S. Army Public Health Command Directorate of Toxicology Health Effects Research Program; 2010.
[28] Levine BS, Furedi-Machacek EM, Rac VS, Gordon DE, Lish PM. Determination of the chronic mammalian toxicological effects of RDX: twenty-four month chronic toxicity/ carcinogenicity study of hexahydro-1,3,5-trinitro-1,3, 5-traizine (RDX) in the Fischer 344 rat. Phase V. Chicago (IL): IIT Research Institute; 1983, Prepared for: U.S. Army Medical Research and Development Command, Fort Detrick, Frederick (MD). Contract No. DAMD 17-79-C-9161. AD A160744.
[29] Lish PM, Levine BS, Marianna-Furedi EM, Sagartz JM, Rac VS. Determination of the chronic mammalian toxicological effects of RDX. Twenty-four month chronic toxicity/carcinogenicity study of hexahydro-1,3,5-trinitro-1,3,5-triazine (RDX) in the B6C3F1 hybrid mouse. Chicago (IL): IIT Research Institute; 1984, Prepared for: U. S. Army Medical Research and Development Command. ADA No. 181766.
[30] Burdette LJ, Cook LL, Dyer RS. Convulsant properties of cyclotrimethylenetrinitramine (RDX): spontaneous audiogenic, and amygdaloid kindled seizure activity. Toxicol Appl Pharmacol 1988;92(3):436–44.
[31] Quinn Jr JM, Bazar MA, McFarland CA, Perkins EJ, Gust KA, Johnson MS. Sublethal effects of subacute exposure to RDX (1,3,5-trinitro-1,3,5-triazine) in the northern bobwhite (*Colinus virginianus*). Environ Toxicol Chem 2009 Jun;28(6):1266–70.
[32] Johnson MS, Quinn MJ, Bazar MA, Gust KA, Escalone BL, Perkins EJ. Subacute toxicity of oral 2,6-dinitrotoluene and 1,3,5-trinitro-1,3,5-triazine (RDX) exposure to the northern bobwhite (*Colinus virginianus*). Environ Toxicol Chem 2007;26(7):1481–7.
[33] Gogal Jr OM, Johnson MS, Larsen CT, Prater MR, Duncan RB, Ward DL, et al. Dietary oral exposure to 1,3,5-trinitro-1,3,5-triazine in the northern bobwhite (*Colinus virginianus*). Environ Toxicol Chem 2003;22(2):381–7.
[34] Ehrich M, Wu X, Werre SR, Major MA, McCain WC, Reddy G. Calcium signaling in neuronal cells exposed to the munitions compound Cyclotrimethylenetrinitramine (RDX). Int J Toxicol 2009;28(5):425–35.
[35] Way M, McCain W, editors. Protective effects of the calcium channel blocker, verapamil, in rats exposed to RDX. Aberdeen Proving Ground (MD): U.S. Army Center for Health Promotion and Preventive Medicine; 2007, USACHPPM Toxicology Study No. 85-XC-08RU-08.

[36] Williams LR, Aroniadou-Anderjaska V, Qashu F, Finne H, Pidoplichko V, Bannon DI, et al. RDX binds to the GABA(A) receptor-convulsant site and blocks GABA(A) receptor-mediated currents in the amygdala: a mechanism for RDX-induced seizures. Environ Health Perspect 2011;119(3):357–63.
[37] Zhang B, Pan X. RDX induces aberrant expression of microRNAs in mouse brain and liver. Environ Health Perspect 2009;117(2):231–40.
[38] Bannon DI, Johnson M, Williams L, Adams V, Perkins E, Gust K, et al. RDX and miRNA Expression in B6C3F1 Mice. Environ Health Perspect 2009;117(3):A98, author reply A98-9.
[39] Gust KA, Pirooznia M, Quinn Jr MJ, Johnson MS, Escalon L, Indest KJ, et al. Neurotoxicogenomic investigations to assess mechanisms of action of the munitions constituents RDX and 2,6-DNT in Northern bobwhite (*Colinus virginianus*). Toxicol Sci 2009;110(1):168–80.
[40] Dilley JV, Tyson CA, Newell GW, editors. Mammalian toxicological evaluation of TNT wastewaters. Volume II. Acute and subacute mammalian toxicity of TNT and the LAP mixture. Menlo Park (CA): SRI International; 1978, Prepared for: U.S. Army Medical Research and Development Command, Washington, DC. Accession Number: AD A080957.
[41] Cholakis JM, Wong LCK, Van Goethem DL, Minor J, Short R. Mammalian toxicological evaluation of RDX. Kansas City (MO): Midwest Research Institute; 1980, Prepared for: U.S. Army Medical Research and Development Command, Fort Detrick; Frederick, MD. AD A092531.
[42] Von Oettingen WF, Donahue DD. Toxicity and potential dangers of cyclotrimethylenetrinitramine. J Ind Hyg Toxicol 1949;31(1):21–31.
[43] French JE, Bradley SL, Schneider NR, Andersen ME, Jenkins Jr LJ. Cyclotrimethylenetrinitramine (RDX)-induced ultrastructural changes in rat liver and kidney. Toxicol Appl Pharmacol 1976;37:122.
[44] Macphail RC, Walker QD, Cook LL. Neurotoxicity of cyclotrimethylenetrinitramine (RDX). Research Triangle Park (NC): U.S. Environmental Protection Agency, Neurotoxicology Division, Health Effects Research Laboratory; 1985, Prepared for: U.S. Army Medical Research and Development Command, Fort Detrick; Frederick, MD. Accession Number: AD A168266.
[45] Fellows RJ, Driver CR, Cataldo DA, Harvey SD. Bioavailability of hexahydro-1,3,5-trinitro-1,3,5-triazine (RDX) to the prairie vole (*Microtus ochrogaster*). Environ Toxicol Chem 2006;25(7):1881–6.
[46] Smith JN, Liu J, Espino MA, Cobb GP. Age dependent acute oral toxicity of hexahydro-1,3,5-trinitro-1,3,5-triazine (RDX) and two anaerobic N-nitroso metabolites in deer mice (Peromyscus maniculatus). Chemosphere 2007;67 (11):2267–73.
[47] Druckrey H, Preussmann R, Ivankovic S, Schmahl D. Organotrope carcinogene Wirkungen bei 65 verschiedenen N-Nitroso-Verbindungen an BD-Ratten. Zeitschrift fuer Krebsforschung. J Z Krebsforsch 1967;69(2):103–201.
[48] Urban H, Danz M. Tumorinduzierende Wirkung von Trinitroso-trimethylen-triamin in Verbindung mit Dimethylsulfoxid (English translation: Tumorigenicity of TTT dissolved in DMSO). Arch Geschwulstforsch 1976;46(8):657–62.
[49] Danz M, Urban H, Stiller KJ, Amlacher E. The proliferative activity of the adrenal cortex as influenced by carcinogenic and noncarcinogenic substances. Exp Pathol (JENA) 1975;11(3–4):142–7.
[50] Danz M, Klinger W, Muller D, Kleeberg U, Glockner R, Ziebarth D, et al. N, N diethyl-4-aminoazobenzene (DEAB): acute actions with respect to possible carcinogenicity as well as the role of solvents. Morphological and pharmacological investigations. Exp Pathol (JENA) 1978;16(1–6):245–53.

[51] Amlacher E, Rudolph C. The thymidine incorporation inhibiting screening system (TSS) to test carcinogenic substances (a nuclear DNA synthesis suppressive short-term test). Arch Geschwulstforsch 1981;51(7):605–10.
[52] Meyer SA, Marchand AJ, Hight JL, Roberts GH, Escalon LB, Inouye LS, et al. Up-and-down procedure (UDP) determinations of acute oral toxicity of nitroso degradation products of hexahydro-1,3,5-trinitro-1,3,5-triazine (RDX). J Appl Toxicol 2005;25 (5):427–34.
[53] Ferguson JW, McCain WC. 14-day feeding study of hexahydro-1,3,5, trinitro-1,3,5-triazine (RDX) in the white-footed mouse, *Peromyscus leucopus*. Aberdeen Proving Ground (MD): U.S. Army Center for Health Promotion and Preventive Medicine; 1999.
[54] Smith JN, Pan X, Gentles A, Smith EE, Cox SB, Cobb GP. Reproductive effects of hexahydro-1,3,5-trinitroso-1,3,5-triazine in deer mice (*Peromyscus maniculatus*) during a controlled exposure study. Environ Toxicol Chem 2006;25(2):446–51.
[55] Jaligamma S, Kale VM, Wilbanks MS, Perkins EJ, Meyer SA. Delayed myelosuppression with acute exposure to hexahydro-1,3,5-trinitro-1,3,5-triazine (RDX) and environmental degradation product hexahydro-1-nitroso-3,5-dinitro-1,3,5-triazine (MNX) in rats. Toxicol Appl Pharmacol 2013;266(3):443–51.
[56] Gust KA, Wilbanks MS, Guan X, Pirooznia M, Habib T, Yoo L, et al. Investigations of transcript expression in fathead minnow (*Pimephales promelas*) brain tissue reveal toxicological impacts of RDX exposure. Aquat Toxicol 2011;101(1):135–45.
[57] Gong P, Basu N, Scheuhammer AM, Perkins EJ. Neurochemical and electrophysiological diagnosis of reversible neurotoxicity in earthworms exposed to sublethal concentrations of CL-20. Enviro Sci Pollut Res Int 2010;17(1):181–6.
[58] Gong P, Guan X, Inouye LS, Deng Y, Pirooznia M, Perkins EJ. Transcriptomic analysis of RDX and TNT interactive sublethal effects in the earthworm *Eisenia fetida*. BMC Genomics 2008;9(Suppl. 1):S15.
[59] Gong P, Inouye LS, Perkins EJ. Comparative neurotoxicity of two energetic compounds, hexanitrohexaazaisowurtzitane and hexahydro-1,3,5-trinitro-1,3,5-triazine, in the earthworm Eisenia fetida. Environ Toxicol Chem 2007;26(5):954–9.
[60] Gong P, Loh PR, Barker ND, Tucker G, Wang N, Zhang C, et al. Building quantitative prediction models for tissue residue of two explosives compounds in earthworms from microarray gene expression data. Environ Sci Technol 2012;46(1):19–26.
[61] Li Y, Wang N, Perkins EJ, Zhang C, Gong P. Identification and optimization of classifier genes from multi-class earthworm microarray dataset. PLoS One 2010;5(10): e13715.
[62] Bannon DI, Dillman JF, Hable MA, Phillips CS, Perkins EJ. Global gene expression in rat brain and liver after oral exposure to the explosive hexahydro-1,3,5-trinitro-1,3,5-triazine (RDX). Chem Res Toxicol 2009;22(4):620–5.
[63] Hart ER. Subacute toxicity of RDX and TNT in dogs. Kensington (MD): Litton Bionetics, Inc; 1974, Prepared for: U.S. Office of Naval Research, Arlington, VA. Accession Number: AD A035717.
[64] Martin DP, Hart ER. Subacute toxicity of RDX and TNT in monkeys. Kensington (MD): Litton Bionetics, Inc; 1974, Prepared for: U.S. Office of Naval Research, Arlington, VA. Accession Number: AD A044650.
[65] Brown D. The acute and chronic biochemical and behavioral effects of cyclotrimethylenetriamine. Baltimore (MD): University of Maryland School of Pharmacy; 1975.
[66] Levine BS, Furedi EM, Gordon DE, Burns JM, Lish PM. Thirteen week toxicity study of hexahydro-1,3,5-trinitro-1,3,5-triazine (RDX) in Fischer 344 rats. Toxicol Lett 1981;8(4-5):241–5.
[67] Levine BS, Furedi EM, Gordon DE, Barkley JJ, Lish PM. Toxic interactions of the munitions compounds TNT and RDX in F344 rats. Fundam Appl Toxicol 1990;15 (2):373–80.

[68] Crouse LCB, Michie MW, Major MA, Johnson MS, Lee RB, Paulus HI. Subchronic oral toxicity of RDX in rats. Aberdeen Proving Ground (MD): US Army Center for Health Promotion and Preventive Medicine; 2006, Toxicology Study No. 85-XC-5131-03.
[69] Hart ER. Two-year feeding study in rats. Kensington (MD): Litton Bionetics, Inc; 1976, Prepared for: U.S. Office of Naval Research, Arlington, VA. Accession Number: AD A040161.
[70] U.S. EPA. Integrated Risk Information System [Internet]. Cincinnati (OH): Office of Health and Environmental Assessment (US); National Center for Environmental Assessment [cited 2000].
[71] Wolford ST, Schroer RA, Gohs FX, Gallo PP, Brodeck M, Falk HB, et al. Reference range data base for serum chemistry and hematology values in laboratory animals. J Toxicol Environ Health 1986;18(2):161–88.
[72] U.S. EPA. Integrated Risk Information System (IRIS) [Internet]. Hexahydro-1,3,5-trinitro-1,3,5-triazine (RDX) (CASRN 121-82-4). Cincinnati (OH): Environmental Protection Agency [cited 1993].
[73] Parker GA, Reddy G, Major MA. Reevaluation of a twenty-four-month chronic toxicity/carcinogenicity study of hexahydro-1,3,5-trinitro-1,3,5-triazine (RDX) in the B6C3F1 hybrid mouse. Int J Toxicol 2006;25(5):373–8.
[74] Sweeney LM, Okolica MR, Gut Jr CP, Gargas ML Cancer mode of action, weight of evidence, and proposed cancer reference value for hexahydro-1,3,5-trinitro-1,3,5-triazine (RDX). Regul Toxicol Pharmacol 2012;64(2):205–24.
[75] Pan X, San Francisco MJ, Lee C, Ochoa KM, Xu X, Liu J, et al. Examination of the mutagenicity of RDX and its N-nitroso metabolites using the Salmonella reverse mutation assay. Mutat Res 2007;629(1):64–9.
[76] Schneider NR, Bradley SL, Andersen ME. The distribution and metabolism of cyclotrimethylenetrinitramine (RDX) in the rat after subchronic administration. Toxicol Appl Pharmacol 1978;46(1):163–71.
[77] Angerhofer RA, Davis G, Balezewski L. Teratological assessment of trinitro-RDX in rats. Aberdeen Proving Ground (MD): U.S. Army Environmental Hygiene Agency; 1986, Accession Number: AD-A166 249.
[78] Danz M, Amlacher E, Urban H, Bolck F. Different action of carcinogenic and noncarcinogenic substances on the proliferative activity of adrenal cortex cells. Exp Pathol (JENA) 1973;8(2):122–7.
[79] Urban H, Amlacher E, Danz M. Tumor-inducing effect of trinitroso-trimethylenetriamine (TTT) when orally applied in dimethylsulphoxide (DMSO) for solvent. Exp Pathol (JENA) 1975;10(5–6):364–5.
[80] Calow P. Physiological costs of combating toxicants: ecological implications. Comp Biochem Physiol C 1991;100(1–2):3–6.
[81] Congdon JD, Dunham AE, Hopkins WA, Rowe CL, Hinton TG. Resource allocation-based life histories: a conceptual basis for studies of ecological toxicology. Environ Toxicol Chem 2001;20(8):1698–703.
[82] Gogal Jr RM Influence of feed exposure to RDX on northern bobwhite (*Colinus virginianus*). Avian Toxicology Study Nos. 031700G and 031300G (draft). Blacksburg (VA): Avian Immunotoxicity Risk Assessment Laboratory, Virginia-Maryland Regional College of Veterinary Medicine, Virginia Polytechnic Institute and State University; 2001.
[83] Baker LM, Larsen CT, Sriranganathan N, Jones DE, Johnson MS, Gogal Jr RM Effects of energetic compounds on the Northern Bobwhite quail and biotransformation applications of the intestinal flora. Bull Environ Contam Toxicol 2004;72(1):1–6.
[84] Sample BE, Opreska DM, Suter GW. Toxicological benchmarks for wildlife. Oak Ridge (TN): Oak Ridge National Laboratory, Risk Assessment Program, Health

Sciences Research Division; 1996, Prepared for: U.S. Department of Energy. Report Number: ES/ER/TM-86/R3.
[85] Johnson MS, Paulus HI, Salice CJ, Checkai RT, Simini M. Toxicologic and histopathologic response of the terrestrial salamander *Plethodon cinereus* to soil exposures of 1,3,5-trinitrohexahydro-1,3,5-triazine. Arch Environ Contam Toxicol 2004;47 (4):496–501.
[86] McFarland CA, Quinn Jr MJ, Bazar MA, Talent LG, Johnson MS. Toxic effects of oral hexahydro-1,3,5-trinitro-1,3,5-triazine in the western fence lizard (*Sceloporus occidentalis*). Environ Toxicol Chem 2009;28(5):1043–50.

CHAPTER 5

Wildlife Toxicity Assessment for High Melting Explosive (HMX)

Mark S. Johnson, Gunda Reddy

Contents

INTRODUCTION

High Melting Explosive (HMX) is one of several compounds that has been released to the environment during the manufacture of explosives and in load, assembly, and pack (LAP) activities at U.S. Army ammunition plants (AAPs) and other military installations. Structurally, the compound (Chemical Abstract Services Registry Number (CAS No.) 2691-41-0) is a completely N-nitrated, eight-member heterocyclic ring compound with the empirical formula, $C_4H_8N_8O_8$. In addition to HMX, it is known by various systematic and trivial names including octogen, cyclotetramethylenetetranitramine, and octahydro-1,3,5,7-tetranitro-1,3,5,7-tetrazocine, among others.

This Wildlife Toxicity Assessment summarizes current knowledge of the likely harmful impacts of HMX on wildlife, emphasizing threshold doses for the onset of toxicological effects, as described in reports of experimental studies of HMX. Surveying the threshold dosimetry of the compound

Wildlife Toxicity Assessments for Chemicals of Military Concern
http://dx.doi.org/10.1016/B978-0-12-800020-5.00005-3 2015 Published by Elsevier Inc.

may support the establishment of Toxicity Reference Values (TRVs) that could serve as protective exposure standards for wildlife ranging in the vicinity of affected sites. The protocol for performing this assessment followed U.S. Army Center for Health Promotion and Preventive Medicine (USACHPPM) Technical Guide 254, *Standard Practice for Wildlife Toxicity Reference Values* [1].

ENVIRONMENTAL FATE AND TRANSPORT

HMX, a more powerful explosive than trinitrotoluene (TNT), has been used as part of the trigger mechanism for atomic (fission) weapons, as a component in plastic explosives, and in rocket propellants [2,3]. The compound's manufacture is limited to a single location in the United States (the Holston Army Ammunition Plant (AAP) at Kingsport, Tennessee), where it has been reported that up to 45 pounds per day (lb/day) have been released to surrounding water bodies in discharged wastewaters from manufacturing and processing. Concentrations of HMX of up to 3.36 milligrams per liter (mg/L) were detected in effluents from the Holston facility [4]. Releases of HMX have also occurred at facilities where munitions are assembled, stored, or tested. For example, compound concentrations of up to 5,700 milligrams per kilogram (mg/kg) were reported in soil at some army sites [4].

Physicochemical properties of HMX relevant to the environmental fate and transport of the compound are listed in Table 5.1.

The vapor pressure and Henry's Law constant are sufficiently low (3.33×10^{-14} mm Hg and 2.60×10^{-15} atm.m^3/mole, respectively) indicating HMX is highly unlikely to enter the air as a vapor. However, aerial dispersion of the compound while adhering to soil, dust particles, or some alternative means of aerosolization is a likely mechanism by which the compound can be released to the atmosphere [2]. HMX has a low log soil organic carbon-water partition coefficient of 0.54, and has the potential for high mobility in soil that could leach to the ground water. For example, HMX was detected in the ground water at the Louisiana AAP at concentrations of up to 4.2 mg/L [4].

Photolysis appears to be the dominant process by which HMX is broken down in the environment, with a reported first order photolytic rate constant of 0.15 days^{-1} [3]. This suggests that an aqueous HMX concentration of 0.5 mg/L will have a half-life of 4 to 5 days when exposed to natural sunlight. Primary products of this process include nitrate, nitrite, and formaldehyde. By

Table 5.1 Summary of the Physical-Chemical Properties of HMX

CAS No.	2691-41-0
Molecular weight	296.16
Color	Colorless
Physical state	Crystalline solid
Melting point	276−280 °C
Boiling point	No data
Odor	No data
Solubility water	5−6.63 mg/L at 20−25 °C; soluble in acetone, cyclohexanone, acetic anhydride, dimethyl sulfoxide
Partition coefficients:	
Log K_{ow}	0.06, 0.26
Log K_{oc}	0.54
Vapor pressure at 25 °C	3.33×10^{-14} mm Hg
Henry's Law constant at 25 °C	2.60×10^{-15} atm.m^3/mole
Conversion factors	1 ppm = 12.11 mg/m^3 1 mg/m^3 = 0.083 ppm

Sources: ATSDR [2], U.S. EPA [3], Talmage *et al.* [4].

contrast, biodegradation/biotransformational processes involving bacteria or other microflora are extremely slow, though the formation of 1,1-dimethyl hydrazine has been demonstrated as a result of anaerobic degradation [3].

TOXICITY PROFILE

Summary of Mammalian Toxicity

Mammalian Oral Toxicity: Acute

In one of a series of studies carried out for the U.S. Army Medical Research and Development Command by Inveresk Research International (IRI), Cuthbert *et al.* [5] reported data obtained from various short-term toxicological tests on HMX. These included guinea pig sensitization, eye and skin irritation studies in rabbits, dermal and intravenous lethality in rats and rabbits, and acute oral lethality studies in rats, mice, and rabbits. In the latter, acute oral lethal dose (LD_{50}) values of 6.5 and 2.0 grams per kilogram (g/kg) were found for male Fischer 344 (F344) rats and B6C3F1 mice, respectively. Female oral LD_{50} values for rats and mice were reported as 7.6 and 3.8 g/kg, respectively. The rodent studies were conducted using five animals/sex/group. However, acute oral Approximate Lethal Dose (ALD) studies using New Zealand white rabbits used one animal/sex/group at 2,000, 1,000, 429, 250, 100, and 50 mg/kg. Females died at every dose level. Males died at the 250, 429, 1,000, and 2,000 mg/kg dose level. Johnson *et al.* [6] conducted an acute oral study using stagewise probit design and determined the LD_{50} to be 93 mg/kg (CI = 82–105, slope = 14.5). Toxic oral exposures resulted in clonic convulsions, to include salivation, more than 21 hours after exposure, suggesting hindgut fermenting species (to include ruminants) as species with gastric physiologies that may enhance absorption of parent compound or metabolites.

Wilson [7] supplemented these findings with toxicokinetic information that had been obtained by administering ^{14}C-labeled HMX by either gavage or intravenous injection to rats and mice. For either experimental animal species, the data indicate a low level of gastrointestinal absorption of unchanged HMX in rodents. For example, in the rats, 85% of the administered dose had accumulated in the feces 4 days after dosing. Furthermore, the comparative levels of radioactivity released to the urine following intravenous versus oral administration of ^{14}C-HMX suggested that less than 5% of the oral dose of the compound had crossed the gastrointestinal absorption barrier. This result is consistent with the low oral lethality reported by Cuthbert *et al.* [5]. According to Wilson [7], the little tissue deposition that had occurred was found in the liver, kidney, and brain. These data with those of Cuthbert *et al.* [5] and Johnson *et al.* [6] suggest that oral absorption is enhanced in hindgut fermenting and possibly ruminant species.

Mammalian Oral Toxicity: Subacute

Greenhough and McDonald [8,9] published two reports in which the 14-day oral toxicity of HMX was determined in F344 rats and B6C3F1 mice. These were essentially range-finding studies for subsequent investigations of the subchronic (90-day) toxicity of this compound in these species. In the first study, the authors exposed six rats/sex/group to HMX for 14 days as a dietary addition at target doses of 0, 333, 1,000, 3,000, and 9,000 mg/kg/day [8]. As tabulated by the authors, the actual achieved average doses equivalent to these levels were 0, 335.2, 957.4, 2,981, and 8,504.3 mg/kg/day in males and 0, 369.2, 1,280, 3,474.25, and 3,055 mg/kg/day in females. In the in-life phase of the study, animals were checked daily for mortality and clinical signs, twice weekly for body weight, and once weekly for food and water consumption. At termination, blood samples were taken from all animals and stored frozen. All carcasses were subjected to a gross necropsy, liver and kidney weights were recorded, while excised pieces of brain, heart, kidney, liver, spleen, and thymus were processed for histopathological examination.

Concomitant with the incremental range of HMX doses, the group-specific incidence of compound-related fatalities was 0/6 (controls), 0/6, 0/6, 0/6, and 5/6 (high-dose) in males and 0/6, 0/6, 1/6, 1/6, and 6/6 in females. These deaths were accompanied by the onset of profound clinical signs characteristic of toxicologically challenged animals, most notably in males at the two highest dose levels, but in all groups of female rats. All HMX-treated male rats displayed dose-related suppression of body weight gain, while the two highest groups showed an actual body weight loss after 4 days of exposure. This food consumption-related deficit had partially rebounded by day 7. All females receiving HMX showed an initial body weight loss to levels that stayed depressed compared to initial values for all but those females receiving 333 mg/kg/day (group 2). Some marginal reductions in relative and absolute liver and kidney weights were observed among the treated groups, although it is unclear to what extent these changes were merely a consequence of dietary fluctuations.

A number of gross pathological findings were described in the report, although some were essentially sporadic in occurrence and, therefore, probably unrelated to dose. However, 4/6 high-dose females displayed smaller than normal spleens and enlarged adrenals—a feature that was also apparent in the single group where four females died prematurely. High-dose male rats displayed centrilobular degeneration of the liver, while hepatocytic hyperplasia and increased cytoplasmic eosinophilia along with lymphocyte depletion in the thymus and spleen were noted in high-dose and other decedent females. However, the extent of these lesions in intermediate groups was not determined, an omission that did not allow a no observed adverse effect level (NOAEL) or lowest observed adverse effect level (LOAEL) based on any observations other than lethality to be established. Based on the findings in the study, these would be (nominally) 3000 and 9000 mg/kg/day, respectively.

A similar protocol to that described above was also used to determine the subacute toxicity of HMX in B6C3F1 mice [9]. Target dietary doses in males (groups 1 − 5) were 0, 100, 300, 900, and 2,700 mg/kg/day, and in females (groups 1 − 5) were 0, 320, 800, 2,000, and 5,000 mg/kg/day. As tabulated by the authors, the actual achieved average doses equivalent to these levels for males (groups 1 − 3) were 0, 119.5, and 383 mg/kg/day, and for females (groups 1 − 4) were 0, 344, 882.7, and 2,045.6 mg/kg/day. The lethality rate for the full sequence of groups was 0/6, 0/6, 5/6, 6/6, and 6/6 in males and 0/6, 0/6, 2/6, 4/6, and 6/6 in females. All HMX-receiving groups displayed clinical signs in response to dosing that were marked by overexcitability in the lower dose groups and by a range of increasingly severe responses leading to death in the higher dose groups. Animals displayed an initial loss of weight that may have been associated with reduced food consumption. For the survivors, these parameters rebounded in parallel during the second week of exposure. This "recovery" was indicated also by similar terminal absolute and relative organ weights between treated and control groups. As described by the authors, the histopathological findings were characterized by a "dose-related increase" in hepatocellular hyperplasia and cytoplasmic eosinophilia, splenic red and white pulp, and thymic cellular depletion. However, by analogy to the 14-day study in rats [8], the absence of any histopathological examinations of HMX-receiving survivors in the intermediate dose groups has forced the selection of an NOAEL and LOAEL based on lethality. These dose values were approximated as 100 and 300 mg/kg/day, respectively, based on the data for male mice, and as 800 and 2,000 mg/kg/day, respectively, based on the data for female mice.

New Zealand white rabbits were exposed via oral gavage to HMX doses of 0, 0.1, 0.5, 1, 5, 10, 25, and 50 mg HMX per kilogram per body weight (HMX/kg/bw) in corn oil comprising four animals per treatment [6]. Mortality or morbidity occurred in rabbits exposed to 25 and 50 mg/kg/day. Seizures began approximately 48 hours following initial dose and were often transient. Incidences of seizure were observed intermittently in rabbits from treatments 10 mg/kg/day or greater. Changes in body mass and feed consumption were also observed relative to dose. No changes were seen in hematology, although there were changes in blood urea nitrate, creatinine, and chloride levels in plasma. The NOAEL and LOAEL were determined to be 5 and 10 mg/kg/day, respectively.

Mammalian Toxicity: Subchronic

Everett *et al.* [10] and Everett and Maddock [11] reported separate 13-week toxicological studies on HMX in F344 rats and B6C3F1 mice. Arising from the range-finding study in rats described earlier [8], 20 rats/sex/group received dietary doses of 0, 50, 150, 450, 1,350, and 4,000 mg/kg/day (males) and 0, 50, 115, 270, 620, and 1,500 mg/kg/day (females) [10]. The actual achieved average doses equivalent to these target levels were 0, 51, 153.5, 461, 1,394, and 4,101 mg/kg/day in males, and 0, 50.3, 115.6, 273.3, 627.7, and 1,511.9 mg/kg/day in females. In addition to a more extensive range of in-life, necropsy, and histopathological observations than in the 14-day study, all rats received an ophthalmic examination before dosing commenced and during week 13 of dosing. Clinical chemistry and hematological analyses were carried out on blood samples taken from the orbital sinus of 10 males and 10 females during weeks 5 and 12 of treatment. Four-hour urine samples were collected from a subset of subjects among the groups during weeks 5 and 12. These samples were monitored for glucose, blood, protein, ketones, color, pH, and specific gravity, among others.

In contrast to the findings of the 14-day study in F344 rats [8], there were no compound-related deaths and few clinical signs in evidence during the 13 weeks of dosing. All ophthalmological observations were unremarkable before and after treatment. However, body weight gain was reduced in a dose-dependent manner with varying degrees of statistical significance in some groups compared to controls. These changes may have been due, at least in part, to fluctuations in food consumption. Some potentially dose-related hematological changes were observed in both sexes of high-dose rats, including reductions in hemoglobin concentration, packed cell volume and erythrocyte count, and increases in methemoglobin levels. Sporadic, statistically significant differences in plasma enzyme activities (e.g., in alkaline phosphatase) were observed in rats exposed to high-dose levels of HMX compared to controls. However, the extent to which these changes were dose-dependent is uncertain, and thus their relationship to HMX treatment cannot be assigned with confidence. Although findings from gross necropsy were benign, some apparent dose-dependent histopathological changes were considered by the authors to be compound-related. These included the appearance of enlarged liver cells featuring large nuclei and granular eosinophilic cytoplasm with associated small necrotic foci, which were most evident in male rats.

The designation of an NOAEL for the histopathological effects of HMX in liver may be controversial. Thus, although the effects were most evident in males receiving the compound at the two highest doses, the Integrated Risk Information System (IRIS) compilers [12] and Talmage *et al.* [4] chose a nominal dose level of 50 mg/kg/day as the NOAEL, based on an incidence of 2/19 in 150 mg/kg/day-receiving males compared to 0/20 in controls. However, since this difference is statistically insignificant by Fisher's exact test, a viable alternative of NOAEL might be the value of 150 mg/kg/day itself, an approach that appears to be more in line with the conclusions of the authors of the study [10]. Regarding the issue of the precise value of the NOAEL, Everett *et al.* [10] submitted that if 150 mg/kg/day were adopted as the NOAEL, the next highest dose (450 mg/kg/day) would be unsatisfactorily high for the LOAEL, since the incidence of histopathological liver lesions was 20/20 at this level. Taking all the incidence data together suggests that the subchronic points-of-departure (NOAEL and LOAEL) for the toxicological effects in F344 rats are likely to exist in a narrow dosimetric region between 100 and 400 mg/kg/day.

Other compound-related histopathological changes were evident in the kidneys of female F344 rats. The incidence of these lesions, characterized by focal atrophy and dilation of the tubules, achieved statistical significance compared to controls at a dose level of 620 mg/kg/day and above (Fisher's exact test from the data in the study). These changes result in nominal NOAELs and LOAELs of 270 and 620 mg/kg/day, respectively.

A 13-week study in B6C3F1 mice featured dietary administration of HMX at target dose levels of 0, 5, 12, 30, 75, and 200 mg/kg/day in males and 0, 10, 30, 90, 250, and 750 mg/kg/day in females [11]. The actual achieved average doses equivalent to these target levels were 0, 5.2, 12.2, 30.5, 75, and 199.8 mg/kg/day in males and 0, 10.5, 30.8, 95.1, 257.1, and 784.5 mg/kg/day in females. A range of toxicological effects was observed similar to those in evidence in the rat study [10]. However, in contrast to the findings in rats, the apparent toxicological consequences of the compound in the mice were profound, with 65% premature deaths observed in high-dose males and 100% deaths in high-dose females. Lower fatality rates were observed at lower dose levels supporting the conclusion that mortality was likely compound-related. However, other than lethality, few if any obvious HMX-related consequences were apparent among the survivors at any dose level, thereby rendering uncertain the causes of death among the high-dose animals and calling into question the utility of the study to delineate a sufficiently discriminating subthreshold point-of-departure for the compound's toxicological consequences. Using mortality as the primary subchronic toxicological effect of HMX from the female mouse data, the nominal NOAEL would be 90 mg/kg/day, with an LOAEL of 250 mg/kg/day. These doses are strikingly similar to those identified for mortality in male B6C3F1 mice in the 14-day subacute toxicity study [9].

Mammalian Oral Toxicity: Chronic

No experimental studies were identified that addressed the chronic toxicity of HMX.

Mammalian Oral Toxicity: Other

No other data relevant to oral exposures for mammals were found.

Studies Relevant for Mammalian TRV Development for Ingestion Exposures

The toxicological database on HMX is limited (Table 5.2). The toxicokinetic findings discussed by Wilson [7] indicate that, typically, only a comparatively small proportion of an orally administered dose of HMX will be absorbed at the gastrointestinal barrier in monogastric species such as rats and mice species used in these assays. In addition, Wilson [7] pointed out that in IRI experiments only small amounts of the compound absorbed survived clearance in the urine where the radioactivity partitioned mostly as polar metabolites. Therefore, since mammals clearly have the capacity to metabolize HMX, the fact that the radioactivity eliminated in the feces was overwhelmingly in the form of unchanged HMX supports the suggestion that this component of the load probably represented unabsorbed substrate rather than HMX that had been absorbed and then undergone hepatobiliary recycling.

As with urinary metabolites, the little amount of compound deposited in the tissues would also change to metabolites of HMX [7]. This implies that the toxicological consequences of HMX, including the hepatic and renal changes seen in histopathological specimens and the compound-induced lethality evident at higher doses in either species of test animal (F344 rats and B6C3F1 mice), will probably have resulted from the biochemical activity of one or more metabolites of HMX rather than the parent compound. Unfortunately, examination of the carcasses of the high-dose mice receiving HMX for up to 13 weeks and the results of the necropsy and histopathological findings in animals treated at lower doses and surviving to term failed to offer any clues as to the causes of the premature deaths induced by HMX. In fact, there is little evidence of a single universally applicable mechanism by which HMX induces toxic effects leading to lethality in rodents. By contrast, histopathological findings in F344 rats were inconsistent since, in the 13-week study [10], sublethal microscopic lesions in the liver were observed primarily in exposed males, while kidney effects were largely restricted to the females. This separate and gender-specific pattern of histopathological lesion formation argues against the existence of a single ubiquitous mechanism by which fatalities such as those observed in both sexes of mice from the 13-week study could have been induced.

Taking the results of all the IRI studies on HMX together suggests that (unknown) metabolites formed from only a very small proportion of the administered load can induce lethality by an unknown mechanism, and that B6C3F1 mice are more susceptible to this effect than F344 rats.

Table 5.2 Summary of Relevant Mammalian Data for TRV Derivation

Test Organism	Test Duration	Test Results: NOAEL (mg/kg/day)	Test Results: LOAEL (mg/kg/day)	Test Results: Effects Observed at the LOAEL	Study
Rabbit	Single acute exposure	100 (m) ? (f)	250 (m) 50 (f)	Mortality, convulsions, miosis, mydriasis, slight hyperkinesias, labored respiration	[5]
Rabbit	LD_{50}	93 mg/kg		Clonic convulsions, salivation	[6]
Rat	LD_{50}	7,360(m and f)			[5]
Mouse	LD_{50}	2,710(m and f)			
Rat (F344)	14 days	2981 (m) 1280 (f)	8504 (m) 3055 (f)	Lethality associated with histopathological liver lesions, lymphocyte depletion, and spleen effects	[8]
Mice (B6C3F1)	14 days	120 (m) 883 (f)	383 (m) 2045 (f)	Lethality associated with histopathological liver lesions, lymphocyte depletion, and spleen effects	[9]
Rabbit	14 days	5	10	Convulsions/seizure, changes in body mass, plasma chloride, BUN, CREA	[6]
Rat (F344)	13 weeks	153 (m) 273 (f)	461 (m) 628 (f)	Histopathological lesions of the liver Focal atrophy and dilation of the kidney tubules	[10]
Mice (B6C3F1)	13 weeks	95 (f)	257 (f)	Lethality	[11]

However, recent toxicity data suggest that the rabbit may be more susceptible to HMX than either B6C3F1 mice or F344 rats [6]. The ALD in rabbits was between 50 and 250 mg/kg reported from Cuthbert *et al.* [5]; Johnson *et al.* [6] reported an LD_{50} of 96 mg/kg (CI = 82–105 mg/kg) in agreement with earlier data. The authors report the manifestation of adverse neurological symptoms more than 20 hours post oral administration, suggesting a much longer period required for absorption. Rabbits are also coprophagic, which may influence the time and surface area required for absorption. The 14-day, repeated-dose study found the NOAEL and LOAEL to be 5 and 10 mg/kg/day, respectively; these levels were much lower than those for rodents. Collectively, these data suggest that hindgut fermenting species (and possible ruminants) are more sensitive to oral exposures of HMX. These data are graphically presented in Figure 5.1.

Mammalian Inhalation Toxicity

No inhalation studies conducted using animals were found.

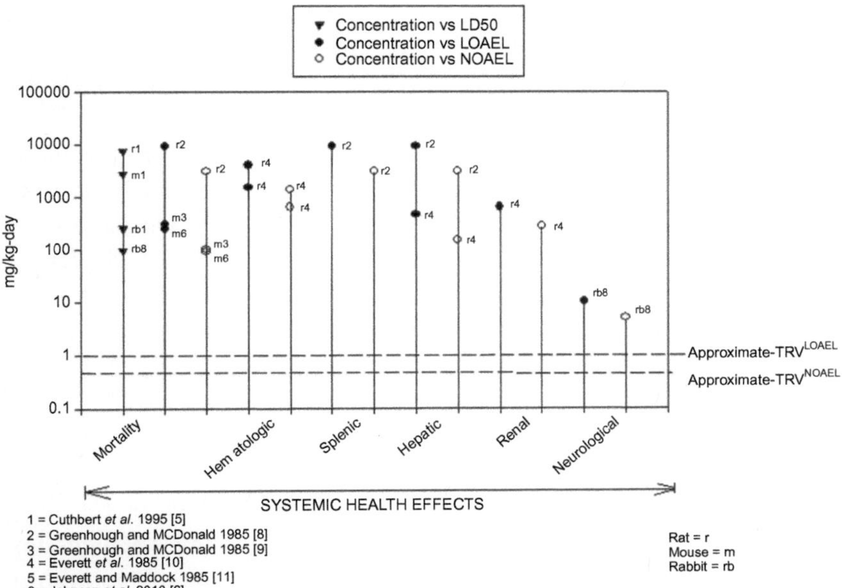

Figure 5.1 HMX Health Effects to Mammals. Letters by symbols represent test species, and numbers represent study.

Mammalian Dermal Toxicity

Cuthbert *et al.* [5] reported data that included dermal toxicity evaluations in rats and rabbits. For rats, the dermal LD_{50} was determined to be greater than 5.0 g/kg/bw. In rabbits, the percutaneous median lethal dose was determined to be 982.03 (861–1,102) mg/kg in abraded/nonabraded skin tests for both sexes. HMX was found to be mildly irritating to the skin of rabbits although not an eye irritant. There was no evidence suggesting that HMX has sensitizing effects using the Magnusson-Kligman Maximisation Test in guinea pigs [5].

Mammalian Toxicity: Other

Cuthbert *et al.* [5] also conducted rat and rabbit intravenous studies using HMX using dimethylsulfoxide (DMSO) as a vehicle. Rat IV LD_{50} was determined to be 25 and 38 mg/kg for males and females, respectively. Rabbit IV LD_{50} was reported to be between 10 and 15 mg/kg for both sexes.

Steinheim *et al.* [13] exposed nine male Norwegian white sheep to 3 days of either clean or contaminated silage as two choices for feed and measured hexahydro-1,3,5-trinitro-1,3,5-triazine (RDX), TNT, or HMX uptake. The explosives were taken from munitions and manually added to the top of the silage during the exposure periods. There was a clear preference for noncontaminated forage; the second most contaminated forage was RDX (eaten half as much as clean forage). HMX was avoided and only minimally ingested, suggesting that food avoidance may occur with wild and/or domestic animals exposed to HMX in field situations.

A companion field study followed the laboratory study to further test whether odor of TNT or HMX affected grazing [14]. Four 12 × 19 m test patches consisting of 3 × 4 m subpatches of soil amended with clean sand, sand with TNT, or sand with HMX, were used to allow eight rams to choose forage areas; observations were recorded. Actual forage was not contaminated. No difference was found in the number of observations of sheep foraging in clean versus contaminated plots.

Summary of Avian Toxicology

Avian Oral Toxicity: Acute

An ALD evaluation was conducted using 16 Northern Bobwhite (*Colinus virginianus*) [15]. Birds were orally gavaged using a water vehicle at eight doses ranging from 125 to 2,125 HMX mg/kg/bw. One bird of each sex was used for each dose group. There was only one death (female; 187 mg/kg) 6 days postexposure. There were no marked signs of overt toxicity.

A subsequent ALD was conducted using eight birds, four groups, at doses ranging from 3188 to 10,760 mg/kg. One female died at 7,173 mg/kg that occurred 12 days postexposure. No dose-related remarkable findings were attributed to exposure. The purity of the compounds was determined to be 98.5%. Additional ALDs were conducted where vehicle (e.g., corn oil) and fasting regime were evaluated, each with no predictable patterns in mortality. The authors report that crop contents consisted of impacted HMX in necropsied birds, suggesting the bolus effect from a nonabsorbable substance. An on-going subchronic study confirms no adverse effects to birds from exposures as high as 10,000 parts per million HMX in feed [15], suggesting that HMX is largely not available for absorption.

Avian Oral Toxicity: Subchronic

Given the lack of toxicity attributed to high acute oral exposures of HMX in birds, *C. virginianus* were exposed to a wide range of HMX concentrations in feed for 28 days [15]. Eight male and eight female Northern Bobwhites were exposed to either 0, 100, 1000, or 10,000 mg HMX/kg feed for 28 days and monitored for gross observations, feed consumption, and changes in body mass. Changes in feed consumption were profound in high exposure groups and lower in the 100 mg/kg group compared to controls. In addition, with these reductions in feed consumption were differential changes in body mass in the 1,000 and 10,000 mg/kg group compared with controls and the 100 mg/kg group. Egg production decreased to 0 in both high exposure groups by week 3 of exposure [15]. No other changes in plasma chemistry or blood parameters were evident. The authors attributed these observations to feed avoidance suggesting the unpalatability of HMX in feed.

Avian Oral Toxicity: Chronic

No data are available.

Avian Oral Toxicity: Other

Brunjes *et al.* [16] conducted a similar study investigating the potential for reproductive, developmental, and egg transfer of HMX in Northern Bobwhite. To determine whether HMX could be deposited in eggs, 11 quail were randomly assigned to one of four groups exposed to either 0, 25, 78, or 137 (analytical) mg HMX/kg feed (period of exposure unknown) presumably until eggs were laid and analyzed for HMX content. HMX was found in all eggs during this part of the study, including controls, and was found 2 days following exposure. No explanation was provided regarding HMX detection in eggs produced by females not exposed to HMX in feed. The authors noted the feed aversion that was consistent with the observations by Johnson *et al.* [15], which indicated concomitant loss in body mass in a dose-response relationship. The concentration of HMX in eggs was greatest at 4 days following the initial exposure and was lowest from 6 to 15 days following the initial exposure.

Reproductive effects were also evaluated in Northern Bobwhite that were exposed to HMX via the feed. Brunjes *et al.* [16] exposed 30 egg-laying females and males in breeding condition to either 0, 12, 53, or 110 mg/HMX/kg feed (analytical), collected the eggs, incubated and hatched young, and recorded growth measurements and hatching rates. Exposure duration was not reported, though it lasted at least 7 weeks when it was reported that a control female was removed from a study for injuries. Although there were no overt effects observed in adults, changes in body mass correlated with reductions in feed consumption were reported. Egg mass did not change relative to treatment, but laying rates were reduced with time of exposure and decreased based on treatment. Hatching rates were low, particularly with eggs laid in the control groups; there were no treatment-related effects. Considering these results, the authors suggested that effects from field exposures to HMX were unlikely given the demonstrative feed aversion observed during both parts of the study [16].

Avian Inhalation Toxicity

No data are available.

Avian Dermal Toxicity

No data are available.

Summary of Amphibian Toxicology

Red-backed salamanders (*Plethodon cinereus*) were exposed to HMX in soil during a range-finding study for 10 days [6]. During exposure, animals were observed for signs of stress (e.g., excess mucus production) and abnormal behavior. Following exposure, salamanders were weighed and euthanized, and blood was taken for enumeration of red and white blood cell counts, hemoglobin content, and five-part leucocyte differential. Animals were exposed to either 0 (<10), 25 (14.8), 100 (94.3), 400 (314.7), 750 (638.7), 1,200 (920.53), 2,500 (1851), or 5,000 (1969) mg HMX/kg Sassafras soil, nominal, and analytical concentrations, respectively. All animals survived the exposure period without any observed adverse effects.

No changes in behavior, body mass, red or white blood cells, hemoglobin, or adverse histology were observed.

Summary of Reptilian Toxicology

Western Fence Lizards (*Sceloporus occidentalis*) were exposed to HMX in corn oil orally via pipette in an acute study using the stagewise probit design [6]. Oral acute does ranged from 200 to 5,000 mg/kg in both sexes. Two of the five male lizards dosed at the 5,000 mg/kg level died within 2 days following exposure. One lizard exhibited seizures and became inactive before death. The cause of death could not be determined although there was evidence of renal inflammation. One female dosed at 500 mg/kg exhibited real limb "paddling" and hyperactivity that was resolved by day 5. No other adverse effects were observed in the other animals with the exception of varying degrees of transient inactivity.

RECOMMENDED TOXICITY REFERENCE VALUES

Toxicity Reference Values for Mammals

TRVs for Ingestion Exposures for the Class Mammalia

The HMX toxicity data are limited and variable. Much of the variation seems attributable to differences between species and study designs. Additionally, the data suggest that HMX is largely not absorbed in monogastric species, such as rodents, birds, and reptiles. Hindgut fermenting species, such as rabbits and possibly ruminants (e.g., cervids), may provide the gastrointestinal retention time needed under specific conditions to enhance absorption. Rabbits were the most sensitive species tested and effects were manifest over 20 hours postexposure [6]. These data are provided with those from other studies in Table 5.2. It is also important to consider that weight loss and feed consumption were inversely proportional and increased relative to HMX concentration. This finding, in addition to that of relatively low acute toxicity, suggested feed aversion, which likely affected other biophysical parameters. Other laboratory food aversion studies support this suggestion [13]; however, in the field, data on food aversion with contaminated soil, not forage, were not predictive for the grazing behavior in sheep [14].

Due to the limited data available, the approximation approach was used to derive the mammalian oral TRV for HMX [1]. An uncertainty factor of 10 was used to derive the NOAEL-based approximate TRV from a 14-day study conducted in rabbits where no adverse effects were observed (lack of convulsions) at 5 mg/kg/day [5]. An uncertainty factor of 10 was used to

Table 5.3 Selected Ingestion TRVs for the Class Mammalia

TRV	Dose	Confidence Level
NOAEL-based	0.5 mg/kg/day	Medium
LOAEL-based	1 mg/kg/day	Medium

Sources: Cuthbert *et al.* [5], Johnson *et al.* [6].

derive the LOAEL-based approximate TRV from the LOAEL reported by Johnson *et al.* [6] at 10 mg/kg/day also using a UF of 10 to derive a LOAEL of 1 mg/kg/day (Table 5.3). These TRVs were given a medium confidence rating since the results are consistent between studies for various species and plausible given the metabolism/absorption studies conducted suggesting that gut physiology is important in absorption.

TRVs for Ingestion Exposures for Mammalian Foraging Guilds

Since the work conducted in rodents has been well documented, TRVs specific to mammalian omnivores could be derived. Using the information from Everett and Maddock [11], mice appear to be more sensitive to the effects of oral HMX exposure than rats. The authors used a 13-week exposure regime where mortality was the only consistent endpoint, possibly due to the low oral bioavailability of HMX. These data are consistent with the findings of other work investigating acute and subchronic exposures in rodents [5,8–11].

An uncertainty factor of 10 was used to derive the NOAEL-based approximate TRV from a subchronic NOAEL. An uncertainty factor of 4 was used to derive the LOAEL-based approximate TRV from a subchronic LOAEL. These TRVs are presented in Table 5.4. Given that these species have been studied extensively, yet there is no other omnivore species evaluated, these TRVs are assigned a confidence rating of Medium.

TRVs for Inhalation Exposures for the Class Mammalia

Not available at this time.

Table 5.4 Selected Ingestion TRVs for Mammalian Omnivores

TRV	Dose	Confidence Level
NOAEL-based	9 mg/kg/day	Medium
LOAEL-based	62.5 mg/kg/day	Medium

Source: Everett *et al.* [11].

TRVs for Dermal Exposures for the Class Mammalia

Not available at this time.

Toxicity Reference Values for Birds

TRVs for Ingestion Exposures for the Class Aves

Two studies were conducted in Northern Bobwhite both yielding similar results. Both were feeding studies where the most significant effect was loss in body mass presumably due to feed avoidance or inappetence [15,16]. Acute toxicity was difficult to demonstrate [15]. Reductions in egg production occurred in both studies concomitantly with reductions in feed consumption and reductions in body mass, suggesting reduction in eggs were likely a consequence of reductions in energy budgets than from a direct toxic mechanism with ova production. Johnson *et al.* [15] found changes in egg production and body mass at 1,000 mg HMX/kg feed or 0.6 mg HMX/kg/bw/day and no adverse effects (NOAEL) at 100 mg HMX/kg feed or 0.1 mg/kg/day. Brunjes *et al.* [16] did not calculate daily oral dose based upon HMX concentration and consumption data and body mass data were not published to allow for a calculation of daily dose. These calculated HMX intakes are considered limited in utility as they are based on marked feed aversion, which, in turn, affected HMX exposure. Given the lack of acute toxicity and results from two studies where authors provide evidence for feed aversion, no TRVs were calculated for birds. However, it should be considered that HMX tainted feed is relatively unlikely to be consumed; therefore, the risk from exposure is expected to be negligible.

Toxicity Reference Values for Amphibians

TRVs for Soil Exposures for the Class Amphibia

Red-backed salamanders were exposed to soil concentrations exceeding 1,969 mg HMX/kg soil for 10 days without observation of adverse effect [6]. No other studies conducted with amphibians from HMX exposure could be found. Therefore, an uncertainty factor of 10 has been applied to the NOAEL value of 1,970 mg/kg resulting in a NOAEL-based value of 197 mg HMX/kg soil. No LOAEL-based value could be derived at this time. These TRVs are presented in Table 5.5. Given the paucity of data for HMX exposures using amphibians, this TRV is assigned a confidence rating of Low.

Table 5.5 Selected Ingestion TRVs for Terrestrial Amphibians

TRV	Soil Concentration	Confidence Level
NOAEL-based	197 mg/kg	Low
LOAEL-based	NA	NA

NA = not applicable.
Source: Johnson *et al.* [6].

Toxicity Reference Values for Reptiles

TRVs for Ingestion Exposures for the Class Reptilia

Western Fence Lizards exposed to acute oral doses on HMX in corn oil provided variable results at the limit dose of 5,000 mg HMX/kg/bw. Two of the five male lizards died and one female exhibited transient neurological effects. No other studies conducted with reptile studies could be found. Given the lack of data, no TRVs for reptiles could be developed for HMX at this time.

IMPORTANT RESEARCH NEEDS

The chemical/physical properties of HMX suggest that systemic exposure will be low for many organisms. The metabolic information from studies conducted in rodents show that most of the ingested HMX is excreted unchanged. However, the data for rabbits suggest differential absorption or biotransformation of HMX in herbivorous animals (e.g., ruminants, hindgut fermenters, etc.). Future work should focus on a greater understanding of the kinetics of HMX in various gastrointestinal physiologies. Additional data should be collected for reptiles and amphibians, though gastrointestinal exposure is likely to be less than that for mammals.

REFERENCES

[1] USACHPPM. Standard practice for wildlife toxicity reference values. Aberdeen Proving Ground (MD): U.S. Army Center for Health Promotion and Preventive Medicine; 2000, Technical Guide 254.

[2] Agency of Toxic Substances and Disease Registry. Toxicological profile for HMX. Atlanta (GA): U.S. Department of Health & Human Services, Public Health Service; 1997.

[3] U.S. EPA. Health advisory for octahydro-1,3,5,7-tetranitro 1,3,5,7-tetrazocine (HMX). Washington (DC): U.S. Environmental Protection Agency Office of Drinking Water; 1988, PB90-273525.

[4] Talmage SS, Opresko DM, Maxwell CJ, Welsh CJE, Cretella FM, Reno RH, et al. Nitroaromatic munition compounds: environmental effects and screening values. Rev Environ Contam Toxicol 1999;161:1–156.

[5] Cuthbert JA, D'Arcy-Burt KJ, Carr SMA. HMX: acute toxicity tests in laboratory animals. Frederick (MD): U.S. Army Medical Research and Development Command, Fort Detrick; 1985, AD A171598.
[6] Johnson MS, McFarland CA, Bazar MA, Quinn Jr MJ, LaFiandra EM, Talent LG. Toxicity of octahydro-1,3,5-tetranitro-1,3,5,7-tetrazocine (HMX) in three vertebrate species. Arch Environ Contam Toxicol 2010;58:836–43.
[7] Wilson AB. Determination of the acute and subchronic mammalian toxicity of HMX. Frederick (MD): U.S. Army Medical Research and Development Command, Fort Detrick; 1985, AD A173743.
[8] Greenhough RJ, McDonald P. HMX: 14-day toxicity study in rats by dietary administration. Frederick (MD): U.S. Army Medical Research and Development Command, Fort Detrick; 1985, AD A171597.
[9] Greenhough RJ, McDonald P. HMX: 14-day toxicity study in mice by dietary administration. Frederick (MD): U.S. Army Medical Research and Development Command, Fort Detrick; 1985, AD A171596.
[10] Everett DJ, Johnson IP, Hudson P, Jones M. HMX: 13-week toxicity study in rats by dietary administration. Frederick (MD): U.S. Army Medical Research and Development Command, Fort Detrick; 1985, AD A171601.
[11] Everett DJ, Maddock SM. HMX: 13-week toxicity study in mice by dietary administration. Frederick (MD): U.S. Army Medical Research and Development Command, Fort Detrick, Frederick; 1985, AD A171602.
[12] U.S. Environmental Protection Agency. Integrated risk information system [Internet]. Cincinnati (OH): Office of Health and Environmental Assessment (US), National Center for Environmental Assessment [cited 2014]. Available at http://www.epa.gov/IRIS.
[13] Steinheim G, Ådnøy T, Voie Ø, Holand Ø, Longva KS. Sheep prefer clean forage over forage contaminated with military explosives TNT, RDX, and HMX. Small Ruminant Res 2011;100:30–3.
[14] Steinheim G, Voie ØA, Ådnøy T, Longva KS, Holand Ø. Effect of contamination of pasture with military explosives on grazing behavior in sheep. Acta Agric Scand Sect A 2011;61:157–60.
[15] Johnson MS, Gogal Jr RM, Larsen CT. Food avoidance behavior to dietary octahydro-1,3,5-tetranitro-1,3,5,7-tetrazocine (HMX) exposure in the northern bobwhite (*Colinus virginianus*). J Toxicol Environ Health 2005;68:1349–57.
[16] Brunjes KJ, Severt SA, Xiaoping JL, Brausch, Cox SA, Conn GP, et al. Effects of octahydro-1,3,5,7-tetranitro-1,3,5,7-tetrazocine (HMX) exposure on reproduction and hatchling development in northern bobwhite quail. J Toxicol Environ Health 2007;70:682–7.

CHAPTER 6

Wildlife Toxicity Assessment for 2,4-Dinitrotoluene and 2,6-Dinitrotoluene

Emily May Lent

Contents

Wildlife Toxicity Assessments for Chemicals of Military Concern
http://dx.doi.org/10.1016/B978-0-12-800020-5.00006-5

INTRODUCTION

The DNTs 2,4-dinitrotoluene (CAS No. 121-14-2) and 2,6-dinitrotoluene (CAS No. 606-20-2) have been associated, either as precursors or by products, with the synthesis of polyurethane foams, coatings, elastomers and explosives. A primary product in propellants, DNTs are also produced as by-products in the manufacture of the common military explosive 2,4,6-trinitrotoluene (TNT). The importance of the DNTs as environmental contaminants is related to their distribution at and around military sites and their potential toxicity to wildlife and other ecological receptors.

Technical grade DNT is composed of approximately 75% 2,4-DNT, 20% 2,6-DNT, and 5% other isomers [1]. It is often found in soils associated with firing points at artillery ranges and with low-order detonations [2]. This Wildlife Toxicity Assessment summarizes the current knowledge of the toxicological impacts of 2,4- and 2,6-DNT on vertebrates. Evaluating the toxicity of 2,4- and 2,6-DNT contributes to the derivation of Toxicity Reference Values (TRVs) that could serve as screening-level benchmarks for wildlife in the vicinity of contaminated sites. The protocol for the development of this assessment is documented in the U.S. Army Center for Health Promotion and Preventive Medicine (USACHPPM) Technical Guide (TG) 254, *Standard Practice for Wildlife Toxicity Reference Values* [3].

TOXICITY PROFILE

Environmental Fate and Transport

The DNTs generated as by-products of the manufacture of 2,4,6-trinitrotoluene are common soil contaminants at munitions manufacturing sites [2]. The discharge to surface water of waste streams generated during munitions manufacturing and processing is the major route of environmental entry for these compounds. Sampling of TNT effluents showed that 2,4-DNT, 2,6-DNT, and 1,3-dinitrobenzene (1,3-DNB) were, in descending order of contribution, the major effluent constituents and accounted for approximately 75% by weight of total organics. Soil contamination by munitions compounds may also occur through open burning, incineration, operational spills, and seepage from landfills and wastewater holding facilities [4]. The 2,4- and 2,6-DNTs have been detected in environmental media (air, surface water, groundwater, soil, and sediment) at 69 and 53 of the 1,467 National Priorities List (NPL) hazardous waste sites, respectively. Where DNTs were detected in some environmental media, 2,4-DNT was detected in air, surface water, groundwater, soil, and sediment at 1, 19, 54, 35, and 16 of 69 sites [5]. The 2,6-DNT isomer was detected in air, surface water, groundwater, soil, and sediment at 1, 18, 32, 20, and 9 of 53 NPL sites where DNTs were detected in some media [5]. A summary of the physical and chemical properties of 2,4- and 2,6-DNT is provided in Table 6.1.

Table 6.1 Summary of Physical-Chemical Properties of 2,4- and 2,6-DNT[a]

	2,4-DNT	2,6-DNT
CAS No.	121-14-2	606-20-2
Molecular weight	182.14	182.14
Color	Yellow	Yellow to red
Physical state	Needles or monoclinic prisms	Solid or rhombic needles from alcohol
Melting point	71 °C	66 °C
Boiling point	300 °C with slight decomposition	285 °C
Density	1.3208	1.2833
Odor	Slight	Slight
Solubility in water	270 mg/L at 22 °C	180 mg/L at 22 °C
Other solvents	Acetone, alcohol, benzene, ethanol, diethyl ether, pyridine, carbon disulfide	Ethanol, chloroform
Partition coefficients:		
Log K_{ow}	1.98	1.72 (estimated)[b]
Log K_{oc}	2.40 (estimated)[b]	1.89 (estimated)[b]
Vapor pressure at 25 °C	1.4×10^{-4}	5.67×10^{-4}
Henry's Law constant at 25 °C	8.79×10^{-8} atm m^3/mole (calculated)	9.26×10^{-8} atm.m^3/mole (estimated)
Conversion factors	1 ppm = 7.4 mg/m^3 1 mg/m^3 = 0.13 ppm	1 ppm = 7.4 mg/m^3 1 mg/m^3 = 0.13 ppm

[a]ATSDR. Agency for Toxic Substances and Disease Registry. Toxicological Profile for 2,4- and 2,6-Dinitrotoluene. Atlanta, GA: U.S. Department of Health and Human Services, Public Health Service, ATSDR; 1998.
[b]Burrows EP, Rosenblatt DH, Mitchell WR, Parmer DL. Organic explosives and related compounds: environmental and health considerations. Technical Report No. AD –8901. Frederick, MD: U. S. Army Biomedical Research and Development Laboratory, Fort Detrick, MD; 1989.

The mobility of 2,4-DNT in soil may be slight to high, depending on the soil type (log $K_{oc} = 1.76$ to 3.3). Based on experiments in sandy loam and sandy silt loam, 2,6-DNT should be fairly mobile despite its estimated log K_{oc} of 1.89 and K_{ow} of 1.72. The low vapor pressures and Henry's Law constants indicate that volatilization from moist soil surfaces and water is not expected for either isomer. Hydrolysis is not expected to occur, and photochemical transformation in soil should not be significant [5–7]. Microbial biodegradation of 2,4- and 2,6-DNT has occurred in soil with half-lives of 25 and 20 days, respectively, at 22 °C when the initial concentration was 0.5 mg/kg [8]. Half-lives of up to 92 days have been measured for 2,6-DNT in soils [9]. Data suggests that if released to soil, 2,6-DNT would degrade, though to a lesser degree than 2,4-DNT.

The fate and distribution of munitions pollutants in the environment are principally driven by microbial and photochemical transformations (photolysis). The photolytic half-lives for 2,4-DNT in river, bay, and pond waters were 2.7, 9.6, and 3.7 hours, respectively [10]. In aqueous biodegradation tests, 2,4-DNT was shown to biodegrade both aerobically and anaerobically. In water, 2,6-DNT should biodegrade (though more slowly than the 2,4 isomer), and photolysis should be rapid (half-life = 12 min) in surface layers [11]. Microbial degradation of 2,4- and 2,6-DNT has been observed for waters and organisms from sites contaminated with DNTs but not for samples taken from other sources [10,12–14]. The 2,4-DNT isomer may have a slight tendency to partition to suspended and sediment organic matter based on the log K_{oc} range. Adsorption of 2,6-DNT to sediments or suspended solids should not be appreciable [5–7]. Volatilization of 2,4- and 2,6-DNT from water should be minimal due to their low vapor pressure and Henry's Law constant [10]. The moderate water solubility and relatively low organic carbon partition coefficients of the two isomers indicate a potential for transport by surface and groundwater [5]. Plant uptake is predicted to occur based on the moderate water solubility. However, uptake of related nitroaromatic compounds (e.g., 2,4,6-TNT) was demonstrated to be inversely related to soil organic carbon content, indicating that plant uptake may be site-specific and limited in carbon-rich environments [15]. Significant bioaccumulation is not expected to occur based on the low lipophilicity of 2,4- and 2,6-DNT (log $K_{ow} = 1.98$ and 1.72, respectively) [5–7].

Given the relatively low affinity to lipids, the DNTs are not expected to appreciably accumulate in animals; therefore, exposure through animal-based food is expected to be minimal. Primary exposures to wildlife are likely to be through consumption of contaminated water and plants, inadvertent ingestion of soils (through grooming and feeding), and potentially through the skin.

MAMMALIAN TOXICITY

Mammalian Oral Toxicity

Following oral administration, DNT isomers are rapidly absorbed, distributed, and eliminated [16–18]. In rats, rabbits, dogs, and monkeys, the bulk of radioactivity from radio-labeled oral DNT exposure was excreted in the urine [5]. In mice, ^{3}H-labeled 2,6-DNT was excreted primarily in the urine while ^{14}C-labeled 2,4-DNT was excreted mostly in the feces [19,20].

Mammalian Oral Toxicity: Acute

Oral LD_{50} values for 2,4-DNT ranged from 240 to 650 mg/kg in rats and 1,340 to 1,954 mg/kg in mice. Toxicity observed in both species included ataxia and cyanosis, with death occurring within the first 24 hours. No treatment-related gross pathology was observed in dead animals, and survivors recovered completely within 48 hours [21–24]. These data are similar to those for TNT exposure in these species.

The oral LD_{50} value for 2,6-DNT ranged from 180 to 795 mg/kg in rats and 621 to 1,000 mg/kg in mice. Toxic symptoms, time to death, gross pathology, and recovery observations were similar to those for 2,4-DNT. Females appeared slightly more tolerant of 2,6-DNT than males [21–23].

Mammalian Oral Toxicity: Subacute

McGown *et al.* (1983) fed rats (5/sex/group) diets containing 0, 0.9, 1.2, 1.9, or 3 g of 2,4-DNT/kg for 14 days. Administration of 2,4-DNT resulted in dose-dependent decreases in food consumption and body weight gain, elevated blood cholesterol and glucose, and oligospermia with degenerative changes of the testes in males [25]. The authors did not estimate average daily intake of 2,4-DNT. However, daily doses have been estimated at 0, 97, 126, 180, and 257 mg/kg/day for males and 0, 96, 121, 186, and 254 mg/kg/day for females [26].

As part of the determination of the Maximum Tolerated Dose for a developmental study, Smith (1983) exposed groups of 10 female CD-1 mice to oral doses of 0, 310, 525, 1,250, 2,500, and 3,500 mg/kg/day of 2,4-DNT for eight days. Mice in the 525 mg/kg/day group lost up to 13%

of their body weight by the eighth day of dosing. By the eighth day, all mice in treatment groups receiving 1250 mg/kg/day or greater died. Adverse observations recorded for the 310 mg/kg/day group included lethargy (10 of 10) and hunched posture (1 of 10) [27].

Lent *et al.* (2012) exposed Sprague Dawley rats to either 2,4-DNT (9, 18, 36, 71, 142, and 284 mg/kg/day) or 2,6-DNT (4, 7, 14, 35, 68, and 134 mg/kg/day) via oral gavage for 14 days. A lowest observed adverse effect level (LOAEL) of 284 mg/kg/day was observed for rats exposed to 2,4-DNT due to mortality, body weight loss, and neurological effects (front limb paralysis); the no observed adverse effect level (NOAEL) for each previously described effect was 142 mg/kg/day. Hepatotoxicity was observed in rats that were dosed with 2,6-DNT. Hepatocellular hyperplasia, oval cell hyperplasia, and hepatocellular hypertrophy were observed in rats in the 35 mg/kg/day and higher dose groups. Testicular atrophy was observed at doses as low as 68 mg/kg/day of 2,6-DNT [28].

Mammalian Oral Toxicity: Subchronic

Bloch *et al.* (1988) administered 0, 0.1, and 0.2% 2,4-DNT in feed to Sprague Dawley rats for three weeks. Average daily intakes were not estimated by the authors but have been estimated by others at 0, 50, and 100 mg/kg/day [26]. Marked changes in Sertoli cell morphology and increases in circulating levels of follicle stimulating hormone and luteinizing hormone were observed in the 0.2% dose group. Vesicles of varying size associated with swollen mitochondria, reduced epididymis weights, and decreased epididymal sperm reserves were observed in 2,4-DNT-treated animals. Rats in both treatment groups had decreased body weight gain and multinucleated spermatids, mild irregularity of basal lamina, vacuolation, and lipid accumulation in Sertoli cells. The authors concluded that 2,4-DNT is capable of testicular injury and altering spermatogenesis; however, the effect of this injury on reproductive function was not evaluated [29].

Lee *et al.* (1978) studied the effects of oral 2,4-DNT administration for 13 weeks in dogs, rats, and mice. The toxic effects observed from 2,4-DNT administration included death, decreased body weight and food consumption, methemoglobinemia, reticulocytosis, hemosiderosis, anemia, hepatocellular changes and altered liver weights/function, severe neurological effects (dogs and rats), and testicular degeneration and decreased spermatogenesis in males. Dogs were the most sensitive species tested, followed by rats and then mice. Reversibility of adverse effects, mutagenicity of the compound, immunologic response, and disposition and metabolism of radiolabeled compound were also determined [19].

A total of 32 beagle dogs were divided into four groups (4/sex/group) and administered 0, 1, 5, or 25 mg/kg/day of 2,4-DNT in capsules for 4 or 13 weeks. One male and one female from each dose group were euthanized at 4 and 13 weeks. An additional male and female from each dose group were removed from treatment at each time point and were euthanized at 8 and 17 weeks, respectively, to study reversibility. Several dogs in the high-dose group were moribund and were euthanized prior to schedule; two dogs in the high-dose, 4-week group were not euthanized until 8 months later to investigate the potential for reversibility over a longer time period. Daily administration of 2,4-DNT at 1 and 5 mg/kg/day resulted in no adverse effects in dogs. Dosages of 25 mg/kg/day resulted in toxicity at 12 to 22 days and were lethal after 22 days. Toxic effects included anorexia and weight loss, neuromuscular incoordination and rigid paralysis (particularly in the hind legs), gliosis and demyelination in the brain, methemoglobinemia and anemia with reticulocytosis and Heinz bodies, hemosiderosis and extramedullary hematopoiesis, and atrophy of the testes with aspermatogenesis. Dogs recovered partially 4 weeks posttreatment and completely after 8 months. Treatment with 2,4-DNT did not alter serum immunoglobin (IgE) levels [19].

A total of 64 male and 64 female CD rats were divided into four equal groups (16/sex/group) and administered diets containing 0, 0.07, 0.2, or 0.7% 2,4-DNT (98% pure) for 4 or 16 weeks in a protocol similar to that used for dogs [19]. Four males and four females from each dose group were euthanized at 4 and 13 weeks. An additional four males and four females from each dose group were removed from treatment at each time point and were euthanized 4 weeks later to study reversibility. The average intake for male and female rats, respectively, was 34.4 and 38.3 mg/kg/day for low-dose, 92.8 and 108.3 mg/kg/day for mid-dose, and 265.6 and 145.2 mg/kg/day for high-dose. The lower intake in high-dose females was the result of a marked decrease in feed consumption.

The only effect observed in the low-dose animals was a slight decrease in weight gain. The mid-dose animals had a greater decrease in weight gain, reticulocytosis, hemosiderosis in the spleen, and decrease or cessation of spermatogenesis. Gliosis and demyelination were observed in the cerebellum of one mid-dose rat. High-dose rats had severe weight loss, anemia with reticulocytosis, greater and earlier onset of splenic hemosiderosis and aspermatogenesis, neuromuscular effects (unusual gait with wide stance and stiff hind legs), and mild to moderate gliosis and/or demyelination in the central nervous system. Eight of 16 high-dose males and one mid-dose male died prior to study termination. Ten of 16 high-dose females died in the first 3 weeks of the study. Survivors across treatments partially recovered 4 weeks after cessation of exposure (recovery of weight gain and anemia, but not aspermatogenesis and splenic hemosiderin incidence). No effects on serum IgE were observed [19,30]. These data are consistent with those effects observed in a 30-day study of tg-DNT in Fischer 344 rats [31].

Sixty-four male and 64 female albino Swiss mice were divided into four equal groups (16/sex/group) and administered 0, 0.07, 0.2, or 0.7% 2,4-DNT in feed for 4 or 16 weeks in an experimental protocol similar to the dog study [19,32]. The average intake in male and female mice, respectively, was 47 and 52 mg/kg/day for the low-dose mice, 137 and 147 mg/kg/day for the mid-dose mice, and 413 and 468 mg/kg/day for the high-dose mice.

Treatment with 2,4-DNT resulted in decreased weight gain in male mice in the low- or mid-dose groups. High-dose mice had weight loss and decreased food consumption, mild anemia, mild depression of spermatogenesis (decreasing fertility), and a few deaths. Surviving mice recovered completely by 4 weeks after cessation of treatment [19,32].

Lee *et al.* (1976) studied the effects of oral 2,6-DNT exposure for up to 13 weeks in dogs, rats, and mice in a protocol similar to that used for 2,4-DNT [33]. The effects of the test article relative to drug metabolizing enzymes, mutagenicity, immunologic response, and reversibility of adverse effects were also evaluated. Treatment-related effects resulting from 2,6-DNT exposure were similar to those observed for 2,4-DNT. Toxic effects included mortality, decreased body weight gain and food consumption, hemosiderosis in spleen and liver, neuromuscular effects in dogs, and degeneration of the testes and decreased spermatogenesis in males. Although dogs remained the most sensitive species, the relative species sensitivities of rats and mice were reversed when compared to 2,4-DNT.

Four groups of eight dogs (4 per sex) were given 0, 4, 20, or 100 mg/kg/day of 2,6-DNT in capsules for 4 or 13 weeks. One male and one female from each dose group were euthanized at 4 and 13 weeks. An additional male and female from each dose group were removed from treatment at 4 and 13 weeks and then euthanized 4 weeks later to study reversibility. Because of severity of symptoms, dogs in the high-dose reversibility group were continued for an additional 2 weeks (19 weeks total) before they were euthanized.

Toxicity occurred in a dose-dependent manner and included many pathological and clinical observations. These included decreased body weight and food consumption, listlessness, incoordination leading to rigid paralysis with occasional tremors, methemoglobinemia with Heinz bodies and anemia with compensatory reticulocytosis and extramedullary hematopoiesis, lymphoid depression with peripheral lymphocytopenia, bile duct hyperplasia, inflammatory and degenerative changes in liver and kidney with elevated serum chemistries, and degeneration and atrophy of the spermatogenic cells of the testes.

Administration of 4 mg/kg/day resulted in no remarkable symptoms, while all dogs given 100 mg/kg/day died between the second and eighth weeks. The toxic effects of oral 2,6-DNT exposure were partially reversed at 4 weeks and completely reversed at 19 weeks posttreatment. Treatment-related effects on serum immunoglobulin E-titers were not observed [33].

Four groups of 32 CD rats each (16 males and 16 females) were given 0, 0.01, 0.05, or 0.25% 2,6-DNT in their feed following the same experimental design as for dogs, except that four animals per sex per group were euthanized at each time point [33]. The average intakes of 2,6-DNT for males and females, respectively, were 7.2 and 7.4 mg/kg/day in the low-dose group, 35.1 and 37.1 mg/kg/day in the mid-dose group, and 144.7 and 155 mg/kg/day in the high-dose group.

No treatment-related effects from 2,6-DNT were observed at the low dose. At the mid-dose, effects observed included decreased weight gain and food consumption, extramedullary hematopoiesis in spleen and/or liver, bile duct hyperplasia, and depression of spermatogenesis and atrophy in the testes. Animals in the high-dose group were more severely affected and exhibited weight loss, decreased erythrocyte count with compensatory reticulocytosis, methemoglobinemia with Heinz bodies, severe testicular lesions, and more prevalent and severe bile duct hyperplasia and extramedullary hematopoiesis. Rats recovered only partially four weeks after treatment cessation [33].

Mice were fed 0, 0.01, 0.05, or 0.25% 2,6-DNT in a protocol similar to the one used for dogs and rats (16 animals per sex per group). The average intakes for male and female mice fed the low-, mid-, and high-dose 2,6-DNT were 11.1 and 11.0, 50.8 and 55.2, and 288.8 and 298.8 mg/kg/day, respectively. Three males from the control group, two males from the low-dose group, eight males and one female from the mid-dose group, and eight males and six females in the high-dose group died on study. The increases in deaths in the mid- and high-dose groups were considered treatment-related. The mid- and high-dose animals also exhibited decreased feed consumption and weight gain, extramedullary hematopoiesis, depression of spermatogenesis and atrophy of the testes, and bile duct hyperplasia. Partial recovery was observed in mice 4 weeks after cessation of treatment. No toxic effects were observed in the low-dose animals. Unlike 2,4-DNT-dosed mice, which appeared to be almost unaffected by doses lethal to rats, mice dosed with 2,6-DNT were more sensitive [33].

Mammalian Oral Toxicity: Chronic

Ellis *et al.* (1979) administered 0, 0.2, 1.5, or 10.0 mg/kg/day of 2,4-DNT (98% pure) in capsules to dogs for up to 24 months (6 per sex per group). After 12 months of dosing, treatment was stopped for two males and two females from each group. One pair per group was euthanized at the time of dosing cessation, while the other pair was allowed to recover for 4 weeks prior to being euthanized. The remaining animals were dosed for 12 more months (24 months total), after which two pairs from each group were euthanized at dosing cessation, and the two remaining pairs from each group were allowed to recover for four weeks prior to their euthanization [34].

No adverse effects from 2,4-DNT were observed in the low-dose (0.2 mg/kg/day) group. Toxicity was observed in some, but not all, dogs in the mid-dose (1.5 mg/kg/day) group, while administration of the high dose (10 mg/kg/day) was toxic to all dogs and lethal to some. Principal target organs were the erythrocytes, nervous system, and biliary tract. Observed effects include methemoglobinemia, anemia, reticulocytosis and Heinz bodies, neuromuscular incoordination and paralysis (especially of the hind legs), degenerative lesions of the cerebellum, and hyperplasia of the biliary tract and gallbladder epithelium. Recovery appeared to occur after dosing cessation; however, animal survival numbers were very low.

The CD rats were administered 0, 0.0015, 0.01, or 0.07% 2,4-DNT in their feed for up to 24 months, with the average intake being 0.57, 3.9, or 34 mg/kg/day for males and 0.71, 5.1, or 45 mg/kg/day for females in low-, mid-, and high-dose groups, respectively. The experimental schedule was similar to that used with dogs, except there were initially 38 animals per sex per group. Four animals per sex per group were euthanized at the interim (12 and 13 months) time points, and the remaining survivors euthanized at 24 and 25 months [34].

No toxicity from 2,4-DNT was noted in rats in the low-dose group. Toxicity occurred in the mid-dose animals and included anemia, decreased weight gain, indicators of liver toxicity, and incidences of mammary tumors. Increased toxicity was observed in the high-dose group and included severely decreased weight gain, shortened life span, livers with hyperplastic foci to hepatocellular carcinoma, decreased spermatogenesis or aspermatogenesis, and increases in background tumors, such as of the connective (fibromas in males) and mammary (fibroadenomas in females) tissues. Occasional neurological symptoms, such as straddling gait, were observed in some high-dose rats but without accompanying histopathological lesions [30,34].

Because earlier studies had shown that 2,4-DNT was less toxic in mice than rats, CD-1 mice were given 0, 0.01, 0.07, or 0.5% in their feed for up to 24 months [34]. There were only small differences in intake between sexes; the average intake for low-, mid-, and high-dose animals was 13.5, 95, and 900 mg/kg/day, respectively. The experimental design was similar to the design used in the dog study, except there were 58 animals per sex per group. Four animals per sex per group were euthanized at 12 and 24 months. Four additional animals per sex per dose group were removed from treatment at each euthanasia point and euthanized 1 month later to study reversibility [34].

Low-dose administration (13.5 mg/kg/day) of 2,4-DNT to mice resulted in toxic nephropathy, excessive pigmentation, liver dysplasia, and renal tumors (in males). The mid-dose (95 mg/kg/day) was very toxic and resulted in a higher incidence and more severe effects than those observed at the low dose, including cystic renal tumors in more than half of the males and atrophy of the testes. High-dose toxic effects were severe and included even greater decreases in weight gain and food consumption, a life span shortened to half that of other dose groups, anemia with many Heinz bodies, and nonfunctioning gonads in both sexes. Generalized pigmentation was observed in many tissues of the high-dose animals. Effects were more severe in males than females [32,34].

One study on the chronic toxicity of 2,6-DNT was located [35]. Although the study was designed to evaluate hepatocarcinogenicity, other parameters were measured. Twenty-eight rats were used in the study with a control and a low (7 mg/kg/day) and high (14 mg/kg/day) treatment of 2,6-DNT in feed. Four rats were euthanized at weeks 4 and 26, while the remaining 20 rats were exposed for 52 weeks total. Throughout the study, concentrations of 2,6-DNT were adjusted based on food consumption and average body weight to maintain target doses. The effects observed for rats chronically exposed to 2,6-DNT were consistent with those observed for subchronic 2,6-DNT administration. These effects included decreased body weights, hepatocellular degeneration and vacuolation, acidophilic and basophilic foci of hepatocellular alteration, cholangiocarcinoma, and hepatocellular carcinoma [35]. The LOAEL for reduced body weight was 7 mg/kg/day. Since this dose was the lowest used, a NOAEL could not be derived.

Mammalian Oral Toxicity: Other

Studies to evaluate developmental effects were conducted by Smith [27] and Hardin *et al.* [36]. In both studies, female CD-1 mice were exposed to 2,4-DNT at 0 and 390 mg/kg/day on GD-6 or 7 through 13. Smith (1987) reported that 10 of 50 mice exposed to 390 mg/kg/day 2,4-DNT died, four of which were pregnant. An additional five died in the postdosing observation period, two of which were pregnant. A total of 16 of the 50 mated female mice were pregnant. Five additional litters were found to be reabsorbed. The reproductive index in the test group (0.68) was significantly lower than that of the control group (0.91). No other measures of reproductive toxicity were altered by 2,4-DNT treatment [27]. Only 1 of 50 mice did not survive following oral doses of 390 mg/kg/day in the Hardin *et al.* study [36], and no other measures of reproductive toxicity were altered by 2,4-DNT treatment in the Hardin *et al.* study [36].

Price *et al.* (1985) and Jones-Price *et al.* (1980) investigated developmental effects of technical grade (tg)-DNT by exposing pregnant rats to oral doses of 14 to 150 mg/kg/day on gestational days 7–20. Overt toxicity included rough coat, lethargy, and hind limb weakness. Blood effects (decreases in red blood cell count and packed cell volume (PCV), and increases in methemoglobin and reticulocyte count) occurred in the 100 mg/kg/day group. No adverse reproductive effects were observed consistent with tg-DNT exposure. The authors concluded that tg-DNT was not teratogenic but may cause embryo/fetal toxicity at levels where maternal mortality occurs [37,38].

In addition to the previously mentioned 2,4- and 2,6-DNT, Lent *et al.* (2012) [28] administered 2,3-, 2,5-, 3,4-, and 3,5-DNT to male Sprague Dawley rats for 14 days. The 3,5-DNT isomer ($LD_{50}=310$ mg/kg/day) was found to be the most toxic as evidenced by death, body weight loss, and neurological effects having been observed at 39 mg/kg/day. Neurological effects included facial twitching, altered head carriage, and paralysis that was restricted to the front limbs. Reduced testes mass and testicular atrophy were noted in the 19 mg/kg/day group (LOAEL); the NOAEL was determined to be 10 mg/kg/day. The 2,3-DNT isomer ($LD_{50}=1{,}100$ mg/kg) caused mortality, body weight loss, and testicular atrophy at 550 mg/kg/day.

A LOAEL of 275 mg/kg/day and a NOAEL of 138 mg/kg/day were determined for 2,3-DNT-based reductions in kidney and liver mass and increased incidence of extramedullary hematopoiesis. Single exposures to 3,4-DNT ($LD_{50}=908$ mg/kg) caused convulsions at 1,000 mg/kg/day; however, neurological effects at 227 mg/kg/day during the 14-day exposure were limited to hypoactivity/staring and facial twitches. The most sensitive response to 3,4-DNT exposure was anemia, which occurred at doses as low as 14 mg/kg/day (LOAEL); the NOAEL for this response was 7 mg/kg/day. The mildest effects were observed in rats exposed to 2,5-DNT ($LD_{50}=616$ mg/kg) with the most sensitive response being trace to mild extramedullary hematopoiesis of the spleen observed in the 39 mg/kg/day and higher dose groups [28].

Studies Relevant for Mammalian TRV Development for Ingestion Exposures

2,4-Dinitrotoluene

The studies selected to derive a toxicity reference value (TRV) for 2,4-DNT were chronic ingestion exposures in several species of mammals. Effects from chronic 2,4-DNT exposures were consistent with those of subchronic exposures and included decreased survival, decreased body weights and food consumption, anemia, hepatocellular dysplasia, kidney tumors in mice, serious neuromuscular effects in dogs, ovarian atrophy in female mice, and decreased spermatogenesis and testicular atrophy in male mice. The most relevant chronic studies were those conducted by Ellis *et al.* (1979 and 1985) [34,39], Lee *et al.* (1985) [30], and Hong *et al.* (1985) [32] in dogs, rats, and mice, respectively. These were well-conducted, of suitable duration, and provided an evaluation of a number of relevant endpoints. Additionally, results from these studies are corroborated by results from a number of other chronic [35,40,41] and subchronic [19,29,42] studies. Acute studies provided additional information but were not considered relevant to the development of TRVs. Table 6.2 summarizes the data from these studies, and Figure 6.1 presents the data in a scatter diagram. All major studies were conducted on contract to the U.S. Army and provided detailed methods and results. These studies are considered high quality and sufficient for use in the derivation of a TRV. Most of these studies have also been presented as peer-reviewed reports.

The USACHPPM TG 254 [3] states that TRVs should be based on ecologically-relevant endpoints. For 2,4-DNT, the neuromuscular effect in dogs [34,39] was chosen as the basis for TRV development. The severe

Table 6.2 Summary of Relevant Mammalian Data for 2,4-DNT

Test Organism/ Route	Test Duration	Test Results				Study
		LD_{50} (mg/kg)	NOAEL (mg/kg/day)	LOAEL (mg/kg/day)	Effects Observed at LOAEL	
Rat/CD/gavage	Single, acute	568 (M) 650 (F)				[22] [21]
Rat/Sprague Dawley/gavage	Single, acute	270 (M)				[23]
Rat/Charles River/ gavage	Single, acute	240 (M)				[24]
Mouse/Swiss albino/gavage	Single, acute	1954 (M) 1340 (F)				[22]
Mouse/CF-1/ gavage	Single, acute	1630 (M)				[23]
Dogs/oral/ capsule	12 days		5	25	Incoordination, stiffness, and abnormal gait	[38]
Rat/Sprague Dawley	5 days		60 (M)	180 (M)	Decreased (though reversible) fertility Oral dog, 0, 0.01, 0.1, or 1 mg/kg-day for 4 weeks, then 0, 0.05, 0.5, and 5 mg/kg-day for 9 more weeks- NOAEL	[19] [24]
Mouse/CD-1	GD 6-13		390 (F)		No change in maternal bw, or developmental indices	[36]
White-footed mouse (*Peromyscus leucopus*)/feed	14 days		158 (M) 74 (F)	286 (M) 158 (F)	Decreased body weight gain, increased liver to body weight and brain/bw ratios	[3]
Rat/CD/feed/ad lib	3 or 6 months			45.3 (F)	Increased incidence of death during parturition	[34]
Dog/oral/capsule	6 months		1.5	10 (M)	4/6 animals died	[34,39]

Dog/oral/ capsule	4 or 13 weeks		5	25	5/8 died	[39] [19]
Rats/CD/feed/ ad lib	3 or 6 months			34 (M) 45 (F)	10–25% Decrease in body weight, decreased fertility (M), difficult parturition (F)	[34]
			34 (M)	45 (M)	Severe atrophy/degeneration of seminiferous tubules	
			5 (F)	45 (F)	Difficult parturition; decreased pup viability	
Rats/CD/feed/ ad lib	4 or 13 weeks			34 (M) 38 (F)	Moderate decrease body weight gain	[19,30]
			34 (M) 38 (F)	93 (M) 108 (F)	Decreased food consumption	
			34 (M) 108 (F)	266 (M) 145 (F)	Reticulocytosis and hemosiderosis. (anemia at 266 (M) and 145 (F))	
			34 (M)	93 (M) 145 (F)	Hepatic and renal effects, demyelination of cerebellum and brain stem, widespread and stiff-legged gait. Death in one male and 100% females	
				93 (M)	Severe decrease in spermatogenesis. Decrease in fertility	
Mice/CD-1/ feed/ad lib	4 or 13 weeks		137 (M) 147 (F)	413 (M) 468 (F)	Mild anemia and reticulocytosis	[32] [19]
			137 (M) 147 (F)	413 (M) 468 (F)	Mild hepatocellular dysplasia 2/16 males and females died	
			47 (M) 52 (F)	413 (M) 468 (F)	Body weight loss with decreased food consumption. Also NOAEL for neurological effects	

Continued

Table 6.2 Summary of Relevant Mammalian Data for 2,4-DNT—cont'd

Test Organism/Route	Test Duration	Test Results: LD_{50} (mg/kg)	NOAEL (mg/kg/day)	LOAEL (mg/kg/day)	Effects Observed at LOAEL	Study
Dogs/oral/capsule	4 or 13 weeks		5	25	Anemia, Heinz bodies, hepatic, renal, and immunological effects	[19,39]
			5	25	Incoordination, abnormal gait, paralysis	
			5(M)	25 (M)	Testicular degeneration/decreased spermatogenesis	
Rats/CD/feed/ad lib	1–2 years		0.57 (M) 0.71 (F)	3.9 (M) 5.1 (F)	Decreased survival	[30]
			3.9 (M)	34.5 (M)	Decreased RBC count	[34]
			5.1 (F)	45.3(F)	Marked anemia (males at 34.5)	
				0.6 (M) 0.7 (F)	Preneoplastic foci of hepatocytes	
			0.51 (M) 0.71 (F)	3.9 (M) 5.1 (F)	Hepatocellular carcinoma, mammary and skin tumors	
			3.9 (M)	34.5 (M)	30% decrease in body weight	
			5.1 (F)	45.3 (F)	27% decrease body weight-both with decreased food consumption	
			0.57 (M) 0.71 (F)	34.5 (M) 45. (F)	Widespread and stiff-legged gait	
			0.57 (M)	0.6 (M) 95 (F)	Atrophy of seminiferous tubules, aspermatogenesis	

Mice/CD-1/ feed/ad lib	2 years		95	900	Decreased survival; anemia, reticulocytosis, Heinz bodies, stiff-legged gait; hyperactivity	[34] [32]
				13.5 (M) 13.5 (F)	Cystic dysplasia; toxic nephropathy	
			95 (F)	13.5 (M)	16% decrease in body weight	
				900 (F)	20% decrease in body weight	
				13.5 (M)	Decreased spermatogenesis and degenerative change; testicular atrophy	
			95 (F)	900 (F)	Ovarian atrophy; nonfunctioning follicles	
				95(M)	Renal solid carcinoma, cystic papillary carcinoma and adenoma, cystic adenoma	
Rat/Fischer-344/ feed/ad lib	1 year			27(M)	Nonneoplastic lesions of hepatocellular degeneration and vacuolation; basophilic and acidophilic foci of cellular alteration. Neoplastic nodule in only 1/20; 25% body weight decrease	[35]
Rat/Fischer-344/ feed/ad lib	78 weeks		8 (M) 8.8 (F)	20 (M) 22 (F)	25% decrease in body weight	[41]
Mouse/C57BL/ 6 N/feed/ad lib	78 weeks		14.4 (M)	72 (M)	18% decrease in body weight	[41]
				15.2 (F)	11% decrease in body weight-24% at 76	
Dog/capsule/1 × daily	2 years		0.2	1.5	Methemoglobinemia, Heinz bodies, loss of hindquarter control, convulsions	[34,39]

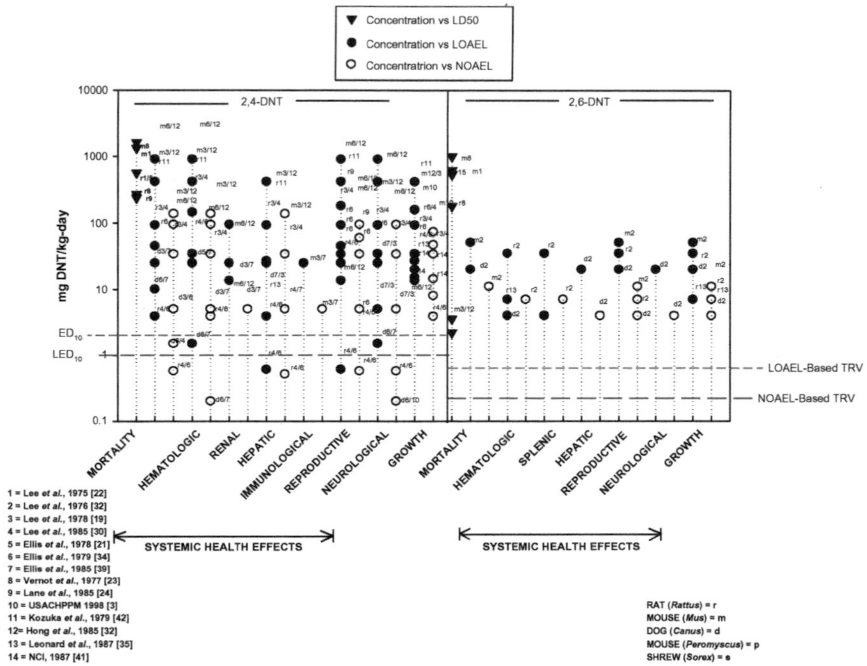

Figure 6.1 2,4- 2,6- DNT health effects to mammals. Letters by symbols represent test species, and numbers represent study.

paralysis and incoordination induced by 2,4-DNT occurred at relatively low concentrations (LOAEL was 1.5 mg/kg/day; NOAEL was 0.2 mg/kg/day) and in some cases, precluded dogs from eating and drinking normally. In less severe cases, hind leg, lip, and tongue control were hampered. These neuromuscular effects would likely lead to difficulties in foraging, avoiding predators, and mating in wild populations. Concomitant with the neuromuscular effects were increased methemoglobinemia and incidence of Heinz bodies. In addition, the onset of these effects in dogs occurred at lower concentrations than the onset of effects such as hepatocellular dysplasia, kidney tumors, ovarian and testicular atrophy, and decreased spermatogenesis seen in other mammalian species. Hence, a TRV based on neuromuscular effects in dogs would likely be protective of other endpoints.

2,6-Dinitrotoluene

One study has characterized the chronic toxicity of 2,6-DNT in rats. Leonard *et al.* (1987) showed that rats dosed with 7 mg/kg/day 2,6-DNT

(LOAEL) for one year were smaller than control rats. Some effects associated with subchronic 2,6-DNT exposures included decreased body weight and food consumption, bile duct hyperplasia, testicular degeneration, decreased spermatogenesis, hepatocellular degeneration, and splenic hemosiderosis. In addition, similar to 2,4-DNT, 2,6-DNT elicited a neuromuscular effect in dogs that manifested as incoordination and lack of balance. The consistency in response among test species and between 2,4- and 2,6-DNT provides a fairly strong case for the toxicity of these compounds. The studies on 2,6-DNT were well-conducted and provided an evaluation of a number of relevant endpoints; however, only one [35] was of suitable duration to derive a TRV with minimal use of uncertainty factors. As outlined in USACHPPM TG 254 [3], TRVs should be derived from ecologically-relevant endpoints. From an ecological perspective, reduced growth and/or associated reductions in food consumption as seen in rats exposed to 2,6-DNT for one year can affect the ecological performance of individuals by causing alterations in energy allocation patterns that could ultimately result in altered reproductive performance [43,44]. On this premise, the Leonard *et al.* (1987) study on rats is most suitable for derivation of the TRV. The TRVs are presented in Tables 6.2 and 6.3.

Mammalian Toxicity: Inhalation

Five rats/sex/group were exposed to measured chamber concentrations of 0.026, 0.196, 0.473, and 0.696 mg 2,6-DNT/L via the nasal passageways for 6 hours. The acute (6-hour) LC_{50} for 2,6-DNT aerosols was 0.24 milligrams per liter (mg/L) for male and 0.66 mg/L for female Fischer 344 rats [45]. Survivors had decreased body weight gain for days 3 through 6, decreased food consumption, and small increases in methemoglobin levels but no macroscopic abnormalities. Animals that died had higher lung-to-body-weight ratios, congested lungs, and a dark appearance to the liver. No longer-term inhalation studies were identified.

Mammalian Toxicity: Dermal

Both isomers failed to demonstrate ocular irritation in rabbits, and both isomers were very mild skin irritants. In guinea pigs, the 2,4-DNT isomer was not a skin sensitizer while the 2,6-DNT isomer was a mild sensitizer [21,22].

Table 6.3 Summary of Relevant Mammalian Data for 2,6-DNT

Test Organism/Route	Test Duration	Test Results: LD 50 (mg/kg)	NOAEL (mg/kg-day)	LOAEL(mg/kg-day)	Effects Observed at LOAEL	Study
Rat/CD/gavage	Single dose	535 (M)			N/A	[22]
		795 (F)				[21]
Rat/Sprague Dawley/gavage	Single dose	180 (M)			N/A	[23]
Mouse/CD	Single dose	621 (M) 807 (F)			N/A	[22]
Mouse/CF-1/gavage	Single dose	1000 (M)			N/A	[23]
White-footed mouse (*Peromyscus leucopus*)/feed	14 days		103 (M) 44 (F)	238 (M) 103 (F)	Increased liver weight and liver to body and brain weight ratios	[3]
Mouse/Swiss-albino/feed	4 or 13 weeks		11 (M)	51 (M)	Death (8/16)	[32]
			11 (F)	55 (F)	Death (6/16)	
			11	51 (M) 55 (F)	Weight loss	
			11 (M)	51 (M)	Decreased spermatogenesis	

Dog/oral/feed	4 or 13 weeks			20 (F)	Death (2/8)	[32]
				4	Mild extramedullary erythropoiesis and lymphoid depletion, mild splenic hematopoiesis	
			4	20	Mortality. Bile duct hyperplasia; degenerative and inflammatory liver changes, dilated tubules, degenerative foci, weight loss with decreased food consumption, incoordination, lack of balance, testicular degeneration	
			20	100	Thymic involution	
Rat/Fischer-344/feed	52 weeks			7 (M)	Hepatocellular degeneration, vacuolation; acidophilic and basophilic foci of cellular alteration, altered serum enzyme activities, 18% decrease in body weight	[35]
				7 (M)	20% decrease in body weight	

SUMMARY OF AVIAN TOXICITY

Avian Toxicity Oral: Acute

2,4-Dinitrotoluene

Northern Bobwhite quail were orally dosed with 96% pure 2,4-DNT in a corn oil vehicle using the up/down procedure [46]. The oral LD_{50} was determined to be 55 mg/kg (20–79 mg/kg 95% confidence interval (CI)). Initial clinical observations included diarrhea and lethargy. Most animals died within 72 hours postdosing.

2,6-Dinitrotoluene

The oral LD_{50} was estimated for Northern Bobwhite quail using the stage-wise probit procedure. Males were orally dosed with 99% pure 2,6-DNT in a corn oil vehicle [47]. The oral LD_{50} was estimated to be 320 mg/kg (195–470 mg/kg 95% CI). Initial observations included loss of body mass, diarrhea, scant white feces, lethargy, and ataxia.

Avian Toxicity Oral: Subacute

2,4-Dinitrotoluene

Northern Bobwhite quail were orally dosed with 2,4-DNT in a corn oil vehicle daily for 14 days [46]. Exposures were confirmed through analytical chemistry including stability and homogeneity in the vehicle. Six treatment groups comprising seven birds of mixed sex were exposed to 0, 0.5, 5, 15, 35, or 55 mg/kg/day. Controls received doses of vehicle equivalent to the amount given to the high-dose group, while doses for other groups were adjusted volumetrically to achieve the desired oral dose of 2,4-DNT. All birds in the 35 and 55 mg/kg/day groups died within 72 hours. Mortality was accompanied by weight loss (emaciation) and diarrhea. Nonsignificant trends in plasma triglyceride and electrolyte levels were suggestive of adverse kidney effects.

2,6-Dinitrotoluene

Northern Bobwhite quail were orally dosed with 2,6-DNT in a corn oil vehicle daily for 14 days [47]. Nine treatment groups comprising at least seven birds of mixed sex were exposed to 0, 50, 100, 190, or 350 mg/kg/day. Oral doses were determined volumetrically. Controls received doses of vehicle equivalent to the amount given to the high-dose group. All birds in the 190 and 350 mg/kg/day groups died or were moribund within 14 days of exposure.

Birds in the 350 mg/kg/day 2,6-DNT treatment were lethargic and had diarrhea, scant white feces, and emaciation of the pectoralis muscle. Mean liver-to-body mass and kidney-to-body mass ratios were increased, relative to the control, in some treatment groups. Spleen-to-body mass ratio was decreased in the 100, 190, and 350 mg/kg/day groups relative to both the control and low-dose treatment groups. Red blood cell counts, hemoglobin content, and packed cell volume (PCV) decreased with increasing dose, although PCV was not significant and hemoglobin did not differ among the 2,6-DNT doses. Nonsignificant trends in reduction of serum glucose, globulin, and total protein were observed. Plasma uric acid levels increased with dose, suggesting kidney-related effects. Six of 10 birds from the high-dose group (350 mg/kg/day) exhibited hepatocellular vacuolation consistent with glycogen accumulation, which corresponded to higher liver weights. A LOAEL of 50 mg/kg/day was established based on anemia in exposed quail. A NOAEL could not be established in this study.

Avian Toxicity Oral: Subchronic

2,4-Dinitrotoluene

Johnson *et al.* (2005) exposed 12 Northern Bobwhite quail per sex per dose to either 0, 1, 5, 15, or 25 mg/kg/day 2,4-DNT in corn oil for 60 days in a manner similar to that described previously. All females and nine of 12 males in the high-dose group died, most within the first week of exposure. Three males and four females died or were moribund in the 15 mg/kg/day groups. Again, lethargy, diarrhea, and weight loss were common accompanying observations. Brain, liver, and kidney-to-body weight ratios were affected by treatment. Increases in brain and liver-to-body weight ratios occurred at levels where mortality occurred. Changes in kidney-to-body weight ratios were most sensitive, occurring at 5 mg/kg/day in females. Increases in triglyceride levels and nonsignificant trends in plasma uric acid levels also occurred at 5 mg/kg/day. Uric acid accumulation was found in kidneys of some birds at levels that caused mortality. Changes in hematological parameters (packed cell volume, hemoglobin, and red blood cell counts) occurred with treatment but were not considered to be biologically significant because they were within normal ranges for this species and occurred predominantly in females. Based on sensitive kidney effects (weight ratios, electrolytes, and trends in plasma uric acid concentrations), the authors reported a NOAEL of 1 mg/kg/day and a LOAEL of 5 mg/kg/day [46].

2,6-Dinitrotoluene

Quinn *et al.* (2007) exposed 12 Northern Bobwhite quail per sex per dose to either 0, 5, 10, 40, or 60 mg/kg/day 2,6-DNT in corn oil for 60 days via oral gavage. Due to excessive loss of body weight, one female and two males in the 40 mg/kg/day group, and one female and three males from the 60 mg/kg/day group either died or were euthanized. Gross abnormalities of the intestinal tract, including enlarged gall bladder, green gizzard contents, and dark, shriveled descending colon and caeca were observed in many of the animals in the 40 and 60 mg/kg/day treatments. There was a non-significant decrease in body weight with increasing exposure level, whereas there were no treatment-related differences in organ weights [48].

Red blood cell count and hemoglobin content were decreased in a dose-dependent manner; however, these values were within normal ranges for this species. Total protein, albumin, globulin, aspartate aminotransferase, and potassium, sodium, and chloride ions were decreased in exposed animals. Levels of circulating uric acid were increased in the high-dose groups. Histopathological abnormalities were observed in the liver (lesions) and kidneys (degeneration and multifocal tubulointerstitial nephritis) of birds exposed to 60 mg/kg/day. Although the ability to lay eggs, date of first egg-laying, average number of eggs, and percentage of abnormal eggs were not affected by treatment, the fewest eggs and the highest percentage of abnormal eggs (42%) were produced by birds in the 60 mg/kg/day group. Based on hematological parameters, the authors reported a NOAEL of 10 mg/kg/day and a LOAEL of 40 mg/kg/day [48].

Avian Toxicity Oral: Chronic

No data were found for birds having had chronic exposures to either DNT isomer.

Studies Relevant for Avian TRV Development for Ingestion Exposures

The data from the Johnson *et al.* (2005) subchronic studies are relevant for TRV derivation for oral 2,4-DNT exposures in birds [46]. The sensitive indicators of effect are those pertaining to the kidney and are corroborated by plasma chemistry, histology, and indicators of dehydration (i.e., weight loss and diarrhea) at higher doses. The most sensitive indicator of adverse kidney and liver effects are changes in organ mass respective to body weight.

The data from the Johnson *et al.* (2007) and Quinn *et al.* (2007) subchronic studies are relevant for TRV derivation for oral 2,6-DNT exposure in birds [47,48]. The TRVs are presented in Tables 6.4 and 6.5, respectively.

Both studies indicated gastrointestinal disturbances, such as diarrhea, as well as adverse kidney and/or liver effects. Hematology and plasma chemistry data were the most sensitive endpoints. Plasma uric acid levels were increased in higher-dose groups in both studies, which was further suggestive of kidney-related effects. The studies for both 2,4- and 2,6-DNT were well-conducted, high-quality studies that assessed a variety of endpoints and thus are appropriate for use in TRV derivation.

Avian Toxicity: Inhalation

No data available.

Avian Toxicity: Dermal

No data available.

SUMMARY OF AMPHIBIAN TOXICITY

Amphibian Toxicity Oral: Acute

2,4-Dinitrotoluene/2,6-Dinitrotoluene

In an acute experiment using the up-and-down (UPD) method, Paden *et al.* (2008) exposed adult male bullfrogs (*Lithobates catesbeiana*) to 2,4-DNT and 2,6-DNT in a polyethylene glycol (PEG) vehicle via oral gavage [49]. Eight adult male bullfrogs for each compound were dosed with 175 to 2,000 mg/kg body weight (bw). Three of three animals died in the highest dose group. The LD_{50} value for 2,4-DNT and 2,6-DNT was determined at 1,098 mg/kg with an approximate 95% CI of 550 to 2,000 mg/kg bw. Clinical signs observed in the high-dose group included changes in respiratory rhythm (dyspnea, cyanosis, and tachypnea), decrease in spontaneous motor activities (i.e., somnolence, loss of righting reflex, prostration, tremors, tonic and clonic convulsion), salivation, muscle tone changes (hypertonia, hypotonia), gastrointestinal changes (vomiting), yellow color changes in the skin and urine, pigmentation of the skin, and ocular signs (relaxation of the nictitating membrane). Toxic signs for animals exposed to 2000 mg/kg bw began within the first 2 hours after dosing; death occurred at 80 hours for 2,4-DNT and 72 hours for 2,6-DNT. Tremors and convulsions appeared after 30 hours and 28 hours for 2,4-DNT and

Table 6.4 Summary of Relevant Avian Data from Oral Exposures to 2,4-DNT

		Test Results				
Test Organism	**Test Duration**	**LD50 (mg/kg)**	**NOAEL (mg/kg/d)**	**LOAEL (mg/kg/d)**	**Effects Observed at LOAEL**	**Study**
Northern Bobwhite (quail)	Single dose	55			Emaciation, diarrhea	[46]
	14 days		15	35	Mortality; changes in kidney/bw, diarrhea	
	60 days		1	5	Increases in relative kidney weight, plasma uric acid trends. Mortality, diarrhea, weight loss $\geq$15 mg/kg-d	

Table 6.5 Summary of Relevant Avian Data from Oral Exposures to 2,6-DNT

		Test Results				
Test Organism	**Test Duration**	**LD50 (mg/kg)**	**NOAEL (mg/kg/d)**	**LOAEL (mg/kg/d)**	**Effects Observed at the LOAEL**	**Study**
Northern Bobwhite (quail)	Single dose	320			Weight loss, diarrhea, scant white feces, lethargy, ataxia	[47]
	14 days			50	Increased uric acid, trends in reduction of serum glucose, globulin, and protein, decreased RBC and hemoglobin	
Northern Bobwhite (quail)	60 days		10	40	Mortality, enlarged gall bladder, edematous gastrointestinal tract, diarrhea, increased plasma protein, albumin, globulin, aspartate aminotransferase, sodium, potassium, chloride, and uric acid	[48]

2,6-DNT, respectively. A loss of appetite was observed in animals dosed at the highest level; however, a moderate increase in body weight, especially in the 2,6-DNT-exposed animals, was due to fluid retention, giving the frogs a bloated appearance. A strong yellow urine color was seen in frogs exposed to both isomers. Liver and kidney enlargement and necrosis (determined using gross morphological changes), particularly in the 2,000 mg/kg-dosed animals, were seen at necropsy. An increase in spleen weight of 1.7- and 5.4-fold was seen in the 2,4-DNT and 2,6-DNT animals, respectively. The authors stated that the LD_{50} results, relative low concentrations typically observed in aquatic environments, and the short half-life of the compounds indicate that adult bullfrogs in the wild are at low risk for either exposure or toxicity.

Amphibian Toxicity Oral: Subchronic

No data were found for amphibians having had subchronic exposures to either DNT isomer.

Amphibian Toxicity Oral: Chronic

No data were found for amphibians having had chronic exposures to either DNT isomer.

Studies Relevant for Amphibian TRV Development for Ingestion Exposures

Few data are available for oral toxicity of DNT isomers to amphibians. The acute LD_{50} study in bullfrogs by Paden *et al.* (2008) represents the only available study [49]. Because the data are relatively similar for both isomers, 2,4-DNT and 2,6-DNT were treated together and used to develop toxicity reference values.

Amphibian Toxicity: Dermal/Oral/Respiratory

2,4-Dinitrotoluene/2,6-Dinitrotoluene

In a chronic aqueous experiment, Paden *et al.* (2011) exposed bullfrog (*Rana catesbeiana*) tadpoles to 2,4-DNT (0, 0.125, 0.25, 0.5, 1, 2, and 4 mg/L) and 2,6-DNT (0, 0.25, 0.5, 1, 2, 4, and 8 mg/L) for 90 days [50]. The static renewal experiment consisted of six replicates of each of the five 2,4- and 2,6-DNT treatment levels and the control in 9 liter aquaria containing 30 tadpoles (Gosner stage 20). The aquaria were placed in a stream to a depth of 12 cm, and the exposure media was replaced every 72 hours. Survival was reduced at all concentrations tested for both isomers. Wet body mass, snout vent length (SVL), and developmental stage were increased in tadpoles surviving exposure to 0.25 and 0.5 mg/L 2,6-DNT for 90 days. Exposure to 0.125 mg/L 2,4-DNT did not affect body mass, SVL, and developmental stage. Developmental and behavioral (swimming) abnormalities were observed in tadpoles exposed to 2,4- or 2,6-DNT at 2 mg/L. The authors concluded that exposure to concentrations of 0.25 mg/L of either 2,4- or 2,6-DNT could impair the growth and survival of larval amphibians [50].

SUMMARY OF REPTILIAN TOXICITY

Reptilian Oral Toxicity: Acute

Western Fence Lizards (*Sceloporus occidentalis*) were orally dosed with 98% pure 2,4-DNT in a 1% methyl cellulose/0.02% Tween 80/deionized water vehicle [51]. The oral LD_{50} was determined, using a stage-wise probit design, to be 380 mg/kg (149–515 mg/kg 95% CI) and 577 mg/kg (406–785 mg/kg 95% CI) for male and female lizards, respectively. Initial clinical signs associated with mortality included lethargy and ataxia.

Reptilian Oral Toxicity: Subacute

Suski *et al.* (2008) assessed the toxicity of 2,4-DNT to Western Fence Lizards (*Sceloporus occidentalis*) in a 14-day oral gavage experiment [51]. Forty-two male lizards were dosed daily with 0, 6.25, 12.5, 25, 50, 100, or 200 mg/kg/day 2,4-DNT in a 1% methyl cellulose/0.02% Tween 80/deionized water vehicle. Doses were adjusted volumetrically to achieve the appropriate oral dose based on individual mass. All lizards in the 200 mg/kg/day and four of six lizards in the 100 mg/kg/day groups were either found dead or were found moribund and were euthanized. No differences in sublethal endpoints (body weight, organ weight, food consumption, hematology, or clinical chemistries) were observed in groups dosed with less than 50 mg/kg/day.

Reptilian Oral Toxicity: Subchronic

Suski *et al.* (2008) assessed the toxicity of 2,4-DNT to Western Fence Lizards (*Sceloporus occidentalis*) in a 60-day oral gavage experiment [51]. Sixty male lizards were dosed daily with 0, 9, 15, 25, 42, or 70 mg/kg/day 2,4-DNT in a 1% methyl cellulose/0.02% Tween 80/deionized water vehicle. Doses were adjusted volumetrically to achieve the appropriate oral dose based on individual mass. Survival was reduced in the three highest dose groups (25, 42, and 70 mg/kg/day), whereas survival in the low-dose groups and the control was 100%. Lizards from the 25, 42, and 70 mg/kg/day dose groups survived an average of 55, 51, and 40 days, respectively.

Body-weight loss occurred early in the study in lizards dosed with 70 and 42 mg/kg/day (days 0 to 7 and 7 to 14, respectively) and occurred late in the study in the 25 and 15 mg/kg/day groups (days 14–21 and 56–60, respectively). The 15 mg/kg/day group also exhibited reduced feeding relative to the controls. The only effect on organ weight was an increase in kidney-to-body weight ratio in the 15 mg/kg/day group relative to the control. Relative to the controls, this dose group also exhibited increased plasma uric acid and phosphorus levels. No other clinical chemistries or hematology measures were affected by 2,4-DNT treatment.

Kidney (tubular degeneration, necrosis, and gout) and liver (cellular necrosis and lipogranulomas) effects were observed in the 9 mg/kg/day group. Mild tubular degeneration was observed in the testes in males exposed to 15 mg/kg/day, and this effect increased in severity in the higher dose groups. Lizards in the three highest dose groups also exhibited altered behavior, including arched posture, hanging, and increased skin pigmentation following dosing. Using the U.S. EPA's Benchmark Dose (BMD) software, the authors derived a BMD of 15.8 mg/kg/day and a benchmark dose lower confidence limit (BMDL) of 12.09 mg/kg/day based on survivability. A LOAEL of 15 mg/kg/day and a NOAEL of 9 mg/kg/day were indicated based on kidney effects (increased plasma uric acid and phosphorus levels and kidney-to-body weight ratio), weight loss, and food consumption.

Reptilian Toxicity Oral: Chronic

No data were found for reptiles from chronic exposures to either DNT isomer.

Studies Relevant for Reptilian TRV Development for Ingestion Exposures

The data from the Suski *et al.* (2008) subacute and subchronic studies are relevant for TRV derivation for oral 2,4-DNT exposures in reptiles [51]. The most sensitive indicators of sublethal effect include weight loss, decreased feed consumption, and effects on the kidney, including increased relative kidney weight, increased plasma uric acid and phosphorus levels, and abnormal kidney histopathology. The TRVs are presented in Table 6.6.

Table 6.6 Summary of Relevant Reptilian Data from Oral Exposures to 2,4-DNT

		Test Results				
Test Organism	**Test Duration**	**LD50 (mg/kg)**	**NOAEL (mg/kg/day)**	**LOAEL (mg/kg/day)**	**Effects Observed at the LOAEL**	**Study**
Western Fence Lizard	Single dose	380 mg/kg (M) 577 mg/kg (F)			Lethargy, ataxia	[51]
	14 days		50 (M)	100 (M)	Mortality, lethargy, ataxia	
	60 days		9 (M)	15 (M)	Increases in relative kidney weight, plasma uric acid, and phosphorus. Renal lesions, degeneration, necrosis, and gout. Weight loss and decreased feed consumption. Mortality $\geq$25 mg/kg-day	

Reptilian Toxicity: Inhalation

No data available.

Reptilian Toxicity: Dermal

No data available.

RECOMMENDED TOXICITY REFERENCE VALUES

Toxicity Reference Values for Mammals

TRVs for Ingestion Exposures for the Class Mammalia

2,4-Dinitrotoluene

Data from at least two Orders and three species have been reported for each of the two isomers of DNT. As described in Section 2.4, the dog appears to be the most sensitive to oral exposures to 2,4-DNT. Adverse neurological effects (loss of hindquarter control, convulsion) were the primary endpoints used to determine the TRV as these effects are likely to be most relevant to the health of mammalian wildlife species. The data from the Ellis *et al.* 1979 dog study were used to derive a TRV using the Benchmark Dose Approach (please refer to Chapter 1 of this book) [34]. A dichotomous, multistage model was used to calculate the TRV. This model was chosen based on several goodness-of-fit tests. The TRVs are presented in Table 6.7. This TRV is given a high confidence level because the study was of suitable duration, high quality, and its results were consistent with those of other studies.

2,6-Dinitrotoluene

Data on the chronic toxicity of 2,6-DNT was limited; only one study, conducted on rats, was located [35]. Rats exposed to 2,6-DNT orally for 1 year exhibited significantly smaller body size compared to control rats. This effect of 2,6-DNT was used to derive the TRV because smaller body size indicates a detrimental effect on the energy budgets of the organism and can lead to reductions in fitness as a result of altered reproductive schedules, increased risk of predation, and overall poor condition. Derivation of the TRV using

Table 6.7 Selected Ingestion TRVs for the Class Mammalia for 2,4-DNT

TRV	Dose	Confidence
LED_{10}	0.67 mg/kg/day	High
ED_{10}	1.4 mg/kg/day	High

Source: [34].

Table 6.8 Selected Ingestion TRVs for the Class Mammalia for 2,6-DNT

TRV	Dose	Confidence
NOAEL-based	0.7 mg/kg/day	Moderate
LOAEL-based	7.0 mg/kg/day	Moderate

Source: [35].

the Benchmark Dose Approach was unacceptable for this data set since it did not meet the necessary criteria described in USACHPPM TG 254 [3]. In the Leonard *et al.* (1987) study on which the TRV is based, all doses of 2,6-DNT showed a significant difference from control, hence a NOAEL was not available while the LOAEL was determined at 7 mg/kg/day [35]. The Approximation Approach was used to obtain the NOAEL-based TRV from the LOAEL using an uncertainty factor of 10. The LOAEL-based TRV is the LOAEL reported from the Leonard *et al.* (1987) study [35]. The TRVs are presented in Table 6.8. These TRVs are given a moderate confidence rating since only one chronic study was located.

TRVs for Inhalation Exposures for the Class Mammalia

Data are insufficient for derivation at this time. However, given the vapor pressure and heterogeneous distribution of these isomers, inhalation exposures are unlikely to be experienced by wildlife.

TRVs for Dermal Exposures for the Class Mammalia

Data are insufficient for derivation at this time.

Toxicity Reference Values for Birds

TRVs for Ingestion Exposures for the Class Aves

2,4-Dinitrotoluene

No chronic avian toxicity studies and only a single subchronic study were identified for 2,4-DNT. The subchronic study in Northern Bobwhite quail [46] was determined to be suitable for TRV derivation. Mortality occurred in Northern Bobwhite quail orally dosed at 15 mg/kg/day or greater. Effects on other endpoints, including egg production, triglyceride levels, and effects observed histologically (e.g., gout tophi of the kidney, hemosiderosis of the spleen), occurred at levels where mortality occurred. The most sensitive indicator of adverse effects was changes in kidney-to-body weight ratio, which was consistent with early onset of disease (uric acid trends, urate accumulation in the kidney, and diarrhea). The NOAEL for this study is 1 mg/kg/day. The study was conducted consistent with Good Laboratory

Table 6.9 Selected Ingestion TRVs for the Class Aves for 2,4-DNT

TRV	Dose	Confidence
NOAEL-based	0.01 mg/kg/day	Low
LOAEL-based	0.13 mg/kg/day	Low

Source: [46].

Practices, and its quality was determined to be sufficient to superior. Therefore, the Approximation Approach was used to derive the TRV [3]. An uncertainty factor (UF) of 10 to extrapolate from a subchronic NOAEL and an additional UF of 10 to account for species differences were used to arrive at the NOAEL-based TRV. The LOAEL-based TRV used a UF of four to extrapolate from a subchronic LOAEL and a UF of 10 was used to account for species differences. The TRVs are presented in Table 6.9. Since this study was conducted with a single species and represents the only data available in birds, the TRV was given a low degree of confidence.

2,6-Dinitrotoluene

Data on the toxicity of 2,6-DNT in birds was limited, as only two studies, both of which conducted on Northern Bobwhite quail, were located [47,48]. Both studies indicated gastrointestinal disturbances such as diarrhea as well as adverse kidney and/or liver effects. Hematology and plasma chemistry data were the most sensitive endpoints. Uric acid levels were increased in higher-dose groups in both studies, further suggestive of kidney-related effects. The LOAEL based on clinical chemistries was 40 mg/kg/day, and the NOAEL was 10 mg/kg/day. These endpoints were used to derive the TRV as they are considered to be early indicators of adverse hepatic and renal effects, which contributed to the observed gastrointestinal distress and ultimately to mortality. Because the two studies identified as relevant to TRV derivation were of subchronic duration and used only a single species, the data fail to meet the minimum requirements. As such, the Approximation Approach was used to obtain the NOAEL- and LOAEL-based TRV [3]. An uncertainty factor of 10 to extrapolate from a subchronic NOAEL and an additional UF of 10 to account for species differences were used to arrive at the NOAEL-based TRV. The LOAEL-based TRV used a UF of four to extrapolate from a subchronic LOAEL, and a UF of 10 was used to account for species differences. Since these subchronic studies represent the only data available in birds and were conducted with a single species, the TRV was given a low degree of confidence. The TRVs are presented in Table 6.10.

Table 6.10 Selected Ingestion TRVs for the Class Aves for 2,6-DNT

TRV	Dose	Confidence
NOAEL-based	0.1 mg/kg/day	Low
LOAEL-based	1.0 mg/kg/day	Low

Sources: [47,48].

Toxicity Reference Values for Amphibians

TRVs for Ingestion Exposures for the Class Amphibia

The acute gavage study with bullfrogs [49] was considered sufficiently unique and able to provide relevant information on the toxicity of 2,4-DNT and 2,6-DNT in amphibians to warrant inclusion and TRV derivation. The reported LD_{50} value of 1,098 mg/kg bw for both isomers was divided by a UF of 1000 to estimate a chronic NOAEL (a UF of 100 for an LD_{50} value to a chronic NOEAL and a UF of 10 for interspecies). This resulted in an approximate NOAEL-based TRV of 1.10 mg/kg bw. To estimate a chronic LOAEL the LD_{50} of 1,098 mg/kg was divided by 200 (a UF of 20 for an LD_{50} value to a chronic LOAEL and a UF of 10 for interspecies). This resulted in an approximate LOAEL-based TRV of 5.49 mg/kg bw. The TRVs determined using acute LD_{50} tests have the least ecological relevance due to their short-term exposures but are provided because the study included important toxicological endpoints and a variety of intraspecific responses. A low level of confidence has been given because data from a single study was utilized, only one species was evaluated, and the test was of short duration. Table 6.11 presents the TRVs for 2,4-DNT and 2,6-DNT for the bullfrog *L. catesbeiana*.

TRVs for Dermal Exposures for the Class Amphibia

The chronic developmental study with bullfrog tadpoles [50] was considered sufficient to provide relevant information on the toxicity of 2,4-DNT and 2,6-DNT in amphibians to warrant inclusion and TRV derivation. The lowest reported LOAEL for both isomers of 0.125 mg/l was divided by a UF of 100 to estimate a chronic NOAEL (a UF of 10

Table 6.11 Selected TRVs for the Class Amphibia for 2,4-DNT/2,6-DNT

TRV	Dose	Confidence
NOAEL-based	1.10 mg/kg/day	Low
LOAEL-based	5.49 mg/kg/day	Low

Source: [49].

Table 6.12 Selected TRVs for the Class Amphibia for 2,4-DNT/2,6-DNT

TRV	Dose	Confidence
NOAEL-based	0.00125 mg/L	Low
LOAEL-based	0.125 mg/L	Low

Source: [50].

for a chronic LOAEL to NOAEL and a UF of 10 for interspecies). This resulted in an approximate NOAEL-based TRV of 0.00125 mg/L. A low level of confidence has been given because data from a single study was utilized and only one species was evaluated. Table 6.12 presents the TRVs for 2,4-DNT and 2,6-DNT for the bullfrog *R. catesbeiana*.

Toxicity Reference Values for Reptiles

TRVs for Ingestion Exposures for the Class Reptilia

2,4-Dinitrotoluene

The subacute and subchronic studies conducted by Suski *et al.* (2008) were the only studies available on the oral toxicity of 2,4-DNT in reptiles [51]. The most sensitive indicators of sublethal effects in the subchronic study were weight loss, decreased feed consumption, and effects on the kidney, including increased relative kidney weight, plasma uric acid and phosphorus levels, and abnormal kidney histopathology. A LOAEL of 15 mg/kg/day and a NOAEL of 9 mg/kg/day are indicated based on these effects. Since this study does not meet the minimum requirements (i.e., single study, single species, subchronic), the Approximation Approach was used to obtain the NOAEL- and LOAEL-based TRV [3]. An uncertainty factor of 10 to extrapolate from a subchronic NOAEL and an additional UF of 10 to account for species differences were used to arrive at the NOAEL-based TRV. The LOAEL-based TRV used a UF of four to extrapolate from a subchronic LOAEL, and a UF of 10 was used to account for species differences. Because the subchronic study had data from only a single reptile species, the TRV was given a low degree of confidence. The TRVs are presented in Table 6.13.

Table 6.13 Selected Ingestion TRVs for the Class Reptilia for 2,4-DNT

TRV	Dose	Confidence
NOAEL-based	0.09 mg/kg/day	Low
LOAEL-based	0.38 mg/kg/day	Low

Source: [51].

IMPORTANT RESEARCH NEEDS

The limited availability of data on the toxicity of 2,6-DNT to wildlife species prevents the development of a high-confidence TRV for mammals. Hence, further chronic studies on the toxicity of 2,6-DNT to mammalian species, particularly mice, are needed. At present, there are few toxicity data for 2,4- and 2,6-DNT in nonmammalian wildlife species such as birds, amphibians, and reptiles. The avian, amphibian, and reptilian TRVs would benefit from data in additional species. Data on the toxicity of these compounds, particularly for 2,6-DNT, are needed for other classes of wildlife like amphibians and reptiles.

REFERENCES

[1] U.S. Environmental Protection Agency (USEPA). Overview of the health effects of selected munition chemicals. EPA 822-R-93-022, Washington, D.C: U.S. Environmental Protection Agency, Office of Water; 1993.

[2] Pennigton JC, Jenkins TF, Ampleman G, Thiboutot S, Brannon J, Hewitt AD, et al. Distribution and fate of energetics on DoD test and training ranges: final report. Washington, D.C: U.S. Army Corps of Engineers, Engineer Research and Development Center; 2006.

[3] USACHPPM. TG 254, Standard practice for wildlife toxicity reference values. U.S. Army Center for Health Promotion and Preventive Medicine: Aberdeen Proving Ground, MD; 2000.

[4] Burrows EP, Rosenblatt DH, Mitchell WR, Parmer DL. Organic explosives and related compounds: environmental and health considerations. Technical Report No. AD–8901, Frederick, MD: U.S. Army Biomedical Research and Development Laboratory, Fort Detrick; 1989.

[5] Agency for Toxic Substances and Disease Registry (ATSDR). Toxicological profile for 2,4- and 2,6-dinitrotoluene. Atlanta, GA: U.S. Department of Health and Human Services, Public Health Service, ATSDR; 1998.

[6] Hazardous Substances Data Bank. 2,4-Dinitrotoluene [Internet]. National Library of Medicine; 2014. [cited June 2014]. Available from, http://toxnet.nlm.nih.gov/cgi-bin/sis/search2/r?dbs+hsdb:@term+@rn+121-14-2:@odiris@/cgi-bin/sis/search2/f?./temp/~PtKAJd:6@.

[7] Hazardous Substances Data Bank. 2,6-Dinitrotoluene [Internet]. National Library of Medicine; 2014 [cited June 2014]. http://toxnet.nlm.nih.gov/cgi-bin/sis/search2/f?./temp/~6WTluH:1.

[8] Jenkins TF, Leggett DC, Miyares PH, Walsh ME, Ranney TA, Cragin JH, et al. Chemical signatures of TNT-filled land mines. Talanta 2001;54(3):501–13.

[9] Loehr RC. Treatability potential for EPA listed hazardous wastes in soil. Washington, D.C: U.S. Environmental Protection Agency; 1989, Contract No.: EPA/600/2-89/011 (NTIS 89-166581).

[10] Spanggord RJ, Mill T, Chou TW, Mabey WR, Smith JH, Lee S. Environmental fate studies on certain munition wastewater constituents. Phase II Laboratory Studies. Fort Detrick, MD: U.S. Army Medical Research and Development Command; 1980, Contract No. DAMD 17-78-8081.

[11] Spanggord RJ, Mabey WR, Mill T, Chou TW, Smith JH, Lee S. Environmental fate studies on certain munition wastewater constituents. Phase III, Part II Laboratory

Studies. Fort Detrick, MD: U.S. Army Medical Research and Development Command; 1981, Contract No. DAMD 17-78-C-8081.

[12] Bausum HT, Mitchell WR, Major MA. Biodegradation of 2,4- and 2,6-dinitrotoluene by freshwater microorganisms. J Environ Sci Health A 1992;27(3):663–95.

[13] Lewis TA, Newcombe DA, Crawford RL. Bioremediation of soils contaminated with explosives. J Environ Manage 2004;70(4):291–307.

[14] Noguera DR, Freedman DL. Reduction and acetylation of 2,4-dinitrotoluene by a *Pseudomonas aeruginosa* strain. Appl Environ Microbiol 1996;62(7):2257–63.

[15] Pennington JC. Plant uptake of 2,4,6-trinitrotoluene, 4-amino-2,6-dinitrotoluene, and 2,amino-4,6-dinitrotoluene using 14C-labeled and unlabeled compounds. Vicksburg, MS: U.S. Army Corps of Engineers, Waterways Experimental Station; 1988.

[16] Rickert DE, Butterworth BE, Popp JA. Dinitrotoluene: acute toxicity, oncogenicity, genotoxicity, and metabolism. Crit Rev Toxicol 1984;13(3):217–34.

[17] Rickert DE, Long RM. Tissue distribution of 2,4-dinitrotoluene and its metabolites in male and female Fischer-344 rats. Toxicol Appl Pharmacol 1980;56(2):286–93.

[18] Rickert DE, Schnell SR, Long RM. Hepatic macromolecular covalent binding and intestinal disposition of [14C] dinitrotoluenes. J Toxicol Environ Health 1983;11 (4–6):555–67.

[19] Lee CC, Ellis HV, Kowalski JJ, Hodgson JR, Hwang SW, Short RD, et al. Mammalian toxicity of munitions compounds. Phase II: effects of multiple doses. Part II: 2,4-Dinitrotoluene. Contract No. DAMD 17-74-C-4073, Kansas City, MO: Midwest Research Institute; 1978.

[20] Schut HA, Loeb TR, Grimes LA, Stoner GD. Distribution, elimination, and test for carcinogenicity of 2,6-dinitrotoluene after intraperitoneal and oral administration to Strain A mice. J Toxicol Environ Health 1983;12(4–6):659–70.

[21] Ellis HV, Hodgson JR, Hwang SW, Halpap LM, Kelton DO, Andersen BS, et al. Mammalian toxicity of munitions compounds Phase I: acute oral toxicity, primary skin and eye irritation, dermal sensitization, disposition and metabolism, and Ames tests of additional compounds. Kansas City, MO: Midwest Research Institute; 1978.

[22] Lee CC, Dilley JV, Hodgson JR, Helton DO, Wiegand WJ. Mammalian toxicity of munitions compounds phase I: acute oral toxicity, primary skin and eye irritation, dermal sensitization, and disposition and metabolism. Report No. 1, Kansas City, MO: Midwest Research Institute; 1975, Contract No. DAMD17-74-c-4073.

[23] Vernot EH, MacEwen JD, Haun CC, Kinkead ER. Acute toxicity and skin corrosion data for some organic and inorganic compounds and aqueous solutions. Toxicol Appl Pharmacol 1977;42(2):417–23.

[24] Lane RW, Simon GS, Dougherty RW, Egle Jr JL, Borzelleca JF. Reproductive toxicity and lack of dominant lethal effects of 2,4-dinitrotoluene in the male rat. Drug and chemical toxicology 1985;8(4):265–80.

[25] McGown EL, Knudsen JJ, Marrs Jr. GE Fourteen-day feeding study of 2,4-dinitrotoluene in male and female rats. San Francisco, CA: U.S. Army Medical Research and Development Command, Division of Research Support, Letterman Army Institute of Research; 1983.

[26] U.S. EPA. Drinking water health advisory for 2,4-dinitrotoluene and 2,6-dinitrotoluene. Washington, D.C: U.S. Environmental Protection Agency, Health and Ecological Criteria Division, Office of Science and Technology, Office of Water; 2008.

[27] Smith KN. Determination of the reproductive effects in mice of nine selected chemicals. Cincinnati, OH: National Institutes of Occupational Safety and Health; 1983.

[28] Lent EM, Crouse LC, Quinn Jr MJ, Wallace SM. Comparison of the repeated dose toxicity of isomers of dinitrotoluene. Int J Toxicol 2012;31(2):143–57.

[29] Bloch E, Gondos B, Gatz M, Varma SK, Thysen B. Reproductive toxicity of 2,4-dinitrotoluene in the rat. Toxicol Appl Pharmacol 1988;94(3):466–72.

[30] Lee CC, Hong CB, Ellis HV, Dacre JC, Glennon JP. Subchronic and chronic toxicity studies of 2,4-dinitrotoluene. Part II. CD rats. J Am Coll Toxicol 1985;4:243–56.
[31] Chemical Industry Institute of Toxicology (CIIT). Thirty-day toxicology study in Fischer-344 rats given dinitrotoluene, technical grade. Research Triangle Park, NC: Chemical Industry Institute of Toxicology; 1977, Contract No.: Final Report. CIIT Docket No. 22397.
[32] Hong CB, Ellis ER, Lee CC, Sprinz JC, Dacre JC, Glennon JP. Subchronic and chronic toxicity studies of 2,4-dinitrotoluene. Part III. CD-l mice. J Am Coll Toxicol 1985;4:257–69.
[33] Lee CC, Ellis HV, Kowalski JJ, Hodgson JR, Short RD, Bhandari JC, et al. Mammalian toxicity of munitions compounds. Phase II: effects of multiple doses. Part III: 2,6-Dinitrotoluene. Kansas City, MO: Midwest Research Institute; 1976, Contract No. DAMD-17-74-C-4073.
[34] Ellis HV, Hagensen JH, Hodgson JR, Minor JL, Hong CB, Ellis ER, et al. Mammalian toxicity of munitions compounds. Phase III: effects of lifetime exposure. Part I: 2,4-dinitrotoluene. Kansas City, MO: Midwest Research Institute; 1979.
[35] Leonard TB, Graichen ME, Popp JA. Dinitrotoluene isomer-specific hepatocarcinogenesis in F344 rats. J Natl Cancer Inst 1987;79(6):1313–9.
[36] Hardin BD, Schuler RL, Burg JR, Booth GM, Hazelden KP, MacKenzie KM, et al. Evaluation of 60 chemicals in a preliminary developmental toxicity test. Teratog Carcinog Mutagen 1987;7(1):29–48.
[37] Jones-Price C, Marks TA, Ledoux TA, Reel JR, Lenghoff-Paschke L, Wolkowski-Tyl R. Teratological and postnatal evaluation of dinitrotoluene in Fischer 344 rats. Research Triangle Park, NC: Chemistry and Life Sciences Group and Chemical Institute of Toxicology; 1980.
[38] Price CJ, Tyl RW, Marks TA, Paschke LL, Ledoux TA, Reel JR. Teratologic evaluation of dinitrotoluene in the Fischer 344 rat. Fundam Appl Toxicol 1985;5(5):948–61.
[39] Ellis ER, Hong CB, Dacre JC, Glennon JP. Subchronic and chronic toxicity studies of 2,4-dinitrotoluene. Part I. Beagle dogs. Int J Toxicol 1985;4(4):233–42.
[40] CIIT. 104-week chronic toxicity study in rats: dinitrotoluene. Research Triangle Park, NC: Chemical Industry Institute of Toxicology; 1982, Contract No.: Final Report, Vol. 1 and 2. CIIT Docket No. 12362.
[41] National Cancer Institute (NCI). Bioassay of 2,4-dinitrotoluene for possible carcinogenicity. CAS No. 121-14-2, Washington, DC: National Cancer Institute, U.S. Department of Health, Education, and Welfare, Public Health Service, National Institutes of Health; 1978, 0163-7185 (Print); 0163-7185. Contract No.: NCI-CG-TR-54.
[42] Kozuka H, Mori M, Naruse Y. Studies on the metabolism and toxicity of dinitrotoluenes. Toxicological study of 2,4-dinitrotoluene (2, 4-DNT) in rats in long term feeding. J Toxicol Sci 1979;4(3):221–8.
[43] Calow P. Physiological costs of combating chemical toxicants: ecological implications. Comp Biochem Physiol C 1991;100(1–2):3–6.
[44] Congdon JD, Dunham AE, Hopkins WA, Rowe CL, Hinton TG. Resource allocation-based life histories: a conceptual basis for studies of ecological toxicology. Environ Toxicol Chem 2001;20(8):1698–703.
[45] Chemical Manufacturers' Association. 2,6-Dinitrotoluene acute (6-hour) inhalation toxicity study in rats. Washington, D.C.: Chemical Manufacturers' Association; 1991.
[46] Johnson MS, Michie MW, Bazar MA, Gogal Jr RM Influence of oral 2,4-dinitrotoluene exposure to the northern bobwhite (*Colinus virginianus*). Int J Toxicol 2005;24(4):265–74.
[47] Johnson MS, Suski J, Bazar MA. Toxicological responses of red-backed salamanders (*Plethodon cinereus*) to subchronic soil exposures of 2,4-dinitrotoluene. Environ Pollut 2007;147(3):604–8.

[48] Quinn Jr MJ, Bazar MA, McFarland CA, Perkins EJ, Gust KA, Gogal Jr RM, et al. Effects of subchronic exposure to 2,6-dinitrotoluene in the northern bobwhite (*Colinus virginianus*). Environ Toxicol Chem 2007;26(10):2202–7.
[49] Paden NE, Smith EE, Kendall RJ. Acute toxicity of 2,4,6-trinitrotoluene, 2,4-dinitrotoluene, and 2,6-dinitrotoluene in the adult bullfrog (*Lithobates catesbeiana*). Bull Environ Contam Toxicol 2008;80(6):487–91.
[50] Paden NE, Smith EE, Maul JD, Kendall RJ. Effects of chronic 2,4,6,-trinitrotoluene, 2,4-dinitrotoluene, and 2,6-dinitrotoluene exposure on developing bullfrog (*Rana catesbeiana*) tadpoles. Ecotoxicol Environ Saf 2011;74(4):924–8.
[51] Suski JG, Salice C, Houpt JT, Bazar MA, Talent LG. Dose-related effects following oral exposure of 2,4-dinitrotoluene on the Western fence lizard, *Sceloporus occidentalis*. Environ Toxicol Chem 2008;27(2):352–9.

CHAPTER 7

Wildlife Toxicity Assessment for Nitroguanidine

Bradley E. Sample, Christine Ann Arenal, Emily N. Reinke, Gunda Reddy

Contents

INTRODUCTION

Nitroguanidine (NQ) is a stable primary explosive compound used as a propellant in cartridges. Combined with nitrocellulose and nitroglycerin, it forms a triple-base propellant mixture [1]. Synonyms for nitroguanidine, CAS No. 556-88-7, include Guanidine, nitro-; Picrite; Picrite (the explosive); α-Nitroguanidine; β-Nitroguanidine; N1-Nitroguanidine; 2-Nitroguanidine; UN 0282; UN1336; and NQ. Environmental releases of NQ are most likely to occur in areas where the munition was manufactured. The detection of NQ in wastewaters from an ammunition plant has raised concerns about its potential effects on ecological receptors and wildlife [1].

This Wildlife Toxicity Assessment (WTA) summarizes current knowledge of the potential harmful impacts on wildlife, and reports Toxicity Reference

Wildlife Toxicity Assessments for Chemicals of Military Concern
http://dx.doi.org/10.1016/B978-0-12-800020-5.00007-7

Values (TRVs) for NQ[1] that are important for the health of wildlife (e.g., mammals, birds, amphibians, and reptiles) exposed to the compound. The toxicity endpoints relate to the likelihood of effects in individual organisms that may be relevant to a population of organisms in the wild, but are not related to demographic rates or outcomes for any particular population. This WTA was developed in accordance with the U. S. Army Center for Health Promotion and Preventive Medicine (USACHPPM) Technical Guide 254, the *Standard Practice for Wildlife Toxicity Reference Values* [2].

TOXICITY PROFILE

Environmental Fate and Transport

Stressing the lack of experimental information, Kenyon [1] reviewed the then-current knowledge of NQ's environmental fate and transport. Much of what is known about the compound's ability to move through environmental media comes from assumptions based upon NQ's physical-chemical properties (Table 7.1; 3, 4). For example, with an estimated vapor pressure of 0.03 mm Hg at 25 °C, NQ should primarily exist in the atmosphere as a vapor, where it will degrade through interaction with photochemically-generated hydroxyl radicals [3]. Kaplan and Kaplan [5] showed that NQ is likely to have a low affinity for soil particles. Microbial degradation of NQ has been reported [5] but requires adequate soil carbon. If soil carbon is insufficient, degradation is limited and leaching to groundwater is possible. Bioassays with soil microorganisms indicate no apparent toxicity with NQ concentrations ranging up to 10 mg/kg soil (1%) for 48 hours [5]. Although NQ is also likely to be highly mobile in aqueous solutions, bioaccumulation in aquatic organisms is expected to be low.

Summary of Mammalian Toxicology

Mammalian Oral Toxicity: Acute

Hiatt *et al.* [6] estimated the median lethal dose (LD_{50}) in ICR mice to be 4,345 mg/kg in females and greater than 5,000 mg/kg in males. Brown *et al.* [7] could not derive a precise LD_{50} in Sprague Dawley rats and estimated the median lethal dose to be greater than 5,000 mg/kg.

[1] Toxicological research on nitroguanidine classification compounds, such as the chloronicotinyl pesticide subset nitroguanidine and those used in pharmaceuticals is not considered relevant to this toxicity profile because of the compounds' diverse chemical structures.

Table 7.1 Summary of Physical-Chemical Properties of Nitroguanidine

CAS No.	556-88-7
Molecular weight	104.07
Color	Colorless
Physical state	Crystals
Melting point	225-250 °C
Boiling point	Decomposes at 232 °C
Odor	No data
Density	1.5 g/mL
Stability	Sensitive to ultraviolet light, absorbs at 264 nm Explosive when shocked or exposed to heat or flame
Solubility in water	4.4 g/L at 25 °C, 90 g/L at 100 °C in water; slightly soluble in methanol and ethanol; insoluble in ether
Partition coefficients:	
Log K_{ow}	–0.89
Log K_{oc}	1.4
Vapor pressure at 25 °C	Extremely low, 0.03 mm Hg
Henry's Law constant at 25 °C	4.5×10^{-12} atm.m^3/mole
Conversion factors	1 ppm = 4.26 mg/m^3 1 mg/m^3 = 0.23 ppm

Sources: Kenyon [1], HSDB [3].

Mammalian Oral Toxicity: Subacute

Morgan *et al.* [8] conducted a 14-day study of NQ administered to Sprague Dawley rats at dietary doses of 0, 100, 316, and 1,000 mg/kg/day (10 rats/ sex per treatment group). A comprehensive suite of parameters was monitored, including clinical signs, behavior, body weights, fcod and water consumption, hematological and clinical chemistry, and gross morphology and histopathology of all organs and tissues. With the exception of an increase in water consumption in rats receiving 316 and 1,000 mg/kg/day, no significant NQ effects were identified, and the investigators concluded that, while NQ appeared to act as an osmotic diuretic with consequent blood serum chemistry alterations, it was essentially non-toxic.

Mammalian Oral Toxicity: Subchronic

Subchronic 90-day studies of NQ toxicity in Sprague Dawley rats [9] and ICR mice [10] used essentially identical protocols. Nitroguanidine was administered in the diet of both test species at concentrations equivalent to doses of 0, 100, 316, and 1,000 mg/kg/day to 15 animals per sex per group. A comprehensive list of possible effects was monitored including clinical signs, body weights, food and water consumption, hematological and clinical chemistry parameters at term, and at necropsy, both gross and histopathology examinations of all organs and tissues.

Female rats in the 1,000 mg/kg/day dose group experienced significant reduction in body weight gain and reduced food consumption rates during the 90-day study [9]. Although not statistically different from controls, male rats in the high-dose group (1,000 mg/kg/day) also displayed reduced weight gain and food consumption. In addition, a dose-related increase in water consumption was observed for both male and female rats. The authors attributed the reductions in body weight gain and food consumption to reduced palatability of feed at the 1,000 mg/kg/day dose level, and increased water consumption due to the diuretic properties of NQ. They concluded that doses of NQ up to 1,000 mg/kg/d did not cause "appreciable toxicological effects." However, it should be noted that IRIS [4] assigned a no observed adverse effect level (NOAEL) of 316 mg/kg/day in the derivation of a reference dose (RfD) based upon the significant reduction in body weight gain observed at the 1,000 mg/kg/day dose level at 5, 6, 8, 9, and 12 weeks. The equivalent lowest observed adverse effect level (LOAEL) was 1,000 mg/kg/day.

The ICR mice exposed for 90 days were also largely unaffected by NQ, although a dose-dependent increase in water consumption was noted [10]. As with rats, the increased water consumption was attributed to the diuretic properties of NQ and was not considered a toxicological effect. All necropsy, histopathological, hematological, and clinical chemistry findings were minimal and apparently incidental to treatment. In high dose (1,000 mg/kg/day) males, a non dose-related increase in relative brain weight was noted, for which the authors assigned little importance [10]. However, IRIS [4] considered these increased brain-to-body weight ratios supportive of the 316 mg/kg/day NOAEL derived from the 90-day study in rats [9].

Mammalian Oral Toxicity: Chronic

No chronic oral mammalian studies were found for NQ.

Mammalian Oral Toxicity: Other

A series of reproductive, developmental, and teratology studies of NQ in Sprague Dawley rats and New Zealand white rabbits were reported. These studies included initial developmental and teratology tests using Sprague

Dawley rats [11] and New Zealand white rabbits [12] dosed during gestation, and a follow-up two-generation reproductive study using Sprague Dawley rats [13].

Coppes *et al.* [11] dosed sperm-positive female Sprague Dawley rats on gestational days (GDs) 6 through 15 at 0 (N=27), 100 (N=27), 316 (N =23), or 1,000 (N=27) mg/kg/day NQ in 1% high viscosity carboxymethylcellulose sodium salt solution by gavage. Surviving rats (N=27, 26, 22, and 21, respectively) were euthanized, ovo-hysterectomized, and the fetuses removed on GD 20. Seven female rats died during the study and one moribund animal was euthanized. Necropsy results for these eight rats indicated that five died of injuries from the dosing method. The remaining three deaths were rats in the high dose (1,000 mg/kg/day) group. However, the authors indicate that these deaths were not likely a direct pharmacological effect, rather they were a result of the high volume required to administer the 1,000 mg/kg/day dose by oral intubation. Evidence included balls of test compound in the stomachs of two dead animals and a small amount of test compound in the stomach of the third. Other observed signs of toxicity in the high dose group (e.g., decreased food consumption and decreased weight gain) were also attributed to interference of the test compound with normal digestion. Fetuses of high-dose dams were significantly lighter in weight and shorter in length with an increased incidence of retarded ossification of the sternebrae, caudal vertebrae, and pubis. Since the fetal malformations were not dose-related, occurred at a low frequency, and were within the range of rates both historical to the lab and published by Palmer [14], the authors considered observed malformations to be spontaneous [11]. Based upon maternal and fetal toxicity observed in the 1,000 mg NQ/kg/day group, Coppes *et al.* [11] reported a NOAEL of 316 mg/kg/day for rats in their protocol. The authors also note that NQ is absorbed with 100% efficiency via oral administration and is excreted unchanged in the urine at a rate of 60 to 80% within the first 8 hours [11]. This suggests that NQ is not easily metabolized by the body and that only 20 to 40% of the administered dose remains after 8 hours.

In a second study, Coppes *et al.* [12] reported results for New Zealand white rabbits that were similar to those observed in rats [11]. Bred rabbits were gavaged on GDs 6 through 18 at 0 (N =17), 100 (N=18), 316 (N=16), or 1,000 (N=22) mg/kg/day NQ in high viscosity 1% carboxymethylcellulose sodium salt solution. On GD 29, surviving rabbits (N=17, 18, 16, and 12, respectively) were euthanized, ovo-hysterectomized, and the fetuses removed. A full range of maternal effects was monitored, and all fetuses were weighed, examined externally, and then processed for either visceral or skeletal examination. A number of clinical signs were observed in the high-dose dams, including reduction in food consumption, failure to gain weight, dehydration, and chromodacryorrhea (the shedding of bloody tears), among others. Fifty percent of the high-dose dams died during the experiment, and survivors consumed less food and gained less weight than controls. Although all 10 deaths occurred in the high dose group, as with the rats, the authors did not consider this a pharmacological effect of NQ. Instead, the high volume necessary to obtain the 1,000 mg/kg/day dose level by oral intubation likely interfered with digestive processes. This resulted in weight loss, and in some cases death. Fetuses from high-dose dams weighed less than controls weighed, and displayed retarded ossification of the sternebrae, olecranon, patellae, and phalanges, but there appeared to be no dose-related visceral malformations and the data did not indicate an NQ-mediated teratological effect.

An increased incidence of fetal resorption in all dose groups was noted (statistically significant in the 100 and 1,000 mg/kg/day groups), but the incidence of resorption in control dams was lower than historically observed in New Zealand white rabbits. Additionally, the resorption rate in NQ-exposed dams was within the normal range [15]. Although the fetal and maternal toxicity evident at the highest dose suggested a NOAEL of 316 mg/kg/day, Coppes *et al.* [12] concluded there were no specific developmental or teratology effects that could be unequivocally attributed to NQ exposure. In this case fetal resorption is questionable as an indication of possible NQ developmental toxicity in rabbits; however, IRIS [4], considering the data equivocal, suggested a developmental toxicity LOAEL of 100 mg/kg/day in rabbits.

Following the rat and rabbit developmental studies, Coppes *et al.* [13] carried out a two-generation reproductive and fertility study to investigate further the potential effects of long-term exposure to NQ. In this study, NQ was mixed in the feed of young adult Sprague Dawley rats at 0, 1.3, 4.0, and 12.7 parts per thousand (ppt), equivalent to daily doses of approximately 0, 100, 316, or 1,000 mg/kg. Parental male and female rats (P_1) received NQ augmented feed starting at 56 to 58 days of age and continuing throughout their lives including through mating, gestation, lactation, and weaning. Progeny of these rats received the same dose levels as their parents through the F_1 and F_2 generations. The authors reported no clinical signs and no NQ-related effects on behavior, mating, fertility, gestation, litter size, pup weights, or survival in animals of either generation. Histopathological examination of adult rat reproductive organs and gross examination of offspring also indicated no toxicological effect of NQ. Some weekly body weights and food consumption rates of adults in the high dose group were significantly lower than controls, but this was not a consistent trend throughout the study. However, terminal body weights for F_1 high dose males and F_1 low and high dose females were significantly lower than controls [13]. The authors concluded that no reproductive or fertility toxicological effects occurred in the study. This suggests a NOAEL of 1,000 mg/kg/day for this study.

Studies Relevant for Mammalian TRV Development for Ingestion Exposures

Acute, subacute, subchronic, and chronic NQ studies representing three species (rats, mice, and rabbits) and two orders (Rodentia and Lagamorpha) were available. The reproductive and developmental studies are considered to be chronic because they were conducted during a critical life-stage (i.e., during gestation) and because the reproductive study was conducted over a period greater than 10% of the lifespan of the test species [2].

These studies indicate that NQ poses few, if any, toxicological hazards to experimental mammals. The most consistent consequence of NQ ingestion was a dose-related increase in water consumption. It was also noted that animals dosed with up to 1,000 mg/kg/day NQ excrete unaltered NQ in the urine; therefore, the authors concluded that NQ acts as an osmotic diuretic. Other effects observed among female test animals on high NQ doses (1,000 mg/kg/day) included death, reduced body weight gain, reduced food consumption and, apparently related to the reduced growth, increased brain-to-body weight ratio at terminal sacrifice. However, in all the studies, these were not considered pharmacological effects of NQ, but rather were attributed to decreased palatability and interference with digestion at the 1,000 mg/kg/day dose level. In the developmental and reproduction studies [11–13], various fetal effects were also noted, including reduced fetal body weights and size and increased incidence of minor developmental alterations (e.g., retarded ossification of the sternebrae).

Coppes *et al.* [12] observed increased fetal resorptions in rabbits at all dose levels. This effect in the 100 and 1,000 mg/kg/day dose groups was significantly different from controls, but was within published normal ranges for the species. Because effects were consistent across studies and species, and NOAELs and LOAELs for several endpoints of effect are available, the subchronic [9,10] and chronic [11–13] studies were considered of sufficient quality and importance to derive class-specific TRVs for NQ. These studies were consistent in quality and reporting of the methods. Data from the acute and subacute studies were not used for TRV derivation purposes. Table 7.2 and Figure 7.1 summarize the results of the relevant NQ studies.

Mammalian Inhalation Toxicity

No relevant studies were identified that addressed NQ inhalation toxicity.

Mammalian Dermal Toxicity

Lewis *et al.* [16] conducted a 24-hour acute dermal toxicity test of NQ on New Zealand white rabbits, and no toxic effects were observed when the compound was applied as a paste at a concentration of 2 g/kg, nor was there any evidence of NQ systemic absorption.

Summary of Avian Toxicology

No relevant avian NQ toxicity studies were identified.

Summary of Amphibian Toxicology

No relevant amphibian NQ toxicity studies were identified.

Summary of Reptilian Toxicology

No relevant reptilian NQ toxicity studies were identified.

RECOMMENDED TOXICITY REFERENCE VALUES

Toxicity Reference Values for Mammals

TRVs for Ingestion Exposures for the Class Mammalia

Based on the information for three species presented earlier, the rabbit was the most sensitive species tested, with a significant increase in fetal resorptions at the 100 and 1,000 mg/kg/day dose levels [12]. An increase in fetal resorptions was also observed in the 316 mg/kg/day dose group, but was not significantly different from controls. Based on this endpoint, IRIS [4] recommended a NOAEL of 100 mg/kg/day. However, the incidence of fetal resorptions in all dose groups was within the normal range for the species [15]. Therefore, this endpoint was not considered useful for TRV derivation. The primary toxicological effects of NQ were consistently observed at the 1,000 mg/kg/day dose level. These effects included

Table 7.2 Summary of Relevant Mammalian Data for TRV Derivation for Nitroguanidine

			Test Results			
Test Type	Test Organism	Test Duration	NOAEL (mg/kg/day)	LOAEL (mg/kg/day)	Effects Observed at the LOAEL	Study
Chronic	Rats (Sprague Dawley)	GD 6-15	316	1,000	Reduced maternal body weight gain. Reduced fetal body weights and lengths and increased incidence of minor malformations.	[12]
	Rabbits (New Zealand)	GD 6-18	316	1,000	Reduced maternal body weight gain. Reduced fetal body weights and lengths and increased incidence of minor malformations.	[13]
			100	316	Increased incidence of fetal resorptions. Results considered equivocal because the number of resorptions was greater than controls, but within published normal limits for the species.	
	Rats (Sprague Dawley)	Two generations	1,000		Though significant decreases in body weights in the F_1 generation occurred at the low and high dose levels, the authors did not consider these appreciable toxicological effects.	[14]
Subchronic	Rats (Sprague Dawley)	90-d	316	1,000	Reduced body weight gain in females. Dose-related increased water consumption at all doses.	[10]
	Mice (ICR)	90-d	316	1,000	Increased relative brain weight in males. Dose-related increased water consumption at all doses.	[11]

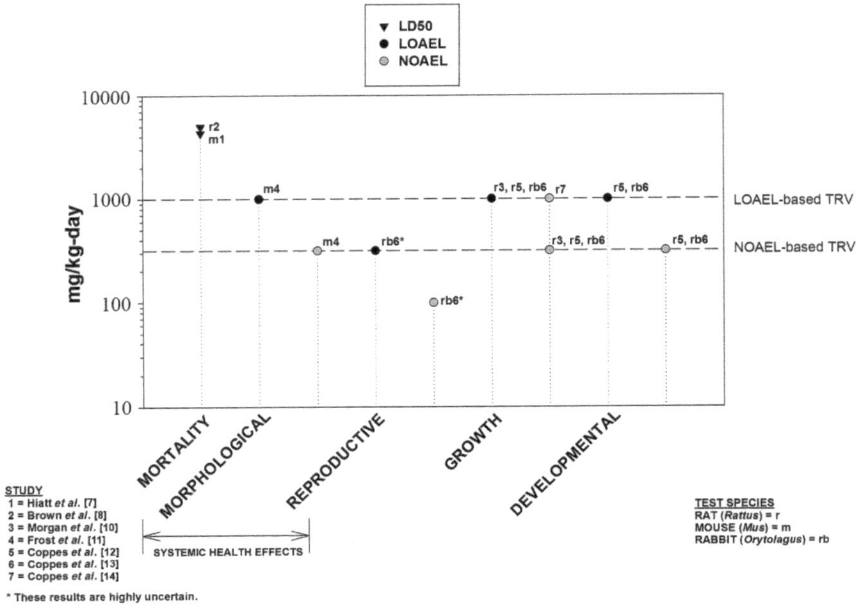

Figure 7.1 Health effects of nitroguanidine in mammals. Letters by symbols represent test species, and numbers represent study.

increased water consumption in both males and females [9,10], reduced body weight and food consumption in females [9,11,12], reduced body weight in fetuses [11] and F_1 generation animals [13], and an increased incidence of minor malformations in fetuses [11,12]. Although the authors did not consider all of these effects to be toxicological in nature (e.-g., reductions in food consumption and body weight were attributed to decreased palatability of the feed or interference of the test compound with digestion), the consistent finding of effects at the 1,000 mg/kg/day dose level in these three species suggest this level as an appropriate LOAEL for NQ. The highest NOAEL within the same endpoints was the 316 mg/kg/day dose level.

Reduced body weights (an indicator of negative effects on growth and/or energy efficiency) and fetal malformations (i.e., developmental toxicity) could adversely affect future fitness. Therefore, these endpoints are ecologically relevant. Since these studies satisfy the minimum data set requirements outlined in USACHPPM [2], uncertainty factors are not needed to derive the TRVs. These data were not sufficient to develop a benchmark dose, but were sufficient for development of a chronic NOAEL and LOAEL. The

Table 7.3 Selected Ingestion TRVs for Nitroguanidine for the Class Mammalia

TRV	Dose (mg/kg/day)	Confidence Level
NOAEL-based	316	Medium
LOAEL-based	1,000	Medium

Source: [12].

class-specific NOAEL and LOAEL are presented in Table 7.3. Although the results were consistent across species and across studies, a medium degree of confidence was assigned to the TRVs based on the equivocal nature of the observed effects.

TRVs for Ingestion Exposures for Mammalian Foraging Guilds

TRVs specific to particular guild associations (e.g., small herbivorous mammals) were not derived. The class-specific TRVs shown in Table 7.3 may be considered to apply to small herbivorous mammals or small omnivorous mammals because rabbits, rats, and mice, respectively, are members of these guilds. As with the class-specific TRVs, confidence in the TRVs is medium. Data to derive TRVs for other guild associations (e.g., carnivorous mammals) are not available at this time.

TRVs for Inhalation Exposures for the Class Mammalia

No data were identified permitting derivation of a mammalian inhalation TRV for NQ.

TRVs for Dermal Exposures for the Class Mammalia

Sufficient data required for development of dermal TRVs for Class Mammalia were not available. However, an acute exposure of 2,000 mg/kg of NQ had no effect on New Zealand white rabbits. This suggests that NQ has a low dermal toxicity to mammals.

Toxicity Reference Values for Birds

No data were identified permitting derivation of an avian TRV for NQ.

Toxicity Reference Values for Amphibians

No data were identified permitting derivation of an amphibian TRV for NQ.

Toxicity Reference Values for Reptiles

No data were identified permitting derivation of a TRV for NQ for reptiles.

IMPORTANT RESEARCH NEEDS

Mammalian TRVs derived for NQ have medium confidence; consequently, mammalian toxicity needs to be further characterized. Specifically, additional species and taxonomic orders should be evaluated to provide a greater breadth of interspecific data. In addition, analyses should be conducted that include dose levels between 316 and 1,000 mg/kg/day to more specifically focus on observed effects at concentrations in this range. Further study of the absorption, retention, and excretion of NQ at high dose levels is also needed. These additional data would increase confidence in the mammalian TRVs, and would enable development of TRVs for specific foraging guilds. Inhalation and dermal studies on mammals were either limited or lacking for NQ. The absence of toxicity data for birds, amphibians, and reptiles precludes TRV derivation for these taxa. Appropriate acute, subacute, subchronic, and especially chronic NQ toxicity data by all exposure routes are needed before reliable avian, amphibian, and reptilian TRVs can be derived. Research studies should include experimental models of species genetically, biologically, and behaviorally similar to wildlife exhibiting the greatest propensity for toxicant exposure. Experimental design should attempt to mimic both exposure type and duration, and include assessments of long-term effects.

REFERENCES

[1] Kenyon KF. A data base assessment of environmental fate aspects of nitroguanidine. Fort Detrick, Frederick (MD): U.S. Army Biomedical Research and Developmental Laboratory; 1982, Technical Report Sep-Oct 1982. Accession Numer: AD-A125591.

[2] USACHPPM. Standard practice for wildlife toxicity reference values. Aberdeen Proving Ground (MD): U.S. Army Center for Health Promotion and Preventive Medicine; 2000, Technical Guide No. 254.

[3] Hazardous Substance Data Bank (HSDB). Nitroguanadine [Internet]. Bethesda (MD): National Library of Medicine (US), National Institute of Health [cited 2001]. Available from: http://toxnet.njm.nih.gov/.

[4] U.S. Environmental Protection Agency (USEPA). Integrated risk information system [Internet]. Cincinnati (OH): USEPA Office of Health and Environmental Assessment, National Center for Environmental, Assessment [cited 2001].

[5] Kaplan DL, Kaplan AM. Degradation of nitroguanidine. Natick (MA): US Army Natick R&D Center; January 1985, Accession Number: AD-A157859. Prepared for: U.S. Army Toxic and Hazardous Materials Agency, Aberdeen Proving Ground, MD.

[6] Hiatt GF, Sano SK, Wheeler CR, Korte Jr DW. Acute oral toxicity of nitroguanidine in mice. Presidio of San Francisco (CA): Letterman Army Institute of Research; 1988 Apr, Accession Number: AD-A194952. Prepared for: U.S. Army Biomedical Research and Developmental Laboratory, Fort Detrick, Frederick, MD.
[7] Brown LB, Wheeler CR, Korte Jr DW. Acute oral toxicity of nitroguanidine in male and female rats. Presidio of San Francisco (CA): Letterman Army Institute of Research; 1988 Mar, Accession Number: AD-A192694. Prepared for: U.S. Army Biomedical Research and Developmental Laboratory, Fort Detrick, Frederick, MD.
[8] Morgan EW, Brown LD, Lewis CM, Dahlgren RR, Korte Jr DW. Fourteen-day subchronic oral toxicity study of nitroguanidine in rats. Presidio of San Francisco (CA): Letterman Army Institute of Research; 1988 Jun, Accession Number: AD-A196666. Prepared for: U.S. Army Biomedical Research and Developmental Laboratory, Fort Detrick, Frederick, MD.
[9] Morgan EW, Pearc MJ, Zaucha GM, Lewis CM, Makovec GT, Korte Jr DW. Ninety-day subchronic oral toxicity study of nitroguanidine in rats. Presidio of San Francisco (CA): Letterman Army Institute of Research; 1988, Accession Number: AD-A203681. Prepared for: U.S. Army Biomedical Research and Developmental Laboratory, Fort Detrick, Frederick, MD.
[10] Frost DF, Morgan EW, Letellier Y, Pearce MJ, Ferraris S, Smith CD, et al. Ninety-day subchronic oral toxicity study of nitroguanidine in rats. Presidio of San Francisco (CA): Letterman Army Institute of Research; 1988, Accession Number: AD-A203757. Prepared for: U.S. Army Biomedical Research and Developmental Laboratory, Fort Detrick, Frederick, MD.
[11] Coppes VG, Orner GA, Korte Jr DW. Developmental toxicity potential of nitroguanidine in rats. Presidio of San Francisco (CA): Letterman Army Institute of Research; 1988, Accession Number: AD-A192383. Prepared for: U.S. Army Biomedical Research and Developmental Laboratory, Fort Detrick, Frederick, MD.
[12] Coppes VG, Gomez CL, Magnuson DK, Korte Jr DW. Developmental toxicity potential of nitroguanidine in rabbits. Presidio of San Francisco (CA): Letterman Army Institute of Research; 1988, Accession Number: AD-A200472. Prepared for: U.S. Army Biomedical Research and Developmental Laboratory, Fort Detrick, Frederick, MD.
[13] Coppes VG, Gomez CL, Clifford CB, Ferraris S, Korte Jr DW. Toxic potential of nitroguanidine on reproductive and fertility in rats. Presidio of San Francisco (CA): Letterman Army Inst of Research; 1990, Accession Number: AD-A224663. Prepared for: U.S. Army Biomedical Research and Developmental Laboratory, Fort Detrick, Frederick, MD.
[14] Palmer AK. Sporadic malformations in laboratory animals and their influence on drug testing. Adv Exp Med Biol 1972;27:45–60.
[15] Korte DW, Coppes VG, Orner GA, Gomez CL. Developmental toxicity potential of nitroguanidine in rats and rabbits. Toxicologist 1990;10:124.
[16] Lewis C, Okerberg CV, Korte Jr DW. Acute dermal toxicity of nitroguanidine in rabbits. Presidio of San Francisco (CA): Letterman Army Institute of Research; 1988, Accession Number: AD-A191796. Prepared for: U.S. Army Biomedical Research and Developmental Laboratory, Fort Detrick, Frederick, MD.

CHAPTER 8

Wildlife Toxicity Assessment for 1,3-Dinitrobenzene

Gunda Reddy

Contents

Wildlife Toxicity Assessments for Chemicals of Military Concern
http://dx.doi.org/10.1016/B978-0-12-800020-5.00008-9 2015 Published by Elsevier Inc.

INTRODUCTION

1,3-Dinitrobenzene (m-DNB; CAS No. 99-65-0) is one of several compounds that have been released to the environment during the manufacture of explosives and in the load, assembly, and pack (LAP) activities at U.S. Army ammunition plants (AAPs) and other military installations. The compound is structurally related to another important military explosive, trinitrotoluene (TNT), of which m-DNB is a manufacturing byproduct and an environmental degradation product. The importance of m-DNB as an environmental contaminant is related to its widespread distribution on military sites and potential toxicity to wildlife and other ecological receptors.

This chapter summarizes current knowledge of the likely harmful impacts of m-DNB on wildlife, with emphasis on identifying the levels at which wildlife species may be adversely affected. Evaluating the toxicity of the compound aims to derive Toxicity Reference Values (TRVs) that could serve as protective exposure standards for wildlife found near m-DNB contaminated sites. The protocol for the performance of this assessment is documented in the U.S. Army Center for Health Promotion and Preventive Medicine (USACHPPM) Technical Guide (TG) 254, the *Standard Practice for Wildlife Toxicity Reference Values* [1].

TOXICITY PROFILE

Environmental Fate and Transport

m-DNB has been used as a component in the chemical synthesis of (a) m-nitroaniline, an intermediate in the production of aniline dyes and (b) m-phenylenediamine, a compound used in the synthesis of aramid fibers and spandex. As with its structural analog 1,3,5-trinitrobenzene (1,3,5-TNB), m-DNB is a manufacturing by-product of the explosive TNT, with the potential for release to the environment in discharged wastewater [2]. Additionally, any 2,4-dinitrotoluene (2,4-DNT) present in the waste stream may be degraded to m-DNB by photolysis under certain pH conditions and organic matter content. Incidental release to the environment of m-DNB might be because of any or all of the manufacturing processes referred to earlier. Soil concentrations of up to 45.2 mg m-DNB/kg soil have been reported at contaminated sites such as AAPs [3]. A list of key physicochemical properties of m-DNB that pertain to the environmental fate and transport of the compound is provided in Table 8.1.

Table 8.1 Summary of Physical-Chemical Properties of 1,3-Dinitrobenzene

CAS No.	99-65-0
Molecular weight	168.11
Color	Yellow-white
Physical state	Crystals/rhombohedral plates
Melting point	89–90 °C
Boiling point	300–303 °C
Odor	No data
Solubility	370–500 mg/L in water at 20–25 °C; Soluble in chloroform, acetone, and ether
Partition coefficients:	
Log K_{ow}	1.49 to 1.62
Log K_{oc}	1.39 to 2.3
Vapor pressure at 25 °C	5.13×10^{-6} mm Hg
Henry's Law constant at 25 °C	2.33×10^{-6} atm.m^3/mole
Conversion factors	1 ppm = 6.88 mg/m^3 1 mg/m^3 = 0.145 ppm

Sources: ATSDR [2], Talmage *et al.* [3], Wentsel *et al.* [5].

The vapor pressure of m-DNB is estimated as 5.13×10^{-6} millimeters of mercury (mm Hg) at 25 °C [3], which is a low value, and implies that partitioning to air is unlikely. Although readily soluble in a variety of organic solvents at ambient temperature, m-DNB is sparingly soluble in water (370–500 milligrams per liter (mg/L)) at 20–25 °C. However, despite its limited solubility, the compound has been identified in both surface water and groundwater. Furthermore, significant concentrations in sediments were identified in streams contaminated with m-DNB. There are no experimental data on the photolysis of m-DNB in aqueous solution but, on theoretical grounds, the process may be expected to occur, given the potential for the compound to absorb light at wavelengths greater than 290 nanometers (nm). Log K_{oc} values have been estimated within the range 1.39–2.3, which indicate a moderate degree of adsorption of m-DNB to suspended sediments, and high to moderate mobility in soil [2].

A variety of microbes subject m-DNB to microbial degradation by using this compound as sources of carbon and nitrogen. This capacity for microbial degradation differs from that of 1,3,5-TNB in which the aromatic six-carbon ring structure is conserved. Various microbial mixed cultures obtained from rivers or sewage sludge effluents break down m-DNB.

Aerobic degradation of m-DNB to carbon dioxide has been demonstrated with the microbial strain *Candida pucherrima* [4] and a variety of microbial genera break down m-DNB [2].

SUMMARY OF MAMMALIAN TOXICITY

Mammalian Oral Toxicity: Acute

The review by Wentsel *et al.* [5] summarizes earlier studies on the acute oral toxicity of m-DNB in experimental animals. The acute oral lethal dose (LD_{50}) of approximately 10 mg of m-DNB/kg was reported in dogs [6]. Cody *et al.* [7] reported an average (male/female) oral LD_{50} for m-DNB of 83 mg/kg for rats. In this study, six male and female Carworth Farm rats were orally administered one of six doses of m-DNB ranging from 36 to 180 mg/kg in corn oil. In another study, the reported average oral LD_{50}s for m-DNB of 59.5 mg/kg in Fischer 344 (F344) rats and 80.4 mg/kg in Swiss mice [8]. Five rats and mice of both sexes were dosed with one of three levels of m-DNB suspended in corn oil. In rats, the doses were 62, 68, and 74 mg/kg while in mice the doses were 50, 70, and 90 mg/kg [8].

Some studies on acute toxicity of m-DNB have focused on endpoints other than lethality. For example, others have demonstrated the formation of methemoglobin in blood samples obtained 5 hours after intraperitoneal injection of male Wistar rats with 100 μmoles/kg of m-DNB in 2 milliliters per kilogram (ml/kg) propylene glycol [9]. This study focused on the methemoglobin-forming potential of 1,3-diamino-2,4,6-trinitrobenzene (DATNB), and the parallel use of other nitrogen-substituted aromatic compounds, such as m-DNB, which served as reference compounds. Of the compounds employed in this survey, m-DNB was one of the most potent inducers of methemoglobin. In another study, the administration of a single oral dose of 50 mg/kg m-DNB by gavage in corn oil to male shrews (*Cryptotis parva*) demonstrated the formation of hemoglobin adducts in blood samples obtained 24 hours after dosing [10]. The potential of nitroaromatic compounds with blood protein and tissue DNA as biological markers of exposure in rat's studies showed that DNB formed stable adducts with blood proteins and tissue DNA after single oral dose of 1.2 mg/kg [11]. *In vitro* incubation of blood with m-DNB also resulted in adduct formation, suggesting a role for cysteine residues in hemoglobin-m-DNB binding. This finding was consistent with an earlier *in vitro* demonstration that m-DNB and other DNB isomers can bind irreversibly to red blood cell macromolecules [12].

A number of single, oral dose studies of m-DNB have provided information regarding the biochemistry and toxicokinetics of neurotoxicological and testicular impairment induced by the compound. Philbert *et al.* [13] used germ-free (GF) and conventional male F344 rats to demonstrate the ability of intestinal microflora to moderate the neuropathological effects of m-DNB. After a single oral dose of 20 or 25 mg/kg m-DNB in polyethylene glycol (PEG), GF male rats (n = 15) displayed a marked ataxic response, in contrast to conventional rats (n = 12) or to GF rats that had been reseeded with a "cocktail" of intestinal microflora.

Conventional rats showed ataxia when dosed with 20 mg/kg m-DNB each day for 5 days [13]. GF rats accumulated approximately 20 and 13 times more radiolabeled m-DNB into the liver and brain, respectively, compared to conventional rats. Light and electron microscopy revealed that histopathological lesions of the brain were limited to the brain stem and inferior colliculus, although animals displaying these features were not necessarily those that had displayed the pronounced clinical signs. The histopathological lesions were similar to those typically seen after thiamine deficiency, a well-documented feature of GF rats. The authors speculated that an observed rise in lactate in the damaged regions of the brains of treated GF rats might reflect interference of oxidative metabolism and pyruvate utilization by m-DNB [13].

Subsequent research by the same group indicated that m-DNB interfered with intracellular redox mechanisms resulting in impaired glucose utilization [14]. The main body of the report contained a detailed histopathological analysis of neurological lesions induced by an experimental protocol similar to that employed in their previous report [13]. Light and electron microscopy revealed the formation of "bilaterally symmetrical vacuolated lesions that involved cerebellar roof, vestibular and superior olivary nuclei and the inferior colliculi." The primary cellular targets were thought to be astrocytes, oligodendrocytes, and vascular elements, with secondary neuronal involvement. However, the precise mechanism of action of the compound has remained obscure [15].

Mammalian Oral Toxicity: Subacute

Subacute studies involve repeated dosing of animals, and parameters are measured at the end of a 14-day study duration. Reddy *et al.* [16] reported a 14-day study on the toxicity of m-DNB in five F344 rats/sex/group that aimed to establish the dosing levels suitable for longer-term studies. The compound was added to the diet at concentrations of 0, 2.5, 10, 25, 75, and 150 mg/kg, yielding respective doses that were calculated as 0, 0.21, 0.8, 1.98, 5.77, and 10.56 mg/kg/day in males and 0, 0.22, 0.87, 2.02, 6.28, and 11.82 mg/kg/day in females [16]. Clinical observations were made twice daily, food and water consumption twice weekly, whole body weights were recorded at the beginning of the study, weekly during the in-life phase of the study, and at termination. A complete profile of hematological and clinical chemistry parameters was assessed in blood samples obtained at necropsy. All tissues and major organs were observed for gross morphological lesions, and the weights of certain organs were recorded. Tissues were sampled, fixed, and processed for histopathology. Slides of sections cut from high-dose and control tissues were examined under a light microscope. In addition, sections of certain potential target organs such as spleen and kidney (male only) were similarly examined for all groups [16].

There were no significant decreases in body weight in any of the treatment groups as compared to control. However, there was a comparative reduction in food consumption in both sexes of animals receiving m-DNB at the highest dose. Some dose-related changes in organ weight/ body weight ratio were seen, including comparative increases in liver, spleen, and kidneys, and reductions in testis. The lowest effective dose for responses such as these was 1.98 mg/kg/day, a level associated with relative increases in kidney weights in male rats, although this particular response may have been incidental to treatment since it did not appear to be part of a dose-response relationship. Though there were no obvious treatment-related changes in clinical chemistry parameters, most hematological indices were reduced compared to controls after 14 days of treatment. For example, a statistically significant reduction in hematocrit and erythrocyte count was seen in female groups at a dose of 2.02 mg/kg/day that appeared to be dose-dependent.

Methemoglobin was significantly increased in females at a dose of 1.98 mg/kg/day and higher, and in males as 0.87 mg/kg/day and higher. Gross pathological signs were seen in the testes of male rats dosed at 5.77 mg/kg/day and higher and in the spleen of females at 2.02 mg/kg/day and males at 1.98 mg/kg/day. There were treatment-related histopathological changes in the bone marrow, spleen, and brain of high-dose rats. In males, histopathological changes were noted in the spleen and kidneys of rats ingesting 1.98 mg m-DNB/kg/day and above and in the testes of rats ingesting 5.77 mg m-DNB/kg/day. These lesions were characterized in the kidneys by the occurrence of tubular degeneration and associated hyaline droplet formation and in the testis by seminiferous tubular degeneration, the appearance of cell debris, and the formation of multinucleated cells.

A number of possible NOAELs and LOAELs were derived from this study. For example, a NOAEL of 0.8 and a LOAEL of 1.98 mg/kg/day would be appropriate, based on the histopathological changes in the kidney in male rats. However, the most sensitive endpoint in the study was the hematological changes and the formation of methemoglobin that were evident in male rats at a dose of 0.8 mg/kg/day that yielded a NOAEL of 0.21 mg/kg/day for a 14 day study [16, Table 8.2].

The subacute toxicity studies were conducted with 1,3-DNB in male and female least shrews (*Cryptotis pavrva*) at equivalent mean daily doses of 0, 0.26, 1.06, 4.26, and 17.0 mg/kg/day for 14 days in the feed. Significant anemia was present in both sexes (hemoglobin and hematocrit at 4.26 mg/kg/day). Significant leukocytosis was observed at the lowest treatment levels of combined sexes at 0.26 and 1.06 mg/kg/day males and granulocytosis and granulocytopenia occurred in female at doses of 0.26 and 17 mg/kg/day group. The LOAEL of 0.26 mg/kg/day was noted in the least shrew for an oral subacute 14-day study [17].

Mammalian Oral Toxicity: Subchronic

Reddy *et al.* [18] conducted a 90-day toxicological study in which 15 F344 rats/sex/group received 0, 1, 6, or 30 mg m-DNB/kg in their diet. These amounts were equivalent to doses of 0, 0.06, 0.35, and 1.73 mg/kg/day in males and 0, 0.07, 0.39, and 1.93 mg/kg/day in females [18]. A full range of in-life, clinical chemistry/hematological (at 45 and 90 days), gross pathological, and histopathological evaluations were performed as described for the 14-day study [16].

Critical findings included an increase in the average relative spleen weight and a reduction in relative testis weights, a reduction in the absolute testis weight and the onset of profound histopathological changes in the spleen, kidney, and testes of high-dose groups. While sporadic changes in clinical chemistry parameters appeared not to be related to treatment, marked dose-dependent changes in hematological parameters were observed, including reductions in hemoglobin, hematocrit, and erythrocyte counts in both sexes of high- and mid-dose rats, increases in platelet counts in high-dose females and in reticulocytes and methemoglobin formation in mid- and high-dose groups. These results indicated a NOAEL of 0.07 mg/kg/day for the hematological responses, with an associated LOAEL of 0.35 mg/kg/day for 90-day study.

Table 8.2 Summary of Relevant Mammalian Data of 1,3-dinitrobenzene for TRV Derivation

		Test Results			
Test Organism	**Test Duration**	**NOAEL (mg/kg/d)**	**LOAEL (mg/kg/d)**	**Effects Observed at the LOAEL**	**Study**
Rat (F344)	5 days	NA	20	Ataxia in 6/6 male rats	[14]
Rat (Sprague Dawley)	12 weeks	0.75	1.5	Reduced spermatid head count	[20]
		NA	0.75	Significant reduction in reproductive performance (pups/litter)	
Rat (F344)	90 days	0.07	0.35	Methemoglobinemia and an increase in reticulocytes	[18]
		0.39	1.73	Reduction in RBCs and in other hematological responses, changes in spleen and testicular histopathology	
Rat (F344)	14 days	0.21	0.8	Methemoglobin formation	[16]
		0.8	1.98	Enlargement of the spleen	
		1.98	5.77	Nephropathy associated with hyaline droplet formation, testicular degeneration	
Rat (Carworth Farms)	8 weeks	NA	4.72	Enlarged spleen, fluctuation in hemoglobin levels, atrophy and histopathological lesions of the testes	[7]
Rat (Carworth Farms)	16 weeks	0.48	1.32	Enlarged spleen	[7]
		1.13	2.64	Depleted spermatogenesis	
Least Shrew	14 days	NA	0.26	Leucocytosis	[17]

The earlier report of Cody *et al.* [7] described subchronic experiments in which m-DNB was administered in drinking water to both sexes of Carworth Farm rats. In one experiment, concentrations of 0, 50, 100, or 200 mg/L were provided to six animals/sex/group for 8 weeks, while in the second group, 0, 3, 8, or 20 mg m-DNB/L were provided to 20/sex/group for 16 weeks [7]. Average dose levels of 0, 4.72, 7.26, and 12.45 mg/kg/day in males and 0, 5.97, 9.0, and 24.43 mg/kg/day in females was calculated for rats in the 8-week study.

In general, dose levels of the 8 week study were associated with profound toxic responses, including four of six fatalities in high-dose (200 mg/L) males, and two of six fatalities in high-dose (200 mg/L) females. In addition, there were reductions in body weight gain compared to the controls were seen, that were probably over and above any anticipated reduction due to curtailed food and water intake. Reduced hemoglobin concentrations, enlarged spleen (except in high dose females), and atrophy of the testes were seen at all m-DNB dose levels.

In the 16-week study, the average doses of 0, 0.4, 1.13, and 2.64 mg/kg/day in males and 0, 0.48, 1.32, and 3.1 mg/kg/day in females was determined for rats, and was based on the data in the report [7]. The compound-related findings were reduced testicular weights with depleted spermatogenesis in the high-dose males and enlarged spleens associated with a number of histopathological manifestations including hemosiderosis in all male dose groups and in the mid- and high-dose female groups. These findings suggested that a NOAEL of 0.48 mg/kg/day (with a related LOAEL of 1.32 mg/kg/day) would be appropriate for the spleen effects, which were the most sensitive responses to m-DNB. A NOAEL of 0.4 mg/kg/day from this study enabled IRIS compilers to derive a human health reference dose (RfD) of 1×10^{-4} mg/kg/day for m-DNB [19].

Testicular impairment and indicators of reproductive success was the subject of a subchronic study on the toxicity of m-DNB in Sprague-Dawley rats [20]. Twelve male Sprague Dawley rats/group were gavaged for 5 days/week for 12 weeks with 0, 0.75, 1.5, 3.0, or 6.0 mg m-DNB/kg/day in a corn oil/acetone mixture. To assess functional male reproduction, each male was mated with two virgin females after 10 weeks, following which pregnant females were terminated on gestation day 21 and evaluated for reproductive parameters. All males were sacrificed after 12 weeks and a battery of sperm parameters were evaluated according to dose, including spermatid count, cauda reserves, percentage motility, morphology, histopathology, and fertility. There was severe toxicity in those animals receiving 6 mg/kg/day m-DNB, as evidenced by ataxia, loss of balance, muscular rigidity, and a lower body weight. These effects became especially marked during the first week of breeding, to such an extent that breeding and m-DNB treatment were discontinued in this group after 4 days. However, there were no clinical signs of toxicity in rats receiving 3 mg/kg/day m-DNB or less. In male rats sacrificed after 12 weeks, there were decreased testicular and epididymal weights at 3.0 mg/kg/day and increased spleen weights in animals that received 1.5 and 3.0 mg/kg/day. The most sensitive parameter of testicular toxicity was the spermatid count, which showed a significant difference (72%) to controls at doses of 1.5 mg/kg/day and above. This observation indicated a NOAEL of 0.75 mg/kg/day and a LOAEL of 1.5 mg/kg/day. In addition to testicular toxicity, male reproductive success was deleteriously impacted by m-DNB. No litters were produced in females that were mated with males receiving 3.0 or 6.0 mg/kg/day, and significantly, fewer pups compared to controls were produced because of matings of females with males receiving 0.75 and 1.5 mg/kg/day. In this study, male rats dosed at 0.75 mg/kg showed disrupted reproductive function (reduction in pups per litter), which suggested a LOAEL of0.75 mg/kg/day [20].

Mammalian Oral Toxicity: Chronic

No studies have explored the toxicity of m-DNB through chronic exposure.

Mammalian Oral Toxicity: Other

A large number of experimental studies have demonstrated that m-DNB is a potent testicular toxicant. A single dose of m-DNB to male laboratory rodents has served as a model for the investigation of testicular impairment. Linder *et al.* [21] administered a single oral dose of 48 mg m-DNB/kg in acetone/corn oil (2% v/v) to eight male Sprague Dawley rats/group and demonstrated a marked seminiferous tubular degeneration, with reduced testicular and epididymal weights, reduced numbers of sperm heads, degeneration of sperm tails, and the appearance of dead and "decapitated" cells. Subsets of treated animals were terminated after different time intervals up to 175 days, throughout which treated males were mated intermittently with untreated females. In line with the reduction in numbers and integrity of the sperm, some of the matings produced no fertilized eggs; for example, at 34–38 days post-treatment. However, some but not all treated males recovered their fertility in subsequent matings. A companion paper examined the detailed histopathological consequences of a single oral dose of

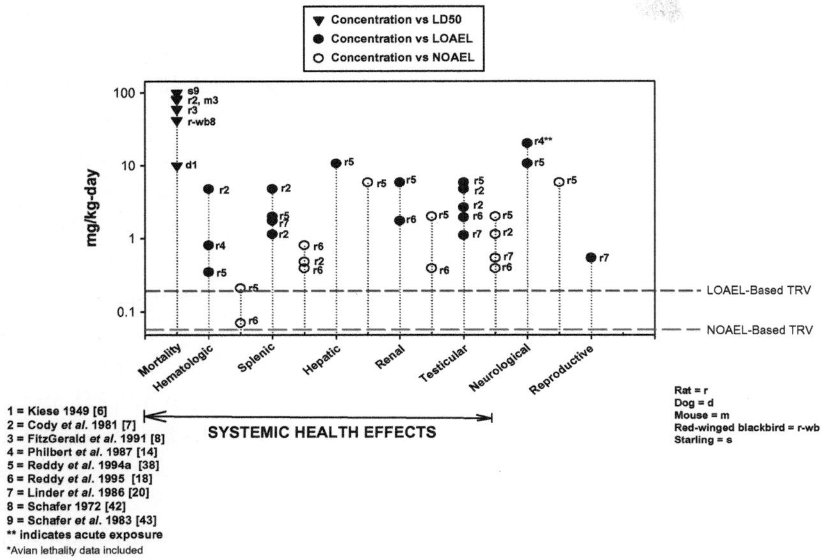

Figure 8.1 1,3-DNB health effects to mammals*. Letters by symbols represent test species, and numbers represent study.

48 mg of m-DNB/kg as previously used [21]. This companion study showed that the testes were severely damaged at 24 hours posttreatment, with the appearance of "increased numbers of regressive seminiferous tubules that exhibited degenerating pachytene spermatocytes, chromatin margination in spermatids, deformed spermatic heads, retained spermatids and reduced numbers of meiotic figures" [22]. The ability of spermatogenesis to recover from this chemical insult was variable, since only three of seven treated males were morphologically and functionally normal at 175 days posttreatment.

Others compared the ability of 1,2-, 1,3-, and 1,4-DNB to induce testicular lesions in four male Wistar rats/group that received a single dose of 50 mg/kg by gavage in PEG. Histopathological lesions of the testis were noted in those animals receiving m-DNB but not 1,2- or 1,4-DNB. This suggests a stereospecificity of the mechanism by which m-DNB induces testicular atrophy and sperm deficits, a feature not shared with the mechanism of methemoglobin induction in which 1,4-DNB was just as effective as m-DNB. In the dose response phase of a study, others found that single doses of 5 and 10 mg m-DNB/kg was ineffective in inducing testicular lesions in Wistar rats [23]. However, testicular lesions that were limited to Stages VIII through XI of the spermatogenic cycle were evident 12 hours after a single oral dose of 25 mg m-DNB/kg and similarly, 48 hours after a single oral dose of 15 mg of 1,3,-DNB/kg. The authors considered the ultrastructural evidence to implicate the Sertoli cells as the prime targets for the toxic action of m-DNB with germ cell damage a secondary event. Others measured a range of testicular and serum hormone concentrations in male Sprague Dawley rats that received a single gavage dose of 32 mg m-DNB/kg in corn oil, and found little if any evidence to support the notion that m-DNB-induced testicular lesions might be secondary to changes in testicular, brain, or pituitary hormone levels [24].

The testicular effects of m-DNB were also apparent in male adult but not prepubertal B6C3F1 mice that were exposed to the compound [25]. In that study, mice were administered a single gavage dose of 0, 8, 16, 32, 40, or 48 mg m-DNB/kg in corn oil/acetone, following which testicular responses were measured at various intervals for the next 25 days. The 48 mg/kg dose was a critical dose for the onset of testicular effects in adult B6C3F1 mice, which resulted in abnormal spermatogenesis, reduced testicular weights, and altered germ-cell type ratios, abnormal chromatin structures, and an increase in abnormal sperm head morphology [25]. Additionally, 16 male Sprague Dawley rats/group were exposed to single doses of 0, 8, 16, 32, or 48 mg m-DNB/kg and the compound's effects on spermatogenesis was determined by flow cytometry, 1, 4, 16, or 32 days postdosing [26]. An increase in the incidence of unusual cell types was observed on day 1 after 48 mg/kg and day 4 after 16 mg/kg or greater. The presence of haploid, diploid, and tetraploid cells signaled a dose-dependent increase in germinal cell types because of treatment, elevated numbers of which persisted to posttreatment day 32 and beyond [26]. This was often accompanied by aberrant sperm effects. Using Sprague Dawley rats, others observed a single-dose NOAEL of 8 mg/kg and a LOAEL of 16 mg/kg for changes seen in epididymis weight, sperm head counts, caudal sperm reserves, and sperm morphology [27].

In vitro fertilization assisted an investigation of the influence of m-DNB on Sertoli cells and thus the functional capacity of developing germ cells [28]. Male Wistar (AP/ALPK) rats were given a single oral dose of 0, 5, 15, or 25 mg m-DNB/kg in PEG, and sperm were collected from the cauda epididymis at various time points thereafter. The sperm was used in an *in vitro* fertilization technique through which reduced sperm fertilizing capacity was observed from 1.5 to 5 weeks and from 7.5 to 8.5 weeks after treatment with

15 and 25 mg m-DNB/kg and for 3, 5.5, 7.5, and 8.5 weeks after treatment with 5 mg m-DNB/kg [28]. Observations indicated that m-DNB did not affect all Sertoli cells equally but acted in a stage-specific manner, with Stages III, IX, XII, and XIV appearing to be especially vulnerable to m-DNB [28]. This stage-specific mechanism of m-DNB on the Sertoli cells was reported [29,30] in which concordant testicular damage was seen in Sprague Dawley rats irrespective of oral or intraperitoneal routes of compound administration. Given the different concentrations of parent compound and metabolites that resulted from intraperitoneal versus oral dosing, it was speculated that only the duration of testicular exposure to the toxicant might govern susceptibility to toxicity once a threshold blood level of m-DNB was reached [30].

Many studies have searched for compounds that might serve as plasma biomarkers of the incipient testicular damage induced by chemical toxicants like m-DNB. Of a number of hormones and enzymes evaluated, the lactate dehydrogenase-C_4 (LDH-C_4) isoenzyme and the androgen binding protein (ABP) were viable candidates when single doses of 0, 10, 20, and 30 mg of m-DNB/kg were administered to male Wistar (Alpk:APfSD) outbred rats and 25 mg m-DNB/kg was administered in a time course experimental protocol [31]. Using the same experimental animal, others have correlated histopathological lesions and ABP production to implicate further Sertoli cell dysfunction as the etiologically significant event in germ cell depletion [32].

Previous reports employed morphometry to visualize the toxicological effects of m-DNB [33,34]. A single oral dose of 0, 15, or 25 mg m-DNB/kg was administered to male Sprague Dawley rats, which were then sacrificed after 22 to 24 days [33]. Collected sperm were measured by light microscopy for total width, interior width, and symmetry using automated sperm morphometry analysis. The studies revealed three subpopulations of m-DNB-affected sperm, two of which were abnormal to varying degrees [33]. Ninety-three percent of sperm that were harvested from the untreated controls had a normal appearance, with dose-dependently lower proportions in m-DNB-treated groups (78% and 66% for the low- and high-dose groups, respectively). Male Sprague Dawley rats were gavaged with a single dose of 0 or 25 mg m-DNB/kg, with groups of four rats being sacrificed 1, 2, 4, or 7 days after administration [34]. Sertoli cells were examined by light microscopy, and the toxicity of m-DNB was evaluated by counting the proportion of germ cells that were lacking abnormalities in each treatment group, and according to four user-defined clusters of stages of the spermatogenic process [34]. A large number of vacuoles were seen in the Sertoli cell cytoplasm one day after m-DNB administration [34], the presence of which was thought to be instrumental in the apoptotic cell death of both pachytene and diplotene spermatocytes, round spermatids, and other germ cells associated with Sertoli cells [34].

STUDIES RELEVANT FOR MAMMALIAN TOXICITY REFERENCE VALUE DEVELOPMENT FOR INGESTION EXPOSURES

Acute, subacute, and subchronic studies have delineated a distinct and well-defined range of toxicological effects of m-DNB. However, the range of animal models employed in these studies, which used predominantly F344 and Sprague Dawley rats, may lend a degree of uncertainty to quantitative estimates of dose levels that are potentially protective of mammalian wildlife. In general, the use of a narrow range of experimental animals reduces confidence in whether the resulting NOAELs and LOAELs have necessarily captured the subthreshold for mammalian wildlife.

IRIS compilers chose splenic enlargement and hemosiderin deposit formation as principal effects of m-DNB for RfD derivation [19]. However, a clear spectrum of other toxicological responses to m-DNB has emerged, including (a) hematological effects (e.g., methemoglobinemia), (b) nephropathy associated with cytoplasmic and/or hyaline droplet formation, (c) structural and functional impairment of brain deficits, and (d) atrophy of the testis with associated degeneration of the seminiferous tubules and sperm.

Methemoglobinemia is reversible and creates a functional hypoxic blood condition. Carbon monoxide poisoning also creates a functionally similar condition. Humans with levels of carboxyhemoglobin (COHb) above 10% have reported symptoms of headache, yet the preponderance of adverse effects occurs when COHb concentrations exceed 2% [35]. Most humans can tolerate under 10% of methemoglobin without any toxic symptoms [36]. Moreover, differences between control and high dose rats of approximately 3% or less following m-DNB exposure were previously reported [18]. Given the uncertainty associated with the reported methemoglobin increase in these studies, increased methemoglobinemia due to m-DNB was likely biologically insignificant [18].

In addition, m-DNB increased splenic erythroid cell hyperplasia and pigmentation [37]. This study reported decreased hematocrit, red blood cell count, and hemoglobin, but an increase in the mean cell volume primarily in the high-dose groups. However, these values were within ranges considered normal for rats of that age group [37]. The combined effects of m-DNB on spleen histopathology and hematological parameters are indicative of anemia, although the biological significance of these effects in these models is questionable.

It is uncertain whether the nephropathic changes and cytoplasmic droplet formation in the kidney following m-DNB exposure represent the same phenomenon as the well-described $\alpha 2_{\beta}$-globulin-mediated hyaline droplet formation that is typical of male F344 rats and those of other strains. For example, in a 90-day study of m-DNB, it was shown that 9 out of 10 high-dose females, but only 1 out of 10 high-dose males displayed cytoplasmic kidney droplets on histopathological examination [18], a finding that contrasts with the current understanding of the $\alpha 2_{\beta}$-globulin-mediated hyaline droplet phenomenon as a characteristic of male rats. For m-DNB, this weakens the analogy with the nephropathy displayed by 1,3,5-TNB in the subacute [38] and subchronic studies [39], which appears to have the "classical" $\alpha 2_{\beta}$-globulin-mediated hyaline droplet etiology. However, the cytoplasmic droplets identified in the kidney of F344 males in a 14-day study of m-DNB toxicity were clearly identified as hyaline [16]. Contrasting these observations with those of a 2-year study of 1,3,5-TNB-induced nephropathy in F344 rats [40] and those of m-DNB-induced nephropathy in a 90-day study [18] (which in each case showed lesions that apparently did not resemble the typical $\alpha 2_{\beta}$-globulin-mediated hyaline droplet histopathology and gender-specific incidence) indicated the possibility of age-related morphological and chemical differences existing between kidney droplets formed in response to shorter periods of nitroaromatic dosing as compared to those manifesting after a longer period of dosing. This response is common following nitroaromatic exposure, and yet the impact of increased incidence of hyaline droplets in the kidney remains unknown. However, in human renal disease patients, hyaline droplets were seen in the cytoplasm of necrotic proximal tubular epithelial cells, and in the absence of necrosis, proximal tubular cells containing hyaline droplets showed microvilli that were degenerating [41]. Hence, given the uncertainty in the etiology related to hyaline droplet formation and the lack of a known biological or ecological significance, this endpoint was not used for derivation of the TRV.

m-DNB appears to be a neurological toxicant, with pronounced histopathological lesions induced in various regions of the brain because of acute dosing. While speculation on the mechanism of these effects has centered on the possibility of impairment of oxidative metabolism and a perturbation of intracellular redox mechanisms and/or changes in the concentrations and activities of various neurotransmitters and their metabolites, evidence is lacking to indicate precisely how the observed effects are brought about [15]. The most pertinent study on the neurological effects of m-DNB in rats showed signs of ataxia although only six conventional (non-GF) rats were dosed with one concentration of m-DNB [14]. The limited data and study design precludes an adequate assessment of the neurological effects of m-DNB, although it is reasonable to infer that ataxia will likely result in decreased survival, the lack of a dose-response regime and endpoints consistent with prior studies suggests the questionable nature of these data for TRV derivation. Given these uncertainties associated with the whole organism effects of the histopathological lesions and the limited availability of data, these studies alone cannot be used for derivation of the TRV.

Impairment of the male reproductive organs with associated decreases in sperm production, motility, and reproductive performance is a consistent response of experimental animals to m-DNB [20,21,23,25,26,28]. As described earlier, a single dose of m-DNB of at least 5 mg/kg/day is sufficient to induce testicular effects and sperm deficits in various strains of experimental animal [28]. In the most pertinent study [20], the NOAEL was 0.75 mg/kg for rats gavaged to m-DNB for 12 weeks. From an ecological perspective, the most significant effect was decreased reproduction in m-DNB dosed males. In fact, no litters were produced from nondosed females mated with males previously dosed with 3 and 6 mg/kg/day [20]. Additionally, the number of pups per litter was significantly less from females mated with males dosed with 0.75 and 1.5 mg/kg/day as compared to females mated with control males [20]. Although the effects of m-DNB on reproduction in males appear to be strong, limited data suggests that this response is somewhat reversible on complete cessation of exposure [20]. Nevertheless, given the severity of reproductive effects of m-DNB, decreased reproduction in males was chosen as the endpoint for derivation of the TRV.

Mammalian Inhalation Toxicity

No inhalation studies conducted using mammals were found.

Mammalian Dermal Toxicity

No dermal studies conducted using mammals were found.

SUMMARY OF AVIAN TOXICITY

Avian Oral Toxicity: Acute

Researchers with the U.S. Fish and Wildlife Service [42,43] reported LD_{50} values for this compound in Red-winged Blackbirds (*Agelaius phoeniceus*) and European Starlings (*Sturnus vulgaris*) of 42 and >100 mg/kg, respectively. No other information was presented.

Avian Oral Toxicity: Subchronic

No data are available.

Avian Oral Toxicity: Chronic

No data are available.

Avian Oral Toxicity: Other

No data are available.

Avian Inhalation Toxicity

No data are available.

Avian Dermal Toxicity

No data are available.

SUMMARY OF AMPHIBIAN TOXICITY

Toxicological data for the effects of m-DNB in amphibian species were not located. Ecotoxicological research on the effects of this compound in amphibians is recommended.

SUMMARY OF REPTILIAN TOXICITY

Toxicological data for the effects of m-DNB in reptilian species were not located. Ecotoxicological research on the effects of this compound in reptiles is recommended.

RECOMMENDED TOXICITY REFERENCE VALUES

Toxicity Reference Values for Mammals

TRVs for Ingestion Exposures for the Class Mammalia

The parameters relevant to m-DNB toxicity involved changes in testicular histopathology and reproductive performance in male rats (Table 8.2). The data relevant for TRV derivation for m-DNB is limited to subchronic oral exposure in rats. No chronic data are available and no repeated dosing studies were conducted on mammals other than rats. Values used to derive the TRV were obtained from 12-week gavage studies in male rats [20].

As outlined in USACHPPM TG254, toxicity values used for derivation of a TRV should be based on potentially ecologically relevant effects. The effects of m-DNB on male reproduction was chosen as the parameter for TRV derivation based on decreased litter production from females mated with males exposed to m-DNB via oral gavage, with a LOAEL of 0.75 mg/kg/day for reproduction [20]. In addition, the level at which reproductive effects occur is protective of the neurological effects, as previously noted [13].

Data on the toxicity of m-DNB is limited to two mammalian species, rats and mice, and no chronic toxicity studies were conducted. Although a study on m-DNB toxicity in shrews was available, it was not useful for TRV development because it provided information only on hemoglobin adduct formation, no controls were used, and subjects were euthanized 24 hours after exposure. Given these data limitations, the approximation approach was the only viable means for deriving the TRV for m-DNB [1]. An uncertainty factor of 20 was used to derive the NOAEL-based approximate TRV from a subchronic LOAEL of 0.75 mg/kg/day for reproduction [20]. An uncertainty factor of four was used to derive the LOAEL-based approximate TRV from the same subchronic LOAEL. Final values for the TRV were rounded. As stated, there are limited data for multiple species and no chronic toxicity data for DNB. However, the reported studies are of relatively high quality with a broad scope of observations that are consistent with each other and of other nitroaromatic compounds. It is for these reasons that the TRV

Table 8.3 Selected Ingestion m-Nitrobenzene TRVs for the Class Mammalia

TRV	Dose	Confidence
NOAEL-based	0.04 mg/kg/d	Medium
LOAEL-based	0.2 mg/kg/d	Medium

Source: [20].

for m-DNB presented below was given a medium confidence value (Table 8.3).

TRVs for Inhalation Exposures for the Class Mammalia

Not available at this time.

TRVs for Dermal Exposures for the Class Mammalia

Not available at this time.

TRVs for Birds

Only acute LD50 data are available for two species of birds, Red-winged Blackbirds, and European Starlings. This information was originally presented [42] and no other supporting information was presented. Although the approximate method could be used to derive a TRV from these data, in the absence of details from the original reports, it is recommended that a TRV not be derived for birds until more information is available.

TRVs for Amphibians

Not available at this time.

TRVs for Reptiles

Not available at this time.

IMPORTANT RESEARCH NEEDS

The limited availability of data on the toxicity of m-DNB to wildlife species precludes the development of a high-confidence TRV. Hence, more studies on the toxicity of m-DNB to wildlife species are required. In particular, chronic toxicity studies on mammals and additional studies on nonmammalian wildlife such as birds, reptiles, and amphibians are particularly warranted. More information regarding the toxicity of m-DNB to wildlife would likely allow the derivation of a high confidence TRV.

REFERENCES

[1] USACHPPM. Standard practice for wildlife toxicity reference values. Aberdeen Proving Ground (MD): U.S. Army Center for Health Promotion and Preventive Medicine; 2000, Technical Guide No. 254.

[2] Agency for Toxic Substances and Disease Registry (ATSDR). Toxicological profile for 1,3-dinitrobenzene/1,3,5-trinitrobenzene. Atlanta (GA): U.S. Department of Health and Human Services, Public Health Service; 1995.
[3] Talmage SS, Opresko DM, Maxwell CJ, Welsh CJ, Certella FM, Reno PH, et al. Nitroaromatic munition compounds: environmental effects and screening values. Rev Environ Contam Toxicol 1999;161:1–156.
[4] Dey S, Godbole SH. Biotransformation of m-dinitrobenzene by *Candida pulcherima.* Indian J Exp Biol 1986;24:29–33.
[5] Wentsel RS, Hyde RG, Jones III WE, Wilkinson MJ, Harward III WE. Problem definition study on 1,3-dinitrobenzene, 1,3,5-trinitrobenzene and di-n-propyl adipate. Frederick (MD): U.S. Army Medical Research and Development Command, Fort Detrick; 1979, Government Accession No. AD A099732.
[6] Kiese M. Pharmacological investigation of m-dinitrobenzene. Arch Exp Pathol Pharmacol 1949;206:361–83.
[7] Cody TE, Witherup S, Hastings L, Stemmer K, Christian RT. 1,3-dinitrobenzene: toxic effects in vivo and in vitro. J Toxicol Environ Health 1981;7(5):829–47.
[8] FitzGerald GB, Austin A, DiGuilio N. Acute toxicity evaluation of nitroaromatic compounds. Frederick (MD): U.S. Army Medical Research and Development Command, Fort Detrick; 1991, Government Accession No. AD A236352.
[9] Watanabe T, Ishihara N, Ikeda M. Toxicity of and biological monitoring for 1,3-diamino-2,4,6-trinitrobenzene and other nitro-amino derivatives of benzene and chlorobenzene. Int Arch Occup Environ Health 1976;37(3):157–68.
[10] Myers SR, Pinorini-Godly MT, Reddy TV, Daniel FB, Reddy G. Gas chromatographic and mass spectrometric determination of hemoglobin adducts of 1,3-dinitrobenzene and 1,3,5-trinitrobenzene in shrew, *Cryptotis parva.* Int J Toxicol 1999;18(5):317–25.
[11] Reddy TV, Daniel FB, Wang H, Wiechman B, Reddy G. Adduction of nitroaromatic compounds with blood proteins and tissue DNA as biological markers of exposures. Cincinnati (OH): USEPA Environmental Exposure Research Laboratory; 1997 Jun, Stock No. AD-A329 242/2. Prepared for the U.S. Army Medical Research and Development Command, Fort Detrick, MD.
[12] Cossum PA, Rickert DE. Metabolism and toxicity of dinitrobenzene isomers in erythrocytes from Fischer-344 rats, rhesus monkeys, and humans. Toxicol Lett 1987; 37(2):157–63.
[13] Philbert MA, Gray AJ, Connor TA. Preliminary investigations into the involvement of the intestinal microflora in CNS toxicity induced by 1,3-dintrobenzaene in male F-344 rats. Toxicol Lett 1987;38(3):307–14.
[14] Philbert MA, Nolan CC, Cremer JE, Tucker D, Brown AW. 1,3-dintrobenzene-induced encephalopathy in rats. Neuropathol Appl Neurobiol 1987;13:371–89.
[15] Philbert MA, Billingsley ML, Reuhl KR. Mechanisms of injury in the central nervous system. Toxicol Pathol 2000;28:43–53.
[16] Reddy TV, Daniel FB, Robinson M, Olson GR, Wiechman B, Reddy G. Subchronic toxicity studies on 1,3,5-trinitrobenzene, 1,3-dinitrobenzene and tetryl in rats. 14-d toxicity evaluation of 1,3-dinitrobenzene in Fischer 344 rats. Cincinnati (OH): USEPA Environmental Monitoring Systems Laboratory; 1944, Stock No. AD-A290641. Prepared for the U.S. Army Medical Research and Development Command, Fort Detrick, MD.
[17] Mock OB, Casteel SW, Darmani NA, Shaddy JH, Besch-Williford C, Towns LC. 1,3-dinitrobenzene toxicity in the least shrew, *Cryptotis parva.* Environ Toxicol Chemist 2005;24(10):2519–25.
[18] Reddy TV, Daniel FB, Robinson M, Olson GR, Wiechman B, Reddy G. Subchronic toxicity studies on 1,3,5-trinitrobenzene, 1,3-dinitrobenzene and tetryl in rats: 90-day

evaluation of 1,3-dinitrobenzene in Fischer 344 rats. Cincinnati (OH): U.S. EPA Environmental Monitoring Systems Laboratory; 1995, Government Accession No. AD A297458. Prepared for the U.S. Army Medical Research and Development Command, Fort Detrick, MD.

[19] U.S. EPA. Integrated risk information system. Online, Cincinnati, OH: Office of Health and Environmental Assessment, National Center for Environmental Assessment; 2000 http://www.epa.gov/iris/subst/0318.htm.

[20] Linder RE, Hess RA, Strader LF. Testicular toxicity and infertility in male rats with 1,3-dinitrobenzene. J Toxicol Environ Health 1986;19(4):477–89.

[21] Linder RE, Hess RA, Perreault SD, Strader LF, Barbee RR. Acute effects and long-term sequelae of 1,3-dinitrobenzene on male reproduction in the rats. I. Sperm quality, quantity, and fertilizing ability. J Androl 1988;9(5):317–26.

[22] Hess RA, Linder RE, Strader LF, Perreault SD. Acute effects and long-term sequelae of 1,3-dinitrobenzene on male reproduction in the rat. II. Quantitative and qualitative histopathology of the testis. J Androl 1988;9(5):327–42.

[23] Blackburn DM, Gray AJ, Lloyd SC, Sheard CM, Foster PM. A comparison of the effects of the three isomers of dinitrobenzene on the testis in the rat. Toxicol Appl Pharmacol 1988;92(1):54–64.

[24] Rehnberg GL, Linder RE, Goldman JM, Hein JF, McElroy WK, Cooper RL. Changes in testicular and serum hormone concentrations in the male rats following treatment with m-dinitrobenzene. Toxicol Appl Pharmacol 1988;95(2):255–64.

[25] Evenson DP, Janca FC, Baer RK, Jost LK, Karabinus DS. Effect of 1,3-dinitrobenzene on prepubertal, pubertal, and adult mouse spermatogenesis. J Toxicol Environ Health 1989;28(1):67–80.

[26] Evenson DP, Janca F, Jost LK, Baer RK, Karabinus DS. Flow cytometric analysis of effects of 1,3-dinitrobenzene on rat spermatogenesis. J Toxicol Environ Health 1989;28(1):81–98.

[27] Linder RE, Strader LF, Barbee RR, Rehnberg GL, Perreault SD. Reproductive toxicity of a single dose of 1,3-dinitrobenzene in two ages of young adult male rats. Fund Appl Toxicol 1990;14(2):284–98.

[28] Holloway AJ, Moore HD, Foster PM. The use of in vitro fertilization to detect reductions in the fertility of male rats exposed to 1,3-dinitrobenzene. Fund Appl Toxicol 1990;14(1):113–22.

[29] Nolte T, Harleman JH, Jahn W. Histopathology of chemically induced testicular atrophy in rats. Exp Toxicol Pathol 1995;47(4):267–86.

[30] McEuen SF, Jacobson CF, Brown CD, Miller MG. Metabolism and testicular toxicity of 1,3-dinitrobenzene in the rat: effect of route of administration. Fund Appl Toxicol 1995;28(1):94–9.

[31] Reader SC, Shingles C, Stonard MD. Acute testicular toxicity of 1,3-dinitrobenzene and ethylene glycol monomethyl ether in the rat: Evaluation of biochemical effect markers and hormonal responses. Fund Appl Toxicol 1991;16(1):61–70.

[32] Suter L, Clemann N, Koch E, Bobadilla M, Bechter R. New and traditional approaches for the assessment of testicular toxicity. Reprod Toxicol 1998;12(1):39–47.

[33] Davis RO, Gravance CG, Teal DM, Miller MG. Automated analysis of toxicant-induced changes in rat sperm head morphometry. Reprod Toxicol 1994; 8(6):521–9.

[34] Matsui H, Takahashi MA. Novel quantitative morphometry of germ cells for the histopathological evaluation of rat testicular toxicity. J Toxicol Sci 1999;24 (1):17–25.

[35] American Conference of Governmental Industrial Hygienists, Inc. (ACGIH). Documentation of the threshold limit values and biological exposure indices. 6th ed. Cincinnati (OH).

[36] Bloom JC, Brandt JT. Toxic responses of the blood. In: Klaassen CD, editor. Casarett and Doull's toxicology. The basic science of poisons. 7th ed. New York (NY): McGraw-Hill; 2008.
[37] Wolford ST, Schroer RA, Gohs FX, Gallo PP, Brodeck M, Falk HB, et al. Reference range database for serum chemistry and hematology values in laboratory animals. J Toxicol Env Health 1986;18(2):161–88.
[38] Reddy TV, Daniel FB, Robinson M, Olson GR, Wiechman B, Reddy G. Subchronic toxicity studies on 1,3,5-trinitrobenzene, 1,3-dinitrobenzene and tetryl in rats: 14-day toxicity evaluation of 1,3,5-trinitrobenzene in Fischer 344 rats. Cincinnati (OH): USEPA Environmental Monitoring Systems Laboratory; 1994, Government Accession No. AD A283664. Prepared for the U.S. Army Medical Research and Development Command, Fort Detrick, MD.
[39] Reddy TV, Daniel FB, Robinson M, Olson GR, Wiechman B, Reddy G. Subchronic toxicity studies on 1,3,5-trinitrobenzene, 1,3-dinitrobenzene and tetryl in rats: Subchronic toxicity evaluation of 1,3,5-trinitrobenzene in Fischer 344 rats. Cincinnati (OH): USEPA Environmental Monitoring Systems Laboratory; 1994, Government Accession No. AD A283663. Prepared for the U.S. Army Medical Research and Development Command, Fort Detrick, MD.
[40] Reddy TV, Daniel FB, Olson GR, Wiechman B, Torsella J, Reddy G. Chronic toxicity studies on 1,3,5-trinitrobenzene in Fischer 344 rats. Cincinnati (OH): USEPA Environmental Monitoring Systems Laboratory; 1996, Government Accession No. AD A315216. Prepared for the U.S. Army Medical Research and Development Command, Fort Detrick, MD.
[41] Sato S, Kitamura H, Ghazizadeh M, Adachi A, Sasaki Y, Ishizaki M, et al. Occurrence of hyaline droplets in renal biopsy specimens: an ultrastructural study. Med Mol Morphol 2005;38(1):63–71.
[42] Schafer EW. The acute oral toxicity of 369 pesticidal, pharmaceutical, and other chemicals to wild birds. Toxicol Appl Pharmacol 1972;21(3):315–30.
[43] Schafer Jr EW, Bowles Jr WA, Hurlbut J. The acute oral toxicity, repellency, and hazard potential of 998 chemicals to one or more species of wild and domestic birds. Arch Environ Contam Toxicol 1983;12(3):355–82.

CHAPTER 9

Wildlife Toxicity Assessment for 1,3,5-Trinitrobenzene (TNB)

Gunda Reddy, Marc A. Williams

Contents

INTRODUCTION

1,3,5-Trinitrobenzene (TNB) is one of several compounds that were released to the environment during explosives manufacture and in load, assembly, and pack (LAP) activities at U.S. Army ammunition plants (AAPs) and other military installations. This nitroaromatic compound has a close structural relationship with the most widely produced military explosive, trinitrotoluene (TNT), of which it is a manufacturing by-product and an environmental degradation product. The importance of TNB as an environmental contaminant relates to its widespread distribution at and around military sites and to its potential toxicity to wildlife and other ecological receptors.

This chapter summarizes current knowledge of the likely harmful impacts of TNB on wildlife, with emphasis on identifying levels at which wildlife species may be adversely effected. Evaluating the compound toxicity intends to contribute to the establishment of Toxicity Reference Values

Wildlife Toxicity Assessments for Chemicals of Military Concern
http://dx.doi.org/10.1016/B978-0-12-800020-5.00009-0

(TRVs) that could serve as protective exposure standards for wildlife ranging near TNB contaminated sites. The assessment performance protocol is documented in the U.S. Army Center for Health Promotion and Preventive Medicine (USACHPPM) Technical Guide 254, *Standard Practice for Wildlife Toxicity Reference Values* [1].

TOXICITY PROFILE

Environmental Fate and Transport

The synthetic compound TNB is a high explosive used for commercial mining and military use, as a narrow-range pH indicator, and as an agent to vulcanize natural rubber [2]. The compound is a manufacturing by-product of the explosive, TNT, and released to the environment in discharged wastewater. Additionally, any TNT present in the waste stream may be degraded to TNB by photolysis under certain conditions of pH and organic matter content [3]. TNB has been released to the environment during Load, Assembly, and Packing (LAP) operations performed at Army Ammunition Plants and from open-pit incineration of waste explosives [4,5]. Military firing ranges may be contaminated by nitroaromatic compounds from ruptured or unexploded ordnance. Soil concentrations of up to 67,000 mg/kg have been detected on military reservations [6].

TNB has an estimated vapor pressure of 3.2×10^{-6} mm Hg at 25 °C [7], a low value implying that partitioning to air is unlikely. TNB is readily soluble in a variety of organic solvents at ambient temperature and has a water solubility of 340 – 385 mg/L at 20 – 25 °C. The compound has been identified in both surface water and groundwater. Furthermore, TNB was previously shown to be present in stream sediments wherein TNB soil concentration data for certain AAPs, depots and arsenals was reported [6,3].

As noted above, TNT will undergo photolysis to produce TNB in aqueous solution, which is resistant to further photolytic degradation. Nitroaromatic compounds, in general, resist hydrolysis under environmental conditions, with TNB conforming to this pattern. The various values derived for a K_{oc} (organic carbon/water partition coefficients) for TNB are estimates, since experimentally determined values are unavailable (Table 9.1). The reported K_{oc} values within the range of 76 – 520 mL/g indicate a moderate degree of adsorption of TNB to suspended sediments, and high to moderate soil mobility [3,9].

TNB is subject to microbial degradation, but the process appears to be limited to the compound's nitrogenous moieties. Thus, aerobic metabolism

Table 9.1 Summary of the Physical-Chemical Properties of 1,3,5-Trinitrobenzene

CAS No.	99-35-4
Molecular weight	213.11
Color	Yellow-white
Physical state	Orthorhombic crystals/rhombic plates
Melting point	122.5 – 125.5 °C
Boiling point	315 °C
Odor	No data
Solubility in water	340 – 385 mg/L at 20-25 °C: soluble in benzene, methanol, ethanol, ether, and carbon disulfide
Partition coefficients:	
Log K_{ow}	1.18
Log K_{oc}	1.88
Vapor pressure at 25 °C	3.2×10^{-6} mm Hg
Henry's Law constant at 25 °C	3.08×10^{-9} atm.m^3/mole
Conversion factors	1 ppm = 8.7 mg/m^3 1 mg/m^3 = 0.115 ppm

Sources: HSDB [2], Talmage *et al.* [3], ATSDR [7], Burrows *et al.* [8].

of TNB by *Pseudomonas* sp. produces 1,5-dinitroaniline, dinitrobenzene, 5-nitrobenzene, and ammonia. TNB is probably not used as a carbon source by this organism, since no further breakdown was observed [3].

Summary of Mammalian Toxicity

Mammalian Oral Toxicity: Acute

The review by Wentsel *et al.* [5] cited the Korolev *et al.* [10] study on the acute oral toxicity of TNB in which respective LD_{50} values of 600, 450, and 730 mg/kg were reported for white mice, white rats, and guinea pigs (strains unknown). Earlier, Fogelman *et al.* [11] determined an LD_{50} of 505 mg/kg for male albino rats (strain unknown) receiving a single gavage dose of 5% weight per volume "TNB" in 0.5% methylcellulose. There were clear-cut indications of cyanosis in addition to clinical and neurological effects. Necropsy of study animals that died during the 7-day observation period revealed hemorrhagic lungs, stained kidneys, and a darkening of the blood that was likely indicative of methemoglobin formation. Some of these features were also evident in survivors sacrificed at term (number unstated). More recently, FitzGerald *et al.* [12,13] conducted a number of acute toxicological tests on military related compounds including TNB. Five male and female Fischer 344 (F344) rats were dosed orally to 185, 260, and 335 mg/kg 1,3,5-TNB in corn oil. Similarly, five male and female Swiss mice were exposed to 500, 700, and 900 mg/kg. In addition to the results derived from primary eye and dermal irritation tests, the average acute oral LD_{50} of TNB for combined sexes was 284 mg/kg in rats and 804 mg/kg in mice [12,13]. TNT, not considered a skin irritant to rabbits, caused eye irritation to rabbits, and was a mild skin sensitizer in guinea pigs [12,13].

The acute studies were also conducted with TNB to investigate endpoints other than lethality. Watanabe *et al.* [14] injected nine male Wistar rats intraperitoneally with 100 μmoles/kg TNB in 2 ml/kg propylene glycol and found formation of methemoglobin in blood samples obtained 5 hours after dosing. In another study, four male F344 rats per group were given a single oral dose of TNB [13]. Blood was collected at 5 hours and 24 hours after dosing. The compound was administered in corn oil via oral gavage at doses of 35.5 or 71 mg/kg. Blood collected 5 hours after TNB administration showed increased methemoglobenemia, although other hematological parameters were unchanged. The study team also found significant anemia with reduced red blood cells (RBC), hemoglobin, and hematocrit in rats that received TNB for 4 or 10 days. Myers *et al.* [16] administered a single oral dose of 50 mg/kg TNB in corn oil to four male shrews (*Cryptotis parva*) and used gas chromatography/mass spectrometry to demonstrate the formation of hemoglobin adducts in blood samples obtained 24 hours after dosing. *In vitro* incubation of blood with TNB also resulted in adduct formation; the data suggested a role for cysteine residues in hemoglobin-TNB binding.

Chandra *et al.* [17,18] administered a single dose of TNB in corn oil by gavage to male F344 rats at 35.5 and 71 mg/kg to examine the acute toxicological effects of TNB on brain and testicular morphology and histopathology. No effects of TNB on these endpoints were seen in the single-dose phase of these studies.

Mammalian Oral Toxicity: Subacute

Subacute studies involve repeated dosing of animals and parameters are measured at the end of a14-day study duration. Reddy *et al.* [19,20], using five F344 rats/sex/group, conducted a 14-day oral toxicity study on TNB to establish suitable dosing levels for longer-term studies. The compound was added to feed to obtain dietary concentrations of 0, 50, 200, 400, 800, and 1,200 mg/kg. The respective doses, as calculated by the authors, were 0, 4.52, 16.85, 33.67, 55.76, and 91.93 mg/kg/day in males, and 0, 4.54, 17.5, 34.14, 59.08, and 79.35 mg/kg/day in females. Clinical signs were monitored twice daily, food and water consumption twice weekly, while body weights were recorded at the beginning, at termination, and weekly during the in-life phase of the study. A full suite of hematological and clinical chemistry parameters was assessed in blood samples obtained at necropsy. All tissues and major organs were observed for gross morphological lesions, and the weights of certain key organs were recorded. Samples from numerous internal organs and tissues were fixed and processed for histopathological examination. Sections of sampled tissues from high-dose and control rats were examined under a light microscope, along with sections from all dose groups for potential target organs such as the spleen and kidney.

After treatment, the body weights of both sexes of high-dose rats were reduced compared to controls, probably associated with a comparative reduction in food consumption in animals receiving TNB. Some dose-related changes in organ weight/body weight ratio were evident, including comparative weight increases in brain, spleen, and kidneys, and reductions in thymus and testes. The lowest dose at which effects were seen was 16.85 mg/kg/day, a level associated with relative increases in kidney weights in male rats. Though there were no obvious, treatment-related changes in clinical chemistry parameters, most hematological indices were altered compared to controls after 14 days of treatment. For example, a statistically significant and dose-dependent reduction in hematocrit and red blood cell count was observed in all female groups and in males that received at least 55.76 mg/kg/day (RBC decrease) and 33.67 mg/kg/day (hematocrit decrease). Reduced hemoglobin concentrations were observed in female rats that received 59.08 and 79.35 mg/kg/day TNB. Conversely, methemoglobin increased in a dose-dependent manner in both sexes at a dose of approximately 34 mg/kg/day and higher.

There were gross pathological signs in the testes of male rats dosed at 55.76 mg/kg/day and higher, effects that were probably associated with the overall reduction in testicular size. The lesions were marked by moderate to severe seminiferous tubular degeneration, fewer mature spermatids and an apparent reduction in spermatogenic cells. Cell debris and some multinucleate cells were observed in a generally restricted tubular lumen. Another consistent histopathological response to TNB was the appearance of hyaline droplets in the cortical tubule cells in the kidneys of male rats receiving 16.85 mg/kg/day and above. As described by the authors, many of these large, irregular-shaped droplets were associated with the onset of tubular degeneration [19,20].

A number of no observed adverse effect levels (NOAELs) and lowest observed adverse effect levels (LOAELs) were evident from the results of this study. For example, a NOAEL of 4.52 mg/kg/day and a LOAEL of

16.85 mg/kg/day corresponded to the histopathological changes in male rat kidneys (cortical tubular hyaline droplet formation) occurring in those animals receiving 200 mg TNB/kg diet and above. However, the most sensitive indices of toxicity were the hematological changes evident in female rats at the lowest dose, which suggests a LOAEL of 4.54 mg/kg/day for these responses (see Table 9.2).

In another study, four male F344 rats/group received gavage doses of 35.5 and 71 mg/kg TNB in corn oil for either 4 or 10 days [13]. Blood was collected daily and showed hematological changes similar to those described by Reddy *et al.* [19,20]. Using the same protocol, Chandra *et al.* [17] also observed a range of histopathological lesions in the brain of high-dose rats, including gliosis, vacuole formation, malacia, demyelination, edema and hemorrhages, and the formation of petechial hemorrhages in the cerebellar peduncles. No such responses were observed in rats receiving TNB at 35.5 mg/kg/day. In addition, histopathological changes and reductions in testis weights were apparent in male F344 rats, with complete cessation of spermatogenesis in those receiving 71 mg/kg for 10 days [18].

Twelve male and female F344 rats/group and 12 male NCI-Black Reiter (NBR) rats/group were gavaged for 10 days with TNB in corn oil at 0, 35.5 and 71 mg/kg. Other groups of male F344 rats received 35.5 mg/kg for 20 or 30 days [21]. The endpoints of the study were the relative incidence of hyaline droplets in the kidney among the groups and an evaluation of the role of the $\alpha 2_{\mu}$-globulin protein in droplet formation. There was dose-dependent hyaline droplet formation in the kidneys of male F344 rats but hyaline droplets were absent from kidney sections of NBR and female F344 rats that had received the same treatments. Immunohistochemical staining with anti-$\alpha 2_{\mu}$-globulin protein anti-serum was apparent in droplet-accumulating regions of the kidneys of affected animals, though mostly confined to the renal proximal convoluted tubular epithelial cells. Hyaline droplets and $\alpha 2_{\mu}$-globulin-positive staining areas were concomitantly absent from reference groups such as treatment-negative controls and female F344 rats and male NBR rats receiving TNB-treatment negative controls. These observations suggest the possible importance of $\alpha 2_{\mu}$-globulin-related hyaline droplet formation in tubular necrosis and subsequent cellular proliferation of kidneys in male F344 rats dosed with TNB. However, they do not address the mechanism by which TNB induces an increase in kidney weight observed in female F344 and male NBR rats.

Immunohistochemical staining of excised brains was used to evaluate the integrity of the blood-brain barrier in F344 rats challenged with TNB [22].

Table 9.2 Summary of Relevant Mammalian Data for TRV Derivation

Test Organism	Test Duration	Test Results: NOAEL (mg/kg/day)	LOAEL (mg/kg/day)	Effects Observed at the LOAEL	Study
Rat (Fischer 344)	2 years	2.68	13.31	Methemoglobinemia, spleen erythroid cell hyperplasia; decreased body weight	[31,32]
Rat (Fischer 344)	90 days	NA	3.91 (m)	Nephropathy/α2u-globulin-associated hyaline droplet formation in males at all doses	[24,25]
		4.29	22.73	Methemoglobinemia, spleen erythroid cell hyperplasia in high-dose and mid-dose groups (both sexes)	
Rat (Fischer 344)	14 days	NA	4.54 (f)	Reduced erythrocyte count and hematocrit in all female groups	[19]
		4.52 (m)	16.85 (m)	Histopathological changes to the kidney in males	
Rat (Sprague Dawley)	90 days	2.0 (m)	9.0 (m)	Sperm motility/seminiferous tubular degeneration of the testes	[35,36]
		NA	2.0 (m)	Nephropathy/hyaline droplet formation in males at all doses	
Rat (Fischer 344)	10, 20, and 30 days	NA	35.5 (m)	Nephropathy/ α2u-globulin-associated hyaline droplet formation in males at all doses	[21]
Rat (Sprague Dawley)	90 days	NA	3.0	Increase in tissue concentration of various neurotransmitters in several regions of the brain, potentially associated with neurological disorders and histopathological lesions	[27]
Rat (Sprague Dawley)	7 weeks (m)	23	51	Testicular degeneration and sperm depletion in males	[34]
	10 weeks (f)	4	23	Encephalitis in females	
Rat (Fischer 344)	10 days	NA	35.5 (m)	Hematological deficits and methemoglobin formation	[15]
Rat (Fischer 344)	10 days	35.5 (m)	71 (m)	Histopathological lesions in the brain	[17]

Continued

Table 9.2 Summary of Relevant Mammalian Data for TRV Derivation—cont'd

		Test Results			
Test Organism	**Test Duration**	**NOAEL (mg/kg/day)**	**LOAEL (mg/kg/day)**	**Effects Observed at the LOAEL**	**Study**
Rat (Fischer 344)	10 days	NA	35.5 (m)	Degenerative changes to the testes	[18]
Rat (Sprague Dawley)	GD6 – 15	45 (f)	90 (f)	Developmental deficits among the pups	[33]
Mouse (*Peromyscus leucopus*)	90 days	67.4 (m)	113.5 (m)	Testicular degeneration in high-dose males	[30]
		23.5 (m)	67.4 (m)	Erythroid hyperplasia, increase in reticulocyte count in mid- and high-dose males	
Shrew (*Cryptotis parva*)	14 days	10.75 (m)	21.60 (m)	Decrease in liver and body weight	[23]
		10.68 (f)	22.24 (f)	Increase in spleen weight	

NA = not applicable; GD = gestational day; f = female; m = male.

Five male F344 rats/group were gavaged with TNB in corn oil at 71 mg/kg for 4, 5, 6, 7, 8, or 10 days with 1 dose per day. The experiments were carried out to determine whether the vascular bed mediates the pathogenesis of TNB-induced encephalopathy by comparing the extent of extravasated plasma albumin between treated and control groups. Animals receiving 10 daily doses of TNB showed an increased intensity and extravascular distribution of immunoreactive albumin in the cerebellum and other parts of the brain, while controls and treated animals granted a recovery period were comparatively unaffected. The authors considered the incidence of vasogenic brain edema to be a critical event in TNB neurotoxicity.

A 14-day toxicity evaluation of TNB was conducted in the shrew, *Cryptotis parva*, by feeding 10 animals/group/sex a diet containing 0, 5, 10, 20, and 40 mg/kg of TNB [23]. The calculated average consumed doses were 0, 10.68, 22.24, 37.79, and 98.27 mg/kg/day for males and 0, 10.75, 21.61, 45.26, and 98.72 mg/kg/day for females. Food and water consumption, body weight and organ weights were measured and hematological and histopathological changes were studied. There were no significant differences in food and water consumption and hematological parameters between control and TNB fed shrews. The NOAEL for both sexes of shrews was 10.68 mg/kg/day for the oral 14-day study. The LOAEL was 21.6 mg/kg/day for the decreased body and liver weights in males and increased spleen weight in females [23].

Mammalian Oral Toxicity: Subchronic

Reddy *et al.* [24,25] described a 90-day toxicological study in 15 F344 rats/sex/group receiving 0, 66.7, 400, or 800 mg TNB/kg diet, which corresponded to doses, as calculated by the authors, of 0, 3.91, 22.73, and 44.16 mg/kg/day in males and 0, 4.29, 24.7, and 49.28 mg/kg/day in females. A full range of in-life, clinical chemistry/hematological, gross pathological, and histopathological evaluations were carried out as previously described for the 14-day study by Reddy *et al.* [19,20].

Critical findings included the formation of kidney lesions characterized by hyaline droplet formation, cortical tubular degeneration, and respondent regeneration of tubular cells. These effects were evident in males at all doses. Other responses included methemoglobinemia and erythroid cell hyperplasia in the spleen of high-dose and mid-dose groups (both sexes) and decreased testicular weight with accompanying seminiferous tubular degeneration. Body weight was also significantly less than corresponding controls in males and females at the two highest doses of TNB. Based on their data the authors considered the most sensitive endpoint to be the kidney lesions evident in male F344 rats and assigned the 3.91 mg/kg/day dose as the subchronic LOAEL for TNB [26].

Narayanan *et al.* [27] investigated the effects of subchronic administration of TNB on the levels of neurotransmitters and their metabolites in the brain of Sprague Dawley rats. In a complex protocol, six animals/sex/group received various dietary amounts of TNB between 0 and 800 mg/kg, equivalent to daily doses of 0, 3, 23, and 51 mg/kg/day in males and 0, 4, 30, and 60 mg/kg/day in females, as calculated by the authors. Exposure duration was 90 days in females and 28 days in males. At necropsy, the authors dissected the excised brains into discrete regions, including brain stem, frontal cortex, cerebral cortex, caudate nucleus, septum, hypothalamus, thalamus, hippocampus, and cerebellum. Among the neurotransmitters measured were norepinephrine, epinephrine, dopamine, 5-hydroxytryptomine, homovanillic acid, 3,4-dihydroxyphenylacetic acid, and 5-hydroxyindoacetic acid.

Compared to controls, levels of some neurotransmitters changed dramatically with treatment. For example, the specific concentration of norepinephrine in the septum increased at least tenfold in both sexes at the lowest dose. Effects such as this suggested a LOAEL of 3 mg/kg/day.

As discussed by Reddy *et al.* [28] and on IRIS [29], Reddy *et al.* [30] published an abstract of a meeting presentation in which a 90-day study of TNB in the White-footed mouse (*Peromyscus leucopus*) was described. Ten animals/sex/group were fed diets of 0, 150, 375, and 750 mg TNB/kg, with corresponding dose levels calculated to be 0, 23.5, 67.4, and 113.5 mg/kg/day in males and 0, 20, 65, and 108 mg/kg/day in females.

Results from this study included erythroid hyperplasia of the spleen, increased reticulocytes in mid- and high-dose males, and testicular degeneration in the high-dose males, among other effects. Based on the statistically significant change in reticulocyte count in the mid- and high-dose males, 23.5 mg/kg/day is a subchronic NOAEL for TNB in *Peromyscus*, as suggested in the review by Reddy *et al.* [28]. This choice differs from a subchronic NOAEL of 67.4 mg/kg/day that was chosen by Talmage *et al.* [3] from data on the TNB-induced testicular effects observed in high-dose animals in the same study. However, both levels of response suggest that the White-footed mouse is less sensitive to the toxic effects of TNB than F344 or Sprague Dawley rats.

Mammalian Oral Toxicity: Chronic

Reddy *et al.* [31,32] used the 14-day and 90-day studies on the toxicity of TNB in F344 rats to determine suitable dose levels of the compound for a

2-year, chronic toxicity study. Using 60 rats/sex for control groups and 75 rats/sex for test groups, the authors carried out a full range of in-life, clinical chemistry/hematological, gross pathological, and histopathological evaluations similar to those described earlier for their 14-day and 90-day studies [19,24,20,25]. However, 10 animals/sex/group were sacrificed after 3, 6, and 12 months for a complete interim histopathological examination of target organs. The levels of TNB in diet were 0, 5, 60, and 300 mg/kg, which corresponded to calculated dose levels of 0, 0.22, 2.64, and 13.44 in males and 0, 0.23, 2.68, and 13.31 mg/kg/day in females.

Among the key findings, Reddy *et al.* [31,32] observed an increased formation of methemoglobin compared to controls in the high-dose groups of either sex. Mid- and high-dose females showed reduced mean corpuscular hemoglobin (MCH), among other hematological fluctuations. While there were no gross necropsy findings related to treatment, histopathological examination of the kidney revealed a myriad of lesions, some probably related to changes in the organ that occur normally in aged F344 rats and some probably related to treatment. For example, necrosis of the cortical tubular cells and formation of tubular cytoplasmic droplets were evident in the mid- and high-dose groups of both sexes. Notwithstanding their appearance in both sexes of F344 rats, these lesions were described by the authors as being "hyaline," large and mostly spheroid in shape, and rendered visible with Mallory's Heidenbain protein stain. Immunohistochemical binding to the $\alpha 2_{\mu}$-globulin protein revealed a positive pattern of staining that had a diffuse pattern across the field of view and not necessarily associated with the droplets. The authors concluded that a diagnosis of $\alpha 2_{\mu}$-globulin-related nephropathy would probably be incorrect since there appeared to be no significant increase in tubular cell necrosis. In addition, there were no signs of the granular casts, linear papillary mineralization, and tubular hyperplasia that are typical manifestations of this condition.

Increased splenic erythroid cell hyperplasia and pigment deposition were present in mid- and high-dose groups in the interim sacrifices but were restricted to high-dose groups in the 2-year survivors. This suggested that, after 2 years, animals had compensated for the regenerative anemia previously noted in the results from interim sacrifices.

In addition to the effects of TNB on hematological and histopathological parameters, adults of both sexes at the highest dose were significantly smaller than controls. Body weight at 2 years was significantly affected by TNB in both males and females. Decreased body weight was likely the result of decreased food consumption seen in both sexes for all doses of TNB.

However, the effect of the compound on food consumption was significant only for males at the intermediate- and high-dose levels. It is also possible that smaller body size in TNB-dosed rats may have been the result of alterations in the energy budget of rats caused by compound-induced toxicity.

The authors considered the methemoglobinemia and splenic erythroid cell hyperplasia with pigment deposition to be the critical effects of TNB and suggested that a NOAEL of 2.68 mg/kg/day would be appropriate [31,32]. This conclusion was set forth in a short review by Reddy *et al.* [28] and adopted by the IRIS compilers in their derivation of an oral toxicity reference dose (human health) for TNB of 3×10^{-2} mg/kg/day [29].

Reddy *et al.* [28] and the U.S. Environmental Protection Agency (U.S. EPA) [29] cite a developmental study in Sprague Dawley rats in which TNB was administered at 0, 11.25, 22.5, 45, and 90 mg/kg/day by gavage in 1% agar to an unstated number of pregnant females during the major period of organogenesis (gestation days 6 – 15) [33]. Adverse clinical signs were apparent among the dams, and a number of developmental effects were noted among the pups of dams receiving the highest dose. These included reduced mean fetal weight and crown-rump length, with skeletal variations noted in one animal. These results suggest a developmental NOAEL of 45 mg/kg/day.

Mammalian Oral Toxicity: Other

Sprague Dawley rats were used to examine the reproductive impacts of dietary exposure to TNB [34]. Ten animals/sex/group received 0, 70, 400, or 800 mg TNB/kg diet, levels calculated by the authors to result in average doses of 0, 3, 23, and 51 mg/kg/day in males and 0, 4, 30, and 60 mg/kg/day in females (0, 8, 55, and 110 mg/kg/day during lactation). Male rats were dosed for 14 days prior to mating, and then throughout the mating period for 35 days. Female rats were dosed for 14 days prior to mating, and then throughout mating, gestation, and lactation, with an additional week post-weaning for 70 days. Pups were maintained on the group-specific treated diet for 7 to 14 days postweaning. F_0 male rats were necropsied at termination with sperm count and morphology evaluated in two or three males/group at sacrifice. Testes and epididymides were weighed and examined histopathologically. F_0 female rats were necropsied at termination with pieces of spleen, kidney, and liver processed for histopathological evaluation. Blood samples were taken from all groups at termination for methemoglobin analysis. Measures of reproductive performance were noted, including the copulation and fertilization indices and the number of live and dead pups at birth.

A number of the effects of TNB in this experiment were consistent with those observed in earlier studies described above. For example, there was evidence of enlarged and discolored spleens with hemisiderosis in high- and mid-dose females, testicular atrophy and spermatogenic deficits in high-dose males, and signs of encephalitis in high- and mid-dose females. Neurological signs, for example, head tilt and loss of equilibrium, were observed in lactating dams. Perhaps the most novel findings of the study related to reproductive performance, which, for copulation and fertility indices, were normal, notwithstanding the spermatogenic deficits referred to earlier. Compared to controls, there were no differences in the F_1 sex ratios and offspring/litter among the groups, although the number of surviving pups after 4 and 21 days were lower than controls in the high-dose group. More than one potential NOAEL can be derived from the results described by Kinkead *et al.* [34]. For example, a NOAEL of 4 mg/kg/day may be protective against TNB-induced encephalitis in female rats (an effect that was evident in mid- and high-dose groups only), while a NOAEL of 23 mg/kg/day may protect against the testicular atrophy that was evident in high-dose males.

In another reproductive study conducted by the same research group, dietary amounts of 0, 30, 150, and 300 mg TNB/kg were administered to 18 male and 12 female Sprague Dawley rats/group for up to 90 days [35,36]. Animals were divided into subgroups to accommodate a number of premating, postmating, and recovery dosing regimens. The doses, as calculated by the authors, were 0, 2, 9, and 19 mg/kg/day for males and 0, 3, 14, and 29 mg/kg/day for females. Subsets of animals were subjected to Opto-Varimex open-field activity evaluation tests, though with no evidence of compound-related depletions in motor skills. At necropsy, samples of blood were taken for clinical chemistry/hematological evaluation. Organ weights were monitored and histopathological evaluations were made of the pituitary, spleen, liver, kidneys, bone marrow, and reproductive organs. Sperm counts and motility were measured in comparison to various indices of reproductive performance.

Among the key findings were decreases in sperm motility and degeneration of the seminiferous tubules of the testes in mid- and high-dose males and nephropathy/hyaline droplet formation in males at all dose levels. Methemoglobin was formed in the mid- and high-dose groups for both sexes, although these levels returned to normal in a subset of males allowed to recover after treatment. Splenic hemosiderosis was noted in mid- and high-dose groups of both sexes. By analogy to the findings of Kinkead *et al.* [34], there were no adverse effects of TNB on mating or fertility, gestation length, sex ratio, or the number of offspring/litter. However, during lactation the body weights of pups borne to high-dose parents were reduced compared to controls. These data revealed that a discriminating index of toxicity appeared to be the changes in sperm morphology and motility to which a NOAEL of 2 mg/kg/day and a LOAEL of 9 mg/kg/day could be assigned. However, 2 mg/kg/day would be a LOAEL for the hyaline droplet-related nephropathy that was evident in male rats at all TNB dose levels.

Studies Relevant to Mammalian TRV Development for Oral Exposures

In the experimental studies described in this review, the overwhelming majority were carried out in laboratory rats, predominantly F344 and Sprague-Dawley strains. Less frequently employed species included the White-footed mouse and the shrew, with no studies identified in birds, reptiles, or amphibians. The principal toxicological effects of TNB are displayed in Table 9.2 and Figure 9.1. Responses to TNB emerging from experimental studies in laboratory animals included [1] nephropathy associated with $\alpha 2_{\mu}$-globulin-associated hyaline droplet formation in male rats, [2] atrophy of the testis with associated degeneration of the seminiferous tubules and sperm deficits, [3] structural and functional impairment of the brain, [4] methemoglobin formation, and [5] reduced body weight. These effects are discussed below, in the context of their relevance to TRV derivation.

Nephropathic changes associated with hyaline droplet formation and $\alpha 2_{\mu}$–globulin formation in male rats appear to be a sensitive indicator of TNB toxicity since no clear-cut NOAEL could be identified for this response in any of the evaluated studies. For example, in Sprague Dawley rats exposed to the compound for up to 90 days, the lowest dose tested (2.0 mg/kg/day) was identified as a subchronic LOAEL for this response [35,36]. This is a lower value than the chronic NOAEL of 2.68 mg/kg/day emerging from the 2-year study on TNB with methemoglobinemia specified as the principal effect [30,27,31,28]. Unfortunately, data from the 2-year study on TNB may be unsuitable to shed light on the issue of a chronic NOAEL for $\alpha 2_{\mu}$-globulin-associated hyaline droplet formation and related nephropathy in males, since Reddy *et al.* [31,32] described incipient nephropathy and droplet formation in the kidneys of *both* sexes of mid- and high-dose F344 rats. They considered this effect inappropriate for a diagnosis of $\alpha 2_{\mu}$-globulin-associated nephropathy in this study, even though

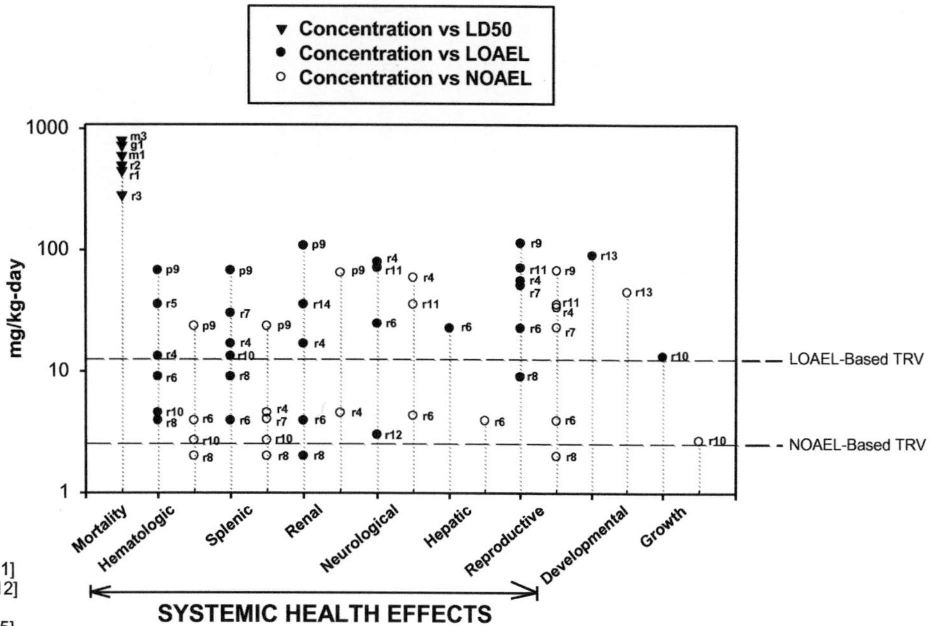

1 = Korolev *et al.* 1977 [10]
2 = Fogelman *et al.* 1955 [11]
3 = FitzGerald *et al.* 1991 [12]
4 = Reddy *et al.* 1994a [19]
5 = Chandra *et al.* 1995a [15]
6 = Reddy *et al.* 1994b [24]
7 = Kinkead *et al.* 1994a [34]
8 = Kinkead *et al.* 1994b [35]
9 = Reddy *et al.* 1995 [30]
10 = Reddy *et al.* 1996, 2001 [31,32]
11 = Chandra *et al.* 1995b [17]
12 = Narayanan *et al.* 1995 [27]
13 = Cooper and Caldwell, 1995 [33]
14 = Kim *et al.* 1997 [21]

RAT (*Rattus*) **= r**
MOUSE (*Mus*) **= m**
GUINEA PIG (*Cavia*) **= g**
New World Mouse (*Peromyscus*) **= p**

Figure 9.1 1,3,5-TNB health effects to mammals. Letters by symbols represent test species, and numbers represent study.

gender-specific hyaline droplet formation had been a consistent feature of the toxicological impacts of TNB in F344 rats [19,24,20,25]. Since the precise nature and biological significance of the droplets remains uncertain, this endpoint cannot be used for the derivation of a TRV for TNB.

Impairment of the male reproductive organs with associated decreases in sperm production and motility is a consistent response of experimental animals to nitroaromatic compounds including TNB. However, similar to the methemoglobinemia, this response appears to be at least partially reversible on cessation of exposure. Moreover, there was no significant effect of TNB on reproductive performance [34–36]. This suggests that these responses may not be critical when deriving quantitative benchmarks for wildlife protection.

Methemoglobinemia can create a functional hypoxic blood condition. Carbon monoxide poisoning also creates a functionally similar condition. Humans with levels of COHb above 10% have reported headaches while other adverse effects, such as decreased psychomotor performance, were reported when COHb concentrations exceeded 2% [36]. Chronic congenital methemoglobinemia in humans was found where 10 to 50% of circulating blood pigment is in the form of methemoglobin with subjects exhibiting no overt signs of toxicity [38]. In chronic toxicity study, Reddy *et al.* [30,31] reported differences between control and high dose rats of less than 2%. Given the uncertainty associated with the reported methemoglobin increase in these investigations, increased methemoglobinemia due to TNB was not considered biologically significant.

Histopathological lesions in various parts of the brain are a consistent feature of responses to TNB on the part of experimental animals. However, the mechanisms by which such responses are brought about are poorly understood, although the induction of an imbalance in neurotransmitters and their metabolites may play at least a partial role in bringing about these effects [26]. No clear links between these histopathological lesions and impaired health or performance were established. Hence, the TRV is not based on this response.

Reduced body weight was found in TNB exposed F344 rats under both subchronic [24,25] and chronic study regimes [31,32]. Both studies were high quality with respect to experimental design and execution. A diminution in growth rate may have ecological ramifications by, for example, causing a delay in time to maturity or increasing risk of predation. Reduced somatic growth may not be a direct effect of exposure to TNB as decreased food consumption was a common response. However, it can be argued that decreased food consumption, as a result of avoidance or direct toxicity, will compromise an organism's function and hence will likely result in altered health or performance. Given the prevalence of reduced body size in TNB exposed mammals and the likely ecological ramifications of this effect, this endpoint was selected for derivation of the TRV based on data from the chronic study by Reddy *et al.* [31,32].

Taken together, the experimental findings summarized in this report present a consistent picture of the overall toxicity of TNB in experimental animals, with dose levels for toxic effects in the 2 to 5 mg/kg/day region. Available evidence suggests that the White-footed mouse may be less sensitive to the toxicological effects of TNB than F344 or Sprague Dawley rats, although Talmage *et al.* [3] used the experimentally derived NOAELs from F344 rats [31,32] and White-footed mice [30] to derive reasonably consistent chronic screening criteria for several wildlife species based on feeding pattern estimates and size comparisons.

Mammalian Inhalation Toxicity

No inhalation studies conducted using mammals were found.

Mammalian Dermal Toxicity

In vitro percutaneous absorption of 14C-TNB was studied using viable skin from hairless guinea pigs (HGP), F344 rats, and human skin in assembled flow-through diffusion cells [39]. Results showed that absorption of TNB in an acetone or water vehicle was rapid, occurring within 24 hours. The absorption of TNB in HGP skin was 72% in acetone and 82% in water. The absorption of TNB in rat skin was 61% in acetone and 66% in water. However, the absorption of TNB in human skin was 38% in acetone and 75% in water. These results show the absorption of TNB in acetone was reduced in human skin when compared to rats and HGP. Although this study provided some insight into the percutaneous absorption of TNB, it did not provide any data on toxicity of this compound when exposure occurs via the dermal route.

Summary of Avian Toxicology

Toxicological data for the effects of TNB in avian species were not located. Ecotoxicological research on the effects of this compound in birds is recommended.

Summary of Amphibian Toxicology

Toxicological data for the effects of TNB in amphibian species were not located. Ecotoxicological research on the effects of this compound in amphibians is recommended.

Summary of Reptilian Toxicology

Toxicological data for the effects of TNB in reptilian species were not located. Ecotoxicological research on the effects of TNB in reptiles is recommended.

RECOMMENDED TOXICITY REFERENCE VALUES

Toxicity Reference Values for Mammals

TRVs for Ingestion Exposures for the Class Mammalia

Decreased body weight, an indication of a lower growth rate, was used to determine the TRV. Growth is an ecologically relevant parameter, which when altered, may affect future fitness. Data on female rats were used because these data were protective of males since females were exposed to slightly lower doses. Moreover, growth as an endpoint is also protective of adverse reproductive effects [33,34]. Growth data, as indicated by body size, also meet the minimum data requirements of the Standard Practice, Section 2.2 [1] and therefore no uncertainty factors are required in the derivation of the TRV. Derivation of the TRV was attempted using the benchmark dose (BMD) approach with data derived from Reddy *et al.* [31,32] with the mean response described as mean body weight at 2 years in Fischer 344 rats. Data from females were used because the doses used for females were slightly lower than for males, and there was a clear dose response. However, Test 2 of the *constant variance model* indicated that variances were heterogeneous and suggested use of a *nonconstant variance model*, which ultimately indicated that the data did not fit this model (Test 3 rejected), and that the variances were poorly modeled (Test 4 rejected). Both variance models were attempted with various distributions. Hence, it was concluded that the BMD approach could not be satisfactorily used. The NOAEL/LOAEL approach was used and the resulting TRVs are shown in Table 9.3. These TRVs are given a medium confidence rating since there was only one chronic study done in F344 rats. Moreover, limited mammalian wildlife toxicity data were available.

Table 9.3 Selected Ingestion TRVs for the Class Mammalia

TRV	Dose	Confidence Level
NOAEL-based	2.68 mg/kg/d	Medium
LOAEL-based	13.31 mg/kg/d	Medium

Source: [30].

TRVs for Inhalation Exposures for the Class Mammalia

Not available at this time.

TRVs for Dermal Exposures for the Class Mammalia

Not available at this time.

Toxicity Reference Values for Birds

Not available at this time.

Toxicity Reference Values for Amphibians

Not available at this time.

Toxicity Reference Values for Reptiles

Not available at this time.

IMPORTANT RESEARCH NEEDS

Data regarding the toxicity of TNB are from laboratory rodents, primarily rats, with a few studies on mammalian wildlife species. Hence, the most obvious research need is for studies on the toxicity of TNB to avian, amphibian, and reptile species. Currently, there are no data on these groups of organisms although they are likely to be present at contaminated sites and important components of local habitats. Toxicological testing on these types of organisms would provide the data needed to develop TRVs for nonmammalian, ecological receptors.

REFERENCES

[1] USACHPPM. Standard practice for wildlife toxicity reference values. Aberdeen Proving Ground (MD): U.S. Army Center for Health Promotion and Preventive Medicine; 2000, Technical Guide 254.

[2] Hazardous Substances Data Bank (HSDB). TNB [Internet]. Bethesda (MD): National Library of Medicine (US), National Institutes of Health [cited 2000]. Available from: http://toxnet.njm.nih.gov/.

[3] Talmage SS, Opresko DM, Maxwell CJ, Cretella FM, Reno PH, Daniel FB. Nitroaromatic munition compounds: environmental effects and screening values. Rev Environ Contam Toxicol 1999;161:1–156.
[4] Simini M, Wentsel RS, Checkai RT, Phillips CT, Chester NA, Major MA. Evaluation of soil toxicity at Joliet Army Ammunition Plant. Environ Toxicol Chem 1995;14:623–30.
[5] Wentsel RS, Hyde RG, Jones III WE, Wilkinson MJ, Harward III WE. Problem definition study on 1,3-dinitrobenzene, 1,3,5-trinitrobenzene and di-n-propyl adipate. Alexandria (VA): Atlantic Research Corporation; 1979, ADA099732.
[6] Hovatter PS, Talmage SS, Opresko DM, Ross RH. Ecotoxicity of nitroaromatics to aquatic and terrestrial species at army Superfund sites. In: Dwyer FJ, Doane TR, Hinman ML, editors. Environmental toxicology and risk assessment: modeling and risk assessment, vol. 6. West Consohocken (PA): American Society for Testing and Materials; 1997.
[7] Agency for Toxic Substances and Disease Registry (ATSDR). Toxicological profile for 1,3-dinitrobenzene/1,3,5-trinitrobenzene. Atlanta (GA): U.S. Department of Health and Human Services, Public Health Service; 1995.
[8] Burrows EP, Rosenblatt DJ, Mitchell WR, Palmer DL. Organic explosives and related compounds: environmental and health considerations. Fort Detrick; Frederick (MD): U.S. Army Medical Research and Development Command; 1989, ADA210554.
[9] Spanggord RJ, Mill T, Chou TW. Environmental fate studies on certain munition wastewater constituents–Laboratory Studies. Final report. Phase II. Menlo Park (CA): SRI International; 1980, Prepared for the U.S. Army Medical Research and Development Command, Fort Detrick; Frederick, MD. ADA099256.
[10] Korolev AA, Voitskhovskaya TV, Bogdanov MV, Arsenieva MV, Zakharova TA. Experimental data for hygienic standardization of dinitrotoluol and trinitrotoluol in surface waters. Gigiena i Sanitariya 1977;10:17–20, Cited in Wentsel et al., 1979.
[11] Fogelman RW, Elsen JR, Paynter OE, Kundzins W. Toxicity of trinitrobenzene-aniline complex, a rodent repellent. Agric Food Chem 1955;3:936–9.
[12] FitzGerald GB, Austin A, DiGuilio N. Acute toxicity evaluation of nitroaromatic compounds. Fort Detrick; Frederick (MD): U.S. Army Medical Research and Development Command; 1991, ADA236352.
[13] FitzGerald GB, DiGuilio N, Desai LS, Reddy G. Acute toxicological evaluation of 1,3,5-trinitrobenzene. J Am Coll Toxicol 1992;Acute Toxicity Data 1: 169–70.
[14] Watanabe T, Ishihara N, Ikeda M. Toxicity of and biological monitoring for 1,3-diamino-2,4,6-trinitrobenzene and other nitro-amino derivatives of benzene and chlorobenzene. Int Arch Occup Environ Health 1976;37(3):157–68.
[15] Chandra AM, Qualls Jr. CW, Reddy G, Meinkoth JH. Hematological effects of 1,3,5-trinitrobenzene (TNB) in rats in vivo and in vitro. J Toxicol Environ Health 1995 Sep;46:57–72.
[16] Myers SR, Pinorini-Godly MT, Reddy TV, Daniel FB, Reddy G. Gas chromatographic and mass spectrometric determination of hemoglobin adducts of 1,3-dinitrobenzene and 1,3,5-trinitrobenzene in shrew Cryptotis parva. Int J Toxicol 1999;18(5):317–25.
[17] Chandra AM, Qualls Jr. CW, Reddy G. 1,3,5-trinitrobenzene-induced encephalopathy in male Fischer-344 rats. Toxicol Pathol 1995;23(4):527–32.
[18] Chandra AM, Qualls Jr. CW, Reddy G. Testicular effects of 1,3,5-trinitrobenzene (TNB). 1. Dose response and reversibility studies. J Toxicol Environ Health 1997;50 (4):365–78.
[19] Reddy TV, Daniel FB, Robinson M, Olson GR, Wiechman B, Reddy G. Subchronic toxicity studies on 1,3,5-trinitrobenzene, 1,3-dinitrobenzene and tetryl in rats: 14-day toxicity evaluation of 1,3,5-trinitrobenzene in Fischer 344 rats. Cincinnati (OH): U.S. Environmental Protection Agency; 1994, U.S. Army Project Order MIPR No. 92MM2525. ADA283664.

[20] Reddy TV, Olson GR, Wiechman B, Reddy G, Robinson M, Torsella JA, Daniel FB. Fourteen-day toxicity study of 1,3,5-trinitrobenzene in Fischer F344 rats. J Appl Toxicol 1996;16(4):289–95.
[21] Kim S, Qualls Jr. CW, Reddy G, Stair EL. 1,3,5-trinitrobenzene-induced alpha-2u-globulin nephropathy. Toxicol Pathol 1997;25(2):195–201.
[22] Chandra AM, Campbell GA, Reddy G, Qualls Jr. CW. Neurotoxicity of 1,3,5-trinitrobenzene: Immunohistochemical study of cerebrovascular permeability. Vet Pathol 1999;36(3):212–20.
[23] Reddy G, Reddy TV, Daniel FB, Leach GJ. Fourteen-day toxicity evaluation of 1,3,5-trinitrobenzene (TNB) in shrew (*Cryptotis parva*), In: Twenty-first meeting of the American College of Toxicology; 2000 Nov 12–15, San Diego, CA. Abstract.
[24] Reddy TV, Daniel FB, Robinson M, Olson GR, Wiechman B, Reddy G. Subchronic toxicity studies on 1,3,5-trinitrobenzene, 1,3-dinitrobenzene and tetryl in rats: subchronic toxicity evaluation of 1,3,5-trinitrobenzene in Fischer 344 rats. Cincinnati (OH): U.S. Environmental Protection Agency; 1994 U.S. Army Project Order MIPR No. 92MM2525.
[25] Reddy TV, Olson GR, Wiechman B, Reddy G, Torsella J, Daniel FB. Subchronic toxicity of 1,3,5-trinitrobenzene in Fischer 344 rats. Int J Toxicol 1998;17(4):393–411.
[26] Bond JA, Chism JP, Rickert DE, Popp JP. Induction of hepatic and testicular lesions in Fischer-344 rats by single oral doses of nitrobenzene. Fundam Appl Toxicol 1981;1:389–94.
[27] Narayanan LD, Caldwell J, Miller CR. Alteration in neurotransmitters and their metabolite levels in 1,3,5-trinitrobenzene-treated Sprague-Dawley rats. Wright-Patterson AFB (OH): U.S. Air Force Armstrong Laboratory; 1995, AL/OET-TR-1995-0133.
[28] Reddy G, Reddy TV, Choudhury H, Daniel FB, Leach GJ. Assessment of environmental hazards of 1,3,5-trinitrobenzene. J Toxicol Environ Health 1997;52(5):447–60.
[29] U.S. EPA. Integrated Risk Information System. 1,3,5-Trinitrobenzene (CASRN 99-35-4) [Internet]. Cincinnati (OH): Office of Health and Environmental Assessment (US), National Center for Environmental Assessment [cited 2000]. Available from http://www.epa.gov/iris/subst/0316.htm.
[30] Reddy TV, Torsella J, Daniel FB, Olson GR, Wiechman B, Reddy G. Ninety-day toxicity evaluation of 1,3,5-trinitrobenzene (TNB) in *Peromyscus leucopus*. Second Society of Environmental Toxicology and Chemistry; 1995 Nov 5–9, Vancouver, British Columbia. Cited in Talmage et al., 1999. Abstract.
[31] Reddy TV, Daniel FB, Olson GR, Wiechman B, Torsella J, Reddy G. Chronic toxicity studies on 1,3,5-trinitrobenzene in Fischer 344 rats. Cincinnati (OH): USEPA Environmental Monitoring Systems Laboratory; 1996 Prepared for the U.S. Army Medical Research and Development Command, Fort Detrick; Frederick, MD. Final Report 1 May 9–30 Apr 95. ADA315216.
[32] Reddy TV, Olson GR, Wiechman B, Reddy G, Torsella JA, Daniel FB, Leach GJ. Chronic toxicity of 1,3,5-trinitrobenzene in Fischer 344 rats. Int J Toxicol 2001;20 (2):59–67.
[33] Cooper KR, Caldwell DJ. Developmental toxicity of 1,3,5-trinitrobenzene in Sprague-Dawley rats. Wright-Patterson AFB (OH): U.S. Air Force Armstrong Laboratory; 1995, Final Report. Cited in Reddy et al., 1997.
[34] Kinkead ER, Wolfe RE, Salins SA, Fleming CD, Caldwell DJ, Miller CR, Latendresse JR. Range-finding study for a reproductive assessment of 1,3,5-trinitrobenzene administered in the diet of Sprague-Dawley rats. Wright-Patterson AFB (OH): U.S. Air Force Armstrong Laboratory; 1994 AL/OE-TR-1994-0072. ADA299032.
[35] Kinkead ER, Wolfe RE, Fleming CD, Caldwell DJ, Miller CR, Marit GB. Reproductive toxicity screen of 1,3,5-trinitrobenzene administered in the diet of Sprague-Dawley rats. Wright-Patterson AFB (OH): U.S. Air Force Armstrong Laboratory; 1994 AL/OE-TR-1994-0144. ADA2980912.

[36] Kinkead ER, Wolfe RE, Flemming CD, Caldwell DJ, Miller CR, Marit CG. Reproductive toxicity screen of 1,3,5-trinitrobenzene administered in the diet of Sprague Dawley rats. Toxicol Ind Health 1995;11(3):309–23.
[37] American Conference of Governmental Industrial Hygienists (ACGIH). 6th ed. Documentation of the threshold limit values and biological exposure indices, vol. 1. Cincinnati (OH): American Conference of Governmental Hygienists, Incorporated; 1991.
[38] Smith RP. Toxic responses of the blood. In: Klaassen CD, editor. Casarett and Doull's toxicology: the basic science of poisons. 5th ed. New York: McGraw-Hill; 1996.
[39] Kraeling ME, Reddy G, Bronaugh RL. Percutaneous absorption of trinitrobenzene: animal models for human skin. J Appl Toxicol 1998;18(6):387–92.

CHAPTER 10

Wildlife Toxicity Assessment for *N*-Methyl-*N*-2,4,6-Tetranitroaniline (Tetryl)

Valerie H. Adams

Contents

INTRODUCTION

Tetryl was described in 1879 and was first used in World War I as a mixture with trinitrotolulene (i.e., Tetrytol) [1]. Tetryl was used until the mid-1970s when concerns over its mutagenicity and efforts to manufacture less sensitive munitions resulted in its discontinuation and replacement with Royal Demolition Explosive (RDX). Although no longer in use, tetryl (CAS No. 479-45-8) is one of several compounds that have been released to the environment during its manufacture, or during load and pack LAP activities at U.S. Army ammunition plants (AAPs) and other military installations. The structure of the compound resembles trinitrobenzene with an additional *N*-methylnitramine substituent group; see Figure 10.1.

Tetryl is also known as 2,4,6-tetryl, tetralite, tetril, *N*-picryl-*N*-methylnitramine, and *N*-methyl-*N*-2,4,6-tetranitrobenzamine, among others [2].

Wildlife Toxicity Assessments for Chemicals of Military Concern
http://dx.doi.org/10.1016/B978-0-12-800020-5.00010-7

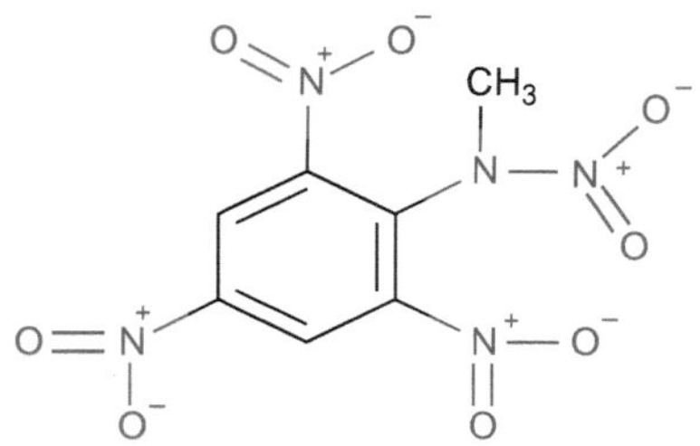

Figure 10.1 Chemical structure of tetryl (molecular formula C7-H5-N5-O8); *Source:* from [2].

Tetryl is also referred to as Nitramine; however, the use of this term is discouraged, as it is a nonspecific term used to describe the chemical functional group. This tetryl wildlife toxicity assessment (WTA) summarizes current knowledge of the likely harmful impacts of this chemical on wildlife and, where possible, identifies exposure levels that may adversely affect wildlife species. Evaluating the toxicity of this compound will facilitate the derivation of TRVs that could serve as protective exposure standards for wildlife species living near affected sites. The protocol for the performance of this assessment is documented in the U.S. Army Center for Health Promotion and Preventive Medicine (USACHPPM) Technical Guide (TG) 254, the *Standard Practice for Wildlife Toxicity Reference Values* [1].

TOXICITY PROFILE

Environmental Fate and Transport

Tetryl is an explosive chemical that was used as a detonator and propellant in military ordnance. While the use of tetryl as a military propellant or detonator has been superseded by other formulations, spills and discharges to waste streams have resulted in contamination of environmental media at or near installations where the compound was manufactured and/or used in LAP activities. For example, the maximum detected concentration of tetryl in soil at Joliet Army Ammunition Plant (AAP) was reported to be 84,000 mg/kg, with concentrations in groundwater up to 67 micrograms per liter (μg/L) [4]. Other locations where tetryl was manufactured or handled are contaminated with the compound to varying extents. Physicochemical data on tetryl that relate to the environmental fate and transport characteristics of the compound are listed in Table 10.1.

The estimated values for vapor pressure and Henry's Law constants are sufficiently low to indicate that tetryl is unlikely to be released to the air as

Table 10.1 Summary of Physical-Chemical Properties of Tetryl

CAS No.	479-45-8
Molecular weight	287.15
Color	Yellow
Physical state	Crystalline solid
Melting point	129–132 °C
Boiling point	187 °C (explodes)
Odor	Odorless
Solubility	Sparingly soluble in water (75–80 mg/L at 20–25 °C), soluble in acetone, alcohol, ether, benzene, glacial acetic acid
Partition coefficients:	
Log K_{ow}	2.4, 2.0, 1.65
Log K_{oc}	3.13–3.47, 2.6, 1.69
Vapor pressure at 25 °C	4.0×10^{-10}, 5.7×10^{-9}, 1×10^{-8} mm Hg
Henry's Law constant at 25 °C	2.0×10^{-12}, 1.0×10^{-11} atm.m^3/mole
Conversion factors	1 ppm = 11.74 mg/m^3 1 mg/m^3 = 0.085 ppm

Sources: Talmage *et al.* [4], ATSDR [5], HSDB [6].

vapor. Aerial dispersion of the compound adhering to soil or dust particles is a possible mechanism by which tetryl can be released to the atmosphere [5]. The experimentally derived K_{OC} values indicate tetryl binds to soil and is influenced by the amount of organic carbon in the soil. Tetryl may be released from soil and enter the groundwater. At some AAPs, tetryl is detected in the parts per billion ranges in groundwater [4]. Tetryl is sensitive to photolysis and hydrolysis; and these processes contribute significantly to tetryl breakdown in the environment [7]. The former process is dominant in sunlit waters, with *N*-methylpicramide, nitrate, picrate, and methylnitramine among the primary products. Formation of most of these products proceeds at a much slower rate in the dark, and *N*-methylpicramide appears to be solely a product of photolysis rather than thermal degradation or abiotic transformation in the absence of light. The polarity (water solubility) of the breakdown products (e.g., picric acid) will influence leaching into groundwater. Tetryl biodegrades under some laboratory reactor conditions [8]. When the soil reactor was amended with molasses, native soil bacteria were able to digest tetryl completely within 3 months. Primary metabolites were transiently detectable in the test system. Pilot scale evaluation of a bioremediation product on tetryl contaminated soil from Joliet AAP found the tetryl starting concentration of 6,000 mg/kg was reduced to 16 mg/kg over a 120-day treatment course [9]. Tetryl in soil may be taken up by plants and metabolized to more polar products with the root being the primary site of accumulation; tetryl was not overtly phytotoxic (highest dose tested 75 micrograms per gram (μg/g) soil; chronic LOEC 25 mg/kg in soil) [10]. Tetryl is considered toxic to aquatic species (average LOEC 0.7 mg/L) [11,12].

Summary of Mammalian Toxicity

Mammalian Oral Toxicity: Acute

In a series of experiments carried out for the U.S. Army Medical Research and Development Command by Toxikon Corporation, FitzGerald *et al.* reported data obtained from various short-term toxicological tests on tetryl [13]. These included eye and skin irritation studies in rabbits, dermal lethality in rabbits, sensitization tests in guinea pigs, and acute oral lethality studies. In the acute study, 10 each of F344 rats and Swiss mice were given 5 grams per kilogram (g/kg) tetryl suspended in corn oil. For both the rats and the mice, only one animal died because of the exposure. Other effects noted in rats included lethargy, piloerection, and loss of body weight. In mice, effects of tetryl exposure were tachypnea, catalepsy, piloerection, and unusual locomotion. FitzGerald *et al.* concluded that tetryl was non-toxic at 5 g/kg in these animal models [13].

Reddy *et al.* conducted an acute toxicity study on male Fischer 344 rats. Four rats/sex/group were given a single oral dose of 0, 500, or 1000 mg/kg body weight tetryl in corn oil via gavage. Twenty-four hours after dosing, drug-metabolizing enzymes and hematological and histological changes were studied [14]. No mortality occurred; however, there were significant effects of tetryl on O-dealkylation activity (ethoxyresorufin and pentoxyresorufin) of liver microsomes. Gross necropsy examination revealed several tetryl-related effects including red to dark-brown blood in the stomach and small intestine, and food containing a yellow granular substance in the stomach. Histological examination showed focal coagulative necrosis at the junction of the stomach and duodenum of tetryl-treated rats.

Mammalian Oral Toxicity: Subacute

Reddy *et al.* conducted a 14-day study originally published as a government report [15]. Both male and female rats (5/sex/group) were fed laboratory chow containing 0, 500, 1,250, 2,000, 2,500, and 5,000 mg tetryl/kg feed and reported that these levels approximated average doses of 0, 31.8, 80,

121, 170.5, and 349.7 mg/kg/day in males and 0, 32.1, 82.5, 130.3, 178.9, and 374.4 mg/kg/day in females [16]. Daily observation of animals for clinical signs of toxicity, food, and water consumption was monitored twice weekly, and body weights were recorded at 7-day intervals. Blood samples for routine clinical chemistry and hematological parameters were obtained immediately prior to termination. Animals were subject to a gross necropsy and the weights of brain, kidneys, liver, spleen, adrenals, lungs, thymus, gonads, and heart were noted. Excised pieces of a wide range of organs and tissues were fixed and processed for histopathological examination. Sections of all tissues from the high-dose and control groups were examined under the microscope, while sections from certain likely target organs such as kidney, spleen, and testis were examined in all groups.

There were few if any clinical signs of toxicity displayed by the animals under test, and all rats survived to term. Similarly, there were essentially no compound-related changes in food and water consumption among the groups. However, the high-dose males displayed a reduction in body weight compared to controls after 14 days. There were no statistically significant changes in organ weights, but some relative organ weights such as liver and spleen in females and kidneys in males were increased in high-dose rats compared to controls. While some sporadic differences in hematological parameters were observed, the only compound-related change appeared to be the dose-dependent increase in the concentration of methemoglobin (metHb), an effect that was statistically significant at doses of 121 mg/kg/day and greater in male rats. Similarly, some clinical chemistry parameters appeared to fluctuate because of tetryl administration. For example, there was a dose-dependent reduction in plasma alkaline phosphatase activity in males receiving 80 mg/kg/day and greater. Necropsy findings and histopathological changes induced by tetryl were sparse other than an increased incidence of cytoplasmic droplets in the proximal renal cortical tubular epithelial cells of male rats. Since these protein-containing droplets were apparent at all dose levels but not controls, this suggests a LOAEL of 31.86 mg/kg/day; see Table 10.2.

Mammalian Oral Toxicity: Subchronic

Reddy *et al.* [16,17] also provided data from a study in F344 rats in which the same parameters were monitored after a 90-day dietary exposure to tetryl. The compound was mixed with laboratory chow to concentrations of 0, 200, 1,000, and 3,000 mg/kg feed, amounts calculated by the authors to be equivalent to average daily doses of 0, 13, 62.43, and 179.63 mg/kg/day in males and 0, 14.2, 68.87, and 199.06 mg/kg/day in females.

Table 10.2 Summary of Relevant Mammalian Data for TRV Derivation

Test Organism	Test Duration	Test Results: NOAEL (mg/kg/day)	Test Results: LOAEL (mg/kg/day)	Effects Observed at the LOAEL	Study
Rat (F344)	14 days	80	121.1	Methemoglobinemia in males.	[15,16]
		NA	31.86	Protein droplet-related nephropathy in males.	
Rat (F344)	90 days	13	62.43	Nephropathy/ $\alpha 2_{\mu}$-globulin-associated hyaline droplet formation in males at all doses. Increased kidney, liver, and spleen weight in both sexes. Methemoglobinemia, lower hemoglobin, and RBC count in both sexes. Decreased adrenal and thymus weights in both sexes.	[16,17]
		14.2	68.87	Reduced body weight in females at mid- and high doses.	

Statistically significant reductions in body weight were evident in both high-dose groups and in the mid-dose females compared to controls, while changes in absolute and relative organ weights were noted, including increases in kidney, liver, and spleen and reductions in brain, adrenals, and thymus. Similar to the body weight changes, these deviations in organ weights were most evident in the high- and mid-dose groups, although the differences in relative kidney weight were also significantly different from controls in low-dose females. Food consumption in all dose groups for both males and females was significantly lower than corresponding controls. Water consumption was significantly higher in females at the high dose. A number of compound-related hematological changes were noted, including reductions in red blood cell count, hematocrit, and hemoglobin concentration, and concomitant increases in reticulocytes, platelets, and (especially) methemoglobin. Dose-related changes in clinical chemistry parameters due to tetryl included reductions in alkaline phosphatase activity and increases in plasma cholesterol levels. The latter was statistically significant in all dosed groups. Necropsy findings were unremarkable, but histopathological changes were observed in stained sections of spleen (probable hemosiderin) and kidney. Thus, both sexes of high-dose rats had excess deposition of pigment in the spleen, where both sexes of high- and mid-dose animals displayed excess pigment deposition in kidneys. There was also some evidence for the formation of $\alpha 2_{\mu}$-globulin-containing hyaline droplets in the kidneys of male rats, a feature that appeared to be dose-related. Reddy *et al.* considered the kidney effects to be the most sensitive indicator of tetryl toxicity and suggested a NOAEL of 13 mg/kg/day and a LOAEL of 62.43 mg/kg/day [16]. In addition, tetryl had a significant deleterious effect on growth rate in female rats with an associated NOAEL of 14.2 mg/kg/day and a LOAEL of 68.87 mg/kg/day (Table 10.2).

The Agency of Toxic Substances and Disease Registry (ATSDR) summarized two Italian studies in which 12 rabbits were orally dosed with 125 mg/kg/day tetryl for 6 and 9 months [5]. Results of the 6-month study indicated some statistically significant changes in blood parameters suggesting a coagulation disorder. However, the effects were inconsistent and did not increase in severity with time. In the 9-month study, renal and hepatic degenerative effects were observed; however, no controls were used and no other doses were administered in this study, precluding statistical analyses. The U.S. Environmental Protection Agency (U.S. EPA) used data from one of the Italian studies to derive human health chronic and subchronic reference doses (RfD) of 0.01 and 0.1 mg/kg/day, respectively, based on the onset of histopathological effects in the kidney, liver, and spleen. However, as pointed out in the ATSDR, the value of these observations is limited by the use of a small number of test animals, a single dose level, and the lack of a control group in the study.

Tetryl was one of a series of compounds tested for tumor induction in the mammary gland of female Sprague Dawley rats [18]. In a nonstandard protocol, 20 animals received 10 doses (40 mg each) of the compound by gavage in sesame oil during a 30-day period at the previously determined maximum tolerated dose (400 mg tetryl). Animals were observed for 9 months, then necropsied, at which time histopathological mounts were made of a number of organs and tissues including the mammary gland. There was no increased incidence of tumor formation due to tetryl, although a number of the other compounds, including 7,12-dimethylbenz(a)anthracene (positive control), were positive for tumor formation.

Mammalian Oral Toxicity: Chronic

Other than the poorly designed rabbit studies reviewed by ATSDR and described in the previous section, no studies were identified that tested the chronic toxicity of tetryl.

Studies Relevant for Mammalian TRV Development for Ingestion Exposures

The data on toxicity of tetryl are primarily limited to acute and subchronic oral exposures in F344 rats and are presented in Table 10.2 and Figure 10.2. Recent toxicological studies involving tetryl have employed either Swiss mice or F344 rats, the latter representing the animal model for which most of the toxicity data on the compound have been derived. Reliable chronic data are not available, and there are no studies relevant to mammalian wildlife. The parameters relevant to tetryl toxicity were changes in hematological, clinical chemistry and histological parameters, and reduction in body weight [16].

As outlined in USACHPPM TG 254, TRV derivation should be based on an ecologically relevant endpoint. Decreased growth indicates an alteration in the overall energy balance of an organism, suggesting that the affected organism obtains and/or allocates energy differently than a control organism [19,20]. Alterations in energy acquisition and/or processing may in turn cause changes in reproductive schedules or leave the affected organism more susceptible to predators for a longer time. Data on the growth effects of tetryl were obtained from Reddy *et al.* [16,17].

Mammalian Oral Toxicity: Other

No other data relevant to oral exposures for mammals were found.

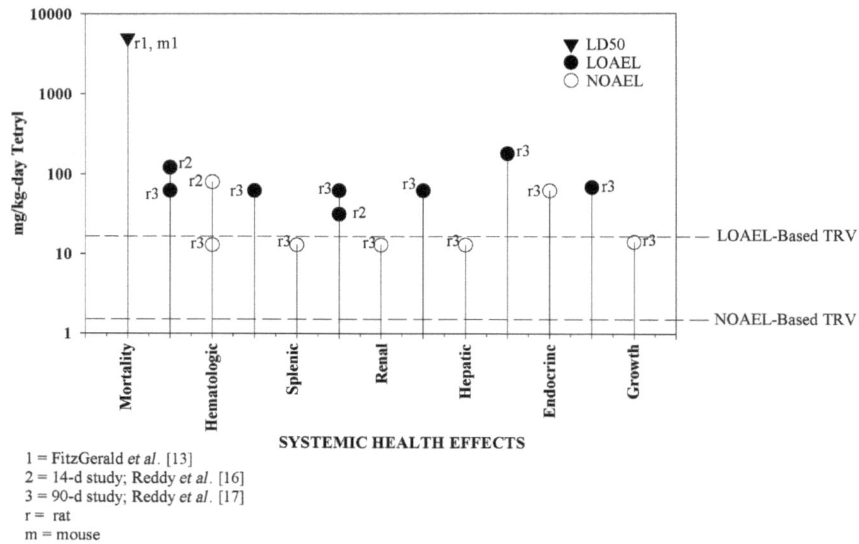

Figure 10.2 Ingested tetryl health effects to mammals. Letters by symbols represent test species, and numbers represent study.

Mammalian Inhalation Toxicity

No inhalation studies conducted using mammals were found.

Mammalian Dermal Toxicity

No dermal studies conducted using mammals were found.

Avian Toxicology

Toxicological data for the effects of tetryl in avian species were not located. Ecotoxicological research on the effects of this compound in birds is recommended.

Amphibian Toxicology

Toxicological data for the effects of tetryl in amphibian species were not located. Ecotoxicological research on the effects of this compound in amphibians is recommended.

Reptilian Toxicology

Toxicological data for the effects of tetryl in reptilian species were not located. Ecotoxicological research on the effects of this compound in reptiles is recommended.

RECOMMENDED TOXICITY REFERENCE VALUES

Toxicity Reference Values for Mammals

TRVs for Ingestion Exposures for the Class Mammalia

Data on reduced female body weight in rats were used to derive the TRV [16,17]. Data on female body weight were used for the TRV, as it was more sensitive and hence protective of males. Decreased body weight, an indication of a lower growth rate, was used to determine the TRV because alterations in this endpoint are likely to affect fitness. For example, alterations in energy acquisition or allocation patterns may impair reproductive function and/or schedules [19,20]. In addition, sustaining a smaller body size for longer times may increase risk of predation. All data concerning the toxicity of tetryl have been generated from studies on laboratory rats; no data from tetryl investigations using wildlife species were found. No reliable chronic studies were identified so the TRV was derived from a subchronic study on F344 rats [16]. Based on the available data and according to procedures outlined in USACHPPM TG 254, the approximation approach was used to derive the TRV for tetryl [3]. An uncertainty factor (UF) of 100 was used to account for interspecific variability (UF = 10) and to derive the NOAEL-based approximate TRV from a subchronic NOAEL (UF = 10) of 14.2 mg/kg/day. An uncertainty factor of 40 (UF = 10 for interspecific variability and UF = 4 to extrapolate from a subchronic LOAEL) was used to derive the LOAEL-based approximate TRV from a subchronic LOAEL of 68.9 mg/kg/day. The minimum data requirements for a high confidence rating were not met; however, sufficient data are available to assign a medium confidence rating to the TRV for tetryl [3] (Table 10.3).

TRVs for Inhalation Exposures for the Class Mammalia

Not available at this time.

TRVs for Dermal Exposures for the Class Mammalia

Not available at this time.

TRVs for Birds

Not available at this time.

TRVs for Amphibians

Not available at this time.

TRVs for Reptiles

Not available at this time.

IMPORTANT RESEARCH NEEDS

The limited availability of data on the toxicity of tetryl to wildlife species precludes the development of a high-confidence TRV. The physicochemical properties of tetryl are such that subsurface soil concentrations of tetryl near the AAP activity may be high; however, exposure to light and water will likely degrade tetryl at the air soil interface. Thus, subterranean or burrowing terrestrial species may have greater exposure risk than the general

Table 10.3 Selected Ingestion TRV for the Class Mammalia

TRV	Dose	Confidence Level
NOAEL-based	0.142 mg/kg/day	Medium
LOAEL-based	1.72 mg/kg/day	Medium

Source: [16].

wildlife populations. Avian species that forage on the ground may have increased exposure risk as well. Hence, more studies on the toxicity of tetryl to wildlife species are required, especially chronic toxicity studies and studies on nonmammalian wildlife such as birds, reptiles, and amphibians. These studies might likely allow the derivation of a higher confidence TRV.

REFERENCES

[1] Zukas JA, Walters W, Walters WP, editors. Explosive effects and applications. New York: Springer; 1998.

[2] ChemIDplus lite. [Internet], National Library of Medicine (US); 2013. [cited 2009 multiple dates, 2013-]. Available from, http://chem.sis.nlm.nih.gov/chemidplus/chemidlite.jsp.

[3] USACHPPM. Standard practice for wildlife toxicity reference values. Aberdeen Proving Ground, MD: U.S. Army Center for Health Promotion and Preventive Medicine; 2000, Technical Guide No. 254.

[4] Talmage SS, Opresko DM, Maxwell CJ, Welch CJ, Cretella FM, Reno PH, et al. Nitroaromatic munition compounds: environmental effects and screening values. Rev Environ Contam Toxicol 1999;161:1–156.

[5] ATSDR. Toxicological profile for tetryl: (2,4,6-Trinitrophenyl-*N*-methylnitramine), Atlanta, GA: U.S. Department of Health and Human Services, Public Health Service, Agency for Toxic Substances and Disease Registry; 1995. Available from, http://www.atsdr.cdc.gov/ToxProfiles/tp80.pdf.

[6] Hazardous Substances Data Bank. Tetryl. National Library of Medicine [Internet] 2013 [cited 2013-2014 multiple dates]. Available from: http://toxnet.nlm.nih.gov/.

[7] Kayser EG, Burlinson NE, Rosenblatt DH. Naval Surface Weapons Center. Kinetics of hydrolysis and products of hydrolysis and photolysis of tetryl. Fort Detrick, Frederick (MD): US Army Medical Research and Development Command; 22 October 1984, Accession Number: AD-A153 144.

[8] Boopathy R, Manning J. Biodegradation of tetryl (2,4,6-trinitrophenylmethylnitramine) in a soil-slurry reactor. Water Environ Res 1998;70(5):1049–55.

[9] Spain JC, Hughes JB, Knackmuss HJ, editors. Biodegradation of nitroaromatic compounds and explosives. Boca Raton (FL): Lewis Publishers; 2000.

[10] Fellows RJ, Harvey SD, Cataldo DA. An evaluation of the environmental fate and behavior of munitions materiel (tetryl and polar metabolites of TNT) in soil and plant systems. Richland (WA): Pacific Northwest Laboratory; MAR 1992, Fort Detrick, Frederick (MD): US Army Medical Research and Development Command. Accession Number:AD-A266548.

[11] Lotufo GR, Rosen G, Wild W, Carton G. Summary review of the aquatic toxicology of munitions constituents. Vicksburg (MS): US Army Engineer Research and Development Center; 2013, Report Number: ERDC/EL TR-13-8 Contract No.: Work Unit 33143.

[12] Nipper M, Carr RS, Biedenbach JM, Hooten RL, Miller K, Saepoff S. Development of marine toxicity data for ordnance compounds. Arch Environ Contam Toxicol 2001;41 (3):308–18.

[13] Fitzgerald GB, Austin A, DiGuilio N. Acute toxicity evaluation of nitroaromatic compounds. Fort Detrick, Frederick (MD): US Army Medical Research and Development Command; MAR 1991.

[14] Reddy G, Qualls CW, Hampton AE G, Yelamanchili A, Kim S. Acute pathological and biochemical effects of 2,4,6-trinitrophenyl-*N*-methylnitramine (Tetryl) in male rats. Res Commun Pharmacol Toxicol 1997;2:1–11.

[15] Reddy TV, Daniel FB, Robinson M, Olson GR, Wiechman B, Reddy G. Subchronic toxicity studies on 1,3,5-trinitrobenzene, 1,3-dinitrobenzene and tetryl in rats. 14-day toxicity evaluation of *N*-methyl-*N*,2,4,6-tetranitroaniline in Fischer 344 rats. Cincinnati (OH): Prepared by the U.S. EPA Environmental Monitoring Systems Laboratory; 1994, Accession Number: ADA 283663.
[16] Reddy TV, Olson GR, Wiechman B, Reddy G, Torsella J, Daniel FB, et al. Toxicity of tetryl (*N*-methyl-*N*,2,4,6-tetranitroaniline) in F344 rats. Int J Toxicol 1999;18(2):97–107.
[17] Reddy TV, Daniel FB, Robinson M, Olson GR, Wiechman B, Reddy G. Subchronic toxicity studies on 1,3,5-trinitrobenzene, 1,3-dinitrobenzene and tetryl in rats. Subchronic (90-day) toxicity evaluation of *N*-methyl-*N*,2,4,6-tetranitroaniline (tetryl) in Fischer 344 rats. Cincinnati (OH): U.S. EPA Environmental Monitoring Systems Laboratory; 1994.
[18] Griswold Jr DP, Casey AE, Weisburger EK, Weisburger JH. The carcinogenicity of multiple intragastric doses of aromatic and heterocyclic nitro or amino derivatives in young female Sprague-Dawley rats. Cancer Res 1968;28(5):924–33.
[19] Calow P. Physiological costs of combating chemical toxicants: ecological implications. Comp Biochem Physiol C 1991;100(1–2):3–6.
[20] Congdon JD, Dunham AE, Hopkins WA, Rowe CL, Hinton TG. Resource allocation-based life histories: a conceptual basis for studies of ecological toxicology. Environ Toxicol Chem 2001;20(8):1698–703.

CHAPTER 11

Wildlife Toxicity Assessment for Nitrocellulose

Michael J. Quinn Jr.

Contents

INTRODUCTION

Nitrocellulose (CAS No. 9004-70-0) is a white, fibrous, pulp-like material that generally conforms to the formula C12H16N4O18 (Figure 11.1). Nitrocellulose is also referred to as pyroxylin, gun cotton, or collodion, although the last term (also listed as CAS No. 9004-70-0) is actually a solution of approximately 6% nitrocellulose, 24% ethanol, and 70% ether [1]. Nitrocellulose can be produced by replacing between two and three hydroxyl groups in the cellulose monomer with nitro groups.

Nitrocellulose is usually kept moist with water or an organic solvent due to the compound's explosive instability when dry [2]. This property has been exploited for military use in explosives, smokeless powder, mortar increments, rocket fuels, and propellants. Nitrocellulose is also used in a number of other commercial applications, such as skin covering to protect wounds from infection, cosmetic formulations, photographic film, inert support for chromatographic separations in biological research, and

Wildlife Toxicity Assessments for Chemicals of Military Concern
http://dx.doi.org/10.1016/B978-0-12-800020-5.00011-9

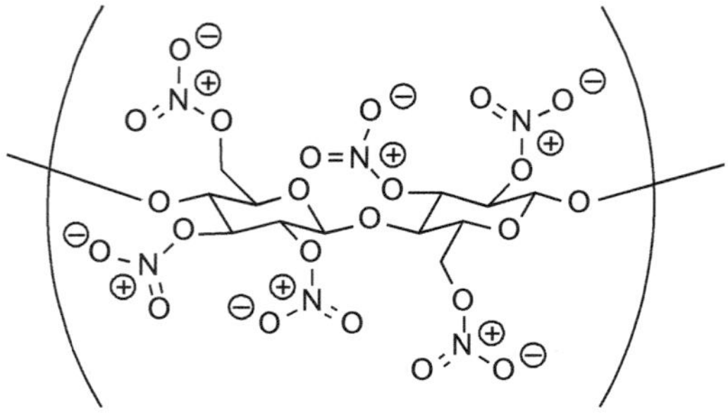

Figure 11.1 Structure of Nitrocellulose.

components in the manufacture of lacquers, patent and artificial leathers, adhesives, and so on. It is a prior-sanctioned food ingredient for use in paper and paperboard products used in food packaging [3,4] and an approved indirect food ingredient [5–7]. Reflecting scant information on the toxic effects of the compound in humans and animals, there is no U.S. Environmental Protection Agency Integrated Risk Information System (U.S. EPA IRIS) record for nitrocellulose. A summary of the physical and chemical properties of nitrocellulose is presented in Table 11.1 and further details can be obtained from available literature on this compound [8–10].

Table 11.1 Summary of Physical-Chemical Properties of Nitrocellulose

CAS No.	9004-70-0
Molecular weight	504.3
Color	White
Physical state	Fibrous
Melting point	160–170 °C (ignites)
Boiling point	34 °C
Odor	Ether-like
Solubility in water	<1 g/L at 20–25 °C: soluble in acetone, esters, ketones, glycol ethers, nitrobenzene, tetrahydrofuran; and ethyl-, butyl-, and amylacetate
Partition coefficients:	
Log K_{ow}	No data
Log K_{oc}	No data
Vapor pressure at 25 °C	1×10^{-5} mm Hg
Henry's Law constant at 25 °C	No data
Conversion factors	1 ppm = 20.63 mg/m^3 1 mg/m^3 = 0.048 ppm

Sources: HSDB [8], ICI [9], Lewis [10], Hartley *et al.* [13].

This chapter summarizes current knowledge of the likely harmful impacts of nitrocellulose on wildlife, emphasizing, where possible, threshold doses for the onset of toxicological effects, as described in reports of experimental studies of the compound.

ENVIRONMENTAL FATE AND TRANSPORT

Little information exists on the environmental fate and transport of nitrocellulose, although some behavior patterns can be inferred from the compound's physical-chemical properties, as listed in Table 11.1. For example, nitrocellulose is virtually insoluble in water, which suggests that atmospheric washout or leaching from landfills and other contaminated sites to streams and groundwater is unlikely. Recognizing that the manufacture of munitions containing the compound will generate substantial amounts of waste nitrocellulose in environmental media, the U.S. Army Environmental Center (U.S. AEC) has published an online information sheet on the potential for safe disposal of the compound by composting [11]. The U.S. AEC carried out a field demonstration at Badger Army Ammunition Plant and identified viable compost mixtures that contained biodegradable substrates and bulking agents to promote microbial catabolic activity. However, the efficiency of the system remains uncertain because of the difficulty of monitoring performance. Sullivan *et al.* [12] suggested that since cellulose triacetate and cellulose trinitrate are resistant to enzymatic digestion, nitrocellulose should be very environmentally persistent. Although biodegradation of nitrocellulose does not appear to be very probable, alkaline hydrolysis of nitrocellulose has been shown to produce metabolites that can be further broken down by microbes [13].

TOXICITY PROFILE

Summary of Mammalian Toxicity

Nitrocellulose is considered to be non-toxic as the only effects seen in mammalian oral toxicity studies (including chronic and multigenerational) have been increased feed consumption, decreased weight gain, and intestinal impaction. The impaction was only observed in mice due to the relatively small gastrointestinal tracts; the nitrocellulose was able to pass freely through the intestines of the larger rats and dogs. The effects on feeding and body weights were due to the nonnutritive bulk of the fibrous material.

Mammalian Oral Toxicity: Acute/Subacute

Few data were identified that address the acute lethality of the compound to mammals. Midwest Research Institute [14] reported a rat and mouse oral median lethal dose (LD_{50}) of a 5% nitrocellulose suspension to be greater than 5,000 milligrams per kilogram (mg/kg). No deaths were reported for the male and female CD rats; however, two of the 10 male albino Swiss mice died with no gross lesions having been observed or reported. This LD_{50} value was also reported for ICR mice [15] and Sprague Dawley rats [16] that received a single dose of Ball Powder® (83.23% nitrocellulose). No mortality was observed during either of these studies.

Mammalian Oral Toxicity: Subchronic

The Midwest Research Institute (MRI) [17] assessed the subchronic toxicity in beagle dogs, rats, and Swiss mice via similar exposure methods; individuals of each species received 1, 3, or 10% nitrocellulose for 13 weeks and were allowed to recover for 4 subsequent weeks. At the end of each study organ weights (heart, liver, spleen, kidneys, adrenal glands, pituitary, thyroid glands, and gonads), gross necropsy, body weight, feed consumption, and hematology (hematocrit, hemoglobin, total and differential erythrocyte and leukocyte counts), and blood chemistry parameters (alkaline phosphatase, cholesterol, glucose, aspartate aminotransferase, and alanine aminotransferase) were measured in all species. The beagle study also performed the bromosulfophthalein (BSP) retention assay (a test for hepatic effects) and measured levels of circulating immunoglobulin-E (IgE) to assess immune system effects. The rat study also measured circulating IgE and also completed a battery of tests to test for mutagenicity.

The only observed change in the beagle study was a 15% increase in feed consumption in dogs from the 10% dose group compared to controls. The authors speculate that net consumption was similar across treatment groups when nitrocellulose is considered to be nonnutritive bulk. No other nitrocellulose induced changes were observed.

A similar increase in feed consumption was observed in the rat 10% nitrocellulose group, with corresponding reductions in body weights that began to approach those of control rats during the recovery period. As with the dog study, no evidence of nitrocellulose toxicity was reported. The mutagenicity tests revealed no changes in chromosome frequency distribution, tetraploid numbers, or in the frequency of chromatid breaks, gaps, or translocation in the peripheral lymphocyte and kidney cultures.

In the mouse study, mortality did occur in the 10% dose group of mice, but this was due to intestinal impaction rather than nitrocellulose toxicity. Fibrous blockages were found in the distal ileum or colon upon necropsy. The actual amount of nitrocellulose ingested cannot be determined since excessive scattering of the fibers in the enclosures was reported; however, the 1 and 3% treatment levels was estimated to be equivalent to 1,690 and 5,062 milligrams per kilogram per day (mg/kg/day) for males and 1,741 and 7,000 mg/kg/day for females, respectively [13]. The authors suggest that no deaths from intestinal impaction were observed in the dogs and rats because the larger sizes of their digestive tracts allowed this material to pass without causing harm. The same effect of increased feeding as seen in the dogs and rats with nitrocellulose exposure was observed in mice, and this is also thought to have occurred to compensate for the nonnutritive fiber in the treated diets. The total and relative (to body weight) spleen weights from the 10% dose group mice were smaller than controls at 13 weeks of exposure, but this effect was no longer apparent following the recovery period. No other effects were observed.

Mammalian Oral Toxicity: Chronic

Additional studies by MRI had exposed beagles, rats, and mice orally to nitrocellulose for 12 and 24 months [18]. No significant treatment effects were observed in dogs or rats exposed to 1, 3, or 10% nitrocellulose for either 12 or 24 months. Many CD-1 mice from the 10% nitrocellulose and 10% cotton linter control treatment groups died within 3 weeks as a result of effects similarly reported in the previously discussed 13-week study. Intestinal impaction occurred at various intestinal sites from the jejunum to the rectum. These mice were replaced, and the next significant cases of mortality occurred at approximately 9 months with about 25% mortality in the 10% nitrocellulose-dosed mice (about three times more deaths than were observed in the cotton control treatment group). No histopathological examinations were performed due to a reported rapid autolysis of tissue. Nitrocellulose toxicity was not attributed to any of the deaths. The absolute intake of nitrocellulose could not be accurately determined for the 10% treatment level however; Hartley *et al.* [13] estimated the daily intake levels for the 1 and 3% doses to have been 1,814 and 4,866 mg/kg/day for the males, and 1,767 and 6,056 mg/kg/day for the females, respectively.

In a three-generation study, no significant differences were observed between treatments in male and female fertility, litter size, live born index, birth weight, viability index, or sex ratios [18]. However, significant reductions in the lactation index and weaning weight were observed in the 10% treatment levels of the F1b and F2b litters. These effects were attributed to the bulky, noncaloric nature of nitrocellulose rather than to a purely chemical one.

Although the actual amount of nitrocellulose ingested by the murine species in these studies could not be accurately measured due to excessive scattering, Hartley *et al.* [13] made a conservative estimate that 10% of the diet approximated 2,800 mg/kg/day in rats and 12,000 mg/kg/day in mice, assuming these animals consumed twice the amount when exposed to 3% treatment levels. Weight loss and an increased consumption were consistently reported across species; weight loss was characterized by a greater proportion of lean body mass than fat tissue. Also, the authors attributed the deaths that occurred in mice, but not rats or dogs, to be attributed to the relatively larger size of the fibers in relation to the size of the murine intestinal lumen. Hartley *et al.* [13] concluded that nitrocellulose, being neither digested nor absorbed, is not toxic to dogs, rats, or mice.

Studies Relevant for Mammalian TRV Development for Ingestion Exposures

One subchronic and one chronic study are relevant for the development of Toxicity Reference Values (TRV) [17,18]. These studies were conducted properly, and contained the information necessary to evaluate the results, and report them in a clear, logical, and concise form. The methods used were consistent with other accepted protocols and contained the proper controls. Thus, these studies met or exceeded the minimum quality requirements. The chronic studies were considered most relevant due to the exposure duration.

Mammalian Toxicity: Inhalation

Inhalation data for mammals were not found.

Mammalian Toxicity: Dermal

An irritation score of less than 0.2 was determined from a modified Draize test in which a 33% aqueous solution of nitrocellulose was applied to intact and abraded skin of 6 NZW rabbits [14]. A mean irritation score of 0 was determined from a study in which 0.5 g Ball Powder® in 0.5 mL isotonic saline was applied to NZW rabbits for 4 hours [15,16]. Taken together, these tests suggest that nitrocellulose is not a primary skin irritant in rabbits.

Summary of Avian Toxicology

Toxicological data for the effects of nitrocellulose in avian species were not located. Ecotoxicological research on the effects of this compound in birds is recommended.

Summary of Amphibian Toxicology

Toxicological data for the effects of nitrocellulose in amphibian species were not located. Ecotoxicological research on the effects of this compound in amphibians is recommended.

Summary of Reptilian Toxicology

Toxicological data for the effects of nitrocellulose in reptilian species were not located. Ecotoxicological research on the effects of this compound in reptiles is recommended.

RECOMMENDED TOXICITY REFERENCE VALUES

Toxicity Reference Values for Mammals

TRVs for Ingestion Exposures for the Class Mammalia

Three studies and three species are available for derivation of TRV values. An acute test in rats [8] determined that nitrocellulose had a low level of acute oral lethality based on an LD_{50} of greater than 5,000 mg/kg. The MRI [17] conducted an acute (4-day) test in rats and two subchronic tests (13 weeks) in dogs and Swiss mice and found no harmful effects in dogs, while in the rats and Swiss mice nitrocellulose appears to cause intestinal problems when it exceeds 10% of the diet and acts as a nonnutritive bulk compound. In the subchronic tests with Swiss mice, lower colon blockage was observed and led to a significant increase in death when compared to controls. Chronic studies [18] with dogs, rats, and CD-1 mice produced similar results with the smaller mice experiencing intestinal impaction and death. A chronic three-generation reproductive toxicity test in rats resulted in reductions in the lactation index and weaning weight of pups when ingested at more than 10% of the diet. Nitrocellulose, as a yellowish-white, matted mass of filaments, appears and behaves similar to raw cotton, and as such is considered toxicologically inert when ingested by dogs, rats, or mice.

Nitrocellulose has been determined to be non-toxic for most monogastric mammalian species, and adverse health consequences from exposure are not expected. However, no data are available that address the potential for

Table 11.2 Selected Ingestion TRVs for the Class Mammalia

TRV	Dose	Confidence Level
NOAEL-based	Non-toxic[1]	Medium
LOAEL-based	Non-toxic	Medium

[1] Nitrocellulose is non-toxic to monogastric mammalian species. However, it may have greater effects on ruminant or hindgut species capable of digesting cellulose.

nitrocellulose toxicity in ruminant or hindgut fermenting species that are capable of digesting cellulose, which could potentially liberate free nitrates upon digestion of nitrocellulose. Therefore, care must be exercised when applying this concept to herbivores. See Table 11.2 for a summary of mammalian TRVs.

TRVs for Inhalation Exposures for the Class Mammalia

At this time, no inhalation TRV can be derived for mammals due to insufficient data.

TRVs for Dermal Exposures for the Class Mammalia

At this time, no dermal TRV can be derived for mammals due to insufficient data.

TRVs for Birds

At this time, no TRVs for birds can be derived due to a lack of data.

TRVs for Reptiles

At this time, no TRVs for reptiles can be derived due to a lack of data.

TRVs for Amphibians

At this time, no TRVs for amphibians can be derived due to a lack of data.

IMPORTANT RESEARCH NEEDS

The limited availability of data on the toxicity of nitrocellulose to birds, reptiles, and amphibians precludes the development of a TRV. Therefore, more studies of this compound are recommended.

REFERENCES

[1] Cosmetic Ingredient Review. 2012 annual report [Internet]. Washington (DC): Cosmetic Ingredient Review [cited 2012]. Available at: http://www.cir-safety.org/sites/default/files/2012%20CIR%20Annual%20Report.pdf.

[2] Budavari S. The Merck index. An encyclopedia of chemicals, drugs, and biologicals. 12th ed. Whitehouse Station (NJ): Merck and Co., Inc; 1996.

[3] U.S. Code of Federal Regulations. U.S. Food and Drug Administration. 21 CFR Part 181. Prior-Sanctioned Food Ingredients. Section 181.30 – Substances used in the manufacture of paper and paperboard products used in food packaging. 1 April 2013.

[4] U.S. Code of Federal Regulations. U.S. Food and Drug Administration. 21 CFR Part 176. Indirect Food Additives: Paper and Paperboard Components. Section 176.170 – Components of paper and paperboard in contact with aqueous and fatty foods. 1 April 2013.

[5] U.S. Code of Federal Regulations. U.S. Food and Drug Administration. 21 CFR Part 175. Indirect Food Additives: Adhesives and Components of Coatings. Section 175.150 – Adhesives. 1 April 2013.

[6] U.S. Code of Federal Regulations. U.S. Food and Drug Administration. 21 CFR Part 175. Indirect Food Additives: Adhesives and Components of Coatings. Section 175.300 – Resinous and polymeric coatings. 1 April 2013.

[7] U.S. Code of Federal Regulations. U.S. Food and Drug Administration. 21 CFR Part 177. Indirect Food Additives: Polymers. Section 177.1200 – Cellophane. 1 April 2013.

[8] Hazardous Substance Data Bank (HSDB). Nitrocellulose [Internet]. Bethesda (MD): National Library of Medicine (U.S.), National Institutes of Health [cited 2002]. Available from: http://toxnet.njm.nih.gov/.

[9] Imperial Chemical Industries, Inc. (ICI). Product safety data sheet for industrial nitrocellulose wetted with alcohol (UN2556) or water (UN2555) [Internet]. London (UK): Imperial Chemical Industries, Inc. [cited 2002]. Accessed at http://www.nitrocellulose.com/nc_english.html.

[10] Lewis Sr. RJ. Sax's dangerous properties of industrial materials. 8th ed. New York (NY): John Wiley & Sons, Inc.; 1992.

[11] U.S. Army Environmental Center (U.S. AEC). Composting of nitrocellulose fines [Internet]. Aberdeen Proving Ground (MD): U.S. Army Environmental Center [cited 2002]. Available from: www.dtic.mil/cgi-bin/GetTRDoc?AD=ADA427022.

[12] Sullivan JH, Putnam HD, Keirn MA, Pruitt BC, Nichols JC. A summary and evaluation of aquatic environmental data in relation to establishing water quality criteria for munitions unique compounds. Part I. Nitrocellulose. Final report; 1987, Contract No. DAMD 17-77-C-7027. Prepared for: U.S. Army Medical Research and Development Command, Fort Detrick, Frederick, MD.

[13] Hartley WR, Glennon J, Gordon L, Normandy J. Nitrocellulose (NC). In: Roberts WC, Hartley WR, editors. Drinking water health advisory: munitions. Boca Raton (FL): Lewis Publishers; 1992. p. 181–99.

[14] Ellis III HV, Hodgson JR, Hwang SW, Halpap LM, Helton DO, Andersen BS, et al. Mammalian toxicity of munitions compound Phase I: acute oral toxicity, primary skin and eye irritation, dermal sensitization, disposition and metabolism, and Ames test of additional compounds. Kansas City (MO): Midwest Research Institute; 1978, Prepared for: U.S. Army Medical Research and Development Command, Fort Detrick, Frederick, MD. Progress Report No. 6, Accession Number ADA 069333.

[15] Hiatt GF, Frost DF, Wheeler C, Kortz DW. Acute oral toxicity of Ball Powder® in ICR mice. Presidio of San Francisco (CA): Letterman Army Institute of Research; 1989, Institute Report No. 351.

[16] Morgan EW, Lollini LO, Wheeler CR, Korte DW. Acute oral toxicity of Ball Powder® in Sprague-Dawley rats. Presidio of San Francisco (CA): Letterman Army Institute of Research; 1989, Institute Report No. 350.
[17] Ellis III HV, Kowalski JJ, Hodgson JR, Bhandari JC, Sanyer JL. Mammalian toxicity of munitions compounds. Phase II: effects of multiple doses. Part IV: nitrocellulose. Kansas City (MO): Midwest Research Institute; 1976, Prepared for U.S. Army Medical Research and Development Command, Fort Detrick, Frederick, MD. Progress Report No. 5, Accession Number: ADA062016.
[18] Ellis III HV, Hagensen JH, Hodgson JR, Minor JL, Hong CB. Mammalian toxicity of munitions compounds. Phase III: effects of life-time exposure. Part III: nitrocellulose. Kansas City (MO): Midwest Research Institute; 1980, Prepared for U.S. Army Medical Research and Development Command, Fort Detrick, Frederick, MD. Progress Report No. 9, Accession Number: ADA079353.

CHAPTER 12

Wildlife Toxicity Assessment for Pentaerythritol Tetranitrate

Michael J. Quinn, Jr., Mark S. Johnson

Contents

INTRODUCTION

Pentaerythritol tetranitrate (PETN) is an explosive chemical that is currently used as the primary ingredient in detonating fuses and as a component (mixed with hexahydro-1,2,5-trinitro-1,3,4-triazine) in "plastic" explosives such as *Semtex* [1]. PETN (CAS No. 78-11-5) is structurally and functionally similar to nitroglycerin in regard to its pharmacology and explosive properties. Thus, in the human health field, both PETN and nitroglycerin are used

Wildlife Toxicity Assessments for Chemicals of Military Concern
http://dx.doi.org/10.1016/B978-0-12-800020-5.00012-0 2015 Published by Elsevier Inc.

medicinally in the treatment of angina because of their shared vasodilatory action. However, for either drug, repeated exposure can establish a sequence of tolerance, then dependence as the body adjusts to the presence of either hypotensive agent. For occupationally exposed subjects, cycles of exposure and withdrawal associated with a 5-day workweek (exposure) followed by a 2-day weekend (withdrawal) can lead to the well-recognized "Monday-morning death" of munitions employees. This phenomenon results from cardiovascular events triggered by unrestrained compensatory vasoconstriction, as the normally high organic nitrate levels in the body become reduced during the weekend [2]. The importance of PETN as an environmental contaminant is related to its distribution at and around military sites and to its potential toxicity to wildlife and other ecological receptors. This chapter summarizes current knowledge of the likely harmful impacts of PETN on wildlife, emphasizing threshold doses for the onset of toxicological effects as described in reports of experimental studies of PETN. Surveying the threshold dosimetry of the compound may point to the establishment of Toxicity Reference Values (TRVs) that could serve as protective exposure standards for wildlife near affected sites.

TOXICITY PROFILE

Environmental Fate and Transport

Pentaerythritol tetranitrate is produced in much smaller quantities than other military explosives. Therefore, there are few data on either the compound's environmental fate or its site-specific concentrations in environmental media. Physicochemical characteristics of PETN that may be relevant to the environmental fate and transport of the compound are listed in Table 12.1.

Layton *et al.* [3] were unable to find any information about PETN's ability to undergo photolysis but speculated that the process is unlikely to be significant since the compound absorbs little ultraviolet or visible light.

The estimated values for vapor pressure and Henry's Law constants are sufficiently low to indicate that PETN is unlikely to be released into the air as a vapor. However, aerial dispersion of the compound adhering to soil or dust particles is a possible mechanism by which PETN can be released to the atmosphere. The low water solubility of PETN coupled with its low K_{ow} suggests a poor bioavailability of PETN to most organisms.

Table 12.1 Summary of the Physical-Chemical Properties of PETN

CAS No.	78-11-5
Molecular weight	316.15[a]
Color	white[a]
Physical state	crystalline solid[a]
Melting point	140 °C[a]
Boiling point	180 °C (decomposes above 150 °C)[a]
Odor	slight[a]
Solubility Water	2.042 mg/L at 30 °C[b]; soluble in benzene, sparingly soluble in alcohol, ether, etc.[a]
Partition coefficients:	
Log K_{ow}	2.04[b]
Log K_{oc}	2.25–3.24[a]
Vapor pressure at 25 °C	1.035×10^{-10} mm Hg[a]
Henry's Law constant at 25 °C	1.2×10^{-11} atm.m^3/mole[a]
Conversion factors	1 ppm = 12.93 mg/m^3; [a]
	1 mg/m^3 = 0.077 ppm[a]

Sources: [a]HSDB [1], [b]Quinn *et al.* [7].

SUMMARY OF MAMMALIAN TOXICITY

Mammalian Oral Toxicity: Acute

Von Oettingen *et al.* [4] administered a 20 milligrams per milliliter (mg/ml) PETN solution by gavage to young female albino rats. Following a 6-hour period, the entire gastrointestinal tract was removed and the contents analyzed. It was determined that only 13% of the compound had been absorbed. Dogs were also given oral doses of 5 milligrams per kilogram (mg/kg) PETN. A gradual decrease in blood pressure corresponding with an increase in respiratory rate and minute volume were reported; no details were presented. No other studies were identified.

Mammalian Oral Toxicity: Subacute

One of the few studies on the toxicity of PETN in laboratory animals was that of the National Toxicity Program (NTP) [5], which addressed the toxicity of the compound mixed in a 1:4 ratio with D-lactose monohydrate as a stabilizer, designated as the mixture PETN, NF. In the various phases of the study, the mixture was administered to F344 rats and B6C3F1 mice in the diet for 14 days, 13 – 14 weeks, or 2 years. In the subacute phase of the study, five rats and five mice/sex/group were exposed to 0, 3,100, 6,200, 12,500,

25,000, or 50,000 mg/kg PETN, NF for 14 days. These concentrations were approximated to doses of 0, 54, 111, 258, 502, and 909 mg/kg/day PETN in male rats; 0, 69, 145, 272, 552, and 1,087 mg/kg/day in female rats; 0, 164, 281, 556, 1,255, and 2,587 mg/kg/day in male mice; and 0, 215, 534, 784, 1,675, and 3,333 mg/kg/day in female mice. No clinical signs or toxicological responses were induced in any of the animals under investigation [5].

Mammalian Oral Toxicity: Subchronic

The same dietary concentrations of PETN, NF that were used in the 14-day study were administered to 10 rats and mice/sex/group for 13 or 14 weeks, resulting in approximately 0,43, 90, 203, 319, or 625 mg/kg/day PETN in male rats; 0, 54, 105, 240, 463, or 931 mg/kg/day in female rats; 0, 131, 319, 450, 1,028, or 2,163 mg/kg/day in male mice; and 0, 780, 336, 694, 1,406, or 3,170 mg/kg/day in female mice. As in the subacute study, few if any overtly PETN-related effects were observed among the responses [5]. There was, however, a tendency for higher-than-normal relative brain and kidney weights in the three high-dose groups in males and the highest dose in female rats, respectively. Nitrite was detected sporadically in the urine of some male and female rats, and a Zymbal gland adenoma was detected in a single high-dose female rat. Only female mice exposed to 3,170 mg/kg/day had lower mean body weights (13%) than controls. No other changes were observed [5].

Kodja *et al.* [6] exposed nine female New Zealand white rabbits/group for 15 weeks to 6 mg/kg/day PETN in either regular chow or chow with added cholesterol. The hypercholesterolemic effects of the cholesterol-spiked diet were marked, although the presence of PETN appeared to significantly reduce the incidence and size of consequent atherosclerotic lesions. Twenty-four hours after the completion of the feeding study, excised cross-sectional pieces of the thoracic aorta were tested for endothelial functional integrity, as indicated by their contractile response to potassium chloride (KCl), phenylephrine, and by their subsequent vasorelaxant response to acetylcholine, measured in each case by a force-displacement transducer. In general, the different diets induced only marginal effects on the contractile response of the thoracic aorta to KCl or to phenylephrine; PETN slightly reduced the vasorelaxing potency of acetylcholine [6].

Mammalian Oral Toxicity: Chronic

The 2-year component of the NTP study on PETN, NF featured the dietary administration of 0, 25,000, and 50,000 mg PETN, NF/kg chow to 50 male

F344 rats/group and to both sexes of B6C3F1 mice. Additionally, 0, 6,200, or 12,500 mg/kg PETN, NF chow was administered to 50 female F344 rats/group. Average doses were equivalent to 0, 300, and 625 mg/kg/day PETN in male rats; 0, 100, and 208 mg/kg/day in female rats; 0, 1,000, and 2,025 mg/kg/day in male mice; and 0, 1,275, and 2,425 mg/kg/day in female mice. For 13 weeks, clinical signs were monitored twice daily, and body weights and food consumption were monitored weekly. Following week 13, all three characteristics were monitored monthly for the remainder of the study. At the study's termination, a full suite of organs and tissues were examined at necropsy and preserved by fixation. Histopathological examinations were performed on sections from all high-dose animals and controls.

In the rats, survival was comparable among the groups, and there were no obvious clinical signs [5]. Body weights and food consumption in PETN-receiving groups were similar to those of the controls, and there were few noteworthy necropsy findings. Mean body weights of high-dose male rats were 2 to 9% lower than the controls throughout the study; body weights of all other groups of rats were similar. Histopathologically, important findings were restricted to the Zymbal gland, in which a small, but increased incidence in adenomas and carcinomas was observed compared to controls. It is noted, however, that the observed incidences of adenomas and carcinomas were within historical data of nontreated rats [5]. This dose-dependent change, while statistically insignificant when compared to controls (Fisher's exact test), was considered by the NTP study scientists and review board to constitute equivocal evidence of the compound's carcinogenicity in this species. However, no such response was evident among the mice, in which few (if any) compound-related toxicological effects of PETN, NF were observed [5]. No nonneoplastic lesions were attributable to PETN exposure in either animal model. Mononuclear leukemia rates in male rats were significantly lower relative to treatment. Hepatocellular adenomas or carcinomas also occurred, with a significant trend observed in female mice [5].

Von Oettingen *et al.* described a one-year study in an unspecified strain of young albino rats [4]. Two groups of 45 rats were given either a control diet or one equivalent to 2 mg/kg/day PETN. Food consumption was measured daily, and animals were weighed weekly. Blood parameters were evaluated at monthly intervals. Other than a tendency for PETN-exposed animals to show slightly higher hemoglobin and erythrocyte values, no other effects were seen [4]. Additional evaluations during this study included a histopathological examination of the brain, heart, lungs, adrenals, liver, spleen, kidneys, testis, and femur.

Quinn *et al.* [7] conducted a study according to the U.S. Environmental Protection Agency (U.S. EPA) Reproduction/Developmental Toxicity Screening Test [8]. Sprague Dawley rats were exposed to oral daily adjusted volumetric doses of 0, 100, 500, or 1000 mg/kg PETN in a corn oil vehicle for up to 56 days. Mating, duration of gestation, body weight, feed consumption, overall condition of adults, and the number, sex, and condition of pups were recorded. Histological examinations were also performed on the ovaries, testes, and epididymides of animals from the control and highest dose groups. Only body weights and feed consumption were affected by treatment; however, these differences may be attributed more to volumetric adjustments of vehicle in the control and high-dose groups than to PETN toxicity. No adverse effects on reproduction or development from PETN exposure were observed [7].

Studies Relevant to Mammalian TRV Development for Oral Exposures

In general, there is a paucity of experimental data that suggest adverse effects from oral exposure to PETN. The incidences of neoplasms associated with the Zymbal gland were within the bounds of historical results and only apparent in rats from the 104-week study [5]. Moreover, there were no differences in the survival of the rats and mice at the end of the 104-week studies, and any ecological impact of cancer in senescent animals is very uncertain and of questionable relevance. In addition, while few reports of studies on the acute lethality of the compound were found, most of the experimental evidence from the NTP suggests the benign nature of the compound when administered to experimental animals (Table 12.2). Thus, when PETN, NF was administered to F344 rats and B6C3F1 mice in dietary concentrations that were at the maximum practical limit (5% wet weight (w/w)), few if any overtly toxicological responses were observed during exposure periods extending up to 104 weeks [5].

Among the few possibly compound-related effects of PETN were the trends toward increased relative kidney weights compared to controls in high-dose female rats after 13 weeks, and the dose-dependent increase in combined adenomas and carcinomas of the Zymbal gland in male and female rats exposed for two years [5]. However, the marginal nature of each of these individual responses raises questions as to whether these findings are incidental or truly compound-related. For example, supporting evidence of the compound's carcinogenicity might be expected to arise from mutagenicity/genotoxicity studies. However, in this instance, the evidence is also

Table 12.2 Summary of Relevant Data for TRV Derivation in the Class Mammalia

		Test Results			
Test Organism	**Test Duration**	**NOAEL (mg/kg/day)**	**LOAEL (mg/kg/day)**	**Effects Observed at the LOAEL**	**Study**
Rat (F344)	14-week	370 (f) 495 (m)	745 (f) ND (m)	Increase in relative brain and kidney weight was observed in females.	[5]
	104-week	240 (m) 166 (f)	500 (m)[a] ND	Males had 2–9% lower body weight than controls.	
Mouse (B6C3F1)	14-week	1125 (f) 1730 (m)	2535 (f) ND (m)	Females had 13% lower body weight than controls.	[5]
	104-week	1620 (m) 1940 (f)	ND	No adverse effects were observed in mice.	
Dog	Not reported	5	ND	Mild changes in blood pressure and respiratory rate. Secondary source: not used in TRV derivation.	[4]
Rat (strain unspecified)	1-year	2	ND	No effects observed. Secondary source: not used in TRV derivation.	

f=female; m=male; ND=not determined.

[a]The incidences of Zymbal gland neoplasms and nonneoplasms were determined to be of "equivocal significance" by the authors on the ability of PETN to be a causative agent. No differences in survival were seen in any of the dose groups (NTP 1989).

equivocal since PETN was negative for gene reversion in the Ames test, negative for the induction of chromosomal aberrations in Chinese hamster ovary cells, but positive for the induction of sister chromatid exchanges in the same experimental system [9,5].

Collectively, these findings speak to the comparatively benign nature of the compound in experimental animals, albeit within the context of a limited information base. Both of the NTP studies (14- and 104-week studies in rats and mice) were conducted, reported, and recorded appropriately and are of sufficient quality to be deemed relevant for TRV derivation.

Mammalian Oral Toxicity: Other

No other data relevant to oral exposures for mammals were found.

Mammalian Inhalation Toxicity

No inhalation studies conducted with animals were found. Von Oettingen *et al.* [4] reported that PETN is absorbed through the lungs based on observations of decreased blood pressure and increased venous and spinal pressures following insufflation of 100 mg PETN in the lower trachea of dogs.

Mammalian Dermal Toxicity

No dermal studies conducted using animals were found. However, a study in which PETN was mixed with acetone and was rubbed onto the hands of a human volunteer resulted in the recovery of the entire amount of compound through washing [4]. This suggests that the probability for dermal absorption is relatively low.

SUMMARY OF AVIAN TOXICITY

Toxicological data for the effects of PETN in avian species were not located. Ecotoxicological research on the effects of this compound in birds is recommended.

SUMMARY OF AMPHIBIAN TOXICITY

Toxicological data for the effects of PETN in amphibian species were not located. Ecotoxicological research on the effects of this compound in amphibians is recommended.

SUMMARY OF REPTILIAN TOXICITY

Toxicological data for the effects of PETN in reptilian species were not located. Ecotoxicological research on the effects of this compound in reptiles is recommended.

RECOMMENDED TOXICITY REFERENCE VALUES

Toxicity Reference Values for Mammals

TRVs for Ingestion Exposures for the Class Mammalia

The suite of studies reported by NTP are comprehensive and of good quality. The NTP report contains subacute, subchronic, and chronic study information for mice and rats. Extremely high doses were used in an effort to bind the effects data. However, no corroborative effects were observed in either target or magnitude of effect [5]. For example, kidney and brain weights were increased in the high dose for female rats only in the 14-week study, yet no other information suggested that adverse effects were observed relative to the kidney and brain function (i.e., histopathology results revealed no effect), nor did the chronic study show the same effects. Male rats in the high-dose group in the 104-week study showed a 2 to 9% lower body mass than controls, yet these effects were not seen in females in the 104-week study or in either sex in the 14-week study [5].

The mouse data were also equivocal. Only the 14-week study showed a significant change in body weight (13%, females only), yet no adverse effects were seen in either sex for the 104-week study [5]. Collectively, these data suggest the possibility of false-positive data rather than a compound-related effect. Additionally, the data reported in von Oettingen *et al.* [4] and Quinn *et al.* [7] suggest no adverse effects from prolonged oral exposure to PETN.

Information is available from three species and two orders (Carnivora and Rodentia). However, only one chronic lowest observed adverse effect level (LOAEL) of questionable relevance is available. Therefore, the no observed adverse effect level (NOAEL)-LOAEL approach was approximated. Since the highest dose tested in rats approximated 166 and 500 mg/kg/day for female and male rats, respectively, the TRVs were calculated from the female NOAEL (166 mg/kg/day) according to the *Standard Practice for Wildlife Toxicity Reference Values* [10]. The NOAEL-based TRV for mammals is therefore 170 mg/kg/day, and the LOAEL-based TRV for mammals is 1,700 mg/kg/day (through the application of a 10-fold uncertainty factor). Since the LOAEL observations do not provide clear indications of an adverse

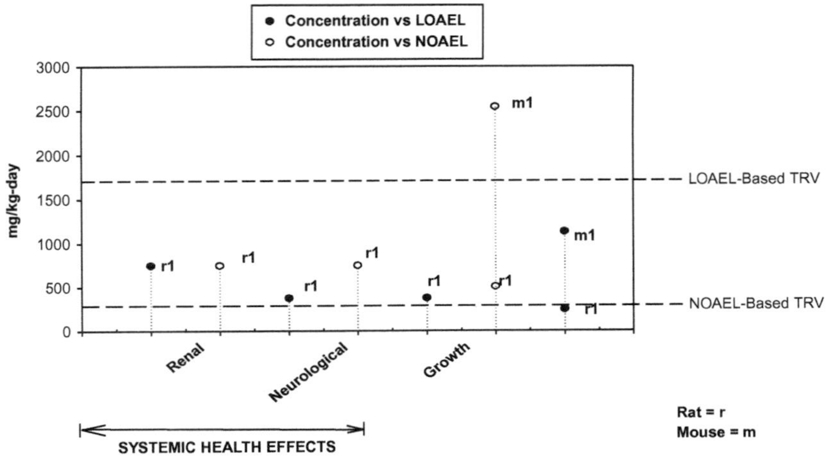

1 = NTP, 1989 [5]

Figure 12.1 PETN health effects to mammals. Letters by symbols represent test species, and numbers represent study.

effect, and only rodents are represented, the LOAEL-based TRV received a confidence rating of "low." However, since no indications of adverse effects were seen at the NOAEL, the NOAEL-based TRV received a "medium" level of confidence. Based on the available data, it is believed that these levels should be protective of species of mammals. Table 12.2 provides a summary of the relevant data for TRV derivation in the Class Mammalia, and Table 12.3 provides the selected ingestion TRVs for same.

TRVs for Ingestion Exposures for Mammalian Foraging Guilds

Toxicity reference values specific to particular guild associations (e.g., small, herbivorous mammals) have not yet been derived. However, these data may be considered as representative of omnivorous rodents. More data for other species would be needed to derive values for other foraging guilds.

Table 12.3 Selected Ingestion TRVs for the Class Mammalia

TRV	Dose	Confidence Level
NOAEL-based	170 mg/kg/day	Medium
LOAEL-based	1700 mg/kg/day	Low

Source: [5].

TRVs for Inhalation Exposures for the Class Mammalia
No data are available at this time.

TRVs for Dermal Exposures for the Class Mammalia
No data are available at this time.

TRVs for Birds
No data are available at this time.

Toxicity Reference Values for Amphibians
No data are available at this time.

Toxicity Reference Values for Reptiles
No data are available at this time.

IMPORTANT RESEARCH NEEDS
Data are lacking indicating clear adverse effects from exposure to PETN. Given the concentrations used in some of the studies reported herein, it is unlikely that further research using traditional laboratory animals would yield different results. However, testing in other mammalian, amphibian, avian, and reptilian species is needed to determine if the effects of PETN on those species differ from those observed in laboratory mammalian models.

REFERENCES
[1] Hazardous Substances Data Bank (HSDB). Pentaerythritol [Internet]. Bethesda (MD) National Library of Medicine (US), National Institutes of Health [cited 2000]. Available from: http://toxnet.njm.nih.gov/.

[2] Abrams J. Nitrate tolerance and dependence. Am Heart J 1980;99:113–23.

[3] Layton D, Mallon B, Mitchell W, Hall L, Fish R, Perry L, et al. Data-base assessment of the health and environmental effects of conventional weapons demilitarization: explosives and their co-contaminants. Livermore (CA): University of California Lawrence Livermore National Laboratory; 1987.

[4] von Oettingen WF, Donahue DD, Lawton AH, Monaco AR, Yagoda H, Valaer PJ. Toxicity and potential dangers of penta-erythritol-tetranitrate (PETN). Washington (DC): U.S. Public Health Service Federal Security Agency; 1944, Public Health Service Bulletin No. 282.

[5] National Toxicology Program (NTP). Toxicology and carcinogenesis studies of pentaerythritol tetranitrate (CAS NO. 78-11-5) with 80% D-lactose monohydrate (PETN, NF) in F344/N rats and B6C3F1 mice. Triangle Park (NC): U.S Department of Health and Human Services; 1989, NTP Toxicology Report 365.

[6] Kojda G, Stein D, Kottenberg E, Schnaith EM, Noack E. In vivo effects of pentaerythritol-tetranitrate and isosorbide-5-mononitrate on the development of atherosclerosis and endothelial dysfunction in cholesterol-fed rabbits. J Cardiovasc Pharmacol 1995;25:763–73.
[7] Quinn Jr MJ, Crouse L, McFarland CA, LaFiandra EM, Johnson MS. Reproductive and developmental effects and physical and chemical properties of pentaerythritol tetranitrate (PETN) in the rat. Birth Defects Res B Dev Reprod Toxicol 2009;86:65–71.
[8] U.S. EPA. Health effects test guidelines, OPPTS 870.3550: reproduction/developmental toxicity screening test. Washington (DC): U.S. Environmental Protection Agency Prevention, Pesticides, and Toxic Substances; 2000. EPA 712-C-00-367.
[9] Mortlemans K, Haworth S, Lawlor T, Speck W, Tainer B, Zeiger E. Salmonella mutagenicity tests: II. Results from the testing of 270 chemicals. Environ Mutagen 1986; 8(Suppl. 7):1–119.
[10] U.S. Army Center for Health Promotion and Preventive Medicine. Standard practice for wildlife toxicity reference values. Aberdeen Proving Ground (MD): U.S. Army Center for Health Promotion and Preventive Medicine; 2000, Technical Guide 254.

CHAPTER 13

Wildlife Toxicity Assessment for Nitroglycerin

Marc A. Williams, Bradley E. Sample, Glenn J. Leach

Contents

INTRODUCTION

Nitroglycerin (NG; also known as trinitroglycerine or TNG, glyceryl trinitrate, and 1,2,3-trinitroxypropane), CAS No. 55-63-0, is produced for use by the military as a component of propellants and explosives and is used commercially as a pharmaceutical agent. Nitroglycerin is a vasodilator widely used by the medical community in the treatment of angina and other cardiovascular pathologies [1,2]. There is an extensive body of literature related to the medical uses of nitroglycerin. However, very few of these studies were found relevant to the development of wildlife toxicity values.

Wildlife Toxicity Assessments for Chemicals of Military Concern
http://dx.doi.org/10.1016/B978-0-12-800020-5.00013-2

This toxicity assessment is based on a thorough review of the scientific literature regarding the toxicological characteristics of nitroglycerin that may pertain to the health of wildlife (mammals, birds, reptiles, and amphibians) exposed to the substance. The protocol for the performance of this assessment is documented in the U.S. Army Center for Health Promotion and Preventive Medicine (USACHPPM) Technical Guide (TG) 254, *Standard Practice for Wildlife Toxicity Reference Values* [3]. This document is designed to support ecological risk assessment activities.

ENVIRONMENTAL FATE AND TRANSPORT

It is possible that nitroglycerin is released to the environment from its production and use as a component of propellants and explosives and as a pharmaceutical compound. Wastewater discharges from the manufacture of commercial dynamite preparations, military explosives, and other sources of production might contain NG. The relatively high solubility of NG in water (i.e., 1,800 milligrams per liter (mg/L) at 20 °C) suggests that environmentally significant concentrations might be dissolved in waste rinse water and be expected to remain in the water column [4]. Neat NG does not polymerize and exists as a stable solid in the form of dipyramidal crystals at temperatures below 13.5 °C. Liquid NG begins to decompose at 50 to 60 °C, and at 145 °C, decomposition is so rapid that the liquid appears to boil. At 256 °C, NG spontaneously explodes [4]. The chemical and physical properties of NG are presented in Table 13.1.

The release of NG to the environment is influenced by both abiotic and biotic processes. Physical and chemical degradation of NG is generally slow [8]. The photolytic half-life of NG in pure water was estimated to be 5 days [9], and the hydrolytic half-life of NG at normal environmental temperatures was predicted to be more than 1 year at pH 3 to 8 [10]. An alkaline environment significantly decreases the hydrolytic half-life of NG [10].

Extensive study was made of the kinetics and products formed on the hydrolysis of NG and each of the isomeric dinitroglycerols (DNGs) and mononitroglycerols (MNGs). Chemical hydrolysis does not follow the successively slower stepwise pathways found for microbial systems. In stark contrast, NG hydrolyzes to a complex mixture of products more slowly than DNGs or MNGs; for example, nitrate, nitrite, and oxalate. Additionally, 1,3-DNG is hydrolyzed to glycidyl nitrate, and 1,2-DNG isomerizes to 1,3-DNG to form a similar product on hydrolysis. Additionally, 2-MNG isomerizes to 1-MNG before hydrolysis, where nitrate is the only product identified [11].

Table 13.1 Physical/Chemical Properties of Nitroglycerin

CAS No.	55-63-0
Molecular weight	227.09
Color	Pale yellow
Physical state	Viscous liquid; triclinic or rhombic crystals below melting point
Melting point	13.5 °C (stable form)/2.8 °C (labile form)
Boiling point	Apparently boils by decomposing rapidly at temperatures of 122–140 °F; explodes at 218–256 °C
Taste	Sweet, burning taste
Solubility	1,800 mg/L @ 25 °C; soluble in acetone, benzene, and chloroform
Partition coefficients	
Log K_{OW}	1.62
Log K_{OC}	1.51 estimated
Vapor pressure, Pa at 20 °C	2.0×10^{-4} mm Hg
Henry's Law constant at 20 °C	Between 3.3 and 0.06 torr-l/mole
Odor	No information
Conversion factors	1 ppm = 9.29 mg/m^3 1 mg/m^3 = 0.108 ppm

Sources: Roberts and Hartley [4], HSDB [5], Lyman *et al.* [6], ACGIH [7].

Nitrate and nitrite are highly mobile in soil and water wherein microbes will convert nitrate to nitrite, which can potentially cause methemoglobinemia in mammals. Plants may accumulate nitrates, and ingestion of plant material by ruminants (rumen microorganisms reduce nitrate to nitrite) may also result in methemoglobinemia [12]. However, the products of NG hydrolysis are not expected to concern wildlife due to a negligible reaction rate outside of an extreme alkaline environment.

Contrary to some earlier reports that NG was resistant to biodegradation, it has proven to be readily biodegradable in batch and continuous tests. Breakdown of NG takes place in stages via the isomeric di- and mononitrates, with each successive step proceeding at a slower rate [13,14]. It was found that NG is not suitable as a source of carbon and nitrogen, so nutrients are essential. Where NG did not biodegrade, it was speculated that earlier experiments had been conducted using NG concentrations that might have been toxic to microorganisms [2,8,13]. In the environment, NG would likely biotransform through a series of successive denitration steps, and the products would be mineralized by biological systems and incorporated into biomass [14].

If released on land, NG may readily leach into the soil. Few experimental data are available on its fate in the soil, but biodegradation and hydrolysis under alkaline conditions might occur. Few data are available on the fate of NG that is released in aquatic environments; however, environmental degradation of NG appears to occur primarily through biodegradation and photolysis. If released to the atmosphere, NG will probably be in aerosol form and subject to gravitational settling and scouring by rain. Photolysis is also a possibility although data are lacking. The Henry's Law constant calculated from reported water solubilities and vapor pressures ranges between 3.3 and 0.06 torr liters per mole (Torr L/mole) resulting in a half-life for volatilization from water of about 3,000 days (or approximately 8 years). Therefore, physical transport from aqueous systems should be relatively unimportant. Because nitroglycerin is relatively soluble in water (1,800 mg/L at 25 °C), its adsorption to soil and sediment and its bioconcentration in aquatic organisms should not be appreciable [5,8,10].

SUMMARY OF MAMMALIAN TOXICITY

Nitroglycerin has profound effects on the systemic and cardiac microcirculation. Its actions are mediated by stimulation of soluble guanylate cyclase in vascular smooth muscle cells. Long-term industrial exposure to NG has been associated with withdrawal symptoms and sudden death from cardiovascular accidents [15].

Nitroglycerin is rapidly absorbed, distributed, metabolized, and eliminated in both laboratory animals and human subjects. Metabolism occurs in both hepatic and extrahepatic tissues via stepwise denitrification, and elimination is primarily in the urine and expired air. Absorption is somewhat less in mice than other species [8].

In most species, urinary metabolites consisted largely of free MNGs, glycerol, and other polar metabolites that included glucuronides; only small amounts of TNG and free DNGs were excreted. Mice excreted only small amounts of free MNG and DNG- and MNG-glucuronides, which indicated the relatively complete biotransformation in mice [4].

Nitroglycerin is absorbed through intact skin in amounts sufficient to cause vasodilation. In humans, the most prominent manifestations of NG toxicity are severe headaches and adverse cardiovascular effects, including organic nitrate dependence in the case of chronic exposure [1]. In animals, the adverse effect most often observed after administration of NG at high dosage levels is decreased weight gain (related to decreased food consumption); effects were also seen in the liver (lesions), blood (methemoglobinemia), and testes (lesions and aspermatogenesis) [4].

There are limited mutagenicity data for nitroglycerin. In bacterial tests with *Salmonella typhimurium*, NG was mutagenic in 1 out of 7 strains tested [16]. It also tested positive for chromosomal aberrations both with and without metabolic activation in at least one test using Chinese hamster ovary cells [17].

Developmental and reproductive studies in animals have failed to demonstrate that NG is a teratogen. However, exposure to high concentrations of NG can result in testicular lesions and male infertility, and delayed development of offspring [8].

Mammalian Oral Toxicity: Acute

The acute toxicity of NG in mammals is moderate. The oral LD_{50} was approximately 500–900 mg/kg in rats and 500–1200 mg/kg in mice [18,19]. Differences in the acute toxicity of NG between genders or among species appeared insignificant or minimal.

Charles River CD rats and Albino Swiss mice were administered NG in lactose/peanut oil by oral gavage, and the survivors were observed for 14 days. The oral LD_{50} was 822 and 884 mg/kg in male and female rats, respectively; and 1,188 and 1,055 mg/kg in male and female mice, respectively. All animals became cyanotic and ataxic within 1 hour of dosing and had very pale extremities and depressed respiration. Animals usually died within 5 to 6 hours of dosing. Additionally, gross pathology was observed in the animals that died. Survivors generally recovered within 24 hours [18]. The number of male and female animals used in these studies was not given.

Oketani *et al.* [19] administered NG in propylene glycol to S1c:dd mice and S1c:SD rats by oral gavage and observed the surviving animals for 7 days. Ten animals of each gender were used in these studies. The LD_{50} was 525 and 540 mg/kg for male and female rats, respectively; and 550 and 500 mg/kg for male and female mice, respectively. Deaths occurred within 48 hours, and survivors recovered within 48 to 72 hours. Autopsy did not reveal any notable findings [19].

Four groups of adult beagle dogs, each consisting of two males and two females, were given 25, 50, 100, or 200 mg/kg/day NG in capsules for

5 days. A transient and dose-related severe methemoglobinemia was observed. Animals that received 100 and 200 mg/kg/day had cyanosis that lasted several hours. High-dose dogs also exhibited decreased activity. Due to the severity of the effects of NG, and to study any protective effect on methemoglobin formation, the high-dose animals were intravenously given 3 mg/kg methylene blue (to induce methemoglobin formation) on the third day, and treatment was discontinued thereafter. The two low-dose groups had transient methemoglobinemia and no adverse clinical effects [20,21].

Mammalian Oral Toxicity: Subchronic

Thirteen-week oral studies were conducted in dogs, rats, and mice as described below [20,21]. Because no adverse effects were observed during the first several weeks of the 13-week studies, the doses were increased 5-fold for dogs at the fifth week, for rats at the sixth week, and for mice at the fourth week. Healthy young adult beagle dogs, four males and four females per treatment group, were given 0, 0.01, 0.1, or 1 mg/kg/day NG in capsules for 4 weeks, then 0, 0.05, 0.5, and 5 mg/kg/day for 9 more weeks. Adverse effects were not seen at any dose [20,21]. The NOAEL was 5.0 mg/kg/day.

CD-1 mice (n = 16 male and n = 16 female per treatment group) were fed 0.001, 0.01, or 0.1% NG in the diet for 3 weeks, after which intakes were calculated as 1.3/1.3, 11.5/10.9, and 107/95 mg/kg/day in males and females, respectively. Next, animals were fed 0.005, 0.05, and 0.5% NG for 10 additional weeks, wherein the intakes were calculated as 6.4/6.9 m/f, 60.2/58.7 m/f, and 608/561 mg/kg/day in males and females, respectively. Treated mice had mild extramedullary hematopoiesis in the liver and spleen that was likely not dose-related. No additional adverse effects were observed at any dose level [20,21]. The NOAEL was 608 and 561 mg/kg/day in males and females, respectively.

CD rats (n = 16 males and n = 16 females per treatment group) were fed 0.001, 0.01, or 0.1% NG in the diet for 5 weeks, after which intakes were calculated as 0.8/0.9, 6.0/6.4, and 59/59 mg/kg/day in males and females, respectively. Next, animals were fed 0.005, 0.05 and 0.5% NG for 8 additional weeks, wherein the intakes were calculated as 2.6/3.1, 24.5/26.5, and 230/234 mg/kg/day in males and females, respectively. Reversible decreases in food consumption, weight gain, and increased SGOT (aspartate aminotransferase) levels (an enzyme that signals and enables detection of liver damage) were observed in the high-dose group after an increased dosage. No additional significant toxicological findings were observed, and no adverse effects were observed in any of the other dose groups. The LOAEL was 230 and 234 mg/kg/day for decreased weight gain and food consumption in males and females, respectively. The NOAEL was 24.5 (males) and 26.5 (females) mg/kg/day. To assess the effects of higher NG doses, CD rats (n = 3 males, and n = 3 females; n = 4 males, and n = 4 female controls) were fed 2.5% NG in the diet (overall average intake of 1,406 or 1,416 mg/kg/day, m/f) for 13 weeks [20,21]. Dosing for males and females was initiated at 1,176 and 1,076 mg/kg/day, respectively, and increased to 1,588 or 1,773 mg/kg/day, respectively, during the last 5 weeks of the study. A lactose control group, composed of five males and five females, was also included.

Animals receiving NG presented with weight loss (at weeks 4–8) and compensated anemia but resumed gaining weight as feeding continued. Treated rats had altered blood chemistries and relative organ weights, hemosiderosis of the liver and spleen, and moderate to severe testicular degeneration and/or atrophy with severe and complete aspermatogenesis.

Mammalian Oral Toxicity: Chronic

Chronic exposure of laboratory animals to high NG concentrations resulted in decreased body weight gain and adverse hematological and liver changes. Hepatocellular carcinomas as neoplastic nodules, and interstitial cell tumors of the testes were frequently observed in the high-dose group of rats after 2-year exposures to NG [21,22]. Four groups of 38 male and 38 female Charles River CD albino rats were fed diets containing 0, 0.01, 0.1, or 1% NG in their diet for 2 years (average intakes were 0, 3.04, and 3.99; 31.5 and 38.1; and 363 and 436 mg/kg/day in males and females, respectively) [19,20]. Adverse effects were not observed in any of the low-dose rats. Mid-dose rats exhibited decreased weight gain in later months, wherein male and female rats were approximately 60 g and 30 g lighter than controls, respectively. Some rats that were fed 0.1% NG (31.0 (males) or 38.1 (females) mg/kg/day) had mild hepatic lesions (foci of hepatocellular alteration that yield hepatocellular carcinomas). High-dose rats had decreased food consumption and dampened weight gain, behavioral effects (decreased activity and failure to groom), compensated anemia with reticulocytosis, elevated serum transaminases, and methemoglobinemia, with some excessive splenic and renal epithelial pigmentation.

After 1 year, eight high-dose rats had cholangiofibrosis, some of which had neoplastic foci in the liver. At 2 years, all 13 surviving high-dose rats and six of 16 mid-dose rats had enlarged and grossly abnormal livers with severe cholangiofibrosis and hepatocellular carcinomas, some of which had metastasized to the lung. There were interstitial tumors of the testes in one-half of the high-dose males, and in some, this led to aspermatogenesis. A decrease in the naturally occurring pituitary chromophobe adenoma and mammary tumors increased life span, especially in females [19–21]. The NOAEL for this study was 3.04 and 3.99 mg/kg/day, and the LOAEL was 31.5 and 38.1 mg/kg/day for decreased weight gain and enlarged, abnormal livers with cholangiofibrosis and hepatocellular carcinomas in males and females, respectively. The LOAEL for male reproductive effects was 363 mg/kg/day. The NOAEL for male reproductive effects was 31.5 mg/kg/day.

Four groups of 58 male and 58 female CD-1 mice were fed diets containing 0, 0.01, 0.1, or 1% NG for 2 years (average intake of 0, 11.1, and 9.7; 114.6 and 96.4; 1,022 and 1,058 mg/kg/day for males and females, respectively) [21–23]. There were no adverse effects in the low- and mid-dose groups during the 2-year study. Decreased feed consumption and weight gain, and behavioral effects (decreased activity and failure to self-groom) were observed in the high-dose mice. After 1 year, the high-dose animals had heme-derived pigment deposits in various organs, and liver dysplasia. At 2 years, there was pigmentation in the liver and to a lesser extent in the spleen and/or kidneys in most high-dose and some mid-dose mice [21–23]. High-dose mice also had decreased weight gain, decreased self-grooming, methemoglobinemia, and its sequelae [21,22]. The NOAEL was 11.1 and 9.7 mg/kg/day, and the LOAEL was 114.6 and 96.4 mg/kg/day in males and females, respectively, for pigment deposits in the liver, spleen, and/or the kidneys.

Four groups of six male and six female beagle dogs were given capsules containing 0, 1, 5, or 25 mg/kg/day NG for 1 year. The only observed effect was an occasional dose-related incidence of transient mild methemoglobinemia in some dogs that were treated for 6 months or more and at all dose levels [21,22]. There were no accompanying effects on body weight, feed consumption, other hematological tests, clinical chemistry, and histological examination. The NOAEL was 25 mg/kg/day [21,22].

There were quantitative and qualitative differential effects between species. Oral doses up to 20 mg/kg/day for 5 days or 25 mg/kg/day for 12 months produced transient methemoglobinemia in dogs. In rats, lifetime feeding of 363 mg/kg/day (males) and 434 mg/kg/day (females) resulted in toxic effects on the liver and blood, and 31.5 mg/kg/day (males) or

38.1 mg/kg/day (females) resulted in mild effects on the liver of some animals. Mice were the least affected. Lifetime feeding of 115 mg/kg/day (males) and 96 mg/kg/day (females) resulted in a compensated anemia and some pigment deposits. The only effect common to all three species was transient methemoglobinemia. Increased interstitial cell tumors of the testes occurred only in rats [21,22]. Suzuki *et al.* [24] exposed C57BL/6 J mice to NG in their drinking water for 18 months at estimated orally administered doses of 0, 1.5, or 6.2 mg/kg/day, and for 12 months at an estimated concentration of 58.1 mg/kg/day. Each treatment group consisted of approximately 50 animals of each gender. There were no treatment-related adverse effects, including body weight changes, in any dose groups. The NOAEL for the 18-month study was 6.2 mg/kg/day. The NOAEL for the 12-month study was 58.1 mg/kg/day.

Mammalian Oral Toxicity: Other

Ellis *et al.* [22] also conducted a three-generation reproductive study in CD rats where the parental generation (F_O) received the same concentrations of NG as those in the chronic study for 6 months prior to mating (i.e., 0, 0.01, 0.1, or 1% NG). Matings consisted of 10 males and 20 females from each group for the F_O generation. Random selection of equal numbers of 20 to 24 pups from the second generation of each treatment group was maintained for each respective treatment. At 3 months old, each male was mated with a female from each group, and again, only the second-generation offspring were selected for continued treatment. This selection process was repeated until the animals from the third generation (F_{3b}) were weaned. Evaluation of all offspring for gross physical abnormalities and for the number of live and dead pups was recorded. Survival and body weights were recorded at 0, 4, and 21 days. Fertility in the F_1 and F_2 generation of high-dose-treated males was severely impacted. These effects appeared to result from decreased feed intake and consequent poor nutritional status of the females and from decreased spermatogenesis (due to interstitial tumors) in the males. No other developmental or reproductive effects were identified [22]. Ellis *et al.* [22] also conducted a series of cytogenic studies in dogs and rats in the chronic study as well as dominant lethal mutation studies. There were no apparent NG-induced mutagenic effects in the cytogenetics analyses of cells from dogs and rats or in the dominant lethal mutation study in rats.

Studies Relevant for Mammalian TRV Development for Ingestion Exposures

Relevant studies for the development of Toxicity Reference Values (TRVs) are the chronic and subchronic studies performed by Ellis *et al.* [21,22] and Lee *et al.* [18,20] in dogs, rats, and mice. These studies were conducted properly and contained the information necessary to evaluate the results and report them in a clear, logical, and concise form. The methods used were consistent with other accepted protocols and contained the proper controls. Thus, these studies met or exceeded the minimum quality requirements. The chronic studies were considered most relevant due to the exposure duration and because the dose levels changed during subchronic studies. This analysis results in NOAEL and LOAEL values that are less certain. Acute studies provided additional information; however, they were not considered applicable to the development of a TRV. The studies by Suzuki *et al.* [24] and Oketani *et al.* [19] were included as informational data points but were not used to derive a TRV because these data were obtained from a secondary reference, and an English version of the original work could not be obtained. Table 13.2 summarizes the data from these studies, and Figure 13.1 presents the data in a scatter diagram.

Mammalian Toxicity: Inhalation

Inhalation data for mammals was not found.

Mammalian Toxicity: Dermal

Nitroglycerin was a very mild skin irritant but not an eye irritant in rabbits and was a moderate sensitizer in guinea pigs [18]. Nitroglycerin was readily absorbed through the skin of rhesus monkeys [25]. No other relevant dermal toxicity data in mammals were found. However, the potential contribution of the dermal pathway to total exposure dose should be considered in the context of any risk assessment.

SUMMARY OF AVIAN TOXICITY

Fink *et al.* [26] conducted an 8-day dietary study in Northern Bobwhite (*Colinus virginianus*). This study was conducted consistent with OECD Guideline 205 [27]. Five pens consisting of 10 birds per pen were randomly sorted to a control, and 50 birds were randomly assigned to 562, 1,000, 1,780, 3,160, or 5,620 parts per million (ppm) NG in feed to achieve 10 birds per pen. All birds were monitored for weight gain and feed

Table 13.2 Summary of Relevant Mammal Data

Test Organism/Route	Test Duration	Test Results: NOAEL	LOAEL	Effects at LOAEL	Study
Mouse/Albino Swiss/Oral in lactose/peanut oil	Single dose	LD_{50} 1188 (m) 1055 (f)		Animals usually died within 5–6 hours; survivors recovered within 24 hours; no gross pathology. N = unknown	[18]
Rat/Charles River/Oral in lactose/peanut oil	Single dose	LD_{50} 822 (m) 884 (f)		Animals usually died within 5–6 hours; survivors recovered within 24 hours; no gross pathology. N = unknown	[18]
Dogs/oral-capsule	5 days	50 mg/kg/day	100 mg/kg/day	Methemoglobinemia; decreased activity. Cyanosis.	[20] [21]
Dog/oral/feed	13 weeks	5.0 g/kg/day	NA	0, 0.01, 0.1, or 1 mg/kg/day for 4 weeks; then 0, 0.05, 0.5, and 5 mg/kg/day for 9 more weeks. No adverse effects seen at any dose. Low number of animals. Oral dog: 0, 0.01, 0.1, or 1 mg/kg/day for 4 weeks; then 0, 0.05, 0.5, and 5 mg/kg/day for 9 more weeks. NOAEL	[20] [21]
Mouse/oral/feed	13 weeks	608 (m) and 561 (f) mg/kg/day	NA	0.001, 0.01, or 0.1% in the diet (intake of 1.3/1.3 m/f, 11.5/10.9 m/f, 107/95 m/f mg/kg/day) for 3 weeks; then 0.005, 0.05, and 0.5% (intake of 6.4/6.9, 60.2/58.7 m/f, 608/561 m/f mg/kg/day) for 10 more weeks. No treatment-related adverse effects at any dose. US EPA (1992) derived LOAEL of 6.7 mg/kg/day based on mild hemosiderosis. Informational NOAEL not considered for derivation of TRVs.	[20] [21]
Rat/oral/feed	13 weeks	24.5 (m) and 26.5 (f) mg/kg/day	230 (m) and 234 (f) mg/kg/day	Decrease in weight gain and food consumption after increase in dosage in high-dose group.	[20] [21]

Continued

Table 13.2 Summary of Relevant Mammal Data—cont'd

Test Organism/Route	Test Duration	Test Results			Study
		NOAEL	LOAEL	Effects at LOAEL	
Rat/oral/feed	13 weeks	NA	NA	Only one dose level (2.5%, 1, 406 or 1, 416 mg/kg-day). Initial weight loss, altered blood chemistries, hemosiderosis in liver and spleen; testicular degeneration and aspermatogenesis.	[20] [21]
Dog/oral/capsule	12 months	5.0 mg/kg/day	NA	Mild methemoglobinemia (less than 3%, observed at all dose levels). USEPA 1992 did not derive NOAEL. Informational NOAEL, not considered for derivation of TRVs.	[22] [23] [21]
Mouse/oral/feed	2 years	11.10 (m) and 9.72 (f) mg/kg/day	114.6 (m) and 96.4 (f) mg/kg/day	Decreased weight gain, decreased grooming, hemosiderosis.	[22] [23] [21]
Rat/Charles River/oral/feed	2 years	3.04 (m) and 3.99 (f) mg/kg/day	31.5 (m) and 38.1(f) mg/kg/day	Reduced growth (weight loss); mild hepatic lesions like those seen in rats fed higher doses. No adverse effects were observed in the low-dose rats.	[22] [23] [21]
Rat/Charles River/oral/feed	3-generation reproduction study	31.5 (m) and 38.1(f) mg/kg/day	363 (m) and 434 (f) mg/k/day	Infertility in high-dose group (no effects in F_O; 1st mating of F_1 produced 3 litters; only 1 F_2 litter). Infertility due to males; testes ¼ size; no sperm in vaginal plugs; severe aspermatogenesis in F_2 males.	[22]
Rat/Charles River/oral/feed	2 years	31.5 mg/kg/day	363 mg/kg/day	Interstitial cell tumors of the testes with aspermatogenesis.	[22] [21]

NA = not applicable.

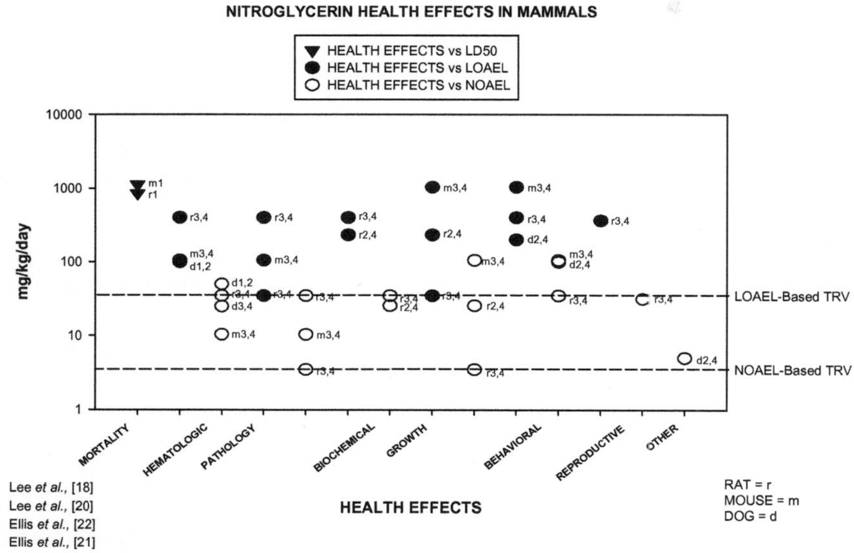

Figure 13.1 Nitroglycerin health effects in mammals. Letters by symbols represent test species, and numbers represent study.

consumption per pen and exposed for 8 days with a 3-day recovery period. No adverse effects were attributed to treatment.

More recently, the developmental effects of nitroglycerin were investigated in Japanese quail [28]. Nitroglycerin was injected into embryos through a small hole drilled into the shell. Embryos were treated at early, middle, and late stages of development with doses of NG ranging from 4.4 μM to 440 μM. The embryos were examined at 24, 48, and 72 hours posttreatment. The NG treatment increased mortality and decreased weight gain in a dose-related fashion with the most dramatic effects seen in the early developmental stage group. The NG treatment also produced a number of malformations, including microphthalmia, craniofacial, heart, and neural tube defects. The authors speculated that the malformations were due to reactive nitrogen species formed from NG and that other organic nitrates might provoke similar responses.

No other toxicological data for the effects of NG in avian species were found.

SUMMARY OF AMPHIBIAN TOXICITY

Toxicological data for the effects of NG in amphibian species were not located. Ecotoxicological research on the effects of this compound in amphibians is recommended.

SUMMARY OF REPTILIAN TOXICITY

Toxicological data for the effects of NG in reptilian species were not located. Ecotoxicological research on the effects of this compound in reptiles is recommended.

RECOMMENDED TOXICITY REFERENCE VALUES

Toxicity Reference Values are used for screening purposes only and are not intended to be predictors of effects in field situations. Site-specific conditions may justify adjustments of these values based on toxicity information that is relevant to specific assessment endpoints.

Toxicity Reference Values for Mammals

TRVs for Ingestion Exposures for the Class Mammalia

Three species of mammals from two different orders were evaluated using dietary exposures of NG. The data set contained studies that were comprehensive in the scope of potential effects and in study duration. Moreover, most studies were well documented and contained refined data that were derived from all observations. Thus, these studies had acceptable quality for TRV derivation.

Most studies had consistent and corroborative findings. Incidences of indicators that suggest liver necrosis, red blood cell lysis, methemoglobinemia, adverse reproductive effects, and weight loss were reported within the same range. Many studies conducted at relatively low levels failed to find adverse effects. Although some findings had questionable biological and ecological significance (e.g., transient and reversible methemoglobinemia), other effects, such as reduced male fertility, were clearly important. Given that most of these data follow a consistent pattern, and that the reproductive data were inappropriate for benchmark dose extrapolation, the NOAEL-LOAEL approach was used.

The NOAEL-based TRV was based on the chronic study by Ellis *et al.* [21] where no adverse effects in weight loss or hepatic lesions were seen. Thus, the NOAEL was derived from the most sensitive species (rat) and sex (males) that were tested at 3 mg/kg/day. The LOAEL-based TRV was derived from the original Ellis *et al.* work [22] where at 31.5 mg/kg/day, mild hepatic lesions and weight loss were found in chronically exposed male rats. Therefore, the LOAEL-based TRV of 32 mg/kg/day was selected. This value was less than those where other endpoints (e.g., reproductive) were observed and was therefore protective of other reported effects. Since the relevant data presented earlier satisfies the minimum data set requirement of the Standard Practice [3], no uncertainty factors were needed to select the TRVs.

The studies used to derive the LOAEL-based TRV were chronic, and the results of those studies were consistent with those of other studies. Thus, this TRV was given a high degree of confidence. The NOAEL-based TRV was derived from a chronic study that evaluated consistent endpoints of toxicity. However, given the information present in other studies during which higher doses than those in the Ellis *et al.* study [22] were tested and no adverse effects were observed, it is likely that the true NOAEL for many species of mammals is higher. Therefore, the NOAEL-based TRV was given a medium degree of confidence. See Table 13.3 for a summary of mammalian TRVs.

TRVs for Ingestion Exposures for Mammalian Foraging Guilds

Toxicity Reference Values specific to particular guild associations (e.g., small herbivorous mammals) have not yet been derived.

TRVs for Inhalation Exposures for the Class Mammalia

Due to insufficient data, no inhalation TRV can be derived for mammals at this time.

TRVs for Dermal Exposures for the Class Mammalia

Due to insufficient data, no dermal TRV can be derived for mammals at this time. However, given that there is evidence that NG has irritant and sensitizing properties, and that dermal exposures were identified as significant, the dermal exposure route should be considered for risk assessment purposes.

TRVs for Birds

Since adverse effects were not found in Northern Bobwhites that were exposed to concentrations of NG greater than 5,000 ppm for 8 days, NG is considered non-toxic for birds in environmental settings.

Table 13.3 TRVs for the Class Mammalia

TRV	Dose	Confidence
NOAEL-based	3.0 mg/kg/d	Medium
LOAEL-based	32.0 mg/kg/d	High

Sources: [21,22].

TRVs for Amphibians

Due to a lack of data, no TRVs for reptiles and amphibians can be derived at this time.

TRVs for Reptiles

Due to a lack of data, no TRVs for reptiles and reptiles can be derived at this time.

IMPORTANT RESEARCH NEEDS

The effects of NG have not been assessed in amphibians or reptiles. It is recommended that these effects be evaluated.

REFERENCES

[1] Gilman AG, Rall TW, Nies AS, Taylor T, editors. The pharmacological basis of therapeutics. 8th ed. New York: Pergamon Press; 1990.
[2] Burrows EP, Rosenblatt DH, Mitchell WR, Parmer DL. Organic explosives and related compounds: environmental and health considerations. Technical Report No. AD–8901. Frederick, MD: U. S. Army Biomedical Research and Development Laboratory, Fort Detrick; 1989.
[3] USACHPPM. TG 254, Standard practice for wildlife toxicity reference values. Aberdeen Proving Ground, MD: U.S. Army Center for Health Promotion and Preventive Medicine; 2000.
[4] Roberts WC, Hartley WR, editors. USEPA drinking water heath advisory: munitions—trinitroglycerol (TNG). U.S. EPA Office of Water. Boca Raton, FL: Lewis Publishers; 1992.
[5] Hazardous Substances Data Bank. Tomes database/micromedix. National library of medicine [Internet]; 2012 [cited November 03, 2014]. http://toxnet.nlm.nih.gov/cgi-bin/sis/search2/f?./temp/~8Z8bV6:1.
[6] Lyman WJ, Reehl WF, Rosenblatt DH. Handbook of chemical property estimation methods. New York: McGraw-Hill; 1985.
[7] American Conference of Governmental Industrial Hygienists (ACGIH). Documentation of the threshold limit values and biological exposure indices. 6th ed. Nitroglycerin. Cincinnati, OH: ACGIH, Inc; 1991.
[8] Smith JG. Water quality criteria for nitroglycerin: final report. Oak Ridge, TN: Oak Ridge National Laboratory; 1986.
[9] Spanggord RJ, Mill T, Chou TW, Mabey WR, Smith JH, Lee S. Environmental fate studies on certain munition wastewater constituents. Phase II Laboratory Studies. Final Report. Contract No. DAMD 17-78-8081. Fort Detrick, MD: U.S. Army Medical Research and Development Command; 1980.
[10] Spanggord RJ, Mill T, Chou TW, Mabey WR, Smith JH, Lee S. Environmental fate studies on certain munition wastewater constituents, Phase I literature review. Final report. Fort Detrick, MD: U. S. Army Medical Research and Development Command; 1980.
[11] Capellos C, Fisco WJ, Ribaudo C, Hogan VD, Campisi J, Murphy FX, et al. Int J Chem Kinet 1984;16:1009–51, as cited in [2].

[12] Cockerham LG, Stone BS, editors. Basic environmental toxicology. Boca Raton, FL: CRC Press; 1994.
[13] Wendt TM, Cornell JH, Kaplan AM. Microbial degradation of glycerol nitrates. Appl Environ Microbiol 1978;36:693–9.
[14] Walker JE, Kaplan DL. Biological degradation of explosives and chemical agents. Natick, MA: U.S. Army Natick Research Development and Engineering Center; 1992.
[15] Lund RP, Häggendal J, Johnsson G. Withdrawal symptoms in workers exposed to nitroglycerine. Br J Ind Med 1968;25(2):136–8.
[16] Maragos CM, Andrews AW, Keefer LK, Elespuru RK. Mutagenicity of glyceryl trinitrate (nitroglycerin) in *Salmonella typhimurium*. Mutat Res 1993;298(3):187–95.
[17] U.S. Environmental Protection Agency (U.S. EPA). Hazard characterization document screening level hazard characterization, nitroglycerin, [Internet][cited November 03, 2014]. Available from, http://www.epa.gov/chemrtk/index/htm.
[18] Lee CC, Dilley JV, Hodgson JR, Helton DO, Wiegand WJ, Roberts DN, et al. Mammalian toxicity of munition compounds. Phase I. Acute oral toxicity, primary skin and eye irritation, dermal sensitization and disposition and metabolism. Report No. 1 AD B011150. Contract No. DAMD-17-74-C-4073. Kansas City, MO: Midwest Research Institute; 1975.
[19] Oketani Y, Mitsuzono T, Ichikawa K, Itono Y, Gojo T, Nagayoshi S, Konoha N. Toxicological studies on nitroglycerin (NK-843). 1. Acute toxicity in mouse and rat. Iyakuhin Kenkyu 1982;13:90–4, as cited in [8].
[20] Lee CC, Ellis III HV, Kowalski JJ, Hodgson JR, Hwang SW, Short RD, et al. Mammalian toxicity of munition compounds. Phase II: effects of multiple doses. Part I: trinitroglycerin. Progress Report No. 2. ADA 047067. Kansas City, MO: Midwest Research Institute; 1977.
[21] Ellis III HS, Hong CB, Lee CC, Dacre JC, Glennon JP. Subacute and chronic toxicity studies of trinitroglycerin in dogs, rats, and mice. Fundam Appl Toxicol 1984;4:248–60.
[22] Ellis HV, Hagensen JH, Hodgson JR, Minor JL, Hong CB, Ellis ER, et al. Midwest Research Institute. Mammalian toxicity of munitions compounds. Phase III: effects of lifetime exposure. Part II: trinitroglycerin. Progress Report No. 8. AD A078746/5. Contract No. DAMD-17-74-C-4073Kansas City, MO: Midwest Research Institute; 1978.
[23] Dacre JC, Lee CC, Ellis III HV, Hong CB. Chronic and carcinogenic study of trinitroglycerin in rats, mice and dogs. Arch Toxicol 1980;(Suppl. 4):88.
[24] Suzuki K, Sudo K, Yamamoto T, Hashimoto K. The carcinogenicity of N-ethocarbonyl-3-morpholinosydnonimine (molsidomine) in comparison with nitroglycerin in C57BL6jms mice. Pharmacometrics 1975;9:229–42, as cited in [8].
[25] Wester RC, Noonan PK, Smeach S, Kosobud L. Pharmacokinetics and bioavailability of intravenous and topical nitroglycerin in the rhesus monkey: estimate of percutaneous first pass metabolism. J Pharm Sci 1983;72:745–8.
[26] Fink R, Beavers JB, Joiner G, Faulcon V, Cadby V, Beskid J, et al. Eight-day dietary LC50-bobwhite quail. Final report. Wilmington, DE: Wildlife International, Limited; 1980.
[27] Organization for Economic Cooperation and Development (OECD). Test No. 205: avian dietary toxicity test. OECD guidelines for the testing of chemicals. Section 2. Éditions OCDE; 1984, http://dx.doi.org/10.1787/9789264070004-en DOI:10.1787/9789264070004-en#10.1787/9789264070004-en#_blank.
[28] Bardai GK, Hales BF, Sunhara GI. Developmental toxicity of glyceryl trinitrate in quail embryos. Birth Defects Res A Clin Mol Teratol 2011;91(Part A):230–40.

CHAPTER 14

Wildlife Toxicity Assessment for 2-Nitrodiphenylamine and 4-Nitrodiphenylamine

Bradley E. Sample, Christine Ann Arenal

Contents

INTRODUCTION

The compound 2-nitrodiphenylamine (2-NDPA) is a reddish-brown crystalline powder used as a stabilizer for propylene glycol dinitrate (PGD) in "Otto Fuel II," an explosive propellant for torpedoes [1]. 2-NDPA constitutes about 1.5% of Otto Fuel II; the other major constituent is dibutyl

Wildlife Toxicity Assessments for Chemicals of Military Concern
http://dx.doi.org/10.1016/B978-0-12-800020-5.00014-4

sebacate (22.5%). The compound is also used as an ingredient in the U.S. Army's double base solid propellants and, in the civilian sector, as a solvent dye. Information concerning the industrial hygiene, toxicology, or environmental fate and transport of 2-NDPA is limited. For example, the Hazardous Substances Data Bank (HSDB) [2] contains no record for the compound. Likewise, 2-NDPA is not included in the U.S. Environmental Protection Agency's Integrated Risk Information System or in publications of the National Institute of Occupational Safety and Health or the Occupational Safety and Health Administration. Toxicological information for an isomer of 2-NDPA, 4-nitrodiphenylamine (4-NDPA) is more readily available and may be a useful surrogate for 2-NDPA. 4-NDPA occurs as dark yellow crystals with a faint aromatic odor [3]. This compound is manufactured at a limited number of sites worldwide and is used solely as a chemical intermediate for industrial uses [4]. Generally, it is fully consumed in chemical reactions during the manufacture of other products.

This Wildlife Toxicity Assessment (WTA) summarizes current knowledge of the potential harmful impacts of 2-NDPA and 4-NDPA on wildlife, emphasizing, where possible, threshold doses for the onset of toxicological effects, as described in reports of experimental studies of these compounds. Available dose-response data for these compounds may support the establishment of Toxicity Reference Values (TRVs), which represent dose levels that could serve as protective thresholds for wildlife exposed to either 2-NDPA or 4-NDPA. It should be noted that environmental exposure to 4-NDPA is currently expected to be negligible due to its limited manufacture and typical disposal methods (i.e., industrial wastes containing 4-NDPA are generally incinerated) [4]. This WTA was developed in accordance with the U.S. Army Center for Health Promotion and Preventive Medicine (USACHPPM) Technical Guide (TG) 254, *Standard Practice for Wildlife Toxicity Reference Values* [5].

TOXICITY PROFILE

Environmental Fate and Transport

Physical-chemical information applicable to the environmental fate and transport of 2-NDPA (Table 14.1) was derived from the Agency of Toxic Substances and Disease Registry (ATSDR) [1] and Acros Organics [6]. In addition, 2-NDPA has low volatility, indicating that it is unlikely to readily partition to the air [1]. Similarly, the solubility in water is low, suggesting that it will more likely absorb to sediment or soil than leach to groundwater.

Table 14.1 Summary of Physical-Chemical Properties of 2-Nitrodiphenylamine

CAS No.	119-75-5
Molecular weight	214.22
Color	Reddish-brown
Physical state	Crystalline
Melting point	75.5 °C
Boiling point	No data
Odor	Faint, ethereal
Solubility in water	<1 mg/L at 20–25 °C: soluble in ethanol, acetone, dimethyl sulfoxide
Partition coefficients:	
Log K_{ow}	4.9×10^{-1}
Log K_{oc}	No data
Vapor pressure at 25 °C	1×10^{-5} mm Hg
Henry's Law constant at 25 °C	No data
Conversion factors	1 ppm = 8.76 mg/m^3 1 mg/m^3 = 0.114 ppm

Sources: ATSDR [1], Acros Organics [6].

The ATSDR [1] reports corroborating evidence for this based on wastewater releases from a U.S. Army ammunition plant. There is some evidence that 2-NDPA may degrade in the environment. For example, photolysis of the compound has been demonstrated [1]. Bacterial degradation of 2-NDPA by *Clostridium* spp. was reported, although the final degradation products was not identified [7].

The physical-chemical data pertaining to the environmental fate and transport of 4-NDPA (Table 14.2) were obtained from three sources [2–4]. Given the limited production, use, and disposal methods, environmental release of 4-NDPA is most likely associated with the manufacturing process [4]. However, these releases are predicted to be negligible due to the physical properties of the compound [4]. Additionally, 4-NDPA is likely to be persistent. The OECD SIDS [4] reports that 4-NDPA is not readily biodegradable (i.e., no biodegradation, no aerobic biodegradability or chemical degradation was observed after 8 weeks in river water, and no ultimate degradation to carbon dioxide (CO_2) after up to 35 days), and is only slowly photodegradable (i.e., photo transformation in water of about 7% in 7 days and a half-life of 70 days). Although it is predicted that 4-NDPA might be persistent in the environment, environmental residues from production activities have not been detected [4]. As with 2-NDPA, the physical properties of 4-NDPA suggest partitioning to soil and sediment rather than

Table 14.2 Summary of Physical-Chemical Properties of 4-Nitrodiphenylamine

CAS No.	836-30-6
Molecular weight	214.23
Color	Dark yellow
Physical state	Crystalline
Melting point	133.5 °C
Boiling point	211 °C (HSDB 2005) or 343 °C (OECD SIDS, 1998)
Odor	Faint, aromatic
Solubility in water	4.1 mg/L at 24 °C: insoluble in water, soluble in alcohol, acetic acid, sulfuric acid
Partition coefficients:	
Log K_{ow}	3.82 at 25 °C
Log K_{oc}	No data
Vapor pressure at 25 °C	<0.15 mm Hg
Henry's Law constant at 25 °C	No data
Conversion factors	1 ppm = 8.76 mg/m^3 1 mg/m^3 = 0.114 ppm

Sources: HSDB [2], Monsanto [3], OECD SIDS [4].

leaching to groundwater [4]. The moderate partition coefficient (Log_{Kow}) of 3.82 suggests a potential for bioaccumulation.

SUMMARY OF MAMMALIAN TOXICITY

Mammalian Oral Toxicity: Acute

Acute oral toxicity data for mammals for both 2-NDPA and 4-NDPA were limited. Whereas ATSDR [1] report an oral median lethal dose (LD_{50}) for 2-NDPA in rats of 6,150 mg/kg, Scharman and Rosencrance [8] report an oral LD_{50} in rats of 12 g/kg (or 12,000 mg/kg). Scharman and Rosencrance [8] further provide a case report of two truck washers who became ill while cleaning out the tank of a vehicle that had been used to transport 2-NDPA. Cyanosis with accompanying methemoglobinemia appeared to be the primary reaction. As with 2-NDPA, 4-NDPA was not observed to be acutely toxic. The oral LD_{50} for rats was reported as greater than 7,940 mg/kg [3,4]. No deaths occurred in three males and two females dosed at 7,940 mg/kg [4].

Mammalian Oral Toxicity: Subchronic

No subchronic data for 2-NDPA were located. However, a 4-week toxicity study using 4-NDPA was conducted by Monsanto Company in 1984 and submitted to the U.S. EPA [9]. Groups of five male and five female Sprague Dawley rats were fed dietary doses of 0, 2,000, 5,000, 10,000, and 20,000 parts per million (ppm) of 4-NDPA for 4 weeks. Using the mean food consumption rate (g/kg bw/day) that was reported for males and females in each dose group, the authors determined that these doses were approximately equal to 0, 195, 469, 943, and 1,761 mg/kg/day. Although no deaths occurred at any of the dose levels, a dose-dependent decrease in body weights of both males and females was observed in the three highest dose groups relative to controls. This difference was statistically significant in the two highest dose groups. Food consumption also was reduced in the 10,000 and 20,000 ppm dose groups, although the body weight to food ingestion ratio was similar to controls. In a paired study, the authors fed control rats the same amount of food as was eaten by rats in the 20,000 ppm group and observed a similar reduction in body weight. Therefore, it was concluded that poor diet palatability, especially among males, was a factor in the observed body weight reductions. Yellow/orange urine and yellow fur were observed in all dose levels, though this was not considered a toxic effect. Finally, renal tubular cysts and casts were observed in males at the 5,000, 10,000, and 20,000 ppm dose levels. No renal effects were noted at the 2,000 ppm treatment. The authors therefore concluded that the kidney is the target organ for 4-NDPA. Because this study occurred over a 1-month period (less than 10% of the life span of the test species, but greater than 14 days), it is considered to be equivalent to a subchronic study [5]. Based on the data presented, a subchronic no observed adverse effect level (NOAEL) of 195 mg/kg/day (2,000 ppm dose level) and a subchronic lowest observed adverse effect level (LOAEL) of 469 mg/kg/day (5,000 ppm dose level) for kidney toxicity (presence of cysts/casts) were derived.

A second subchronic study conducted by Monsanto is reported in OECD SIDS [4] and in the Material Safety Data Sheet (MSDS) for 4-NDPA [3]. In this study, 30 male and 30 female Sprague Dawley rats were given 0, 800, 2,000, and 5,000 ppm 4-NDPA for 90 days. No data on the body weights and food consumption rates were provided in this secondary source; however, it is noted that the 800 ppm level is equivalent to 57 mg/kg/day [4]. Based on these data, a food consumption rate of 71.25 g food/kg bw/day was calculated and used to determine daily doses of 143 and 356 mg/kg/day for the 2,000 and 5,000 ppm dose groups, respectively. Observed effects included decreased body weight gain, histopathological effects on kidneys and spleen, slight anemia, and increased methemoglobin values. (Note: OECD SIDS [4] does not indicate at which dose levels each of these effects occurred). No toxic effects were observed at the 800 ppm exposure level. Therefore, the 800-ppm level (57 mg/kg/day) represents a subchronic NOAEL and the 2,000-ppm dose group (143 mg/kg/day) represents a subchronic LOAEL.

Mammalian Oral Toxicity: Chronic

No chronic mammalian oral toxicity data were located for either 2-NDPA or 4-NDPA.

Mammalian Oral Toxicity: Other

Although additional oral toxicity data for 2-NDPA were not available, data were identified for a developmental study using 4-NDPA [10,11]. Groups of 25 female rats were dosed with 250, 1,000, and 3,000 mg/kg/day of 4-NDPA and a dose volume of 10 milliliters per kg (mL/kg) during days 6 through 15 of gestation. A control group of 25 female rats was dosed with corn oil at the same volume during the same gestation period. Females were observed for clinical signs of toxicity, and body weights and food consumption were measured on gestation days 0, 6, 9, 12, 16, and 20. Fetuses were removed via cesarean section on gestation day 20 and were weighed, sexed, and examined for external, skeletal, and visceral abnormalities. Maternal toxicity occurred in the two highest dose groups, with eight females (32%) in the 3,000 mg/kg/day dose group and one female (4%) in the 1,000 mg/kg/day dose group found dead during the treatment period. Clinical signs of toxicity (e.g., no or few feces) and dose-dependent reductions in body weight gain and food consumption were observed at these two dose levels. No signs of maternal toxicity were observed at the 250 mg/kg/day dose level. Fetal weights were significantly reduced within the 3,000 mg/kg/day dose group. There was also a significantly increased frequency of litters with extra vertebrae in the 3,000 mg/kg/day dose group, and a slight, but not statistically significant, increase in the number of fetuses/litters with extra ribs in the 1,000 and 3,000 mg/kg/day dose groups. Based on these findings, the authors considered the 250 mg/kg/day dose level to be the NOAEL for a lack of effects to the dams or fetuses.

Studies Relevant for Mammalian TRV Development for Ingestion Exposures

Since the developmental study reported in U.S. EPA [10] and Bannister *et al.* [11] was conducted during a sensitive life stage (gestation) and adverse effects (i.e., decreased fetal body weights, skeletal anomalies) were observed, this study is considered to be equivalent in value to data collected using a chronic study design [5]. Thus, the 250 mg/kg/day dose group represents a chronic NOAEL, and the 1,000 mg/kg/day dose level would represent the chronic LOAEL. Chronic studies for 2-NDPA were not available; however, given the low acute toxicity of 2-NDPA, it is unlikely that further studies would provide useful results (i.e., beyond the limit value).

Based on the available studies, the primary target organs for 4-NDPA are the kidneys and blood (4 and 8, respectively). Additionally, effects were found at high exposures in fetuses from dams dosed with 4-NDPA (Table 14.3, Figure 14.1). These studies represent acute, subchronic, and chronic exposure, although the Sprague Dawley laboratory rat is the only test species that has been studied. Effects considering the magnitude and means of exposure are consistent among the three studies. Therefore, all three studies are considered sufficient to derive class-specific TRVs for 4-NDPA. No data on subchronic or chronic exposures to 2-NDPA were located. Acute studies alone were not considered relevant for TRV development. However, these studies indicated that neither 2-NDPA nor 4-NDPA were highly toxic.

Mammalian Inhalation Toxicity

One dust inhalation study using 4-NDPA is reported in Monsanto [3] and OECD SIDS [4]. In this study, 10 male and 10 female Sprague Dawley rats were exposed to 4-NDPA by whole body dust inhalation at 8.5, 29, and 52 milligrams per cubic meter (mg/m^3) for 6 hours per day, 5 days per week for 21 days. Increased methemoglobin, absolute and relative liver weights, and hematological changes were observed at the 29 and 52 mg/m^3 dose levels. No adverse effects were noted at the 8.5 mg/m^3 dose level. Although OECD SIDS [4] reports this as a subacute study, treatment occurred over less than 10% of the life span of the test species but for greater than 14 days. Therefore, this study is considered to be subchronic [5], with the 8.5 mg/m^3 dose level representing the subchronic NOAEL and the 29 mg/m^3 dose level representing the subchronic LOAEL.

Mammalian Dermal Toxicity

A dermal LD_{50} of greater than 7,940 mg/kg has been reported [3,4]. No deaths occurred at this level; only one male and one female were tested [4].

Table 14.3 Summary of Relevant Mammalian Data for TRV Derivation (4-NDPA)

Test Type	Test Organism	Test Duration	Test Results: NOAEL (mg/kg/day)	Test Results: LOAEL (mg/kg/day)	Test Results: Effects Observed at the LOAEL	Study
Developmental	Rat (Sprague Dawley)	10 days (exposure during gestation)	250 (♀)	1,000 (♀)	Maternal mortality, reduced body weight gain, reduced food consumption, and clinical signs of toxicity (e.g., no or few feces).	[10]
	Rat (Sprague Dawley)	10 days (exposure during gestation)	250 (fetal)	1,000 (fetal)	Increased incidence of 14th full ribs. Decreased body weights at 3,000 mg/kg/days.	
Subchronic	Rat (Sprague Dawley)	30 days	195 (♂) 469 (♀)	469 (♂) 943 (♀)	Renal tubular cysts and casts in males. Dose-dependent decrease in body weights of both males and females.	[9]
	Rat (Sprague Dawley)	90 days	57 (♂ + ♀)	143 (♂ + ♀)	Decreased body weight gain, histopathological effects on kidneys and spleen, slight anemia, and increased methemoglobin values.	[3] [4][a]

[a]Secondary sources.

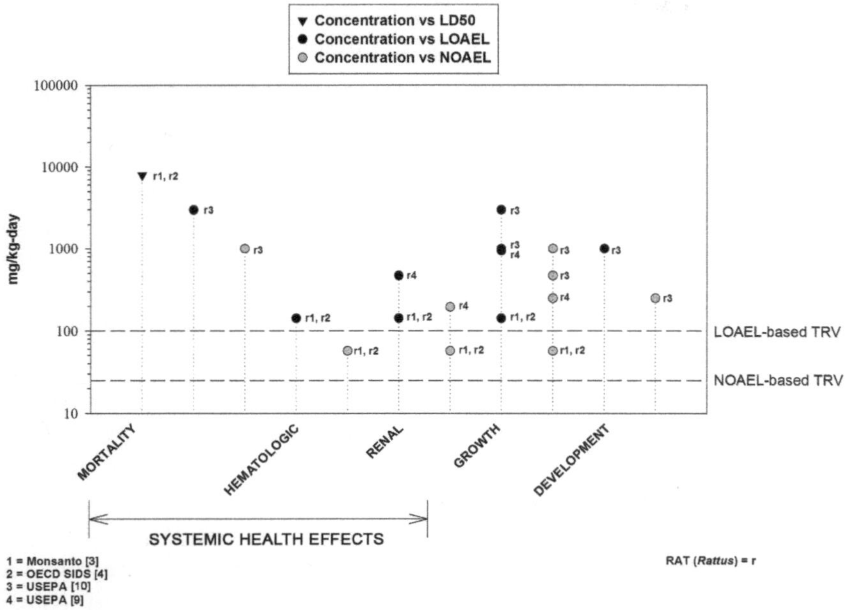

Figure 14.1 Health effects of 2-Nitrodiphenylamine in mammals. Letters by symbols represent test species; numbers represent study.

SUMMARY OF AVIAN TOXICITY

No relevant toxicity studies for either 2-NDPA or 4-NDPA in birds were identified.

SUMMARY OF AMPHIBIAN TOXICITY

No relevant toxicity studies for either 2-NDPA or 4-NDPA in amphibians were identified.

SUMMARY OF REPTILIAN TOXICITY

No relevant toxicity studies for either 2-NDPA or 4-NDPA in reptiles were identified.

RECOMMENDED TOXICITY REFERENCE VALUES

Toxicity Reference Values for Mammals

TRVs for Ingestion Exposures for the Class Mammalia

Available data suggests that both 2- and 4-NDPA are relatively non-toxic. Limit test criteria were exceeded for both compounds (limit test values are 2–5 g/kg; if met, further studies are not normally conducted [12]. However, further data were collected for the more toxic of the two isomers (4-NDPA), with most findings in general agreement. Feeding studies suggest 4-NDPA reduces palatability, and therefore exposures that may cause adverse effects are highly unlikely. Nevertheless, data collected from the developmental study [10] can be used to derive a TRV for chronic exposures for application in risk assessments. Study exposure occurred during a sensitive life stage (gestation), and maternal toxicity was evaluated. Adverse effects on maternal survival, maternal and fetal growth (body weight gain, food consumption, and fetal body weight), and fetal development (reduced growth and increased skeletal anomalies) of rats exposed to 4-NDPA were observed. Decreased growth in rats fed 4-NDPA was also observed in the subchronic studies; however, this effect may be a consequence of reduced palatability. Both decreased maternal survival and reduced maternal and fetal growth were used to determine the TRV because these endpoints have the potential to be ecologically important. As indicated in USACHPPM TG 254 [5], mortality, reproduction, development, and growth are toxicological endpoints that are relevant to the health and ecology of the whole organism. Therefore, a LOAEL for rats of 1,000 mg/kg/day was identified based on the U.S. EPA study [10]. The highest NOAEL within the same species and endpoints was 250 mg/kg/day, reported by the same authors [10].

Despite the relevancy of the toxicity endpoints, the three studies identified as suitable for TRV derivation do not meet the minimum data set requirement outlined by USACHPPM TG 254 [5]. Namely, the studies do not provide data for three or more mammalian species and do not represent two different taxonomic orders. Therefore, TRVs based on an approximation of the NOAEL and LOAEL were developed for Class Mammalia using uncertainty factors (UFs). Because only one species is represented, an UF of 10 was applied to the ingestion NOAEL and LOAEL (250 and 1,000 mg/kg/day, respectively) to account for interspecific variability. Although available data suggests that both compounds are relatively non-toxic in monogastric rodents, there is medium confidence in the resulting ingestion TRVs (Table 14.4) due to the absence of toxicity data for hindgut fermenting or ruminant species that may respond differently from exposure. Given that the data from the 4-NDPA studies seem to be more toxic than 2-NDPA, the TRVs derived for 4-NDPA may be used as a surrogate until additional 2-DNPA toxicity data become available.

Table 14.4 Selected Ingestion TRVs for the Class Mammalia

TRV	Dose (mg/kg/day)	Confidence Level
NOAEL-based	25	Medium
LOAEL-based	100	Medium

Source: [10].

TRVs for Ingestion Exposures for Mammalian Foraging Guilds

TRVs specific to particular guild associations (e.g., monogastric rodents) have not yet been derived. However, the class-specific TRVs shown in Table 14.4 are applicable to the small herbivorous mammal guild, which includes rats. As with the class-specific TRVs, only one species is represented, so the confidence level for the TRVs is medium. Data to derive TRVs for other guild associations (e.g., carnivorous mammals) are not available at this time.

TRVs for Inhalation Exposures for the Class Mammalia

One study [3,4] evaluated the inhalation toxicity of 4-NDPA to mammals. This study is considered subchronic, and does not meet the minimum data set requirements (i.e., only one species and one taxonomic order represented and no chronic NOAEL or LOAEL determined). Consequently, UFs were applied to the NOAEL and LOAEL from the study. For the NOAEL (8.5 mg/m^3), an interspecific UF of 10 and a subchronic NOAEL-to-chronic NOAEL UF of 10 were applied, whereas an interspecific UF of 10 and a subchronic LOAEL-to-chronic LOAEL UF of 4 were applied to the LOAEL (20 mg/m^3). Since the study was subchronic, only one species was evaluated, and there is uncertainty associated with the relevance of the toxicity endpoint to the reproduction, growth, or survival of the organism. Confidence in the TRVs derived is very low. As with oral toxicity, no data for 2-NDPA were available; thus, TRVs developed for 4-NDPA (Table 14.5) may be used as surrogates pending further toxic evaluation of 2-NDPA.

Table 14.5 Selected Inhalation TRVs for the Class Mammalia

TRV	Exposure (mg/m^3)	Confidence Level
NOAEL-based	0.085	Low
LOAEL-based	0.725	Low

Sources: [3,4].

TRVs for Dermal Exposures for the Class Mammalia

Sufficient data required for development of dermal TRVs for the Class Mammalia are not available. However, the high reported dermal LD_{50} (>7,940 mg/kg) for 4-NDPA suggests that this compound has a low dermal toxicity to mammals. No dermal toxicity data were available for 2-NDPA.

Toxicity Reference Values for Birds

No data were identified to permit the derivation of an avian TRV for either 2-NDPA or 4-NDPA.

Toxicity Reference Values for Amphibians

No data were identified to permit the derivation of an amphibian TRV for either 2-NDPA or 4-NDPA.

Toxicity Reference Values for Reptiles

No data were identified to permit the derivation of a reptilian TRV for either 2-NDPA or 4-NDPA.

IMPORTANT RESEARCH NEEDS

Toxicological data for both 2- and 4-NDPA are minimal but suggest both compounds are relatively non-toxic, although limit-test requirements are exceeded [12]. Additional subchronic data in other species would increase confidence in TRV estimates. No data are available for hindgut fermenting or ruminant species. These species may have a different capacity to absorb and assimilate NDPA. However, given the observed reduction in palatability in previous NDPA studies, potential for significant exposure may be low.

No data were available for other classes of wildlife (birds, amphibians, and reptiles). Research is desirable for testing in these species, first in regards to acute toxicity, and second to corroborate these findings with mammalian data to determine the need for further studies. Given that the use of NDPA isomers is limited and highly controlled, with no reported environmental residues, the development of a robust, multi taxa toxicity database may not be justified. Until a greater need is established, the development of additional toxicity studies for either 2-NDPA or 4-NDPA is not recommended.

REFERENCES

[1] ATSDR. Toxicological profile for otto fuels II. Atlanta (GA): U.S. Department of Health and Human Services; 1995.

[2] Hazardous Substance Data Bank (HSDB) Propylene. [Internet]. Bethesda (MD): National Library of Medicine (US), National Institute of Health [cited 2013]. Available from: http://toxnet.njm.nih.gov/.

[3] Monsanto, Incorporated. Material safety data sheet for 4-Nitrodiphenylamine. In: U.S. EPA, editor. Follow-up information: teratology study in rats with 4-nitrodiphenylamine (Final report) with attachments and cover letter dated 042391. Washington (DC): U.S. Environmental Protection Agency; 1991 EPA/OTS: Doc #89-910000119.

[4] OECD. Screening Information Data Set (SIDS). Benzenamine, 4-nitro-N-phenyl. CAS No: 836-30-6. Paper presented at: SIDS Initial Assessment Meeting (SIAM) 8; 1998 Oct 28–30; Paris, France. Available from: www.inchem.org/documents/sids/sids/836306.pdf.

[5] USACHPPM. Standard practice for wildlife toxicity reference values. Aberdeen Proving Ground (MD): U.S. Army Center for Health Promotion and Preventive Medicine; 2000 Technical Guide No. 254.

[6] Acros Organics. Material safety data sheet for 2-nitrodiphenylamine [Internet] [cited 2009]. Available from: http://www.acros.com/ http://www.fishersci.com/ecomm/servlet/msdsproxy?productName=AC415940100&productDescription=2-NITRODIPHENYLAMINE+99%25+10G&catNo=AC415940100&vendorId=VN00032119&storeId=10652.

[7] Powell S, Franzmann PD, Cord-Ruwish R, Toze S. Degradation of 2-nitrodiphenylamine, a component of Otto Fuel II, by *Clostridium* spp. Anaerobe 1998;4(2):95–102.

[8] Scharman EJ, Rosencrance JG. Exposure to 2-nitrodiphenylamine. Vet Hum Toxicol 1994;36:345.

[9] U.S. EPA. Initial submission: four week study of 4-Nitrodiphenylamine administered in the diet to sprague-dawley rats with cover letter dated 080792. Washington (DC): U.S. Environmental Protection Agency; 1992 EPA/OTS: Doc#88-920007690.

[10] U.S. EPA. Follow-up information: teratology study in rats with 4-Nitrodiphenylamine (Final report) with attachments and cover letter dated 042391. Washington (DC): U.S. Environmental Protection Agency; 1991 EPA/OTS: Doc #89-910000119.

[11] Bannister RM, Brewster DW, Rodwell DE, Schroeder RE, Barnett Jr. JW. Developmental toxicity studies in rats with 4-aminodiphenylamine (4-ADPA) and 4-nitrodiphenylamine (4-NDPA). Toxicologist 1992;12:103.

[12] U.S. EPA. Health effects test guidelines: acute oral toxicity. Washington (DC): U.S. Environmental Protection Agency Office of Prevention, Pesticides, and Toxic Substances; 2002 OPPTS 870.1100, EPA 712-C-02-190.

CHAPTER 15

Wildlife Toxicity Assessment for Picric Acid (2,4,6-Trinitrophenol)

Rachel M. Hebert, Allison M. Jackovitz

Contents

INTRODUCTION

Picric acid (2,4,6-trinitrophenol) was once an extensively used high explosive for military applications that was once used as a component of rocket fuel [1]. The compound also has nonmilitary uses, both in research and in industry, as a sensitizer in photographic emulsions, a chemical intermediate in the production of picramic acid and chloropicrin, and a component in matches and batteries [2]. Picric acid (CAS No. 88-89-1) is a severe explosion hazard due to its sensitivity to friction, heat, and shock [1]. Additionally, picric acid reacts vigorously with a variety of materials and is toxic

Wildlife Toxicity Assessments for Chemicals of Military Concern
http://dx.doi.org/10.1016/B978-0-12-800020-5.00015-6

to humans through all routes of entry [1]. Given its historical and current widespread application, it is likely that the compound was released into the environment during the manufacture of explosives and in load, assembly, and pack (LAP) activities at U.S. Army ammunition plants and other military installations [2,3].

This chapter summarizes the current knowledge of potential impacts of picric acid on wildlife, emphasizing threshold doses for the onset of toxicological effects, as described in reports of experimental studies of the compound. Surveying the threshold dosimetry of the compound may point to the establishment of Toxicity Reference Values (TRVs), which could serve as protective exposure standards for wildlife ranging near affected or contaminated sites.

ENVIRONMENTAL FATE AND TRANSPORT

As summarized in the Hazardous Substance Data Bank (HSDB), the widespread use of picric acid in explosives and other military applications has likely resulted in the compound's release into the environment in various waste streams [1]. The amount of picric acid in these waste streams has not been quantitated; however, certain aspects of the compound's likely disposition into the environment can be inferred from its physical and chemical characteristics. As shown in Table 15.1, the low vapor pressure of 7.5×10^{-7} millimeters of mercury (mm Hg) at 25 °C suggests that picric acid may exist in both vapor and particulate phases in the ambient atmosphere [1].

Atmospheric products of photolysis are possible since picric acid absorbs light at wavelengths greater than 290 nanometers (nm). Soil-borne picric acid might be expected to have high mobility due to the compound's high solubility in aqueous solution and estimated K_{oc} of 180. In fact, picric acid's low pKa of 0.42 also indicates that the compound will probably exist in the anionic form in soil and therefore, display relatively limited mobility. With a Henry's Law constant of 1.7×10^{-8} (atm m^3/mol), the compound's capacity to volatilize from moist soil surfaces is unlikely to be an important fate process. Among biotic processes, cultures of anaerobic Pseudomonads, such as *Pseudomonas aeroginosa*, can transform the compound to the degradation product, 2-amino-4, 6-dinitrophenol [1]. In an *in vitro* study, the concentration of picric acid was decreased by 22% during 30 days of incubation with *Pseudomonas aeroginosa* [4]. The authors attributed 1.7% of the decrease to conversion of picric acid to picramic acid; the remaining 20.3% decrease

Table 15.1 Summary of the Physical-Chemical Properties of Picric Acid

CAS No.	88-89-1
Synonyms	2,4,6-trinitrophenol, 2-hydroxy-1,3,5-trinitrobenzene, melinite, picral, carbazotic acid
Molecular weight	229.10
Color	Yellow
State	Crystals
Melting point	122–123 °C
Boiling point	Explodes at 300 °C
Odor	Odorless
Solubility	Water: 12.7 g/L at 25 °C; soluble in benzene, chloroform, ether, alcohol, ethanol, and acetone
Partition coefficients:	
Log K_{OW}	1.44180
K_{OC}	180
pKa	0.42
Vapor pressure (at 25 °C)	7.5×10^{-7} mm Hg
Henry's Law constant (at 25 °C)	1.7×10^{-8} atm m^3/mole

Source: HSDB [1].

was attributed to adsorption onto bacterial surfaces, bacterial accumulation, and degradation into unidentifiable products [4].

SUMMARY OF MAMMALIAN TOXICITY

Mammalian Toxicity: Acute

There are very few data on the toxicity of picric acid in experimental studies; however, one study investigated its acute toxicity, cause of death, metabolites, and toxicokinetics in rats [5]. Oral LD_{50} values of 290 ± 57.5 and 200 ± 42.9 mg/kg were determined for male and female F344 rats receiving the compound via gavage [5]. Doses ranged from 50 to 800 mg/kg. Blood gas analysis was performed to determine the cause of death in a separate group of male rats administered between 100 and 400 mg/kg. Additionally, Wyman *et al.* [5] identified metabolites via urine collection and analysis, and examined toxicokinetics and tissue distribution of picric acid. In the metabolite portion of the study, rats were administered 100 mg/kg picric acid. Rats used in the toxicokinetics portion were given either 100 mg/kg picric acid via gavage or 50 mg/kg via injection in the penile vein; radioactivity of all samples was assayed by liquid scintillation spectrometry.

No rats died of oral exposures ≤ 100 mg/kg picric acid in the acute toxicity experiment [5]. The authors noted the steep dose mortality curve and suggested that the likely cause of death was saturation of the buffering capacity of the blood at the higher doses of picric acid. Blood gas analyses in male rats after acute exposure showed that death was a result of acidosis. Urine analysis displayed products of *N*-acetylisopicramic acid, picramic acid, *N*-acetylpicramic acid, and picric acid. The urinary excretion of radioactivity of 100 mg/kg picric acid was 60% at 24 hours post administration. Blood was the principal depot of radioactivity after oral administration of picric acid in the toxicokinetics and tissue distribution study. The highest concentration of radioactivity was found in the spleen, followed by kidney, liver, lung, and testes. Overall, picric acid has low levels of retention in the body for long periods of time [5].

In a different study, respiratory depression and other nonspecified respiratory changes were reported. A lowest lethal dose (LDLo) of 60 mg/kg was determined for dogs after a single subcutaneous injection [6].

Mammalian Toxicity: Subacute

Takahashi *et al.* [7] reevaluated the toxicity of picric acid in young rats and determined the toxicity in newborn rats, for comparison. For the dose-finding in newborn rats, four pups/sex/dose were given picric acid by gavage at 0, 16.3, 81.4, or 407 mg/kg/day on postnatal days (PNDs) 4–17 (14 days) and euthanized on PND 18. Low body weight behavioral changes, and death were seen at 81.4 and 407 mg/kg/day in the dose-finding study [7]. Additionally, six pups/sex were given picric acid by gavage at 0, 4.1, 16.3, or 65.1 mg/kg/day on PNDs 4–21 (18 days) and euthanized after the last treatment. In the main study, no deaths were observed, but lower body weights were noted at 65.1 mg/kg. Because of this pattern of lower body weight, the NOAEL for newborn rats was considered 16.3 mg/kg/day.

A dose-finding study and main study were then performed with young rats (5 weeks old). Initially, three rats/sex/dose were given picric acid by gavage at 0, 20, 100, or 500 mg/kg/day for 14 days. All males and one female rat given 500 mg/kg/day died by day 2 of the dosing period. In the main study, six rats/sex/dose received picric acid by gavage at 0, 4, 20, or 100 mg/kg/day for 28 days. No deaths were observed throughout the 28 day experimental period [7]. Significantly higher values of relative liver and spleen weights and hemolytic anemia were observed in the 100 mg/kg/day dose group. Hemolytic anemia was not seen in newborn rats receiving

81.4 mg/kg picric acid. Males in the 100 mg/kg/day dose group also had small testes at the end of the recovery period and lower values of relative epididymides weight, indicating testicular toxicity [7]. The revised NOAEL for young rats was determined as 20 mg/kg/day. Taken together, data from Takahashi *et al.* indicate that picric acid was more toxic to newborn rats, since newborns are in a sensitive life stage.

Mammalian Toxicity: Subchronic

No relevant subchronic mammalian toxicity studies were identified.

Mammalian Toxicity: Chronic

Van Esch *et al.* [8] exposed groups of Wistar rats to 500 ppm of picric acid, picramide, or hexanitrodiphenylamine in feed for up to 2.5 years. The main purpose of this study was to evaluate the carcinogenicity of hexanitrodiphenylamine; since the products of hydrolysis were picric acid and picramide, they were included in parallel. No adverse effects were reported [8].

Studies Relevant for Mammalian TRV Development for Ingestion Exposures

Few studies were available that evaluated the effects of oral picric acid exposure to mammals.

Mammalian Oral Toxicity: Other

No other data relevant to oral exposures for mammals were found.

Mammalian Inhalation Toxicity

No inhalation studies conducted with mammals were found.

Mammalian Dermal Toxicity

No dermal studies conducted with mammals were found.

SUMMARY OF AVIAN TOXICITY

An LDLo was reported for Pigeons resulting from a single subcutaneous injection of 200 mg/kg [6]. No other data were available.

SUMMARY OF AMPHIBIAN TOXICITY

No relevant amphibian picric acid studies were identified.

SUMMARY OF REPTILIAN TOXICITY

No relevant reptilian picric acid studies were identified.

RECOMMENDED TOXICITY REFERENCE VALUES

Toxicity Reference Values for Mammals

TRVs for Ingestion Exposures for the Class Mammalia

Few relevant studies were conducted with picric acid. Acute data included four species from at least two orders. The LD_{50} values for male and female rats are 290 ± 57.5 and 200 ± 42.9 mg/kg, respectively [5]. Blood gas analysis suggests severe acidosis as the cause of death from these acute exposures.

The approximation must be used since data are few. This method requires that uncertainty factors (100 for the female LD_{50} for the NOAEL-based TRV and 20 for the LOAEL-based TRV) be applied to these acute data [9]. Therefore, the mammalian NOAEL-based TRV is 2 mg/kg/day and the LOAEL-based TRV is 10 mg/kg/day (Table 15.2). Both are given a low degree of confidence based on the scarcity of data for picric acid. Given the structural similarities between picric acid and other nitroaromatics (e.g., 2,4,6-trinitrotoluene, 1,3,5-trinitrobenzene), these values are within similar ranges and seem reasonable.

TRVs for Ingestion Exposures for Mammalian Foraging Guilds

Toxicity reference values specific to particular guild associations (e.g., small herbivorous mammals) have not yet been derived. However, these data are representative of omnivorous rodents and may be considered as such. More data for other species would be needed to derive values for other foraging guilds.

Table 15.2 Selected Ingestion TRVs of Picric Acid for the Class Mammalia

TRV	Dose mg/kg/day	Confidence Level
NOAEL-based	2	Low
LOAEL-based	10	Low

Source: [5].

TRVs for Inhalation Exposures for the Class Mammalia

No data were identified to derive a mammalian inhalation TRV for picric acid.

TRVs for Dermal Exposures for the Class Mammalia

No data were identified to derive a mammalian dermal TRV for picric acid.

Toxicity Reference Values for the Class Aves

No data were identified to derive an avian TRV for picric acid.

IMPORTANT RESEARCH NEEDS

To develop useful TRVs, long-term repeated dosing studies are needed for mammals and other vertebrates. The chemical/structural similarities between picric acid and other nitroaromatics may make using surrogates useful in the interim.

REFERENCES

[1] Hazardous Substance Data Bank (HSDB). Picric acid [Internet]. Bethesda (MD): National Library of Medicine (US), National Institute of Health [cited 2012]. Available from: http://toxnet.njm.nih.gov/.

[2] Cooper KR, Burton DT, Goodfellow WL, Rosenblatt DH. Bioconcentration and metabolism of picric acid (2,4,6-trinitrophenol) and picramic acid (2-amino-4,6-dinitrophenol) in rainbow trout *Salmo gairdneri*. J Toxicol Environ Health 1984;14 (5–6):731–47.

[3] Roberts WC, Hartley WR. Toxicological profiles of munitions and other environmental contaminants in drinking water. In: Roberts WC, Hartley WR, editors. Drinking water health advisory: munitions. Boca Raton (FL): Lewis Publishers; 1992.

[4] Wyman JF, Guard HE, Won WD, Quay JH. Conversion of 2,4,6-trinitrophenol to a mutagen by *Pseudomonas aeruginosa*. Appl Environ Microbiol 1979;37(2):222–6.

[5] Wyman JF, Serve MP, Hobson DW, Lee LH, Uddin DE. Acute toxicity, distribution, and metabolism of 2,4,6-trinitrophenol (picric acid) in Fischer 344 rats. J Toxicol Environ Health 1992;37(2):313–27.

[6] Registry of Toxic Effects of Chemical Substances (RTECS). Picric acid [Internet]. Washington (DC): National Institute for Occupational Safety and Health (US), U.S. Department of Health and Human Services [cited 2001].

[7] Takahashi M, Ogata H, Izumi H, Yamashita K, Takechi M, Hirata-Koizumi M, Kamata E, Hasegawa R, Ema M. Comparative toxicity study of 2,4,6-trinitrophenol (picric acid) in newborn and young rats. Congenit Anom (Kuoto) 2004;44(4):204–14.

[8] van Esch GJ, Vink HH, van Genderen H. Influence of hexanitrodiphenylamine on the incidence of neoplasms in the mammary tissue of rats. Nature 1957;180(4584):509–10.

[9] USACHPPM. Standard practice for wildlife toxicity reference values. Aberdeen Proving Ground (MD): U.S. Army Center for Health Promotion and Preventive Medicine; 2000, Technical Guide No. 254.

CHAPTER 16

Wildlife Toxicity Assessment for 2-Amino-4,6-Dinitrotoluene and 4-Amino-2,6-Dinitrotoluene

Marc A. Williams, Stephen W. Rice

Contents

INTRODUCTION

This Wildlife Toxicity Assessment (WTA) is based on a thorough scientific literature review of the toxicological characteristics of nitroglycerin and relevant to the health of wildlife (e.g., mammals, birds, reptiles, and amphibians) following exposure to the compound. The protocol for the performance of this assessment is documented in the U.S. Army Center for Health Promotion and Preventive Medicine (USACHPPM) Technical Guide (TG) 254, *Standard Practice for Wildlife Toxicity Reference Values* [1]. This chapter aims to support ecological risk assessment activities.

Wildlife Toxicity Assessments for Chemicals of Military Concern
http://dx.doi.org/10.1016/B978-0-12-800020-5.00016-8

The compounds 2-amino-4,6-dinitrotoluene (2A-DNT; CAS No. 35572-78-2) and 4-amino-2,6-nitrotoulene (4A-DNT; CAS No. 19406-51-0) are the primary reduction products of 2,4,6-trinitrotoluene (TNT). Both occur from the reduction of one of the nitro groups on the benzene ring. Hovatter *et al.* [2] included 2A-DNT in their survey of the ecological toxicity of nitroaromatic compounds released from U.S. Army Superfund sites, because the compound is a microbial and environmental degradation product of the military explosive, TNT. Although less well characterized, it is possible that the compound's structural isomer, 4A-DNT, can be produced from TNT under similar circumstances following the release of TNT from Army Ammunition Plants (AAPs) in large quantities (as "pink water"). In addition, 2A-DNT occurs at sites within and around army munitions factories and during load, assembly, and pack (LAP) activities at AAPs and other military installations. The chemical structure of 2-amino-4,6-dinitrotoluene and 4-amino-2,6-nitrotoluene are shown in Figure 16.1. Additionally, a concise summary of the physical and chemical properties of 2A-DNT is described in Table 16.1.

This WTA summarizes the current state of knowledge of the likely harmful impacts of 2A-DNT and 4A-DNT on wildlife, emphasizing threshold doses for the onset of toxicological effects, as described in reports of experimental studies of the compound. Surveying the threshold dosimetry of the compound may point to the establishment of Toxicity Reference Values (TRVs) that could serve as protective exposure standards for all

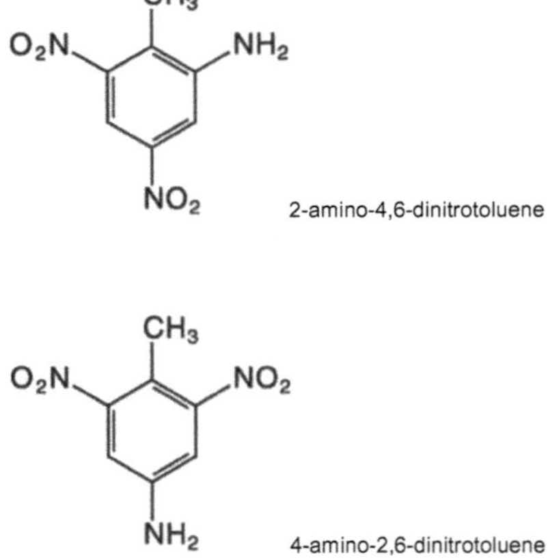

Figure 16.1 Chemical structure of 2-amino-4,6-dinitrotoluene and 4-amino-2,6-dinitrotoluene.

Table 16.1 Summary of the Physical-Chemical Properties of 2-Amino-4,6-Dinitrotoluene and 4-Amino-2,6-Dinitrotoluene

	2-Amino-4,6-Dinitrotoluene	4-Amino-2,6-Dinitrotoluene
CAS No.	35572-78-2	19406-51-0
Molecular weight	197.17 g/mol	197.15 g/mol
Color	NA	NA
State	Crystals	NA
Melting point	173–176 °C	NA
Boiling point	NA	NA
Odor	NA	NA
Solubility	38 mg/L at 20 °C*	43 mg/L at 20 °C*
Partition coefficients:		
Log K_{OW}	1.06, 1.94, 0.5 (all estimated)	NA
K_{OC}	NA	NA
Vapor pressure (at 20 °C)	4×10^{-5} mm Hg (estimated)	NA
Henry's Law constant (at 25 °C)	3×10^{-3} L-torr/mole (estimated)	NA
Conversion factors	1 ppm = 8.06 mg/m^3 1 mg/m^3 = 0.124 ppm	NA

NA = not available.
Sources: [3], *empirically determined [18].

wildlife ranging approximately close to the affected sites. Since both 2A- and 4A-DNTs are often found *in vivo* and are the primary reduction products in the metabolism of TNT, more information on the WTA summarizing the effects of TNT can be found in Chapter 3 of this book.

TOXICITY PROFILE

Environmental Fate and Transport

Talmage *et al.* [3] discussed the widespread occurrence and distribution of 2A-DNT at military sites since TNT is produced and released to the environment. Thus, where TNT is found, 2A-DNT, and its structural isomer 4A-DNT, are also likely to be present since it is one of the microbial degradation products of TNT. Additionally, 2A-DNT was detected in the soil at Joliet Army Ammunition Plant, Illinois, at concentrations of up to 19 mg/kg, and up to 37 mg/kg at Raritan Arsenal, New Jersey. Although 2A-DNT has a limited capacity to bioaccumulate, soil ingestion models, as described for example, for the Northern Bobwhite, suggest that soil ingestion represents a major exposure pathway [4].

Amino dinitrotoluenes are produced relatively rapidly on release of TNT to the soil, and one of the most significant challenges with regard degradation of TNT is the persistence of 2A-DNT and 4A-DNT in the soil and environment [3]. Others [5] have also outlined degradation pathways of TNT in soil. Moreover, prior empirical evidence suggested that environmental release and subsequent degradation of 2A-DNT and 4A-DNT is a key determinant in the rate of biodegradation and mineralization of TNT in the soil [6]. Also, some metabolic information supports this contention in animal systems [7,8].

Additionally, photolytic degradation of 2A-DNT might also occur since it is photosensitive to solar ultraviolet (UV) at wavelengths greater than 290 nm and highly susceptible to photoreduction by sodium borohydride. The biotic degradation of 2A-DNT has been demonstrated in the presence of mixed microbial isolates [9], undifferentiated sludge samples, and in the presence of pure cultures of *Pseudomonas* spp, and particularly *Pseudomonas savastoni*, that can metabolize 2A-DNT as a rich nitrogen source [10]. Since prolonged incubation with these preparations can result in a mixture of degradation products including triaminotoluene, it is possible that microbial degradation is a relatively nonstereospecific enzymatic reduction process. This would suggest that 4A-DNT might also undergo a similar range of interactions. For example, the addition of glucose to pure cultures of *Pseudomonas savastoni* appeared to augment 2A-DNT and 4A-DNT synthesis and dampened TNT denitration [9,10]. Further, a strain of *P. aeruginosa* was found to aerobically metabolize 2A-DNT and 4A-DNT in an oxygen-dependent process to polar products that were not identified [11]. Authors concluded that oxidative metabolism of the 2A-DNT and 4A-DNT products might occur through oxidation of the methyl group and decarboxylation and/or via the action of oxidative deaminases [11].

Moreover, *Clostridium bifermentans* CYS-1 was found capable of degrading TNT to aliphatic polar compounds via degradation to 2A-DNT and 4A-DNT in the first instance [12]. Under anoxic conditions, *Desulfovibrio* spp. are capable of metabolizing TNT [13,14]. One isolated strain called *Desulfovibrio* sp. strain B could metabolize TNT as the sole source of nitrogen, wherein metabolism of the nitro groups of TNT to the corresponding amines formed triaminotoluene (TAT), which then underwent reductive elimination of the amino groups [13]. Others found that a pure culture of *Desulfovibrio* catalyzed the total reduction of TNT to TAT in liquid culture in the absence of a reducing agent, wherein TNT was the sole nitrogen source, pyruvate provided carbon [14].

Summary of Mammalian Toxicity

Mammalian Oral Toxicity: Acute

There are very few data on the toxicity of the amino-dinitrotoluenes in experimental studies, and the only studies located addressed the acute lethality of the compound. Ellis *et al.* [15] reported data on the acute toxicity testing of a number of substituted and unsubstituted dinitrotoluenes including 2A-DNT and 4A-DNT in CD rats and Swiss or B6C3F1 mice. For 2A-DNT, LD_{50} values ranged from 1,394 (female) to 2,240 (male) mg/kg in the rats and 1,522 (female) to 1,722 (male) mg/kg in the mice. For 4A-DNT, the values ranged from 939 (female) to 1,360 (male) mg/kg in the rats and from 1,342 (male) to 1,495 (female) mg/kg in the mice (Table 16.2).

Ellis *et al.* [15] also carried out a toxicokinetic study in which male CD rats were given a single oral dose of ^{14}C-labeled 4A-DNT at 1/10 the LD_{50}. Animals were kept in metabolic cages for the collection of expired air, and specimens of feces and urine. At termination, major organs were excised to measure the radiological activity. The data suggested an absorption factor of approximately 50%, with most of the radioactivity being cleared to the urine. By contrast, there was very little tissue deposition of radioactivity and very little was expired on the breath.

Mammalian Oral Toxicity: Subchronic

No data relevant to subchronic oral exposures for mammals were found.

Mammalian Oral Toxicity: Chronic

No data relevant to chronic oral exposures for mammals were found.

Mammalian Oral Toxicity: Other

No data relevant to other oral exposures for mammals were found.

Table 16.2 Acute Data of Oral Exposures from 2A-DNT and 4A-DNT

Compound	**LD_{50} Mouse***	**LD_{50} Rat***
2A-DNT	$1,722 \pm 154$ (m)	$2,240 \pm 85$ (m)
	$1,522 \pm 71$ (f)	$1,394 \pm 191$ (f)
4A-DNT	$1,342 \pm 107$ (m)	$1,360 \pm 53$ (m)
	$1,495 \pm 90$ (f)	939 ± 76 (f)

* mg/kg ± S.D.
Source: [15].

Studies Relevant for Mammalian TRV Development for Ingestion Exposures

Few data are available for the monoamine DNTs. These data represent studies in rats and mice from acute exposures only [15]. Since these data are relatively similar, in both species, 2A- and 4A-DNTs were treated together and used to develop values.

Mammalian Inhalation Toxicity

No inhalation study data conducted using mammals were found.

Mammalian Dermal Toxicity

No dermal study data conducted using mammals were found.

Summary of Avian Toxicology

Quinn *et al.* [4] conducted an evaluation of the oral toxicity of 2A-DNT in the Northern Bobwhite (*Colinus virginianus*). Subacute (14 days) and subchronic (60 days) oral gavage exposure studies were conducted following determination of the median acute lethal dose (LD_{50}=1,167 mg/kg). Dose levels for the subacute study were 0, 50, 125, 625, 550, and 1,000 mg/kg body weight (bw)/day. All birds in the 125 mg/kg and above treatments died or were moribund within 14 days of exposure, and only one male from the 50 mg/kg dose group died within the 14 days. These mortality data were used to help determine dose groups for the subchronic (60 days) study: 0, 0.5, 3, 14, and 30 mg 2A-DNT/kg bw/day in corn oil.

Only four females (3 at the 14 mg/kg/day dose, and 1 at the 30 mg/kg/day dose) failed to survive throughout the 60 days of subchronic exposure. One female from the 14 mg/kg/day dose failed to survive at day 49 of exposure, and two females from the 14 mg/kg/day dose were euthanized at days 35 and 48 of exposure to 2A-DNT due to an inability to stand. The female bobwhite that was euthanized at day 35 had an edematous gastrointestinal tract with milky, green food contents in the gizzard and pale kidneys. The fourth female that died by day 48 following exposure to a dose of 30 mg/kg/day presented with an edematous gastrointestinal tract and a small, white patch on one of the liver lobes. Histological findings from this bird were limited by moderate tissue autolysis. All males survived throughout the 60 days in all treatment groups.

Hepatocellular vacuolization increased minimally in the highest exposure group compared to controls; however, this was not believed to be related to 2A-DNT exposure due to moderate to severe vacuolization also being observed in control bird livers. Feed consumption in the control and 30 mg/kg/day treatment groups was lower than was found in the other dose groups, perhaps due to the larger volumes of corn oil that these birds received with each daily gavage. These effects on feeding resulted in no apparent differences in body weights among the treatment levels. Plasma triglyceride levels from birds exposed to 0.5, 3, and 14 mg/kg/day 2A-DNT were significantly higher than birds from the control and 30 mg/kg/day dose groups; these elevated levels were most likely due to the increased feed consumption.

The lowest observed adverse effects level (LOAEL) was determined to be 14 mg/kg based on mortality, and the no observed adverse effects level (NOAEL) was determined to be 3 mg/kg based on lack of effects at this oral exposure level of subchronic (60 day) study.

Summary of Amphibian Toxicology

Johnson *et al.* [16] exposed nine tiger salamanders (*Ambystoma tigrinum*) to soil and earthworms containing TNT and 2A- and 4A-DNT. Initial soil concentrations (at time of initial exposure) were 280, 39, and 62 mg/kg soil of TNT, 2A- and 4A-DNT, respectively. Appropriate controls were used. At the termination of exposure (14 days later) soil concentrations were 59, 58, and 78 mg/kg soil of TNT, 2A- and 4A-DNT, respectively. Salamanders were fed earthworms in identical soil preparations. Earthworm concentrations ranged from 0.25 to 0.79 μg/g TNT, 2.1 to 2.6 μg/g 2A-DNT, 2.1 to 2.5 μg/g 4A-DNT, and had trace amounts of 2,4-diamino-6-nitrotoluene. Salamanders were evaluated for immunological indicators of effects (phagocytosis, radical oxygen intermediate production), blood parameters (5-part cellular differentials, total protein, hematocrit, and other endpoints), and for histopathological indicators of the liver and kidney. No adverse effects were reported for any endpoint, and it was remarked that the animals appeared healthy and maintained an appetite. No adverse effects were reported from 14-day exposure to these conditions.

Studies Relevant for Amphibian TRV Development for All Exposures

This study used a microcosm design that considered all pathways of exposure and potential variation in feeding regimes [16]. Since soil concentrations of 2A- and 4A-DNT were monitored, these data are used to derive a NOAEL for the terrestrial salamander *Ambystoma tigrinum*. A minimum soil concentration of 39 mg/kg was selected, which reflects all exposure pathways. Since adverse effects were not observed in the study, a LOAEL is unavailable.

Summary of Reptilian Toxicology

McFarland *et al.* [17] studied the acute, subacute, and subchronic toxicological effects of 2A-DNT under laboratory conditions in the Western Fence Lizard (*Sceloporus occidentalis*) to identify the potential of this compound for reptile toxicity. Lizards lost an average of 12.4% body weight across all doses tested (15 to 150 mg/kg/day), and survivors at day 14 of the study improved by day 5 posttreatment.

RECOMMENDED TOXICITY REFERENCE VALUES

Toxicity Reference Values for Mammals

TRVs for Ingestion Exposures for the Class Mammalia

Only acute data were available for 2A- and 4A-DNT [15]. The data are fairly consistent and the studies were reported and evaluated properly. Given these data, a TRV may be approximated through the application of uncertainty factors prescribed by TG 254 [1]. Using the lowest LD_{50} value (i.e., female rat LD_{50} of 4A-DNT, see Table 16.2) and applying an uncertainty factor of 100 and 20, respectively, a NOAEL- and LOAEL-based TRV was derived with values of 9.4 mg/kg/d and 47.0 mg/kg/d, respectively (see Table 16.3). Since these values are based upon acute data only, a low degree of confidence is given to these TRVs.

Both isomers have relatively low water solubility. This suggests that the bioavailability of these compounds is low. These facts coupled with the relative toxicity of TNT suggest that these values should be protective for mammals and that incidence of toxicity is considered low.

TRVs for Inhalation Exposures for the Class Mammalia

This assessment is not yet complete pending available data.

Table 16.3 Selected Ingestion TRVs for Combined 2A- and 4A-DNTs: Class Mammalia

TRV	Dose	Confidence Level
NOAEL-based	9.4 mg/kg/day	Low
LOAEL-based	47.0 mg/kg/day	Low

Source: [15].

TRVs for Dermal Exposures for the Class Mammalia

This assessment is not yet complete pending available data.

Toxicity Reference Values for Birds

Only one study that assessed the toxicity of 2A-DNT following ingestion in birds [4] could be located (see Table 16.4). The approximation approach is required for TRV derivation as the data do not fit the minimum data requirements of TG 254 [1] since the study is subchronic and is the only study available for this taxonomic class. Using the NOAEL of 3 mg/kg/day and the LOAEL of 14 mg/kg/day from the subchronic study, and applying uncertainty factors of 10 and 4, respectively, a NOAEL-based and a LOAEL-based TRV was calculated (Table 16.4). A low confidence level was assigned to the TRV due to limited exposure length and lack of available avian data for 2A-DNT toxicity.

Toxicity Reference Values for Reptiles

Only McFarland *et al.* [17] have produced a relevant reptilian study assessing 2A-DNT in the Western Fence Lizard (60-day subchronic). As the only study in this taxonomic class, the approximation approach is required for TRV derivation as outlined by TG 254 [1]. McFarland *et al.* produced a NOAEL of 5 mg/kg/day and a LOAEL of 15 mg/kg/day based on the critical endpoints of survival and loss of body weight. Applying the respective uncertainty factors of 10 and 4 to these values, a NOAEL-based and a LOAEL-based TRV were approximated and listed in Table 16.5. A low

Table 16.4 Selected Ingestion TRVs for Class Aves

TRV	Dose	Confidence Level
NOAEL-based	0.3 mg/kg/day	Low
LOAEL-based	3.5 mg/kg/day	Low

Source: [4].

Table 16.5 Selected Ingestion TRVs for Reptiles

TRV	Dose	Confidence Level
NOAEL-based	0.5 mg/kg/day	Low
LOAEL-based	3.75 mg/kg/day	Low
BMDL	3.38 mg/kg/day	-
BMD	5.47 mg/kg/day	-

Source: [17].

confidence level was assigned to the TRV as the study was of limited length of exposure and no other reptilian data is available for 2A-DNT toxicity.

BMD Approach for TRV of Reptiles

The benchmark dose (BMD) approach was also performed for the McFarland *et al.* [17] subchronic (60 day) study in male Western Fence Lizards. The endpoint used in derivation was the effect of 2A-DNT on the mortality of the lizards after 60 days. Mortality showed a clear dose response and is a critical effect toward the survival of the species, making this a relevant endpoint for BMD analysis. Four dichotomous models (the Gamma, Logistic, Probit, and Weibull models) fit the data with overall goodness of fit p-values greater than 0.1. The Probit model (Figure 16.2) yielded the lowest and thus most sensitive BMDL and BMD values of 3.38 and 5.47, respectively.

Toxicity Reference Values for Terrestrial Amphibians

Johnson *et al.* [16] identified values of 2A- and 4A-DNT in soil where no adverse effects were observed in tiger salamanders. Since the exposures were relatively brief, considering the average life span of *Ambystoma* salamanders (>10 years), these were classified as acute exposures and a NOAEL was identified [16]. In addition, since dermal exposures to TNT were reported to be considerable, a pathway-specific (i.e., oral) TRV would not be

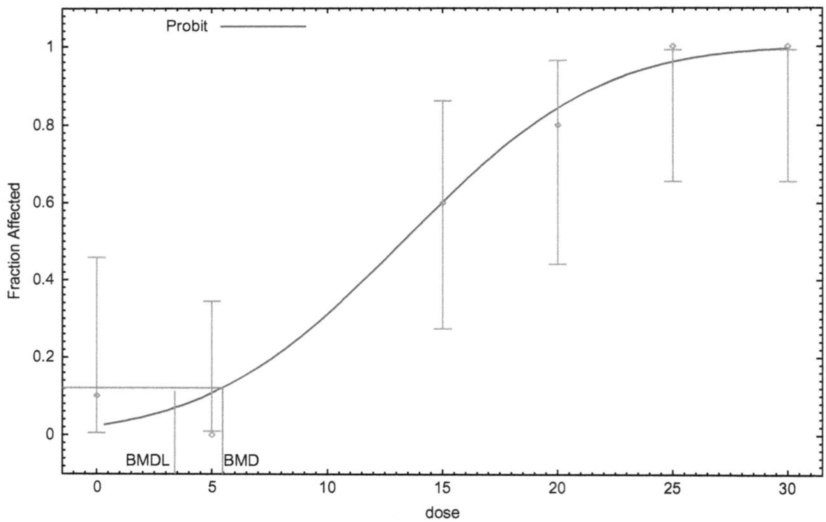

Figure 16.2 Probit model for mortality of Western Fence Lizards, with BMR of 10% risk for the BMD and 0.95 lower confidence limit for the BMDL.

Table 16.6 Selected Soil TRVs for Terrestrial Amphibians

TRV	Dose	Confidence Level
NOAEL-based	0.13 mg/kg soil (dry weight)	Low
LOAEL-based	Not available	-

Source: [16].

appropriate. Since this study used a holistic exposure regime, a media-based value for soil could be derived.

The acute (14-day) NOAEL of TNT in soil (39 μg/g) was divided by a UF of 300 to approximate a chronic NOAEL for terrestrial amphibians (a UF of 30 for an acute NOAEL to a chronic NOAEL and a UF of 10 to extrapolate across multiple species). This resulted in an approximation of a NOAEL-based TRV of 0.13 mg 2A-DNT or 4A-DNT/kg soil dry weight intended to be protective of terrestrial amphibians. However, a LOAEL was not identified, and thus an approximation of a LOAEL-based TRV could not be derived. Table 16.6 presents the selected TRVs. A low confidence level has been assigned to the available TRV because a study observing adverse effects was not available, the only study is of limited length of exposure, and no other terrestrial amphibian data are available.

IMPORTANT RESEARCH NEEDS

Only acute and subacute toxicity data are available for mammals and for salamanders. No data are available for other species of amphibians. Long-term (i.e., subchronic to chronic) oral toxicity testing is needed for all vertebrates, and acute data are needed for birds and other species of mammals (non-rodent). In addition, metabolic data that provides evidence of bioavailability of both compounds is also needed.

REFERENCES

[1] USACHPPM. Standard practice for wildlife toxicity reference values. Aberdeen Proving Ground (MD): U.S. Army Center for Health Promotion and Preventive Medicine; 2000, Technical Guide No. 254.

[2] Hovatter PS, Talmage SS, Opresko DM, Ross RH. Ecotoxicity of nitroaromatics to aquatic and terrestrial species at army superfund sites. In: Doane TR, Hinman ML, editors. Environmental Toxicology and Risk Assessment: Modeling and Risk Assessment. West Consohocken (PA): American Society for Testing and Materials; 1997, Sixth Vol.

[3] Talmage SS, Opresko DM, Maxwell CJ, Welsh CJ, Cretella FM, Reno PH, et al. Nitroaromatic munition compounds: environmental effects and screening values. Rev Environ Contam Toxicol 1999;161:1–156.

[4] Quinn MJ, McFarland CA, LaFiandra EM, Bazar MA, Johnson MS. Acute, subacute, and subchronic exposure to 2A-DNT (2-amino-4,6-dinitrotoluene) in the northern bobwhite (*Colinus virginianus*). Ecotoxicology 2010 Jun;19(5):945–52.

[5] Walsh M. Environmental transformation products of nitroaromatics and nitramines. U.S. Army Corps of Engineers Cold Regions Research & Engineering Laboratory: Hanover (NH); 1990, Special Report No. 90.

[6] Bumpus JA, Tatarko M. Biodegradation of 2,4,6-trinitrotoluene by Phanerochaete chrysosporium: identification of initial degradation products and the discovery of a TNT metabolite that inhibits lignin peroxidases. Curr Microbiol 1994;28(3):185–90.

[7] Yinon J. Toxicity and Metabolism of Explosives. Boca Raton (FL): CRC Press; 1990.

[8] Johnson MS, Vodela JK, Reddy G, Holladay SD. Fate and the biochemical effects of 2,4,6-trinitrotoluene (TNT) exposure to tiger salamanders (*Ambystoma tigrinum*). Ecotoxicol Environ Saf 2000 Jun;46(2):186–91.

[9] Esteve-Nunez A, Caballero A, Ramos JL. Biological degradation of 2,4,6-trinitrotoluene. Microbiol Mol Biol Rev 2001 Sep;65(3):335–52.

[10] Martin JL, Comfort SD, Shea PJ, Kokjohn TA, Drijber RA. Denitration of 2,4,6-trinitrotoluene by Pseudomonas savastanoi. Can J Microbiol 1997 May;43(5):447–55.

[11] Alvarez MA, Kitts CL, Botsford JL, Unkefer PJ. Pseudomonas aeruginoasa strain MA01 aerobically metabolizes the aminodinitrotoluenes produced by 2,4,6-trinitrotoluene nitro group reduction. Can J Microbiol 1995 Nov;41(11):984–91.

[12] Lewis TA, Ederer MM, Crawford RL, Crawford DL. Microbial transformation of 2,4,6-trinitrotoluene. J Ind Microbiol Biotechnol 1997;18(2,3):89–96.

[13] Boopathy R, Kulpa CF. Trinitrotoluene (TNT) as a sole nitrogen source for a sulfate-reducing bacterium *Desulfovibrio* sp. (B strain) isolated from an anaerobic digester. Curr Microbiol 1992 Oct;25(4):235–41.

[14] Preuss A, Fimpel J, Dickert G. Anaerobic transformation of 2,4,6-trinitrotoluene (TNT). Arch Microbiol 1993;159(4):345–53.

[15] Ellis HV, Hong C-B, Lee C-C. Mammalian toxicity of munitions compounds. Summary of toxicity of nitrotoluenes. Kansas City (MO): Midwest Research Institute; 1980, Prepared for the U.S. Army Medical Research and Development Command, Fort Detrick, Frederick, MD. Accession Number: AD A080146.

[16] Johnson MS, Holladay SD, Lippenholz KS, Jenkins JL, McCain WC. Effects of 2,4,6-trinitrotoluene in a holistic environmental exposure regime to a terrestrial salamander: *Ambystoma tigrinum*. Toxicol Pathol 2000 Mar;28(2):334–41.

[17] McFarland CA, Quinn Jr MJ, Boyce J, LaFiandra EM, Bazar MA, Talent LG, et al. Toxic effects of oral 2-amino-4,6-dinitrotoluene in the western fence lizard (Sceloporus occidentalis). Environ Pollut 2011 Feb;159(2):466–73.

[18] Allen NA, Major MA. The re-evaluation of water solubility and octanol/water partition values for military explosives commonly found in soil contaminants; 2001, USACHPPM Report (draft manuscript in preparation).

CHAPTER 17

Wildlife Toxicity Assessment for Triacetin

Michael J. Quinn, Jr., David Ziolkowski Jr.

Contents

INTRODUCTION

Triacetin is a triester of glycerin and acetic acid. It has been used for over 75 years for a wide range of uses, including cosmetic biocide (most often as a fungicide), plasticizer, solvent in cosmetic formulas, food additive (as a flavoring agent and adjuvant), and as a binder for combustible material in solid-rocket propellants. It is also known as glyceryl triacetate, glycerol triacetate, glycerin triacetate, glycerine triacetate, triacetyl glycerine, acetin-tri, 1,2,3-triacetoxypropane, 1,2,3-propanetriol triacetate, 1,2,3-propanetriyl triacetate, and acetic-1,2,3-prepanetriyl ester [1]. Common trade names include Enzactin, Fungacetin, Glyped, Kesscoflex TRA, and Vanay. It is a colorless, oily liquid that, although is most often synthesized, can be found naturally in cod-liver oil, butter, and other fats [2]. Reflecting

Wildlife Toxicity Assessments for Chemicals of Military Concern
http://dx.doi.org/10.1016/B978-0-12-800020-5.00017-X

its chemical nature and its widespread use as a Food and Drug Administration generally recognized as safe (FDA, GRAS) food additive, there is no U.S. Environmental Protection Agency Integrated Risk Information System (U.S. EPA, IRIS) record on triacetin, and no threshold limit values have been established to protect occupationally exposed workers to the compound. The only limitation on triacetin is Good Manufacturing Practice (GMP) requirements set by the FDA [3].

The potential environmental effects of triacetin are of particular interest because it is a high production volume chemical. Japan, the leading producer of triacetin, was estimated to produce 5,000 tons per year (t/yr), compared to the global production of about 10,000 to 50,000 t/yr [4]. One Japanese production site that produced 2,000 t/yr was estimated to release about 1,440 kilograms per year (kg/yr) through wastewater [4]. In the United States, Shackelford and Keith [5] detected triacetin in samples collected from the Tennessee River, but no concentration levels were reported.

This chapter summarizes current knowledge on potentially harmful impacts to wildlife from triacetin exposure. Where possible, emphasis is placed on threshold doses for the onset of toxicological effects, as described in reports of experimental studies. Surveying the threshold dosimetry of the compound may help to establish Toxicity Reference Values (TRVs). These levels could serve as protective exposure standards for all wildlife that may become exposed to triacetin while ranging near affected sites. The protocol for the performance of this assessment is documented in the U.S. Army Center for Health Promotion and Preventive Medicine (USACHPPM) Technical Guide (TG) 254, Standard Practice for Wildlife Toxicity Reference Values [6]. Table 17.1 summarizes the chemical and physical properties of triacetin.

TOXICITY PROFILE

Environmental Fate and Transport

Table 17.1 summarizes the physical-chemical information applicable to the environmental fate and transport of triacetin. The body of available evidence suggests that triacetin is unlikely to persist in environmental media. When triacetin is released into the environment, it is believed to degrade rapidly.

Due to its low vapor pressure of 0.0033 hectopascal (hPa) at 25 °C, triacetin most likely exists entirely in the vapor phase in the ambient atmosphere [4] where it is degraded photochemically to hydroxyl radicals. An estimated K_{oc} of 10.5 suggests that the compound should readily leach from soil to surface water or groundwater. When in water, degradation is expected to

Table 17.1 Summary of Physical-Chemical Properties of Triacetin

CAS No.	102-76-1
Molecular weight	218.20
Color	Colorless
Physical state	Oily liquid
Melting point	–78 °C
Boiling point	259 °C
Density	1.1562 g/cm^3 at 25 °C
Odor	Faint, fatty odor with bitter taste
Solubility in water	58 g/L at 20–25 °C. 70 g/L at 25 °C. Soluble in acetone, benzene, ethanol, ether, chloroform
Partition coefficients:	0.21 (25 °C)
Log K_{ow}	2.5×10^{-1}
Log K_{oc}	1.02
Vapor pressure at 25 °C	2.48×10^{-3} mm Hg; 3.306×10^{-3} hPa at 25 °C
Henry's Law constant at 25 °C	No data
Conversion factors	1 ppm = 8.92 mg/m^3 1 mg/m^3 = 0.11 ppm

Sources: OECD [4], HSDB [7], CRC Handbook of Chemistry and Physics [40].

occur rapidly and almost entirely by hydrolysis, producing glycerol and acetic acid, which are further broken down to carbon dioxide. Because of its high solubility and low vapor pressure in a water medium, 99% of the initial parent compound is believed to remain until hydrolysis is complete [4]. A linear relationship between degradation rates and increasing alkalinity has been demonstrated for triacetin hydrolysis [7]. Although biodegradation is generally accepted to be sufficiently rapid not to cause significant environmental effects, the previous detection of triacetin in the Tennessee River suggests that there may be limited capacity for persistence in surface water [7].

Deposition of triacetin in surface water could result in direct exposure of wildlife via dermal contact and ingestion. Evidence suggests, however, that triacetin is rapidly hydrolyzed by a number of enzymes in most tissues, including lipase in pig [8] and rat [9] small intestines, carboxyesterases in the rat liver and plasma [10], electric organ cholinesterase [11], intact cells from the electric eel's organ of Sachs [12], and nonspecific tissue esterases [13]. It is readily metabolized in the gastrointestinal tract where it is treated similar to a small triglyceride. Triacetin is broken down more efficiently than butterfat, hydrogenated cottonseed oil, and coconut oil [14]. As such, triacetin has been shown to have more of a nutritive than toxic effect, as evidenced by being able to be a source of liver glycogen [15]. It has also been demonstrated to be used as efficiently as glucose when fed in amounts equal in caloric value to 15% glucose [13].

Summary of Mammalian Toxicology

Mammalian Oral Toxicity: Acute

Most of the data concerning acute oral exposure of triacetin in mammals are reported from nonpublished sources, with very little, if any, information available regarding methodology. At present, the only reported effect from these studies is mortality expressed as lethal dose $(LD)_{50}$ values. A single LD_{50} value from a peer-reviewed journal has been calculated as 9,300 milligrams per kilogram (mg/kg) rat body weight (bw) by the Organisation for Economic Co-operation and Development (OECD) [4] from a given literature value of 8.0 milliliters per kilogram (ml/kg) bw and density of 1.16 grams per cubic centimeter (g/cm^2) [16]. The remaining LD_{50} values that come from unpublished reports are continually reported as secondary references in review papers. Of these secondary referenced reports, the single acute exposure test with reported methods states an LD_{50} of 6,400 to 12,800 mg/kg bw for rats and 3,200 to 6,400 mg/kg bw for mice [17,18, and as reported in 19]. The lowest LD_{50} value from a secondary source with no described methods is less than 2,000 mg/kg bw for rats [20, as reported in 4]. This dose limit caused no mortality and, although the signs measured were not reported, no signs of systemic toxicity were observed during the 14-day observation period.

Mammalian Oral Toxicity: Subacute

No subacute triacetin mammalian oral toxicity data were identified.

Mammalian Oral Toxicity: Subchronic

A rigorous study conducted by the Japanese Ministry of Health and Welfare [21] administered triacetin by daily oral gavage at 40, 200, or 1,000 mg/kg BW to male Sprague Dawley rats for 44 days and females for 41 to 48 days [as reported in 4]. In agreement with most other triacetin toxicity studies, no effects on body weight, food consumption, necropsy findings, histopathological observations, or blood chemistries were observed. What makes this study unique, however, is that this is the only study known to assess reproductive effects of triacetin. Male and female rats were dosed from 2 weeks prior to mating, with females dosed to day 3 post-partum. No adverse effects were observed in the mating or fertility indices, gestation length, number of corpea lutea and implantations, implantation, gestation or delivery indices, parturition, maternal behavior at delivery and lactation, number of offspring or live offspring, sex ratio, offspring body weight, and live birth and viability indices. In addition, no adverse toxic effects were found in the offspring. Consequently, the no observed adverse effect level (NOAEL) is considered 1,000 mg/kg bw/day for this study.

Mammalian Oral Toxicity: Chronic

The only chronic triacetin mammalian oral toxicity study identified was a study attempting to find methods to produce a synthesized diet composed of glycerol, formose sugars, and triacetin for use on long-duration space missions [22]. Growing rats (age not specified) were fed diets containing between 20 and 60% by weight for 90 days. A large loss in weight and considerable mortality was observed at 60%, while no adverse effects were observed at 20%, approximating a daily dose of 10,000 mg/kg bw. No other details relating to the toxicity of triacetin were reported.

Mammalian Oral Toxicity: Other

Many LD_{50} values are reported in Fiume [1] and OECD [4]; however, the details of the unpublished reports' methods are often either unclear or completely missing. Since these studies are often reported in triacetin toxicity summary reports, they warrant some mention here. Gast [23] reports an LD_{50} for male Swiss mice to be 1,800 mg/kg bw, and 1,100 mg/kg bw for females [reported in 4]. Mice were also reported to have a range of LD_{50} values from 3,200 to 6,100 mg/kg bw [24] and 3,200 to 6,400 mg/kg bw [25, reported in 1]. The LD_{50} values for rats are reported as 3,480 mg/kg bw [26, reported in 27] and from 6,400 to 12,800 mg/kg bw [25, reported in 1]. The only report to have determined an LD_{50} for rabbits reports this value at over 2,000 mg/kg bw [28, reported in 4].

Two subchronic oral toxicity studies were found, however limitations in study design prevent them from being appropriate for TRV derivation. Lynch *et al.* [29] fed Sprague Dawley rats a diet in which 28.5% of the total calories were supplied by triacetin for 30 days. The rats had *ad libitum* access to food, but because consumed food amounts were not reported, the actual dose could not be determined. No overt signs of toxicity were reported in these rats. In another triacetin feeding study, Cox [30] stated that the small sample size (N=4) prevented any definitive conclusions when triacetin (55% w/w) was fed *ad libitum* to weanling rats for 60 days. Triacetin dose was unable to be determined because food consumption was unreported.

Though no overt signs of toxicity were observed, triacetin was reported to promote survival and fair growth when compared to the control diet of coconut oil.

The focus of many of the studies was an attempt to modulate nutritional parameters and responses through the repeated oral administration of triacetin to mammals. Imoto and Namioka [31] supplemented the basal diet of pigs with triacetin over a 47-day period; the animals increased bw gain compared to controls and demonstrated a 56 to 59% efficiency. Triacetin administration was associated with a decrease in blood glucose and an increase in lactate and ketone bodies; glycogen levels increased in liver, heart, and femoral muscle [32].

A series of reports addresses the comparative nutritive effects of dietary triacetin versus broadly isocaloric diets containing longer-chain triglycerides [29,33]. When the longer-chain triglyceride diets were made available to male Sprague Dawley rats for 30 days, bodily fat and protein levels were increased. Fat cell diameters, however, were smaller in animals receiving triacetin compared to those receiving longer-chain triglycerides [33]. Some fluctuations were observed in plasma lipid parameters, and the amounts of deoxyribonucleic acid (DNA) in intestinal mucosal cells were elevated in rats receiving triacetin compared to either controls or those receiving longer-chain triglycerides, however no overt toxic effects were observed [29].

Studies Relevant for Mammalian TRV Development for Ingestion Exposures

Lawrence *et al.* [16] determined an LD_{50} of 9,300 mg/kg bw in rats (Table 17.3) and Shapira *et al.* [22] reported a 90-day NOAEL of 10,000 mg/kg bw in rats, and reported a large loss in weight and considerable mortality when rats received a diet containing 60% triacetin (approximately 30,000 mg/kg bw). The only study to observe reproductive effects reported a NOAEL at the highest treatment level of 1000 mg/kg BW over a period of 41 to 48 days.

Mammalian Inhalation Toxicity

A lethal concentration $(LC)_{50}$ for rats following a 4-hour inhalation exposure to triacetin was determined to be greater than 1.721 milligrams per liter (mg/L) [28, as reported in 1]. Studies conducted by Fassett and Roudabush [34] reported no adverse toxic effects observed in rats exposed to vapor concentrations of 2.2 mg/L for 6 hours/day, 5 days/week, for 90 days or

72.8 mg/L for 6 hours/day for 5 days [19]. Because triacetin photodegrades rapidly when in the vapor phase, exposure through environmental inhalation is unlikely.

Mammalian Dermal Toxicity

An unpublished report by Bailey [35] is the only report of a dermal LD_{50} in mammals; LD_{50} in rabbits was determined to be over 5,000 mg/kg bw [cited in 4]. All other data concerning dermal irritation due to triacetin is from unpublished letters or reports. Previous toxicity reviews recategorize triacetin from that of a nondermal irritant [36, citing 37] to a slight dermal irritant [1,19], citing minimal edema and erythema in two guinea pigs receiving triacetin patches of either 5 or 20 cubic centimeters per kilogram (cc/kg) for 24 hours [25].

Avian Toxicology

The only study to assess triacetin toxicity in birds [38] reported little irritation in male Hubbard Broiler Chickens immediately following intramuscular injection with 0.5 mL triacetin, with complete disappearance by day 7 [36].

Amphibian Toxicology

A low-density lipoprotein $(LDL)_0$ of 150 mg/kg bw in frogs was reported [39, cited in 4]; however, neither the study details nor citation date are provided.

Reptilian Toxicology

No information was identified on triacetin toxicity in reptiles.

RECOMMENDED TOXICITY REFERENCE VALUES

Toxicity Reference Values for Mammals

TRVs for Ingestion Exposures for the Class Mammalia

Triacetin is a high-volume production chemical (triglyceride) that has been used for a variety of purposes for over 75 years. Although one source estimates an approximate annual release rate of 7% from a production site, most toxicity reports and reviews agree that no significant environmental exposure is likely.

The historic doses (oral) used for median lethal exposures suggest that triacetin is a comparatively safe compound regarding its effects on experimental animals. This has prompted the FDA to regard triacetin as a GRAS ingredient with no limitations on its conditions of use other than those in the current Good Manufacturing Practice (GMP), found in Title 21, Part 184.1901 of the Code of Federal Regulations (21 CFR 184.1901) [3]. In addition, the Cosmetic Ingredient Review (CIR) expert panel reviewed the safety of triacetin and concluded that the compound is safe as used in cosmetic formulations [1].

Therefore, triacetin is considered to be non-toxic and of no significant threat to wildlife. This determination is given a medium level of confidence due to a lack of current rigorous studies. Table 17.2 provides a summary of the relevant data for TRV derivation in the Class Mammalia, and Table 17.3 provides the selected ingestion TRVs for same.

TRVs for Inhalation Exposures for the Class Mammalia

Triacetin has been assessed as non-toxic to mammals. Adverse effects could not be observed as a result of high acute oral exposures, and given the low vapor pressure of the compound it is unlikely to be encountered via the inhalation route at exposures exceeding the oral limit dose.

TRVs for Dermal Exposures for the Class Mammalia

Not available at this time.

Table 17.2 Summary of Relevant Mammalian Data for TRV Derivation

		Test Results				
Test Organism	**Test Duration**	**NOAEL (mg/kg/day)**	**LOAEL (mg/kg/day)**	**LD_{50} (mg/kg)**	**Effects Observed at the LOAEL**	**Study**
Rat	Single exposure	NA	NA	9,300	NA	[16]
Rat	41–48 days	1,000	NA	NA	NA	[21]
Rat	90 days	10,000	30,000	NA	Weight loss, considerable mortality	[22]

NA, not applicable.

Table 17.3 Selected Ingestion TRVs for the Class Mammalia

TRV	Dose	Confidence
NOAEL-based	Non-toxic	Medium
LOAEL-based	Non-toxic	Medium

Sources: [16,22].

Toxicity Reference Values for Birds

Not available at this time.

Toxicity Reference Values for Amphibians

Not available at this time.

Toxicity Reference Values for Reptiles

Not available at this time.

IMPORTANT RESEARCH NEEDS

The limited, substandard data available on the toxicity of triacetin indicate the need for more research on mammalian oral, dermal, and inhalation routes of exposure. Chronic mammalian studies, especially low-dose chronic exposure studies, are particularly necessary. No studies with birds, amphibians, or reptiles could be found that could be used to generate reliable toxicity reference values; any studies testing the toxicity of triacetin in these classes of animals would be useful. However, despite the lack of quality information available, there is general agreement that triacetin is a low priority chemical.

REFERENCES

[1] Fiume MZ. Cosmetic Ingredients Review Expert Panel. Final report on the safety assessment of triacetin. Int J Toxicol 2003;22(Suppl. 2):1–10.

[2] Grant J, editor. Hackh's chemical dictionary. 4th ed. New York (NY): McGraw-Hill; 1972.

[3] FDA. Triacetin. Washington (DC): Federal Register; 2005, 21 CFR 184.1901.

[4] OECD. Screening information data set (SIDS). Triacetin. Paper presented at: SIDS Initial Assessment Report Meeting (SIAM) 15; 2002 Oct 22–25; Boston, MA.

[5] Shackelford WM, Keith LH. Frequency of organic compounds identified in water. Athens (GA): Environmental Research Laboratory, Office of Research and Development, U.S. Environmental Protection Agency; 1976, Environmental monitoring series. EPA-600/4-76-062.

[6] USACHPPM. Standard practice for wildlife toxicity reference values. Aberdeen Proving Ground (MD: U.S. Army Center for Health Promotion and Preventive Medicine; 2000, Technical Guide No. 254.

[7] Hazardous Substance Data Bank (HSDB). Triacetin [Internet]. Bethesda (MD): National Library of Medicine (US), National Institute of Health [cited 2012]. Available from: http://toxnet.njm.nih.gov/.

[8] Serrero G, Negrel R, Ailhaud G. Characterization and partial purification of an intestinal lipase. Biochem Biophys Res Commun 1975;65(1):89–99.

[9] Barry RJ, Jackson MJ, Smyth DH. Handling of glycerides of acetic acid by rat small intestine in vitro. J Physiol 1966;185(3):667–83.

[10] Murphy SD, Cheever KL. Effect of feeding insecticides. Inhibition of carboxyesterase and cholinesterase in rats. Arch Environ Health 1968;17(5):749–58.
[11] Mounter LA, Cheatham RM. Specificity of electric organ cholinesterase. Enzymologia 1963;25:215–24.
[12] Rosenberg P, Dettbarn WD. Ester-splitting activity of the electroplax. Biochim Biophys Acta 1963;69(1):103–14.
[13] World Health Organization (WHO). Toxicological evaluation of some food colours, thickening agents, and certain other substances. Geneva, Switzerland: World Health Organization; 1975, WHO Food Additives Series No. 8.
[14] Deuel Jr HJ, Hallman L. The rate of absorption of synthetic triglycerides in the rat. J Nutr 1940;20(3):227–32.
[15] Deuel Jr. HJ, Butts JS, Blunden H, Cutler CH, Knott L. Studies on ketosis. IX. Glycogen formation from various purified and natural fats. J Biol Chem 1937;117–9.
[16] Lawrence WH, Malik M, Autian J. Development of a toxicity evaluation program for dental materials and products. II Screening for systemic toxicity. J Biomed Mater Res 1974;8(1):11–34.
[17] Fassett DW. Corporate health and environment laboratories. Report No. 900256006, Rochester (NY): Eastman Kodak Company; 1948, Unpublished data.
[18] Fassett DW. Corporate health and environment laboratories. Rochester (NY): Eastman Kodak Company; 1955, (1). Unpublished data.
[19] Eastman Chemical Company. Toxicity summary for Eastman triacetate, glyceryl triacetate. Kingsport (TN): Product Safety & Stewardship; 2004.
[20] Unichema Chemie BV. RCC NOTOX 0831/1056. 1988(1). Unpublished data.
[21] Ministry of Health and Welfare (MHW), Japan. Toxicity testing reports of environmental chemicals, vol. 6; 1988, 127-147.
[22] Shapira J, Mandel AD, Quattrone PD, Bell NL. Current research on regenerative systems. Life Sci Space Res 1969;7:123–9.
[23] Gast JH. Some toxicity studies with triacetin. (Abstract). Fed Proc Fed Am Soc Exp Biol 1963;22:368.
[24] Bisesi MS. Glyceryl triacetate (Triacetin). In: Clayton GD, Clayton FE, editors. Patty's industrial hygiene and toxicology, 4th ed., vol. II, part D. New York (NY): John Wiley & Sons; 1994.
[25] Anstadt GW. Information on the toxicity of triacetin. Letter to D.R. Mathieu, Flavor and Extract Manufacturers' Association; 1976 Sep 13.
[26] Li RC, Sah PPT, Anderson HH. Acute toxicity of monoacetin, diacetin, and triacetin. Proc Soc Exper Biol & Med 1941;46:26.
[27] Von Oettingen WF. The aliphatic acids and their esters: toxicity and potential dangers. AMA Arch Ind Health 1960;21:28–65.
[28] Unichema Chemie BV. Summary of information on triacetin. Cosmetic, Toiletry, and Fragrance Association; 1994, Unpublished data.
[29] Lynch JW, Miles JM, Bailey JW. Effects of the short-chain triglyceride triacetin on intestinal mucosa and metabolic substrates in rats. JPEN J Parenter and Enteral Nutr 1994;18(3):208–13.
[30] Cox Jr WM. The nutritive value of pure fatty acid esters. J Biol Chem 1933;103:777–90.
[31] Imoto S, Namioka S. Nutritive value of acetate in growing pigs. J Anim Sci 1983;56 (4):858–66.
[32] Imoto S, Namioka S. Acetate-glucose relationship in growing pigs. J Anim Sci 1983;56 (4):867–75.
[33] Lynch JW, Bailey JW. Dietary intake of the short-chain triglyceride triacetin vs. long-chain triglycerides decreases adipocyte diameter and fat deposition in rats. J Nutr 1995;125(5):1267–73.

[34] Fassett DW, Roudabush RL. Inhalation of saturated vapor of triacetin. Laboratory of Industrial Medicine. Rochester (NY): Eastman Kodak Company; 1955 Jul 28.
[35] Bailey DE. Report to Research Institute for Fragrance Materials (RIFM); 1976 May 21.
[36] Opdyke DL. Fragrance new materials monographs: triacetin. Food Cosmet Toxicol 1978;16:879–82.
[37] Fassett DW. 1 Esters. In: Patty FA, editor. Industrial hygiene and toxicology. 2nd ed. New York (NY): Interscience Publishers; 1963.
[38] Hem SL, Bright DR, Banker GS, Pogue JP. Tissue irritation evaluation of potential parenteral vehicles. Drug Dev Ind Pharm 1974;1(5):471–7.
[39] National Institute for Occupational Safety and Health (NIOSH). 1976.
[40] Haynes WM, editor. CRC handbook of chemistry and physics. 94th ed. London (UK): CRC Press; 2013.

CHAPTER 18

Wildlife Toxicity Assessment of White Phosphorus

Marc A. Williams, Stephen W. Rice

Contents

INTRODUCTION

Phosphorus has three allotropes: white, red, and black. White phosphorus is pyrophoric, spontaneously igniting in the presence of oxygen until complete combustion or no more oxygen is available. The pyrophoric nature of white phosphorus, also known as Willy Pete, makes it an ideal signaling, screening, and incendiary material for military purposes. Red phosphorus is created by heating white phosphorus to 250 °C or by exposing white phosphorus to sunlight. Applications of red phosphorus include safety matches, fireworks, and pesticides. As seen with safety matches, frictional heating is enough to convert it back to self-igniting white phosphorus. Black phosphorus is created by heating white phosphorus with a mercury catalyst. It is the safest allotrope and has no commercial use. White phosphorus is the most reactive and most commercially and militarily important allotrope, therefore, this wildlife toxicity assessment will focus solely on white phosphorus.

Wildlife Toxicity Assessments for Chemicals of Military Concern
http://dx.doi.org/10.1016/B978-0-12-800020-5.00018-1

White phosphorus has the appearance of a waxy solid that is colorless-to-white when pure, although the technical grade is yellow. Elemental white phosphorus exists as a tetramer, P_4, which is produced by heating phosphate rock in a furnace and condensing the resulting vapor. White phosphorus spontaneously ignites when exposed to air, producing a thick, white smoke. In the presence of oxygen, white phosphorous reacts to form phosphorus oxides; however, an oxide-passivating layer may form, retarding oxidation and allowing the phosphorus to reach land and water in its elemental form [1]. Elemental P_4 entering the environment may be increased when white phosphorus is fired over wetlands and particles enter the water or sediments before oxidation is complete. Because white phosphorus has very low water solubility and is highly lipophilic, it can remain inert in anoxic aquatic environments for years [2]. Releases of white phosphorus to the environment can also occur during the production of white phosphorus and related munitions. Effluent from white phosphorus production facilities, known as "phossy water," contains both dissolved and suspended white phosphorus [1]. A summary of the chemical and physical properties of white phosphorus is shown in Table 18.1.

According to the U.S. Environmental Protection Agency's (U.S. EPA) most recent reports from their Toxics Release Inventory Program, over 89,800 lbs of white phosphorus were released into the surface waters of

Table 18.1 Summary of Physical-Chemical Properties of White Phosphorus

CAS No.	7723-14-0
Molecular weight	124 g/mol
Color	Colorless/white to yellow
Physical state	Waxy, flammable solid
Melting point	44.1 °C (111.4 °F)
Boiling point	280.0 °C (536.0 °F)
Odor	Garlic-like
Solubility in water	3 mg/L; soluble in alkali, ether, chloroform, benzene, and toluene
Partition coefficients:	
Log K_{ow}	3.08
Log K_{oc}	3.05
Vapor pressure at 25 °C	0.025 mm Hg at 20 °C
Henry's Law constant at 25 °C	$1.36–2.11 \times 10^{-3}$ atm·m³/mole
Conversion factors	1 ppm = 5.07 mg/m³ 1 mg/m³ = 0.2 ppm

Sources: ATSDR [2], Budavari *et al.* [3], HSDB [6].

the United States alone between 1988 and 2007 [4]. Since the annual discharge of P_4 into surface waters has remained relatively constant for the past 9 years, and its half-life in water or wet sediments is long enough to be described as "indefinite" [5], there is a high potential for exposure of wildlife to this chemical. As pelleted P_4 that remains in surface waters or anoxic sediments is the most persistent and environmentally relevant form of white phosphorus, the scope of this Wildlife Toxicity Assessment (WTA) will be limited to the pelleted form and will not include discussions of the toxicity of white phosphorus smoke.

The toxicity of P_4 to human beings has been recognized for many years. For example, in the 19th century, persons employed in making matches developed chronic ailments because of exposure to the element. Perhaps the best known of these conditions was "phossy jaw," a disease caused by absorption of P_4 fumes through dental cavities [6]. Case studies of accidental poisonings suggest a median lethal dose (LD_{50}) of about 1 mg/kg in human beings.

Given the potential for occupational exposure, agencies charged with setting protective standards for the workplace have designated exposure limits for the element. For example, the National Institute for Occupational Safety and Health (NIOSH) and the Occupational Safety and Health Administration (OSHA) set time-weighted average exposure limits of 0.1 mg/m^3 for P_4. The Integrated Risk Information System (IRIS) of the U.S. EPA reports a reference dose (RfD) for oral exposure to the element of 2×10^{-5} mg/kg/day [7].

Most studies concerning toxic effects of P_4 on wildlife have focused on avian species. This research was conducted in an effort to determine the cause of the mass die-offs of waterfowl occurring at Eagle River Flats (ERF), Alaska, U.S. First recorded in 1982, up to 2,000 birds, mostly bottom-feeding ducks and swans, have died annually at Eagle River Flats due to unknown causes [8]. P_4 had been used for military practice in that area since the 1940s and was found in both the sediment where the ducks and swans feed, and in the gizzards of the carcasses collected from the site. The P_4 that entered one of the main salt marshes in ERF was identified as the main cause of waterfowl deaths, and was implicated in potential poisoning of predators due to its ability to bioaccumulate up the food chain [9]. Particle size varies, but often is similar to seeds or aquatic invertebrates that is food for water birds, and is also small enough to allow for significant accidental exposure through sediment ingestion [8]. This WTA summarizes available information on the likely effects of P_4 on wildlife. Toxicity Reference Values (TRVs) are derived when suitable data are available. These TRVs are intended to facilitate screening-level ecological risk assessments for wildlife in contaminated areas. The method used in this assessment is documented in the U.S. Army Center for Health Promotion and Preventive Medicine Technical Guide 254, *Standard Practice for Wildlife Toxicity Reference Values* [10].

TOXICITY PROFILE

Environmental Fate and Transport

The release of elemental phosphorus to the air during its production or use is well documented. In addition, deposition of P_4 to soil and sediment was demonstrated at military firing ranges, and P_4 was released to environmental water bodies from its production sites. For example, according to the Agency for Toxic Substances and Disease Registry (ATSDR) [2], a production facility in Canada released 68 to 91 kg/day of colloidal P_4 into a nearby inlet of Placentia Bay, in Newfoundland. Effluents and landfills at U.S. Army ammunition plants also contain the element, and P_4 that has been deposited in the soil may leach to surface water and groundwater.

Elemental phosphorus is unlikely to persist in environmental media as P_4 for more than a short time since it undergoes rapid oxidation under aerobic conditions and spontaneously combusts between 30 and 40 °C [11]. Conversely, persistence of the element is likely where prevailing conditions are anaerobic, such as in some sediment. Washout and rainout from the air will transfer any unreacted P_4 to water or land. However, reaction products such as phosphine or phosphates may persist for longer periods than the elemental form. ATSDR [2] has identified a number of competing processes through which P_4 partitions in soils, sediments, and surface water. These include volatilization, leaching, and the possibility that the element may be taken up and bioconcentrated in aquatic organisms. These processes will take place concurrently with P_4's chemical modification to such compounds as phosphine, which itself is highly reactive. Physical and chemical characteristics of each contaminated water body can affect the rate of P_4 disappearance from streams, lakes, and rivers. The pH and oxygen content of the water are important factors to consider in the depletion of P_4 from waterways; equally important is whether P_4 is present in the dissolved, sorbed, colloidal, or particulate state. While the production of phosphine is favored in conditions of low oxygen pressure, oxidation products of P_4 include hypophosphorus acid, phosphorus acid, and phosphoric acid.

Some striking levels of P_4 have been reported at sensitive environmental sites. For example, the concentration of the contaminant in the overflow water of a settling pond at a P_4 filling station in Arkansas was reported to be approximately 84 mg/L [2]. Sediments near a phosphorus manufacturing plant were reported up to 3,000 ppm. Eagle River Flats, an estuarine salt marsh in Alaska that is part of a U.S. Army firing range, has elemental phosphorus concentrations up to 10 ppm [12]. As further discussed in this chapter, such levels are associated with the demise of large numbers of estuarine birds. These species forage in the anoxic sediments and ingest unspent pellets of P_4 with their food. Since P_4 readily accumulates in tissues, making biomagnification a likely event, predators and scavengers may be exposed through ingestion of prey items.

Summary of Mammalian Toxicity

Mammalian Oral Toxicity: Acute

White phosphorus oral LD_{50} values of 3.03 mg/kg were reported in rats and 4.82 mg/kg in mice, respectively [13]. Additionally, LD_{50} values of 4.2 and 6.6 mg/kg were reported in Sprague Dawley rats using tricaprylin and corn oil as the gavage vehicle, respectively [14]. A number of single-dose studies have examined toxicological and physiological endpoints other than lethality. For example, in researching the mechanism of P_4-induced deposition of lipid in the liver (of female Wistar rats), the incorporation of radiolabeled DL-leucine, orthophosphate, and acetate into the liver and plasma lipoprotein fractions of rats was charted following a single oral dose of 1 mg P_4 in olive oil [15]. While the incorporation of phosphate into lipid was the same in treated animals and controls, P_4-treated rats incorporated less DL [1-^{14}C]-leucine into the liver and plasma lipoprotein fractions than did controls. Simultaneously, the specific concentrations of free and esterified fatty acids in the liver were markedly increased. This suggested that the formation of a fatty liver in response to P_4 might occur through inhibition of protein synthesis, which in this case resulted in reduced production of apolipoproteins. Others reached a similar conclusion after dosing female Sprague Dawley rats with 10 mg/kg P_4 by gavage in 0.1% mineral oil [16]. The researchers had pretreated the rats with a number of antioxidants that, in other experiments, had blocked the induction of a fatty liver by carbon tetrachloride. However, several of the compounds were ineffective in blocking the P_4-fatty liver response. Supporting the concept of a perturbation of protein synthesis by P_4, the authors prepared subcellular fractions by differential ultracentrifugation and showed a disaggregation of polyribosomes because of P_4

administration. These results are consistent with the concept that an inhibition of protein synthesis mediates the P_4-induced onset of hepatic fat deposition.

Mammalian Oral Toxicity: Subacute

The literature searches on P_4 identified a study in which female Wistar rats were given feed spiked with yellow phosphorus for 16 days [17]. The study design included 16 animals. Four animals were placed on the P_4-amended diet for 16 days, and four control animals were fed stock diets for the same period. Four of the remaining animals were fed the P_4 amended diet of the same dose for 8 days followed by 16 days of stock diet, and the final four animals were used as controls for this group. P_4 was combined with feed by melting 65 mg of yellow phosphorus in 95 mg of soybean oil; it was permanently suspended by the addition of 5 ml of diethyl ether. As calculated by the authors, the daily dose of P_4 was 0.065 mg/rat/day, equivalent to 0.186 mg/kg/day if a default body weight of 350 g is assumed. After the treatments were terminated, all animals were euthanized by diethyl ether inhalation. The primary focus of the experiment was to determine the effects of P_4 on bone. Dissected tibiae-fibulae were examined by radiography, demineralized using a vacuum pump, and then sectioned for microscopic evaluation. The observed principle effect of P_4 was reversible blockade of bone resorption. Morphological changes associated with dietary P_4 included wider trabeculae, the retention of the chondroid core, and, histologically, the enlargement of deep-seated osteocytes with the formation of rounded nuclei. The study was too short to determine any growth effects from P_4 exposure; no statistical analyses were done [17].

Mammalian Oral Toxicity: Subchronic

No studies were identified that address the toxicology of P_4 under a subchronic dosing regimen.

Mammalian Oral Toxicity: Chronic

No studies were identified that address the toxicology of P_4 under a chronic dosing regimen.

Mammalian Oral Toxicity: Other

International Research and Development Corporation carried out a one-generation reproductive study on behalf of Monsanto Inc. that was reported to the U.S. EPA as a Toxic Substances Control Act Test Submission

(TSCATS) [14]. The U.S. EPA IRIS reported an RfD of 2×10^{-5} mg/kg/day for P_4 [18]. Fifteen male and 30 female Charles River COBS® CD® rats per group were gavaged for 80 days at 0, 0.005, 0.015, or 0.075 mg/kg/day P_4 in corn oil, and then throughout a complete reproductive cycle of mating, gestation, parturition, and weaning. The experimental protocol was unusual in that the F_0 generation then entered another complete reproductive cycle such that exposure to P_4 continued until termination of survivors at 30 weeks. Clinical signs were observed twice daily, food consumption and body weights were monitored weekly, and some survivors received a postmortem examination and histopathological work-up.

Monitored reproductive effects included male and female fertility indices, duration of gestation, the number of live and stillborn pups/litter, pup survival through weaning, and pup weights. Generally, few observed effects could be specifically ascribed to the treatments given. For example, there was no apparent incidence of fatty liver at the employed dose levels. However, an important toxicological effect of P_4 was the death of a substantial portion of the high-dose dams close to parturition, on gestation days (GDs) 21 and 22. This unusual phenomenon was observed in both reproductive cycles and was recognized by the U.S. EPA's Office of Toxic Substances as a striking treatment response [14]. The high-dose treatment may also have been associated with a reduction in pup viability, since the proportion of stillborn pups in high-dose dams was increased in both matings. However, this trend was not statistically significant. In agreement with the conclusions presented in the IRIS database, the no observed adverse effect level (NOAEL) for parturition mortality was 0.015 mg/kg/day, with a lowest observed adverse effect level (LOAEL) of 0.075 mg/kg/day.

The Monsanto Company followed up with another one-generation reproductive study in which the influence of gavage vehicle on P_4 was investigated [19]. Fifteen male and 30 female Sprague Dawley rats/group were gavaged for an 11-week premating period with either corn oil alone (group I), 0.075 mg P_4/kg/day in corn oil (group II), 0.075 mg P_4/kg/day in corn oil (group III), tricaprylin alone (group IV), or 0.075 mg P_4/kg/day in tricaprylin (group V). This regimen was maintained through mating, gestation, parturition, and weaning, except for group III in which the dams were terminated on GD 15. As with the previous study, a full suite of systemic and reproductive/developmental parameters was investigated, but all toxicological effects were confined to the dams of groups II, III, and V, who displayed mortality in late gestation in a similar manner to those described previously [14]. Respective percentage mortality among the dams in these groups was 46.7,

30, and 36.7% compared to 0 and 13.3% in groups IV and I (vehicle controls). While there were few, if any, necropsy findings in dams, pups, or sires in this study, most of those dams that died at parturition exhibited severe necrosis of the liver. These findings confirmed an earlier reproductive study [14] and suggested parturition mortality to be an effect of P_4 rather than of an interaction between P_4 and corn oil.

Others discussed how a single oral exposure of 0.75 mg yellow phosphorus/100 g bw in rats caused peroxidative decomposition of microsomal lipid, which might have been a possible first step in abnormal phosphorus-induced hepatic lipid accumulation [20]. In addition, some vesiculation and dispersion of ribosomal granules on hepatic rough endoplasmic reticulum were observed 12 h after the exposure that was described above. However, the significance of this observation is unknown. Finally, Ashburn *et al.* [21] reported to have dosed guinea pigs with 0.75 mg/kg for 4 days each week or 1.5 mg/kg for 2 days each week. However, for reasons that might have been due to the appearance of the animal, or to obtain a "more satisfactory weight curve," dosing was not accurately quantified because two or more successive doses were occasionally omitted [21]. The main objective of this study was to produce liver cirrhosis; however, this was not achieved and the various types of hepatic lesions that were identified were observed in animals from both treatment groups.

Studies Relevant for Mammalian TRV Development for Ingestion Exposures

Bone, liver, and the near parturition uterus appear to be the primary target organs for the toxicological impacts of P_4. Unfortunately, insufficient data exist to compare site-specific threshold doses for the onset of the toxicological impacts to bone, liver, and the uterus when close to parturition. Nonetheless, it possibly can be inferred from the longer-term studies in rats [14,19] that the late gestational mortality of dams receiving 0.075 mg P_4/kg/day is likely to be the most sensitive effect of P_4, since no formation of lipid deposits (the other major response to P_4 exposure) was evident in the liver at this dose level. In fact, the precise hepatic status of the dams in late gestation remained uncertain since identical doses induced few histopathological liver responses in one study [14], but significant necrosis was observed in affected dams in the other study [19].

As listed in Table 18.2 (and shown in Figure 18.1), other effects of P_4 in experimental animals of possible relevance to TRV derivation included a failure of bone resorption in female Wistar rats [17]. Additionally, there

Table 18.2 Summary of Relevant Mammalian Data for TRV Derivation

		Test Results			
Test Organism	**Test Duration**	**NOAEL (mg/kg/day)**	**LOAEL (mg/kg/day)**	**Effects Observed at the LOAEL**	**Study**
Rats (Wistar)	16 days	ND	0.186	Perturbation of bone, failure of resorption	[17]
Rats (Charles River COBS CD)	30 weeks	0.015	0.075	Parturition mortality in dams, growth retardation in males, forelimb hair loss	[18]
Rats (Sprague Dawley)	20 weeks	ND	0.075	Late gestational mortality in dams, liver necrosis in affected dams	[19]

ND = No data.

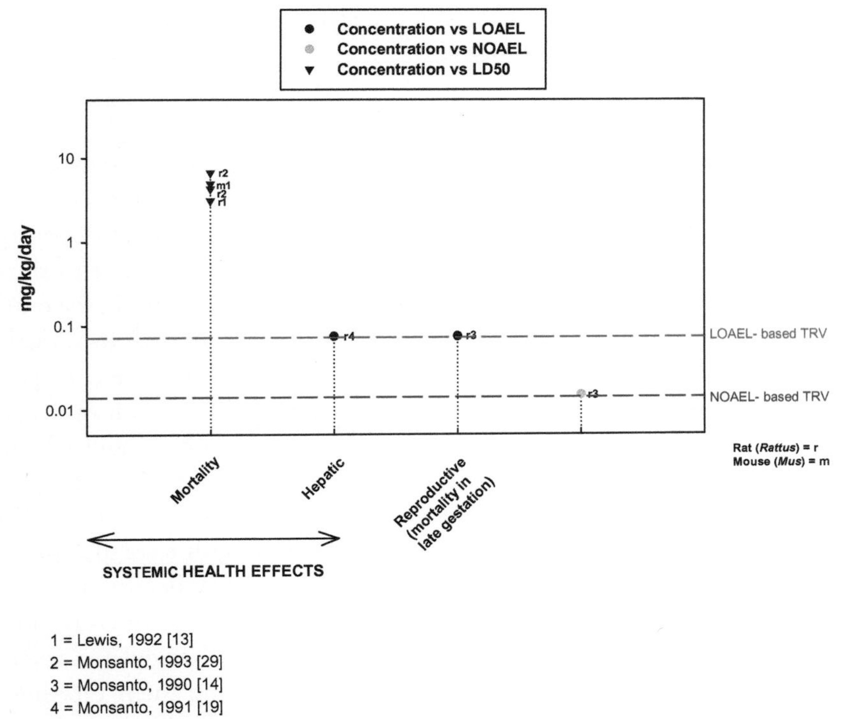

Figure 18.1 White Phosphorus Health Effects to Mammals. Letters by symbols represent test species, and numbers represent study.

was marginal growth retardation seen in male Charles River COBS® CD® rats following subchronic exposure to white phosphorus at a dose of 0.075 mg/kg/day.

Mammalian Inhalation Toxicity

No studies were located that examined the toxicological effects of P_4 via inhalation exposure.

Mammalian Dermal Toxicity

No studies were located that examined the toxicological effects of P_4 via dermal exposure.

Summary of Avian Toxicology

There is an extensive bibliography of the effects of P_4 on the health of birds, and much of our current knowledge has come from studies of the die-offs of waterfowl that have occurred in recent years at the Eagle River Flats artillery range in Alaska. The implication of P_4 as the causative agent has led to experimental studies in which the effects of the element were monitored under more closely controlled conditions.

Avian Oral Toxicity: Acute

Racine *et al.* [12] were among the first groups of scientists who systematically examined the die-offs of waterfowl at Eagle River Flats. Addressing the yearly death rate of more than 1,000 dabbling ducks and swans over an extended period, these researchers showed that mortality was due to their ingestion of unspent P_4 particles located in the bottom sediments of the estuarine salt marsh. Affected species included Mallards, Pintails, Green-Winged Teal, Greater Scaup, and Trumpeter and Tundra Swans. Racine *et al.* [12] observed the behavior of birds that were affected in the wild and showed that it could be repeated in six adult Mallards gavaged once with 12 mg P_4/kg in tricaprylin. The presence of P_4 was apparent in the sediments and in the gizzard contents of some exposed animals. Analysis of fat tissue confirmed that P_4 had been absorbed. Furthermore, since sick or dead birds were susceptible to predation and scavenging, secondary transmission of P_4 occurred in species such as Bald Eagles and Gulls [22]. A census of affected species was taken and it was determined that Mallards, Pintails, and Green-Winged Teals were far more likely to die than other on-site species, among them Northern Shovelers and American Wigeons [23]. The reasons for this discrepancy were unclear but might be related to their feeding behavior, the size, the texture of the grit, or the shape and size of their beaks.

Penned Mallards were studied to obtain a median LD_{50} by gavaging a single dose of P_4 dissolved in corn oil [24]. In this study, values of 6.46 mg/kg were obtained in adult males and 6.96 mg/kg in adult females. However, P_4 was more toxic in the pelletized form, for which a median LD_{50} of 4.05 mg/kg was derived [24]. Among the pathological responses were enlarged spleens, effects on the kidneys (e.g., formation of hyaline droplets), and hepatic effects including necrosis, fat deposition, and the appearance of hepatic foci.

An experimental study exposed 30 Mallards to a single dose or two doses of 5.4 mg P_4/kg in corn oil [8]. The animals were divided by sex, such that five females and five males were assigned to one of the following groups: control, single dose, or double dose. Samples of blood were taken pretreatment and 2 days posttreatment. Immediately following the posttreatment blood draw, the double dose group received their second P_4 dose. Two days after the second administration of P_4 a final blood draw was taken [8]. This group measured clinical chemistry and hematological parameters in the blood samples. They noted consistent decreases in hemoglobin (Hb) and hematocrit as compared to controls, and concomitant increases in the levels of blood urea nitrogen and potassium ions. Additionally, the activities of lactate dehydrogenase (LDH), aspartate aminotransferase, and alanine aminotransferase were all increased [8].

Another study reported the toxicological consequences of P_4 in the Mute Swan, *Cygnus olor* [25]. Thirty male and female birds were gavaged with single doses of pelleted P_4 at 2.98, 3.64, 4.40, or 5.28 mg/kg of body weight. The LD_{50} was determined at 3.65 mg/kg, where death occurred 4.5 days following the administration. A substantial number of clinical chemistry changes were observed as a result of P_4 treatment, accompanied by morphological signs of toxicity, such as erosion of the gizzard lining, fatty liver, petechial hemorrhages in the pancreas, "smoking gizzard," hepatomegaly, and hemorrhaging in the liver. Unlike mallards, there appeared to be no effects of P_4 to the kidneys of mute swans. An additional study was conducted to examine the effect of a single dose of P_4 on 15 egg-laying hens, *Gallus domesticus* [26]. Three equal groups of the hens received a single dose of 1, 3, or 5 mg/kg P_4 that was dissolved in tricaprylin by gavage. They monitored some behavioral and physiological parameters while collecting the eggs and measuring P_4 in extracts of yolk and egg white. Body weights, egg weights, and shell thickness were unaffected by treatment, but the frequency of laying declined compared to controls in those receiving single doses of 3 or 5 mg/kg by approximately 30% from predose rates. P_4 was detected in egg yolk for all treatment groups, as early as 1 to 2 days postdosing; however, not in the egg white. The hens were euthanized when egg-laying had ceased for 7 days or when P_4 was not detected in eggs for 2 to 3 days. The researchers also noted the presence of P_4 in the fat and skin of the hens in the 5 mg/kg treatment group.

Avian Oral Toxicity: Subacute

Sparling *et al.* [8] exposed female Mallards to 0.0, 0.5, 1.0, or 2.0 mg P_4/kg/day, and male mallards to 0.0 or 1.0 mg P_4/kg/day, for 7 days. All birds were phlebotomized 1 week prior to the study and on the first and eighth day following dosing. All the same clinical chemistry and hematological changes that were previously mentioned in Sparling *et al.* [8] were observed again in this experiment. In addition, an inverse correlation between Hb concentrations and plasma LDH activities suggested that the ratio between the two parameters might serve as a non-invasive "bioindicator" of the extent of P_4 exposure in ducks at contaminated environmental sites such as Eagle River Flats.

Nam *et al.* [9] exposed 15 American Kestrels to white phosphorus for 7 days. The chemical had been mixed in the diet via tricaprylin, to concentrations of either 6.4 or 0.7 ppm. Initially, all birds were fed a diet contaminated with P_4 at the 6.4 ppm concentration. After 2 days, the birds were divided into two groups, the accumulation group, and the cyclic dose group. The first group continued on the high dose diet and the second group was fed a lower dose of 0.70 μg of P_4 per gram of diet for 3 days, followed by the higher concentration diet (6.4 ppm) for an additional 2 days. The authors plotted a timed course of uptake and demonstrated the appearance of P_4 in fat and skin deposits within 24 hours of the onset of exposure. However, no signs of toxicity were apparent at any exposure levels.

In another study, Nam *et al.* [26] dosed five hens consecutively for 5 days with P_4 at 1 mg/kg in tricaprylin by gavage. Hens were monitored for behavioral and physiological changes, while collecting the eggs to measure P_4 content. Similar to the earlier reported results, P_4 content was confirmed in the yolk of all eggs collected. In addition, egg production was suppressed by approximately 20% when compared to predose conditions, but no effects on egg weight or shell thickness were observed.

Avian Oral Toxicity: Subchronic

No studies were identified that addressed the toxicology of P_4 under a subchronic dosing regimen.

Avian Oral Toxicity: Chronic

No studies were identified that address the toxicology of P_4 under a chronic dosing regimen.

Avian Oral Toxicity: Other

Vann *et al.* [11] studied the sublethal effects of P_4 on reproduction in mallards. Five mating pairs received up to seven daily doses of 2.6 mg P_4/kg in deionized water via gavage. By contrast, other mallard mating pairs received 1.3 mg/kg of P_4, and kept over winter to determine persistence of the reproductive effects. Three of 15 females receiving 2.6 mg/kg died, and the number of eggs laid by survivors over a 2-week period was markedly reduced in this group when compared to controls. Hatched ducklings and unhatched embryos from birds receiving 1.3 mg/kg were found to have abnormalities, among them spinal defects, small eyes, hydrocephaly and twisted legs and feet. Another phase of the study administered up to 2.0 mg P_4/kg to laying female mallards, which resulted in an apparent dose-dependent increase in the time it took the Mallards to lay 12 eggs. This effect is most likely due to P_4, which can cause liver perturbation and affects the synthesis and organization of lipoproteins. Egg weights were broadly similar among groups. Some fluctuations in hatchling weights were observed, and deformities and abnormalities were apparent in embryos that failed to hatch. A similar experiment where only male birds were dosed with P_4 resulted in no adverse effects on fertility [11].

Sparling and Federoff [27] dosed chickens with P_4, then, in turn, fed the chickens to captive Kestrels. To ensure a wide divergence of doses, the researchers implanted an extra pellet of P_4 in some of the chicken carcasses. This allowed the Kestrels to be sorted into control, "PELL," and "NOPEL" groups. The PELL group was thought to be receiving P_4 at a dose of approximately 8.8 mg/kg/day while the NOPEL group was exposed to 0.62 mg/kg/day. Both exposure groups displayed fatty livers, but 53% of the PELL group died, as compared to 20% of the NOPEL group and none of the controls. A number of clinical chemistry, hematological, and pathological changes were observed (Table 18.3). At necropsy, the PELL group of Kestrels had fatty livers, hepatomegaly, and mild hyaline droplet-related nephrosis.

Roebuck *et al.* [28] determined if P_4 toxicity varied by form (dissolved or particulate) or particulate size in ducks. Birds were given a one-time oral dose of either 1.57 mg/ml in oil or small (0.5–1.6 mm) or large (1.1–2.2 mm) particles. The expected signs of intoxication were observed: extensive drinking, head shaking, lethargy, poor motor control, and convulsions.

Table 18.3 Clinical Chemistry, Hematological, and Pathological Changes in Captive Kestrels Fed Chickens Dosed with Spiked P_4

Parameter	Control (n = 10)	0.62 mg/kg/day (n = 15)	8.8 mg/kg/day (n = 15)
Mortality (%)	0	20	53*
Weight loss (%)	3 ± 10.1	8.7 ± 9.4	25.0 ± 11.3*
Final Weight (g)	135.7 ± 13.9	130.9 ± 17.2	116.8 ± 17.2*
Liver/body weight	2.3 ± 0.4	2.4 ± 0.5	2.7 ± 0.3*
Hematocrit (% PCV)	47.9 ± 4.5	47.1 ± 4.4	38.8 ± 4.1*
Hemoglobin (mg/dL)	15.8 ± 2.1	15.7 ± 2.1	12.6 ± 2.0*
Lactate dehydrogenase (U/L)	552 ± 47	1,646 ± 1,555*	1,693 ± 836*
Alanine aminotransferase (U/L)	258 ± 264	605 ± 685	766 ± 723
Glucose (g/dL)	2,514 ± 2,609	2,421 ± 1,922	4,917 ± 4,053
Triglycerides (mg/dL)	2,805 ± 2,968	1,565 ± 1,031	1,974 ± 1,683
Inorganic P (mg/dL)	49.5 ± 22.3	67.8 ± 41.5	89.6 ± 56.6

*Indicates significant differences.
Source: [27].

Neither form of phosphorus (dissolved or particulate) nor particulate size, however, had a differential effect on toxicity; dose ingested was the only factor that affected degree of toxicity.

Studies Relevant for Avian TRV Development for Ingestion Exposures

Avian studies, in both field and controlled environments, have shown severe toxic effects from P_4 exposure in avian species. Most field information relevant for the TRV derivation is from Eagle River Flats, a U.S. Army base near Anchorage, Alaska. Racine *et al.* [12] were among the first to discover that P_4 is ingested in the particulate form as the birds feed from bottom sediments of the wetland. White phosphorus was detected in the sediment by gas chromatography–mass spectrometry at concentrations ranging from 0.0025 to 10.2 μg/g wet weight. Racine *et al.* [12] produced evidence for P_4 toxicity by conducting a study where Mallards were dosed with 12 mg/kg and observed the same lethal responses in the controlled setting as in the field. Toxicity was also confirmed by the presence of P_4 in the fat and skin of all exposed birds.

There are a few controlled experiments relevant for TRV development. First, Sparling *et al.* [24] conducted an oral toxicity study in which 20 juvenile Mallards were given a single dose of P_4 dissolved in oil, where 10 birds were given 2.6 mg/kg and the remaining birds were given 3.7 mg/kg. In addition, the researchers orally gavaged P_4 in pellet form at concentrations of 1, 2, 4, and 6.4 mg/kg to male Mallards. The observed effects were more severe for Mallards dosed with P_4 pellets, as opposed to dissolved P_4, causing the mortality-based lowest observed adverse effect level (LOAEL) to be below 1.0 mg/kg for pelletized P_4 compared to the LOAEL for P_4 in oil, which is reported to be 2.0 mg/kg. The observed effects associated with P_4 in oil are hepatic or renal damage, and with palletized P_4, the observed effects are hemorrhagic and necrotic foci. Within the same study, the researchers reported a median LD_{50} of 4.05 mg/kg of P_4 in pellet form. In addition, Sparling *et al.* [25] conducted a study of the toxicity of white phosphorus in the Mute Swan. They calculated a LD_{50} of 3.65 mg/kg of P_4 when gavaged in pellet form. Vann *et al.* [11] evaluated P_4 toxicity and its effect on reproduction in female Mallards. They discovered that a continuous dose (7 days) of 0.5 mg/kg/day of P_4 significantly reduces reproduction and clutch size. Of the studies mentioned, Vann *et al.* [11] is most suitable for TRV derivation since it incorporated one of the longer exposure durations and effects were related to reproduction, which has high ecological relevance. The data are summarized in Table 18.4.

Table 18.4 Summary of Relevant Avian Data for TRV Derivation

		Test Results			
Test Organism	**Test Duration**	**NOAEL (mg/kg/day)**	**LOAEL (mg/kg/day)**	**Effects Observed at the LOAEL**	**Study**
Mallards *(Anas platyrhynchos)*	1 day	ND	1.0	Pellet form: hemorrhagic and necrotic foci	[24]
		ND	2.0	Liquid form: hepatic or renal damage	
Mallards *(Anas platyrhynchos)*	7 days	NA	NA	4.05 mg/kg/d	[24]
Swans *(Cygnus olor)*	7 days	NA	NA	3. 65 mg/kg/d	[25]
Mallards *(Anas platyrhynchos)*	7 days	NA	0.5	Took hens longer to lay same clutch size as control; also noted were regions of hepatic necrosis, enlarged spleens, and gallbladders	[11]

NA = Not applicable.
ND = No data.

Avian Inhalation Toxicity

No studies were located that examined the toxicological effects of P_4 via inhalation exposure.

Avian Dermal Toxicity

No studies were located that examined the toxicological effects of P_4 via dermal exposure.

Amphibian Toxicology

No data are available on the toxicity of P_4 in amphibians.

Reptilian Toxicology

No data are available on the toxicity of P_4 in reptiles.

RECOMMENDED TOXICITY REFERENCE VALUES

Toxicity Reference Values for Mammals

TRVs for ingestion Exposures for the Class Mammalia

The endpoint chosen for the mammalian TRV is late gestational mortality in the Charles River COBS® CD® rats [14]. Late gestational mortality would likely affect the reproduction and the survival of a population. The researchers concluded that continuous exposure of P_4 by oral gavage at a dose of 0.075 mg/kg/day adversely affected the dams at parturition and decreased the mean number of viable pups at birth. Therefore, they established a LOAEL of 0.075 mg/kg/day. The NOAEL was established as 0.015 mg/kg/day, where none of the previously mentioned effects occurred. Although this study cannot be considered "chronic" in that it did not cover greater than 10% of the organism's lifetime, it may serve as a surrogate chronic study because it occurred over sensitive life stages. Initially, the benchmark dose approach was attempted to derive the TRV. This attempt failed because the doses derived were higher than the experimental NOAEL and LOAEL. Therefore, the TRV was derived using the NOAEL/LOAEL approach, and the values are presented in Table 18.5. Since dosing duration ranged from 80 days to 30 weeks, a medium confidence level was assigned. A summary of the data is shown in Table 18.5.

TRVs for Inhalation Exposures for the Class Mammalia

Not available at this time.

Table 18.5 Selected Ingestion TRVs for the Class Mammalia

TRV	Dose	Confidence
LOAEL-based TRV	0.075 mg/kg/day	Medium
NOAEL-based TRV	0.015 mg/kg/day	Medium

Source: [14].

TRVs for Dermal Exposures for the Class Mammalia

Not available at this time.

Toxicity Reference Values for Birds

TRVs for Ingestion Exposures for the Class Aves

Decreased fecundity was selected as the endpoint on which to base the P_4 TRV in birds since impaired reproduction is likely to contribute negatively to population sustainability or growth. Vann *et al.* [11] found that a continuous dose of 0.5 mg/kg/day of P_4 significantly reduced reproduction and clutch size in mallards. These effects can have a direct impact on population persistence. Although this study incorporated a subacute exposure regime, it is relevant since some bird species are migratory and may feed from P_4 polluted areas for a short amount of time, as seen at the Eagle River Flats site. Since this study was subacute, the TRV was derived using the approximation approach. An uncertainty factor of 40 was used to obtain the NOAEL-based TRV from the acute LOAEL (0.5 mg/kg/day) and an uncertainty factor of eight was used to derive the LOAEL-based TRV from the acute LOAEL (0.5 mg/kg/day). A low confidence rating was assigned to this avian TRV because few laboratory toxicity studies have been conducted. Similar to the mammalian TRV, this avian TRV was based on oral gavage exposure, and incompletely simulated wildlife exposure. The benchmark dose (BMD) approach was considered for all relevant studies but the appropriate data were not available at this time. A summary of the information is found in Table 18.6.

Table 18.6 Selected Ingestion TRVs for the Class Aves

TRV	Dose	Confidence
NOAEL-based	0.0125 mg/kg/day	Low
LOAEL-based	0.0625 mg/kg/day	Low

Source: [11].

Toxicity Reference Values for Amphibians

Not available at this time.

Toxicity Reference Values for Reptiles

Not available at this time.

IMPORTANT RESEARCH NEEDS

The limited availability of data on the toxicity of white phosphorus to wildlife species precludes the development of high-confidence TRVs. Hence, there is a need for additional studies of the toxicity of white phosphorus to wildlife species, including studies of particle size and form, which will ultimately assist in simulating true exposures. Particularly warranted are long-term, chronic toxicity studies on mammals and additional studies on non-mammalian wildlife such as birds, reptiles, and amphibians. Measures of reproductive and developmental impacts involving one or more species and generations would be especially useful in TRV derivation.

REFERENCES

[1] Davidson KA, Hovatter PS, Sigmon CF. Water quality criteria for white phosphorus. AD-A186613; ORNL-6336, Ft. Detrick, Frederick, MD: U.S. Army Medical Research and Development Command; 1987.

[2] Agency for Toxic Substances and Disease Registry (ATSDR). Toxicological profile for white phosphorus. Atlanta, GA: U.S. Department of Health and Human Services; 1997.

[3] Budavari S, O'Neill MJ, Smith A, Heckelman PE, Kinneary JF. The merck index. An encyclopedia of chemicals, drugs, and biologicals. 12th Edition. Whitehouse Station, NJ: Merck and Co., Inc; 1996.

[4] U.S. Environmental Protection Agency (U.S. EPA). Toxic releases inventory 2007 public data release. Toxic Releases Inventory Program; 2009.

[5] Nam S-I, Walsh MR, Collins CM, Thomas L. Eagle river flats remediation project: comprehensive bibliography – 1950 to 1998. U.S. Army Corps of Engineers, Cold Regions Research, and Engineering Laboratory; 1999, Special Report 99-13.

[6] Hazardous Substances Databank (HSDB). On-line Database. Washington, DC: National Library of Medicine; December, 2006. http://toxnet.nlm.nih.gov/cgi-bin/sis/search2/f?./temp/~RmdY6H:1.

[7] U.S. Environmental Protection Agency (U.S. EPA). Integrated risk information system. Online. Office of health and environmental assessment. Cincinnati, OH: National Center for Environmental Assessment; October, 2014. http://www.epa.gov/iris/subst/0460.htm.

[8] Sparling DW, Vann S, Grove RA. Blood changes in mallards exposed to white phosphorus. Environ Toxicol Chem 1998;17:2521–9.

[9] Nam S-I, Roebuck BD, Walsh ME. Uptake and loss of white phosphorus in American kestrels. Environ Toxicol Chem 1994;13:637–41.

[10] U.S. Army Center for Health Promotion and Preventive Medicine (USACHPPM). Standard Practice for Wildlife Toxicity Reference Values; Technical Guide 254.
[11] Vann SL, Sparling DW, Ottinger MA. Effects of white phosphorus on mallard reproduction. Environ Toxicol Chem 2000;19:2525–31.
[12] Racine CH, Walsh ME, Roebuck BD, et al. White phosphorus poisoning of waterfowl in an Alaskan salt marsh. J Wildl Dis 1992;28:669–73.
[13] Lewis RJ. Sax's dangerous properties of industrial chemicals. 8th ed.; 1990, New York, NY: Van Nostrand Reinhold; 1992.
[14] Monsanto. One-generation reproductive study in rats with elemental phosphorus with attachments and cover letter dated 010990. EPAOTS0518525-1.
[15] Seakins A, Robinson DS. Changes associated with the production of fatty livers by white phosphorus and by ethanol in the rat. Biochem J 1964;92:308–12.
[16] Pani P, Gravela E, Mazzarino C, Burdino E. On the mechanism of fatty liver in white phosphorus-poisoned rats. Exp Mol Pathol 1972;16:201–9.
[17] Whalen JP, O'Donohue N, Krook L, Nunez EA. Pathogenesis of abnormal remodeling of bones: effects of yellow phosphorus in the growing rat. Anat Rec 1973;177:15–22.
[18] Condray JR. Elemental yellow phosphorus one-generation reproduction study in rats. IR-82-215; IRD No. 401-189 St. Louis, MO: Monsanto Company; 1985 (As cited, U.S. EPA 2002).
[19] Monsanto. Follow-up Information: the one-generation reproduction study in rats with elemental phosphorus with attachments and cover letter dated 080591. Monsanto Co; 1991, EPAOTS0518525-4.
[20] Ghoshal AK, Porta EA, Hartcroft WS. The role of lipoperoxidation in the pathogenesis of fatty livers induced by phosphorus poisoning in rats. Am J Pathol 1969;54:275–91.
[21] Ashburn LL, McQueeney AJ, Faulkner RR. Scarring and precirrhosis of the liver in chronic phosphorus poisoning of guinea pig. Proc Soc Exp Biol Med 1948;67:351–8.
[22] Roebuck BD, Walsh ME, Racine CH, Reitsma L, Steele B, Nam S-I. Predation of ducks poisoned by white phosphorus: exposure and risk to predators. Environ Toxicol Chem 1994;13:1613–8.
[23] Steele BJ, Reitsma LR, Racine CH, et al. Different susceptibilities to white phosphorus poisoning among five species of ducks. Environ Toxicol Chem 1997;16:2275–82.
[24] Sparling DW, Gustafson M, Klein P, Karouna-Renier N. Toxicity of white phosphorus to waterfowl: acute exposure in mallards. J Wildl Dis 1997;33:187–97.
[25] Sparling DW, Day D, Klein P. Acute toxicity and sublethal effects of white phosphorus in mute swans, *Cygnus olor*. Arch Environ Contam Toxicol 1999;36:316–22.
[26] Nam S-I, MacMillan DL, Roebuck BD. The translocation of white phosphorus from hen (*Gallus domesticus*) to egg. Environ Toxicol Chem 1996;15:1564–9.
[27] Sparling DW, Federoff NE. Secondary poisoning of kestrels by white phosphorus. Ecotoxicology 1997;6:239–47.
[28] Roebuck BD, Nam S-I, MacMillan DL, Baumgartner KJ, Walsh ME. Toxicity of white phosphorus (P_4) to ducks and risk for their predators: effects of particle size. Environ Toxicol Chem 1998;17:511–8.
[29] Monsanto. Elemental phosphorus: acute oral toxicity study in rats with cover letter dated 061093. Monsanto Co.; 1993, EPAOTS0537723.

PART IV

Wildlife Toxicity Assessments of Selected Metals

CHAPTER 19

Wildlife Toxicity Assessment for Antimony

Bradley E. Sample, Christine Ann Arenal

Contents

INTRODUCTION

Antimony (Sb, stibium) is a semimetallic element that belongs to the vanadium family (Group VA) of the periodic table and shares certain chemical and physical properties with lead, arsenic, and bismuth [1]. In nature, antimony is associated with sulfur as stibnite and occurs in ores with arsenic. A silvery white metal, antimony is too brittle to be used alone and is commonly alloyed with lead and copper to be used in the manufacture of batteries, sheet and pipe metal, ammunition, and pewter [2]. Antimony trioxide (ATO) is added to textiles, plastics, rubbers, paper, and paints as a fire retardant [3]. Antimony compounds were historically used therapeutically in

Wildlife Toxicity Assessments for Chemicals of Military Concern
http://dx.doi.org/10.1016/B978-0-12-800020-5.00019-3

antihelminthic and antiprotozoic treatments; such use is largely discontinued due to antimony's toxicity [4]. Major sources of environmental antimony contamination are smelters, coal combustion, and the incineration of waste and sewage sludge [4]. Antimony exists in valences of −3, 0, +3, and +5. Dissolved antimony in natural waters is found mostly as hydroxides in the trivalent and pentavalent states [5]. The tri- and pentavalent forms are the most stable forms of antimony [1] and are of the most interest in biological systems. Although the Chemical Abstracts Service (CAS) number (7440-36-0) refers to the elemental form of antimony, this Wildlife Toxicity Assessment (WTA) considers the potential toxicity of antimony compounds most often encountered in the environment.

This WTA summarizes the toxicity of antimony with emphasis on identifying its adverse effects and developing Toxicity Reference Values (TRVs). The TRVs are intended to serve as benchmarks for screening-level ecological risk assessments. This WTA was developed in accordance with the U.S. Army Center for Health Promotion and Preventive Medicine (USACHPPM) Technical Guide (TG) 254, *Standard Practice for Wildlife Toxicity Reference Values* [6].

TOXICITY PROFILE

Environmental Fate and Transport

A major source of antimony released into the environment is combustion at smelting operations [2]. Antimony volatilizes during combustion, resulting in small, submicron particles that condense when cooled. These particles are dispersed by wind and removed by gravitational settling and by wet and dry deposition. Antimony that results from smelting is most frequently deposited when dissolved in rain.

The fate of antimony in soil is highly variable and dependent upon soil characteristics and the particular form of antimony [2]. Whereas its nonmineral forms may bind to inorganic and organic ligands, mineral forms may be unavailable for binding. Similarly, although some research has observed moderate mobility of antimony over a range of soil conditions, other studies find that antimony is strongly adsorbed to soil [2]. Antimony forms coprecipitates with hydrous iron, manganese, and aluminum oxides, which may complex with organic matter in soil and sediment.

In sediment, antimony release is bimodal, increasing at both low and high pH, with the dominant species varying by pH. At pH 2.7, Sb (III) dominates; at pH 4.3, Sb (III) and Sb (V) concentrations are equivalent; and at pH 6.3 and above, Sb (V) is the dominant species [2]. Although sewage sludge increases the mobility of many trace elements through complexation with dissolved organic compounds or increased ionic strength, this did not appear to be the case with antimony [7]. Oxidation of saline sediment containing pyrite and other readily oxidizable sulfur compounds may produce sulfuric acid and overwhelm the buffering capacity of the sediment, thus lowering the pH and releasing antimony and other toxic metals [2]. Antimony released into waterways is generally particle-bound, is transported as suspended sediment, and settles out in depositional areas.

The mobilization of antimony from sediment is dependent on organic matter, salinity, and pH [2]. Antimony absorbs strongly to colloidal materials and can be transported to groundwater or surface water with colloidal material. Industrial inputs into water commonly contain Sb (III), which is oxidized to Sb (V) under natural conditions. Chemical and biochemical transformations of antimony in water are not well described; however, methylated antimony species (e.g., methylstibonic acid and dimethlystionic acid) may be present throughout the water column [2]. Table 19.1 presents a summary of the physical and chemical properties of antimony.

Table 19.1 Summary of Physical-Chemical Properties of Antimony

CAS No.	7440-36-0
Molecular Weight (MW)	121.75
Color	Silvery white
Physical State	Solid
Melting Point	630.5 °C
Boiling Point	1,750 °C
Density at 25 °C	6.684 g/cm^3
Odor	No data
Water Solubility at 25 °C	Insoluble
Log K_{ow}	0.73 (estimated)
Log K_{oc}	1.2 (estimated from MCI) 0.63 (estimated from Log K_{ow})
Vapor pressure	1 mmHg at 886 °C
Henry's Law Constant	No data
Flashpoint	No data
Conversion factors:	Since antimony exists in the atmosphere in the particulate state, the concentration is expressed as mg/m^3.

Sources: ATSDR [2], HSDB [29], U.S. EPA [30].

SUMMARY OF MAMMALIAN TOXICITY

Mammalian Oral Toxicity

Mammalian Oral Toxicity—Acute

Ainsworth *et al.* [8] conducted several laboratory experiments to investigate whether elevated antimony concentrations previously found in the organs of wild, short-tailed field voles (*Microtus agrestis*) at a contaminated grassland site near an antimony smelter were sufficient to cause harmful effects. In one experiment, 10 voles, aged 35 to 40 days, were fed *ad libitum* a control diet of 6,700 mg antimony/kg for 7 days; in another experiment, six voles were fed with 20,000 mg antimony/kg for 12 days. No adverse effects were observed in either experiment; however, observational data such as body weight and food consumption were not measured. In a study of acute antimony toxicity in rats (*Rattus norvegicus*), young female and male albino rats were dosed once by gavage with up to 34,600 mg ATO/kg and observed for 14 days [9]. While rats exhibited ruffed fur and hypoactivity during the first 2 days of observation, no mortality was reported, and the acute oral LD_{50} was greater than 34,600 mg ATO/kg, or 14,451 mg Sb/kg.

Antimony potassium tartrate (APT) at doses up to 168 mg ATP (or 30.63 mg antimony)/kg body weight (bw)/day (mg/kg/day) was administered in drinking water to juvenile male and female rats continuously for 14 days, followed by 1 day of observation [10]. A concurrent study using the same dosing regimen exposed juvenile male and female mice (*Mus musculus*) up to 407 mg APT (or 74.2 mg antimony)/kg/day. Whereas 30.63 mg antimony/kg/day had no adverse effects on rats, mice that received 74.2 mg antimony/kg/day displayed treatment-related liver and forestomach lesions, dose-related increases in relative liver weight, and decreased body and organ weights.

The National Toxicology Program (NTP) [11] also evaluated effects from APT on mice and rats. Although there are slight differences between the two studies and a more extensive presentation of the results in Dieter *et al.* [10], the data presented in NTP [11] consisted of identical APT dose levels, duration, test subjects, and methodology as those reported in Dieter *et al.* [10]. No references were identified that would indicate whether NTP [11] and Dieter *et al.* [10] represent the same study. These two studies are therefore assumed to be separate, independent studies.

The clinical signs of toxicity observed in NTP [11] included rough haircoat, emaciation, abnormal posture, hypoactivity, and decreased fecal material, and were consistent with avoidance of APT-dosed water. Water consumption was reduced in all mouse treatment groups in proportion to dosage, with some dose-related increases in relative liver weight observed. Body weight gain was significantly reduced in male mice dosed with 273 or 407 mg APT/kg/day (49.77 or 74.2 mg antimony/kg/day) and in females dosed with 74.2 mg antimony/kg/day for 8 days. With the exception of males in the 74.2 mg antimony/kg/day dose group, all animals recovered by day 16 of the experiment.

In male rats, 17.14 and 30.63 mg Sb/kg/day in drinking water reduced the water consumption by 20 to 30%, and a 10 to 40% reduction in water consumption was observed in female rats dosed with 5.1 to 30.63 mg Sb/kg/day. Increases in relative liver weight occurred in both male and female rats, while increases in relative kidney weight were observed in female rats in the 30.63 mg Sb/kg/day water ingestion dose group [11].

Mammalian Oral Toxicity: Subchronic

Ainsworth *et al.* [8] did not observe any adverse effects to voles supplied ad libitum with a 500 mg Sb/kg diet for 30, 40, 50, or 60 days. The number of 35- to 40-day-old voles per test group was not reported, and no observational data, such as body weight and food ingestion, were recorded.

Gurnani *et al.* [12] studied the clastogenic effects of antimony on Swiss albino male mice (*Mus musculus*). Three doses (400, 667, and 1,000 mg/kg bw) of aqueous suspensions of ATO, corresponding to 1/50, 1/30, and 1/20 of the oral LD_{50} (>20,000 mg/kg/day), were fed by gavage daily to five 8-week-old male mice for 7, 14, and 21 days. A dose of 1,000 mg ATO/kg bw/d (415.1 mg Sb/kg/day) for 21 days was reported to be lethal; however, no other information on mortality (e.g., percentage of mortality) was provided in the study.

Hext *et al.* [13] fed male and female rats a diet containing 1,000, 5,000, or 20,000 mg ATO/kg continuously for 90 days, resulting in a dose of 84, 421, and 1,686 ATO/kg/day for male rats and 97,494, and 1,879 mg ATO/kg/day for female rats [13]. No ocular toxicity, no significant intergroup differences in rate of body weight gain or total weight gained, and no difference in food consumption were observed. The only noteworthy effect observed in the study was an increase in absolute and relative liver weights by about 10% in male and female rats fed 20,000 mg ATO/kg (307.36 mg Sb/kg/day for males and 342.54 mg/kg/day for females) compared to controls. However, in the absence of any histological effects, these observations were considered minor and incidental to treatment.

Ainsworth *et al.* [14] fed mice and voles laboratory diets containing 500 and 6,700 mg Sb/kg for 18 and 21 days, respectively. While the mice remained healthy, the kidney and spleen dry weights were significantly lower for both dose levels as compared to mice in the control group [14]. Among voles, no changes in external or internal appearance were observed. In addition, antimony concentrations in the liver of voles given 6,700 mg Sb/kg were much lower than in mice given the same diet.

Poon *et al.* [15] administered drinking water containing 0.5, 5, 50, and 500 mg Sb/L (as APT) to male and female rats continuously for 13 weeks. Animals in the 500 mg/L Sb dose group (with doses estimated to be 42.17 and 45.69 mg Sb/kg/day for males and females, respectively) consumed about 35% less water and 12% less food than rats in the control and lower dose groups, which resulted in a significant depression in weight gain. Although significant effects were not observed at lower dose levels, the NOAEL for antimony in drinking water was 0.06 mg/kg/day based on biochemical and histological changes and the retention of antimony in red blood cells and the spleen [15]. For the purposes of this WTA, the 42.17 and 45.69 mg Sb/kg/day doses were identified as the LOAELs for male and female rats, respectively.

Mammalian Oral Toxicity Chronic

Rossi *et al.* [16] exposed pregnant rats to 1 and 10 mg ATO/L in drinking water ad libitum from conception to weaning and continued the exposure of male and female pups from birth through 60 days of age. Whereas antimony exposure at both dose levels significantly decreased maternal body weights, pup body weights were significantly reduced only at the 10 mg/L dose. No significant difference in length of gestation, number of newborns per litter, or teratogenic effects was observed for either dose. Based on these data, chronic LOAELs of0.05 and 0.57 mg Sb/kg/day were calculated for maternal and pup body weight loss, respectively.

To evaluate effects of antimony on longevity and tumor production, Kanisawa and Schroeder [17] provided male and female white Swiss mice with drinking water containing 5 mg Sb/L (as APT) from the time of weaning to natural death (>1.5 years). Using body weights that were previously reported [17], this concentration translates to a daily antimony dose of 1 mg/kg/day. Although the treated mice weighed less than the controls, and females had a shorter lifespan, the observations were not statistically significant. Therefore, the single dose was considered to be a NOAEL.

Schroeder *et al.* [18] administered 5 mg Sb/L (as APT) in drinking water to male and female mice from the time of weaning to natural death (~540 days). Assuming a mature mouse ingests 7 grams (g) of water and 6 g of food per 100 g body weight per day, Schroeder *et al.* [18] estimated the daily antimony dose of 0.35 mg Sb/kg bw/day. No demonstrable effects in males were observed. However, some decrease in the lifespans of most females and some suppression of growth in older mice were reported, although no differences were statistically significant. Further, no definitive pathological lesions were reported following antimony accumulation in tissues. In another experiment [19], male and female rats were provided drinking water containing 5 mg Sb/L (the calculated dose was 0.64 mg Sb/kg bw/day) in basal drinking water from the time of weaning to natural death (>540 days). Negligible effects on growth and mature weight were observed, but lifespan was reduced in both males and females. Based on both studies, the authors [19] concluded that innate toxicity from antimony, expressed as a decrease in survival and significantly reduced longevity (i.e., lifespan), were greater in rats than in mice.

Studies Relevant for Mammalian TRV Development for Ingestion Exposures

Only some of the studies identified were retained for TRV derivation. For example, data generated by intraperitoneal (i.p.) injection are not representative of field exposure routes and thus were not used in TRV derivations. Note that in studies where other methods of exposure were used in parallel with i.p. injection, only the data generated by i.p. injection were summarily excluded. For example, data from Dieter *et al.* [10] that were generated through drinking water ingestion were included, while data generated by ip injection were not. In another instance, Gurnani *et al.* [12] reported mortality in rats at the highest dose level (1,000 mg/kg/day) over a 21-day exposure period. However, the study did not report specifics on percent mortality within this dose group, which precluded its use in TRV development. The quality and reporting of the methods of the remaining studies met the criteria for inclusion as appropriate data for TRV derivation.

The clinical signs of antimony toxicity observed in acute studies include rough haircoat, hypoactivity, abnormal posture, emaciation, and decreased fecal material [9,11]. The general toxic effects on growth observed in most studies were a decrease in body weight and an increase in liver and kidney weights. Chronic exposure of rats and mice to antimony decreased longevity and suppressed growth in older mice [17–19]. While the effects from exposure are consistent in all studies regardless of duration, only the chronic studies examine long-term effects and recovery from antimony exposure and are the most relevant to TRV development.

The relevant studies provide data for three species (vole, mouse, and rat) within the same taxonomic order (Rodentia), thereby not meeting the minimum data requirements of three species from two taxonomic orders for use of either the benchmark dose or NOAEL/LOAEL approach. Therefore, the approximation approach was used for the development of the TRVs [6]. The LOAELs and NOAELs for the studies relevant to TRV derivation are listed in Table 19.2 and displayed graphically in Figure 19.1.

Mammalian Inhalation Toxicity

Mammalian Inhalation Toxicity: Acute

No relevant studies that addressed the acute inhalation toxicity of antimony were found.

Mammalian Inhalation Toxicity: Subchronic

A single subchronic antimony mammalian inhalation toxicity study was identified in the literature [20,21]. Newton *et al.* [20] exposed 7-week-old male and female rats to 0.25, 1.08, 4.92, and 23.46 mg/m^3 ATO dust for 6 hours/day, 5 days/week for 13 weeks [20]. Corneal irregularities were observed with approximate equal incidence (30%) in all exposed groups. The irregularities appeared after 2 weeks of exposure and did not abate during the 27-week recovery period. Assuming a body weight of 0.35 kg, these exposures equate to a subchronic inhalation LOAEL for corneal effects of 0.013 mg Sb/kg/day. Mean body weight gain for males in the 23.46 mg/m^3 group was significantly lower than the control males (6%), and it did not recover during the recovery period. Mean and relative lung weights were increased in the 4.92 and 23.46 mg/m^3 groups. Although the lung weights in the 4.92 mg/m^3 group returned to normal during the recovery period, the lung weights in the 23.46 mg/m^3 group remained elevated.

Mammalian Inhalation Toxicity: Chronic

Eight-week-old male and female rats were exposed by inhalation to ATO dust at exposure levels of 0, 0.06, 0.51, and 4.50 mg/m^3 for 6 hours/day, 5 days/week for 12 months [20,22]. Dose-related increases in chromodacryorrhea (red or bloody tears) in males were reported. However, chromodacryorrhea, observed in both sexes, is often a secondary effect related to dental abnormality and, therefore, may be explained by periodontal disease. Exposure-related ocular opacities observed in rats were found to be more

Table 19.2 Summary of Relevant Mammalian Oral Toxicity Data for TRV Derivation for Antimony

			Test Results					
Test Type	**Test Organism**	**Test Duration**	**Dietary NOEC (mg/kg)**	**Dietary LOEC (mg/kg)**	**NOAEL (mg/kg/day)**	**LOAEL (mg/kg/day)**	**Effects Observed at the LOAEL**	**Study**
Acute	Short-tailed field vole	7 days	6,700		1,382[a]			[8]
	Short-tailed field vole	12 days	20,000		4,124[a]			[8]
	Albino rats - Charles River strain	1 dose;14 days of observation			14,451			[9]
	Rat (F344/N)	14 days			30.63	—	—	[10]
	Mouse (B6C3F)	14 days			—	74.20	Liver and forestomach lesions in 3/10 females and 1/10 males; increase in liver weight; decrease in heart, spleen, thymus, kidney, and body weight for high-dose groups.	[10]
	Mouse (B6C3F)	14 days			—	10.76	Reduced water consumption; some dose-related increases in relative liver weight; clinical signs of toxicity included rough haircoat, emaciation, abnormal posture, hypoactivity, and decreased fecal material.	[11]
	Mouse (B6C3F)	14 days			31.72	49.77	Significantly reduced body weight gain in males, which recovered by day 16; reduced water consumption; some dose-related increases in relative liver weight; same clinical signs of toxicity.	[11]

Continued

Table 19.2 Summary of Relevant Mammalian Oral Toxicity Data for TRV Derivation for Antimony—cont'd

			Test Results					
Test Type	**Test Organism**	**Test Duration**	**Dietary NOEC (mg/kg)**	**Dietary LOEC (mg/kg)**	**NOAEL (mg/kg/day)**	**LOAEL (mg/kg/day)**	**Effects Observed at the LOAEL**	**Study**
	Mouse (B6C3F)	14 days			49.77	74.20	Significantly reduced body weight gain for males and females in which females recovered by day 16 and males did not recover. Treatment-related lesions in liver and forestomach in males and females; reduced water consumption; some dose-related increases in relative liver weight; same clinical signs of toxicity.	[11]
	F344/N rat	14 days			2.92	5.10	10–40% reduced water consumption in females.	[11]
	F344/N rat	14 days			2.92	30.63	20–30% reduced water consumption in males; 10–40% reduced water consumption in females; increase relative liver weight in males and females; increase in relative kidney weight in females.	[11]

Subchronic	Short-tailed field vole	30 to 60 days	500	—	103.1[a]		—	[8]
	Wistar rats Alpk: APSD strain	90 days			Male: 76.75 Female: 90.06	Male: 307.36 Female: 342.54	Absolute and relative liver weights were increased by about 10% in both sexes.	[13]
	Laboratory-bred vole	18 days	—	500		103.1[a]	Dry weights of kidney and spleen were significantly lower than control.	[14]
	Laboratory-bred vole	21 days	6,700	—	1382[a]		—	[14]
	Sprague Dawley rat	90 days			Male: 5.58 Female: 6.13	Male: 42.17 Female: 45.69	Consumed 35% less water and 12% less food than all other test groups but returned to control levels during recovery period; significant depression in weight gain in males started at week 6 and in females at week 12.	[15]
Chronic	NOS albino normotensive rat	Adult: conception, 60 days after delivery			—	Calculated: 0.05	Significant decrease in body weight on 20th day of gestation.	[16]
	NOS albino normotensive rat	Pup: conception, 60 days after delivery			Calculated: 0.055	Calculated: 0.57	Significantly reduced pup body weight from 10th to 60th day of age.	[16]
	White Swiss mouse of Charles River strain (CD-1)	From weaning to natural death			1		Treated mice weighed less than controls, and females had a shorter life span (observations were not statistically significant).	[17]

Continued

Table 19.2 Summary of Relevant Mammalian Oral Toxicity Data for TRV Derivation for Antimony—cont'd

			Test Results					
Test Type	Test Organism	Test Duration	Dietary NOEC (mg/kg)	Dietary LOEC (mg/kg)	NOAEL (mg/kg/day)	LOAEL (mg/kg/day)	Effects Observed at the LOAEL	Study
	Charles River CD stain mouse	From weaning to natural death			0.35		Some decrease in life spans of most females; some suppression of growth of older male and female mice (observations were not statistically significant).	[18]
	Long-Evans strain rat	From weaning to natural death			—	0.64	Significant reduction in longevity later in life; disturbances in glucose and cholesterol metabolism associated with ingestion.	[19]

[a]NOAEL calculated assuming bw = 25.7 g using dry weight allometric food ingestion model.

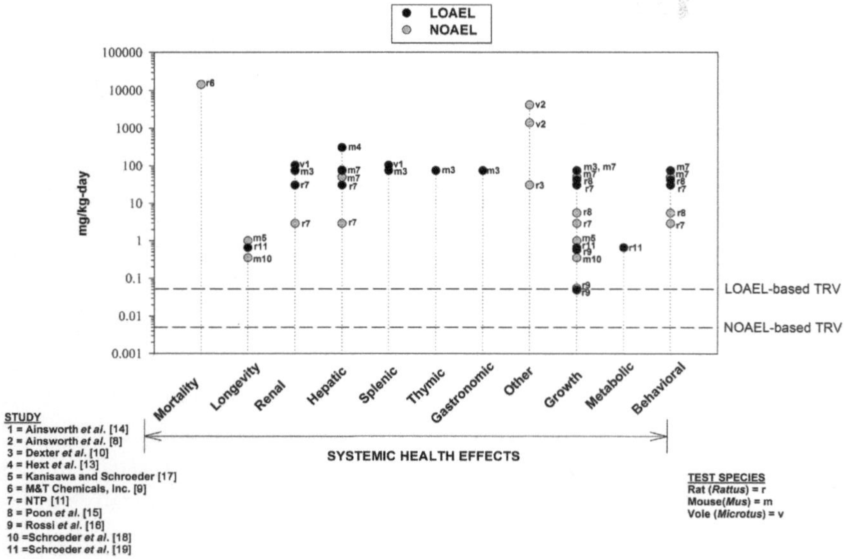

Figure 19.1 Health effects of dietary exposure of mammals to antimony. Letters by symbols represent test species, and numbers represent study.

severe in females. There were no significant differences in body weight gain among the groups. While lung weights were unaffected, clearance of ATO from the lung was reduced by 80% in the 4.50 mg/m^3 group.

Eight-month-old Wistar rats (90 males and 90 females per group) were exposed by inhalation to either ATO (45 mg/m^3 average concentration), or antimony ore concentrate, principally Sb_2S_3 (36 to 40 mg/m^3 average concentration), for 7 hours/day, 5 days/week, for up to 52 weeks and sacrificed 20 weeks after terminating exposures [23]. Serial sacrifices were performed at 6, 9, and 12 months. Mortality did not differ significantly in either of the treatment groups. A significant increase of lung neoplasms was observed in 27% of all exposed females in the ATO treatment groups. After 9 and 12 months of exposure, males weighed significantly less than controls although the maximum differences were no greater than 6.2%. Exposure to the antimony ore concentrate also resulted in a significant increase of lung neoplasms in 25% of all exposed females and a significant reduction in the weight of females exposed for 9 and 12 months.

Several other chronic inhalation studies were identified, but due to methodological or reporting issues, they were not suitable for TRV development. Watt *et al.* [24] exposed rats and miniature swine to ATO dust concentrations of 1.6 and 4.2 mg/m^3 for 6 hours/day, 5 days/week for one year. Rats and swine of the same treatment group were exposed in the same chamber. Because study design details such as the number of animals per group and age were not presented, a reliable effects threshold could not be determined. Gross *et al.* [25] exposed male rats to ATO dust (100 to 125 mg/m^3) for 25 hours/week for 14.5 months and found no differences in spontaneous pneumonia or bronchiectasis in experimental groups. Control rats were exposed to uncontaminated "dust" at a concentration of 25 mg/m^3 air. This control dust was later determined to contain 1% ATO. In addition to this discrepancy, neither the age and sex of the rats tested nor the data for all animals tested were presented.

Dernehl *et al.* [26] exposed guinea pigs to ATO as dust for 2 hours/day, 7 days/week for the first 3 weeks, after which the exposure time increased to 3 hours/day up to week 27. The exposure concentration ranged from 13 to 424 mg/m^3. (Note: Although the concentrations were reported as "mgm," mg/m^3 was the assumed unit of measurement.) A significant increase in lung and liver weights, uniform thickening of the alveolar wall (indicative of the formation of pneumonitis), and fatty degeneration of the liver were reported in animals exposed to 77 mg/m^3 [2] and above. However, the usefulness of this study is limited due to the lack of details provided in the experimental design. Specifically, the change in the duration period and exposure concentrations was not clearly defined, and the age, sex, and number of animals in treatment and in the control groups were not presented.

Studies Relevant for Mammalian TRV Development for Inhalation Exposures

The study designs and data quality in Watt *et al.* [24], Gross *et al.* [25], and Dernehl *et al.* [26] did not meet the requirements for inclusion as studies relevant for the development of TRVs. The remaining studies did meet the criteria and the data were included in the TRV derivation. The relevant studies provide data for only one species (rat) within the same taxonomic order (Rodentia), thereby not meeting the minimum data requirements for either the benchmark dose or NOAEL/LOAEL approach. Therefore, the approximation approach was used for the development of the TRVs [6]. The LOAELs for studies relevant to the TRV derivation are listed in Table 19.3 and displayed graphically in Figure 19.2.

The general toxic effects observed in the antimony inhalation studies were corneal irregularities, ocular opacities, chromodacryorrhea, an increase in lung weight and lung neoplasms, and a decrease in body weight. Although corneal irregularities, ocular opacities, and cataracts were related to inhalation of antimony, chromodacryorrhea was not conclusively determined to be a toxic effect because it is often a secondary effect related to dental

Table 19.3 Summary of Relevant Mammalian Inhalation Toxicity Data for TRV Derivation for Antimony

Test Type	Test Organism	Test Duration	Test Results: NOAEL[a] (mg/kg/day)	Test Results: LOAEL[a] (mg/kg/day)	Test Results: Effects Observed at the LOAEL	Study
Subchronic	Fischer 344 rat	6 hr/day, 5 days/wk for 13 wks followed by a 27-wk observation period	—	0.013	Corneal irregularities appeared after about 2 weeks (treatment related, but not dose related) in 30% exposed animals and did not abate during observation period.	[20,21]
	Fischer 344 rat	6 hr/day, 5 days/wk for 13 wks followed by a 27-wk observation period	0.247	1.17	Mean body weight gain for males in the 23.46 mg/m^3 group was significantly lower than the control males (6%) and did not recover during the recovery period.	[20,21]
	Fischer 344 rat	6 hr/day, 5 days/wk for 13 wks followed by a 27-wk observation period	0.054	0.247	Mean and relative lung weights increased and returned to normal during recovery period; same corneal irregularities.	[20,21]
Chronic	Fischer 344 rat	6 hr/day, 5 days/wk for 12 months followed by a 12-month observation period	—	0.003	Dose-related increase in chromodacryorrhea in males, and found in females—may be explained by peridontal disease. Increases in ocular opacitites in both sexes were more severe in females.	[20,22]
	Fischer 344 rat	6 hr/day, 5 days/wk for 12 months followed by a 12-month observation period	0.025	0.226	Lung clearance of antimony trioxide was reduced by 80%; same ocular effects.	[20,22]
	Wistar-derived rat	7 hours/day, 5 days/week for 12 months with 20-week observation period	—	2.6	Significant increase of lung neoplasms in 27% of exposed females.	[23]
	Wistar-derived rat	7 hours/day, 5 days/week for 12 months with 20-week observation period	—	2.36	Males weighed significantly less than controls; same effects to lung.	[23]

[a]Doses expressed as mg antimony/kg/day.

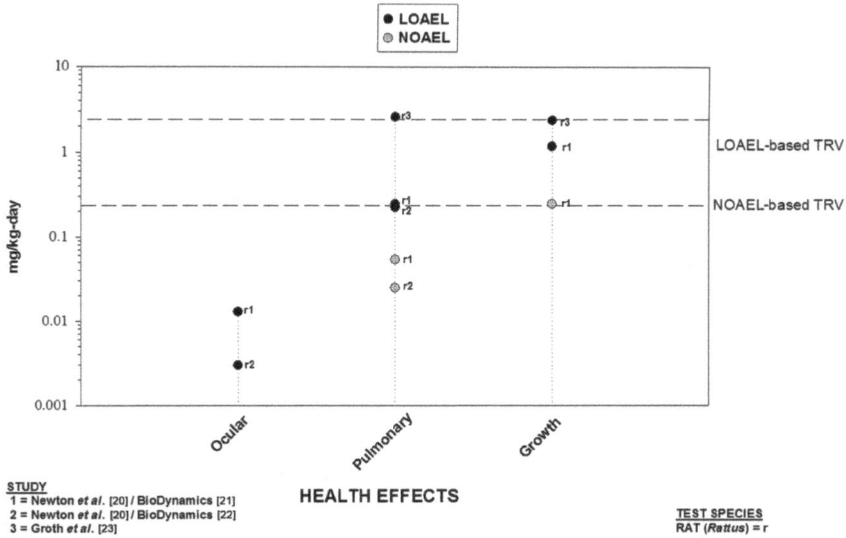

Figure 19.2 Health effects of inhalation exposure of mammals to antimony. Letters by symbols represent test species, and numbers represent study.

abnormality or periodontal disease [20,21]. Groth *et al.* [23] was the only chronic study to show a significant presence of lung neoplasms and a significant increase in weight loss. Specifically, a significant increase in lung neoplasts were observed in 27% of female rats exposed by inhalation to ATO, and males weighed significantly less than controls; however, the maximum differences were no greater than 6.2%. The antimony ore concentrate treatment groups displayed lung neoplasms in 25% of all exposed females, and weighed significantly less than controls, where the maximum differences were no greater than 6.4%. Statistically significant data provided from this chronic exposure study is more substantive and shows a clear impact on growth compared to the chronic exposure data obtained from the Newton *et al.* [20] and BioDynamics [21,22] studies.

Mammalian Dermal and Ocular Toxicity

Mammalian Dermal and Ocular Toxicity—Acute

Horton *et al.* [27] applied 100 mg of finely ground antimony thioantimonate (ATA) once to one eye of each of the 12 New Zealand white rabbits. Treated eyes of six of the rabbits remained unwashed for 13 days and the treated eyes of the remaining six rabbits were irrigated 20 to 30 seconds after ATA application. Superficial injury appeared in four rabbits with unwashed eyes when examined with ultraviolet light at day 7, but all signs of irritation were cleared by 13 days posttreatment. Similar but slightly less severe reactions were observed in the rabbits with eyes irrigated after application of ATA. The mild eye irritation observed was reversible, and the corneal effects seen could possibly have been due to mechanical abrasions by the dry powder.

Antimony triacetate at a dose of 12,800 mg/kg was applied to the shaved trunks of young male and female New Zealand albino rabbits for 24 hours and observed for 21 days [28]. All animals had severe skin irritation; extensive sloughing of the skin was observed in three-quarters of the animals tested. No mortality was observed and, although there was no control group, the dermal LD_{50} was reported to be in excess of 12,800 mg antimony triacetate/kg. In a subsequent study, M & T Chemicals, Incorporated [9] conducted several additional acute toxicity and sensitization studies using rabbits and guinea pigs. Antimony trioxide (1,000 and 2,000 mg/kg) was applied to the shaved backs of young adult male and female New Zealand albino rabbits per test group for 24 hours and observed for 14 days. Minimal irritation to the skin and an acute dermal $LD_{50} > 2{,}000$ mg/kg were reported. In another skin irritation experiment with no control group, two test patches were shaved on the back of young New Zealand albino rabbits; one patch was abraded by epidermal incisions, and the other site remained intact. ATO (500 mg moistened with water) was applied to both test patches on each individual for 24 hours with a 3-day observational period; minimal irritation to the abraded patches and no irritation to the intact patches were observed. Albino guinea pigs (sex and age not reported) were also used to evaluate the skin-sensitizing properties of ATO (500 mg moistened with water) when applied to the shaved back of each animal for a period of 6 hours. Although there was no control group, ATO was determined not to be a sensitizer to albino guinea pigs.

Eye irritation tests were conducted with young New Zealand albino rabbits, during which ATO (100 mg) was instilled into the conjunctival sac of the right eye of each individual, leaving the left eye as the control. One-half of the test subjects' eyes were rinsed 4 seconds after application, while the remaining animals' eyes were not rinsed during the 14-day observational period. Whereas the unwashed eyes displayed "extreme" irritation, the washed eyes displayed "severe" irritation, a somewhat lower level of effect.

Mammalian Dermal and Ocular Toxicity: Subchronic

Horton *et al.* [27] conducted a dermal toxicity study in which three groups of male and female New Zealand white rabbits received 2,000 mg/kg bw of calcium cup grease, 5% ATA/calcium cup grease, or 0.5% ATA/calcium cup grease for 6 hours daily during a 13-week period. Both ATA grease treatments, as well as the calcium grease treatment, resulted in mild to moderate dermal edema and no significant mortality. Significantly increased liver/body ratios were observed in females in all treatment groups, but the authors attributed this effect to calcium cup grease exposure rather than ATA exposure.

Mammalian Dermal and Ocular Toxicity: Chronic

No studies on chronic dermal and/or ocular toxicity of antimony were found.

Studies Relevant for Mammalian TRV Development for Dermal and Ocular Exposures

Although several dermal and ocular toxicity studies of antimony in mammals were identified, they are generally not representative of likely field exposures and thus were not appropriate for use in TRV derivation.

SUMMARY OF AVIAN TOXICITY

Toxicological data for the effects of antimony on avian species were not located.

SUMMARY OF AMPHIBIAN TOXICITY

Toxicological data for the effects of antimony on amphibian species were not located.

SUMMARY OF REPTILIAN TOXICITY

Toxicological data for the effects of antimony on reptilian species were not located.

RECOMMENDED TOXICITY REFERENCE VALUES

Toxicity Reference Values for Mammals

Toxicity Reference Values for Ingestion Exposures for the Class Mammalia

Based on the information provided on mammalian oral toxicity (earlier), the rat appears to be more sensitive to antimony, in terms of growth, than mice.

However, Ainsworth *et al.* [14] observed antimony to be more toxic to mice than to voles. Specifically, antimony concentrations in the liver of voles given 6,700 mg/kg/day were much lower than in mice given the same diet, illustrating the desirability of working with a species known to be exposed in the field rather than a laboratory species. However, these data suggest that laboratory rodents are likely a sensitive species and may represent a conservative assessment of antimony toxicity. Although the data may indicate antimony to be more toxic to rats in terms of growth, this conclusion is limited because only three different species have been evaluated. From the available, appropriate, and relevant data, the lowest LOAEL is 0.05 mg Sb/kg bw/day [16], a concentration that resulted in a significant decrease in adult body weight on the twentieth day of gestation. Because a decrease in body weight can affect potential fitness, this growth endpoint is considered to be ecologically relevant and appropriate for the derivation of the TRV. This value is somewhat more protective than the mammalian TRV for antimony (0.059 Sb mg/kg/day) derived by the U.S. EPA [4] using the adult vole, shrew, and weasel as the surrogate receptors. The lowest NOAEL is 0.055 mg/kg/day, as reported by Rossi *et al.* [16], based on reduced pup body weight from the tenth to sixtieth day of age. However, the NOAEL value is a magnitude higher than the LOAEL value, which may be due to exposure to a less critical developmental life stage (adults). As such, the NOAEL was not used for TRV derivation.

The LOAEL dose of 0.05 mg Sb/kg bw/day was derived from pregnant female NOS albino normotensive rats exposed to ATO (1 mg ATO/L in drinking water ad libitum) from conception through the twentieth day of gestation. The average rat body weight at 20 days of gestation was 0.330 kg and was used to calculate the average daily intake. According to Sample *et al.* [29], average body weights of the animals in the study should generally be used when calculating NOAELs and LOAELs. Since the minimum data set requirements of providing data for three species and at least two different taxonomic orders were not met, the approximation approach was used to derive the TRV [6]. The LOAEL is from a chronic exposure; therefore, no uncertainty factors were applied, and a LOAEL-based TRV of 0.05 mg Sb/kg bw/day was selected. An uncertainty factor of 10 was applied to this LOAEL-based TRV to obtain the NOAEL-based TRV of 0.005 mg Sb/kg bw/day [6]. These TRVs are presented in Table 19.4 and are given a low confidence rating due to a lack of adequate chronic toxicity studies on other mammalian classes.

Table 19.4 Selected Ingestion TRVs for the Class Mammalia

TRV	Dose	Confidence Level
NOAEL-based	0.005 mg/kg/day	Low
LOAEL-based	0.05 mg/kg/day	Low

Source: [6]

Toxicity Reference Values for Inhalation Exposures for the Class Mammalia

There are comparatively fewer studies on the toxic effects to mammals from inhalation exposure to antimony than from ingestion. The toxic effects observed in the antimony inhalation studies were generally consistent and included ocular impairment, increased lung weight and lung neoplasms, and decreased body weight, all of which can impact the health and growth of the animal, thereby adversely impacting the ecology or health of the population. Only one species (laboratory rat) was represented in both subchronic and chronic exposures, and NOAELs were not reported in any of the studies. Although the data may be limiting, Groth *et al.* [23] reported significant toxic effects on rats exposed by inhalation to either ATO (45 mg/m^3 average concentration) or antimony ore concentrate, principally Sb_2S_3 (36 to 40 mg/m^3 average concentration). In the ATO treatment groups, a significant presence of lung neoplasms was observed in 27% of all exposed females.

After 9 and 12 months of exposure, males weighed significantly less than the controls, among which the maximum differences were no greater than 6.2% as compared controls. In the antimony ore concentrate treatment groups, a significant presence of lung neoplasms was observed in 25% of all exposed females, and females exposed for 9 and 12 months weighed significantly less than controls, among which the maximum differences were no greater than 6.4% as compared to controls.

The lowest statistically significant chronic inhalation LOAEL is 2.36 mg Sb/kg bw/day dose associated with reduced growth in male rats [23]. Because a decrease in body weight can affect potential fitness, this growth endpoint is considered to be ecologically relevant and appropriate for the derivation of the TRV. The average male rat body weight for the entire exposure period to ATO, 0.585 kg, was used to calculate the average daily intake. The minimum data set requirements of providing data for three species and at least two different taxonomic orders were not met; therefore, the approximation approach was used to derive the TRV [6]. Because a NOAEL was not provided in Groth *et al.* [23], an uncertainty factor was used

Table 19.5 Selected Inhalation TRVs for the Class Mammalia

TRV	Dose	Confidence Level
NOAEL-based	0.236 mg/kg/d	Low
LOAEL-based	2.36 mg/kg/d	Low

Sources: [6,23].

to calculate the value. An uncertainty factor of 10 was used to calculate the NOAEL-based TRV (0.236 mg Sb/kg bw/day), and an uncertainty factor of 1 was used to calculate the LOAEL-based TRV (2.36 mg Sb/kg bw/day) from the chronic LOAEL [6]. These TRVs are presented in Table 19.5 and are given a low confidence rating due to the lack of adequate chronic toxicity studies on other mammalian classes.

Toxicity Reference Values for Dermal and Ocular Exposures for the Class Mammalia

The dermal exposure data suggests a mild to moderate sensitization reaction to severe durations of topical applications of various materials containing antimony, although the test animals recovered quickly. In addition, these data are most likely representative of physical reactions to the application material (grease or calcium cup grease) rather than of toxicological effects related to the concentration of antimony. The limited similarity between laboratory exposures and field exposure, the short study duration, and the animals' quick recovery from the effects show the data has nominal relevance. Therefore, these data were not considered suitable for deriving a dermal exposure TRV at this time.

Toxicity Reference Values for Birds

No data were identified that would permit the derivation of an avian TRV for antimony.

Toxicity Reference Values for Amphibians

No data were identified that would permit the derivation of an amphibian TRV for antimony.

Toxicity Reference Values for Reptiles

No data were identified that would permit the derivation of a reptilian TRV for antimony.

IMPORTANT RESEARCH NEEDS

Although several studies were available and are consistent among growth effects, mammalian TRVs derived for antimony have been assigned low levels of confidence. This is due to the lack of a sufficient number of species, families, and orders that would provide a greater breadth of taxonomic diversity. The oral toxicity data represented three species within the same taxonomic order, and the inhalation toxicity data represented only one species, resulting in low confidence in the TRVs. Dermal and ocular toxicity data, although available, were not representative of likely field conditions, and the observed toxic effects may be attributed to the application material rather than to antimony exposure. Therefore, TRVs could not be calculated. In addition, none of the available mammalian studies evaluated reproductive or developmental effect endpoints. Such data would be especially applicable to ecological risk evaluations.

There were no studies of the toxic effects from antimony exposures to birds, reptiles, and amphibians that were appropriate for TRV development. Before reliable avian, amphibian, and reptilian TRVs can be derived, antimony toxicity in these wildlife classes needs to be adequately characterized. Appropriate acute, subacute, subchronic, and especially chronic toxicity data derived through biologically relevant exposure routes are needed. Research studies should include experimental models of species that are genetically, biologically, and behaviorally similar to the wildlife that exhibits the greatest propensity for toxicant exposure. In addition, toxicity studies that examine demographic factors such as birth, death, and recruitment would have much greater ecological significance. In general, more research on the toxicity of antimony to wildlife species is needed. Particularly warranted are long-term, chronic toxicity studies on birds, reptiles, and amphibians.

REFERENCES

[1] U.S. Environmental Protection Agency (U.S. EPA). Final drinking water criteria document for antimony. Washington DC: Office of Science and Technology, Office of Water; 1992, Report Number: EPA/920/5-00372.

[2] Agency for Toxic Substances and Disease Registry (ATSDR). Toxicological profile for antimony. Atlanta, GA: U.S. Department of Health and Human Services, Public Health Service; 1992.

[3] U.S. Geological Survey (U.S. GS). Mineral commodity profiles antimony, Washington, DC: U.S. Department of the Interior; 2004. Open-File Report 03-019. Available from, http://pubs.usgs.gov/of/2003/of03-019/of03-019.pdf.

[4] U.S. Environmental Protection Agency (U.S. EPA). Ecological soil screening levels for antimony. Interim final. Washington, DC: Office of Solid Waste and Emergency Response; February 2005, OSWER Directive 9285.7-61.
[5] Mok WM, Wai CM. Distribution and mobilization of arsenic and antimony species in the Coeur d'Alene River, Idaho. Environ Sci Technol 1990;24:102–8.
[6] USACHPPM. Standard practice for wildlife toxicity reference values. Aberdeen Proving Ground, MD: U.S. Army Center for Health Promotion and Preventive Medicine; 2000, Technical Guide No. 254.
[7] Gerritse RG, Vriesema R, Dalenberg JW. Effect of sewage sludge on trace element mobility in soils. J Environ Qual 1982;11:359–64.
[8] Ainsworth N, Cooke JA, Johnson MS. Behavior and toxicity of antimony in the short-tailed field vole (*Microtus agrestis*). Ecotoxicol Environ Saf 1991;21(2):165–70.
[9] M&T Chemicals, Inc. Acute toxicity studies with antimony oxide 2996-30. Prepared by Industrial BIO-TEST Laboratories, Inc. Northbrook, IL.
[10] Dieter MP, Jameson CW, Elwell MR, et al. Comparative toxicity and tissue distribution of antimony potassium tartrate in rats and mice dosed by drinking water or intraperitoneal injection. J Toxicol Env Health 1991;34(1):51–82.
[11] National Toxicology Program (NTP). NTP report on the toxicity studies of antimony potassium tartrate in F344/N Rats and B6C3F Mice (Drinking Water and Intraperitoneal Injection Studies). NTP TOX II, NIH Publication No. 92-3130. U.S. Department of Health and Human Services, Public Health Service, National Institutes of HealthResearch Triangle Park, NC: National Institutes of Health; 1992.
[12] Gurnani N, Sharma A, Talukder G. Comparison of clastogenic effects of antimony and bismuth as trioxides on mice in vivo. Biol Trace Elem Res 1993;37(2–3):281–92.
[13] Hext PM, Pinto PJ, Rimmel BA. Subchronic feeding study of antimony trioxide in rats. J Appl Toxicol 1999;19(3):205–9.
[14] Ainsworth N, Cooke JA, Johnson MS. Biological significance of antimony in contaminated grassland. Water Air Soil Pollut 1991;57–58(1):193–9.
[15] Poon R, Chu I, Lecavalier P, Valli VE, Foster W, Gupta S, Thomas B. Effects of antimony on rats following 90-day exposure via drinking water. Food Chem Toxicol 1998;36(1):21–35.
[16] Rossi F, Acampora R, Vacca C, Maione S, Matera MG, Servodio R, Marmo E. Prenatal and postnatal antimony exposure in rats: effect on vasomotor reactivity development of pups. Teratog Carcinog Mutagen 1987;7(5):491–6.
[17] Kanisawa M, Schroeder HA. Life term studies on the effect of trace elements on spontaneous tumors in mice and rats. Cancer Res 1969;29(4):892–5.
[18] Schroeder HA, Mitchener M, Balassa JJ, Kanisawa M, Nason AP. Zirconium, niobium, antimony and fluorine in mice: effects on growth, survival and tissue levels. J Nutr 1968;95(1):95–101.
[19] Schroeder HA, Mitchener M, Nason AP. Zirconium, niobium, antimony, vanadium and lead in rats: life term studies. J Nutr 1970;100(1):59–68.
[20] Newton PE, Bolte HF, Daly IW, Pillsbury BD, Terrill JB, Drew RT, Ben-Dyke R, Sheldon AW, Rubin AW. Subchronic and chronic inhalation toxicity of antimony trioxide in the rat. Fund Appl Toxicol 1994;22(4):561–76.
[21] Bio/Dynamics, Inc. (BioDynamics). A three month inhalation toxicity study of antimony trioxide in the rat followed by a recovery period – with cover letter dated 11/27/85. Washington, DC: Covington & Burling; 1985.
[22] Bio/Dynamics, Inc. (BioDynamics). A one year inhalation toxicity study of antimony trioxide in the rat (with a one year recovery period). Washington, DC: Antimony Oxide Industry Association (AOIA); 1990.
[23] Groth DH, Stettler LE, Burg JR, et al. Carcinogenic effects of antimony trioxide and antimony ore concentrate in rats. J Toxicol Env Health 1986;18(4):607–26.

[24] Watt WD, Perrin EV, Kerfoot EJ. Chronic inhalation toxicity of antimony trioxide; Validation of the TLV. Detroit, MI: Wayne State University; 1982, Accession Number OTS0206195.
[25] Gross P, Brown JH, Hatch TF. Experimental endogenous lipoid pneumonia. Am J Pathol 1952;28:211–21.
[26] Dernehl CU, Nau CA, Sweets HH. Animal studies on the toxicity of inhaled antimony trioxide. J Ind Hyg Toxicol 1945;27:256–62.
[27] Horton JR, Gaworski CL, Newton PE, Bruner RH. Evaluation of the acute toxicity, irritation, sensitization, and subchronic dermal toxicity of antimony thioantimonate lubricant. Wright-Patterson Air Force Base, OH: Department of the Navy, Naval Medical Research Institute; March 1986, AAMRL-TR-86-010; NMRI-86-25.
[28] M & T Chemicals, Incorporated. Acute dermal toxicity study with antimony triacetate. Princeton, NJ: AME Associates; 1967.
[29] Hazardous Substances Data Bank. National Library of Medicine [Internet], Available from: http://toxnet.nlm.nih.gov/cgi-bin/sis/search2/f?./temp/~lFCQgU:1.
[30] United States Environmental Protection Agency (U.S. EPA). Estimation programs interface suite (EPISuite) [Internet] Version 4.0 [cited January 2009]. Available from: www.epa.gov/oppt/exposure/pubs/episuite.htm.

CHAPTER 20

Wildlife Toxicity Assessment for Thallium

Desmond I. Bannon

Contents

INTRODUCTION

Thallium, found between mercury and lead in the periodic table, is a metallic element that occurs in the earth's crust at a concentration of 0.3–0.6 ppm. Thallium is frequently found in sulfur-containing ores of zinc, lead, and copper [1], as well as potassium minerals, often forming salts with bromine,

Wildlife Toxicity Assessments for Chemicals of Military Concern
http://dx.doi.org/10.1016/B978-0-12-800020-5.00020-X

chlorine, fluorine, and iodine [2]. Although some pure thallium mines exist, the metal has also been recovered as a by-product from dust accumulated in the flues of zinc, lead, and copper smelters [3]. When refined, thallium is a soft bluish-white heavy metal that oxidizes rapidly in air to form a coating of thallium oxide. Elemental thallium is highly reactive and exists in two oxidation states, thallous (I) and thallic (III).

In natural waters, thallium occurs primarily as the thallous ion [4]. The thallous ion (Tl^+) is relatively insensitive to pH and has an ionic radius and charge similar to monovalent cations such as potassium (K^+), allowing it to enter cells easily and compete in K-dependent biological systems [5]. Tl^+ ion binds particularly well to Na/K-ATPase [6] and this may account for some of its toxic properties. Thallium binds to cysteine residues, preventing cross-linking bond formation critical to the keratinization of hair, a possible mechanism for the alopecia associated with acute thallium poisoning [7]. Thallium disrupts mitochondria [8] and interferes with energy production [9], including glycolysis, the Krebs cycle, and oxidative phosphorylation [9,10].

Thallium was primarily used as a rodenticide and insecticide until 1972, when it was banned because of its potential harm to humans [11]. Currently, thallium compounds are used in photoelectric cells, highly reflective and low melting glasses, electrical switches and closures that will work at subzero temperatures, low range thermometers, fireworks, in the manufacture of synthetic gems, and as catalysts in chemical syntheses [11,12]. Like other metals, elemental thallium cannot be further degraded or metabolized, and as such, can persist where there has been high levels of contamination. Potential exposure of wildlife receptors to thallium can occur via wet and/or dry deposition [11].

Since the United States now imports all of its thallium, overall levels of thallium in the environment are relatively low but it can still be released from coal operations, cement manufacture, and metal smelters. Due to its low annual production (<1 ton per year) [3], the pollution potential of thallium may be best judged on production of thallium wastes or by-products [13].

This Wildlife Toxicity Assessment (WTA) summarizes the toxicity of thallium with emphasis on identifying adverse effects and developing Toxicity Reference Values (TRVs). The TRVs are intended to serve as benchmarks for screening-level ecological risk assessments. The protocol for TRV development is documented in the U.S. Army Center for Health Promotion and Preventive Medicine (USACHPPM) Technical Guide 254, *Standard Practice for Wildlife Toxicity Reference Values* [14]. The CAS number, 7440-28-0, refers to the elemental form of thallium, although this WTA also considers the potential toxicity of thallium compounds that are most often encountered in the environment.

TOXICITY PROFILE

Environmental Fate and Transport

The annual use of thallium in the United States is less than 1 metric ton (compared to ~1.4 million metric tons for lead) [3]. Primary uses include radioactive thallium in cardiovascular imaging, as an activator in gamma ray detection scintillators, in high temperature superconductors, and in lenses and prisms. When 63 minable metals were compared on the basis of their global annual production and their environmental impact; iron and aluminum displayed the largest impacts, with thallium and tellurium the lowest [15]. While the potential for pollution from thallium, based on its annual usage, is much less than that from lead or other toxic metals, thallium is more acutely toxic than other metals.

From an environmental perspective, the combustion of coal, cement production, and smelting operations are the predominant sources of atmospheric release of thallium into the atmosphere [16], at least in the United States. Significant amounts of thallium release into the water occurs as a consequence of various mining and refining operations, iron and steel production, and ore processing releases [17]. Soil contamination occurs mainly from solid wastes from coal combustion and smelting [18]. Table 20.1 presents the physical-chemical properties of elemental thallium. Once deposited on land or water, thallium compounds are unlikely to volatilize, although wind dispersion from soil

Table 20.1 Summary of the Physical-Chemical Properties of Thallium

CAS No.	7440-28-0
Molecular weight	204.38
Color	Bluish-white
Physical state	Soft heavy metal
Melting point	304 °C
Boiling point	1473 °C
Odor	None
Solubility in water	Insoluble; slightly soluble in nitric acid
Partition coefficients:	NA
Log K_{ow}	NA
Log K_{oc}	NA
Vapor pressure at 25 °C	1.412×10^{-23} mm Hg
Henry's Law constant at 25 °C	NA
Conversion factors	1 ppm = 8.36 mg/m^3 1 mg/m^3 = 0.12 ppm

Sources: HSDB [2], ATSDR [11].

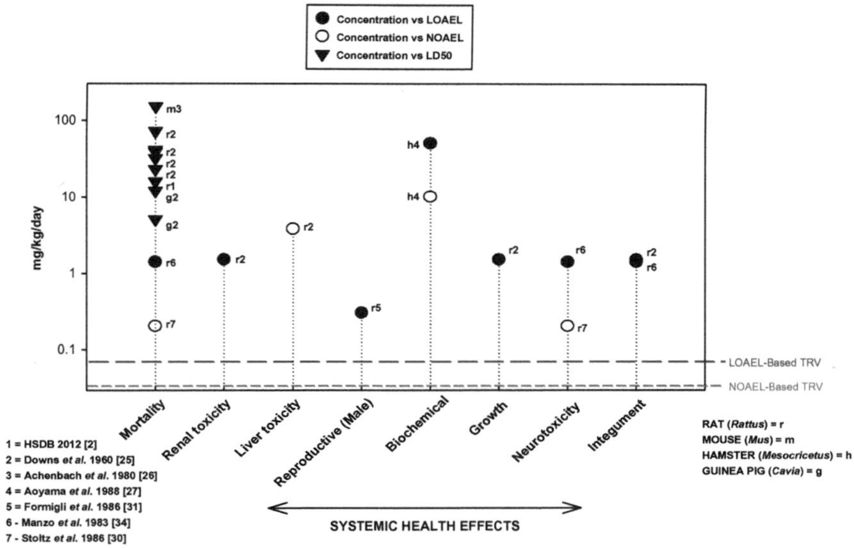

Figure 20.1 Thallium: Health Effects to Mammals. Letters by symbols represent test species, and numbers represent study.

surfaces in dry conditions is possible. Although elemental thallium is insoluble in water, thallium compounds possess a range of solubility in water, where they exist mostly as monovalent (thallous) ions. As such, thallium can readily leach to groundwater, and/or plants take them up from the soil. The hazardous substances data bank [2] identifies a number of field experiments that have quantified the thallium content of various food crops. In addition, some data are available that indicate that thallium can bioaccumulate in aquatic and terrestrial food chains [18–21].

Thallium adsorbs to soils containing clays, organic matter, and iron oxides [22–24] and is not transformed or biodegraded [22]. No scientific studies have been found regarding the abiotic or biotic transformation of thallium compounds in water. Thallium has been found to oxidize slowly in the atmosphere, and although it is not degraded by photochemical reactions, thallous chloride has been shown to be photosensitive [5,11].

SUMMARY OF MAMMALIAN TOXICITY

Mammalian Oral Toxicity: Acute

Acute thallium toxicity in mammals is characterized by the following: anorexia, vomiting, diarrhea, skin changes including hair loss (alopecia), dyspnea and nervous disorders, and respiratory failure leading to death [24]. A number of LD_{50} values for thallium compounds are documented in the secondary toxicological/pharmacological literature, including 41.2 mg/kg for thallium acetate in rats and 16 mg/kg for thallium sulfate in brown rats [2].

Downs *et al.* [25] studied the effects of thallium acetate (TA) and thallic oxide (TO) in Wistar-strain albino rats, albino rabbits, guinea pigs, and beagle dogs. The study evaluated the effects of acute TA and TO exposure in four types of mammals and subchronic exposures in rats. The acute exposure analysis consisted of a single oral dose of TA or TO to each of the four mammalian species. Details were not provided as to how TA and TO were orally administered, and treatments generally consisted of one individual animal/ per dose. As another assessment of acute toxicity, single doses of TA and TO were administered to rabbits, guinea pigs, and rats via intravenous and/or intraperitoneal injections [25]. Thallium acetate was administered as an aqueous solution with a concentration of 1, 2, or 5%. Five percent TO was suspended in a solution containing distilled water and 0.8% carboxy methylcellulose. Results indicated that the lowest lethal dose for a mammal (guinea pig) was 5 mg TO/kg and 12 mg TA/kg via single-dose oral administration. Downs *et al.* had also calculated a 7-day LD_{50} value for female rats for intraperitoneal and oral routes of exposures. The LD_{50} values were 23 mg TA/kg and 72 mg TO/kg for intraperitoneal exposures and 32 mg TA/kg and 39 mg TO/kg for oral exposures.

Using SWS mice, Achenbach *et al.* [26] determined the oral LD_{50} of thallium (as thallium sulfate) as a function of time, and obtained values ranging from 150 mg T1/kg (36 hours) to approximately 2,000 mg T1/kg (1 hour).

Aoyama *et al.* [27] evaluated the possible association of thallium toxicity with peroxidation processes in male golden hamsters. Lipid peroxidation eliminates lipids in cellular membranes and generates endogenous toxicants (free radicals and electrophiles). Free radicals and electrophiles may react with nearby molecules such as membrane proteins, or disperse and react with molecules such as DNA. Glutathione (GSH) protects lipids against peroxidation via electron transfer from glutathione to the peroxidase free radical. Glutathione peroxidase (GSH Px) is the enzyme used in the oxidation reaction of glutathione [28]. The amount of GSH decreases when thallium is bound to sulfhydryl groups [27], which increases lipid peroxidation, and dampens protection against free radicals. One dose of thallium malonate was administered orally via gavage to male golden hamsters at 10 mg/kg and 50 mg/kg doses. The study found that there was an increase in lipid peroxidation in the kidney 1 day after the 10 mg/kg dose. A significant decrease in the nonprotein sulfhydryls (NPSH) and GSH Px was observed in the kidney at 10 mg/kg. NPSH and GSH Px concentrations had also significantly decreased in the liver and kidney at 50 mg/kg. Indicators of renal and liver damage were also noted. Renal damage was shown in other studies [29].

Mammalian Oral Toxicity: Subacute

No studies on subacute oral toxicity of thallium were found.

Mammalian Oral Toxicity: Subchronic

Downs *et al.* [25] conducted a range-finding study and a subchronic study to investigate the subchronic oral toxicity of TA and TO. Initially, seven groups of five rats/group were exposed to dietary concentrations of 0, 0.0002, 0.001, 0.005, 0.01, 0.05, and 0.50% TA for 1 month. Dietary levels of 0.005% or greater resulted in death rates from 60 to 100% within 10 days. No effects were seen at the 0.001% and 0.0002% levels.

Ten weanling Wistar rats (five rats/sex/group) received 0, 0.0005, 0.0015, 0.003, or 0.005% TA, or 0, 0.002, 0.0035, 0.005, 0.01, or 0.05% TO in the diet for 15 weeks. Growth, survival, and organ weights were assessed. Statistical analyses were conducted to compare organ weights between the control and experimental groups. A decrease in growth was observed in males and increased mortality was noted in males and females exposed to 0.003% TA. Eighty percent of males and 60% of females died between 4 and 8 weeks of the study. After 2 weeks, alopecia was found in rats fed 0.0015% TA and 0.003% TA. At the end of the 15-week study, surviving rats from all dose groups were sacrificed for histopathological analysis. No significant histological changes were found in the lungs, liver, kidneys, heart, spleen, brain, stomach, and skin. A considerable decrease (up to 180 g) in body weight was observed in males and females at 0.005% TO, 0.01% TO, and 0.05% TO and a moderate reduction at 0.0035% TO and 0.002% TO in males. Increased mortality occurred in rats of both sexes at $\geq$0.005% TO. At the end of the 15-week study, one male rat and three female rats from the 0.0035% TO group and all males and three females from 0.002% TO group had survived. It was noted by the researchers that the deaths observed in the females at the 0.0035% TO level might not have been due to thallium exposure since the same number of deaths were observed in the females fed 0.002% TO. Alopecia was observed in rats at 0.002% TO and 0.0035% TO. At the end of the 15-week study, surviving rats from all dose groups were sacrificed for histopathological analysis of the organs. Significant histological changes were noted in the kidneys but not in the lungs, liver, and brain. Effects were noted at 0.002% TO (1.51 mg/kg/day), which is the lowest observed adverse effect level (LOAEL) and the lowest dose used in this study.

In a 90-day study [30], rats were exposed by gavage to levels of 0, 0.01, 0.05, and 0.25 mg/kg/day thallium (as thallium sulfate). No deaths were reported. Increased rate of alopecia and observations of lacrimation and rough coat was found in all thallium treatments with increasing incidence. No significant treatment-related changes were reported in hematology parameters, though blood sugar content was reduced with moderate increases in serum glutamic-oxaloacetic transaminase (SGOT) and lactate dehydrogenase (LDH), and sodium levels in rats from all treatment groups. No changes in body weight, feed consumption, or organ weights attributable to treatment were found.

Formigli *et al.* [31] studied the toxicological effects to the male reproductive system by exposing 10 male Wistar rats to 10 ppm thallium (as thallium sulfate) in drinking water for 60 days, a dose equivalent to about 0.3 mg thallium/kg/day. The control group consisted of 10 male Wistar rats. The researchers evaluated sperm motility and conducted histopathological examinations of the testis and epididymis using electron microscopy. A histopathological examination found that there were changes to the tubular epithelium and structural changes in the Sertoli cells. The study found a significant increase in the number of immature sperm in the epididymis and a significant decrease in sperm motility. In an evaluation of various reproductive endpoints, Chapin *et al.* [32] found that fertility had decreased when sperm motility in Swiss mice was 37% or less. Based on the findings in Formigli *et al.* [31] and Chapin *et al.* [32], it can be concluded that the LOAEL for likely fertility decreases in male rats was 0.3 mg/kg/day thallium. Since the rats were not given thallium doses lower than 0.3 mg/kg/day, a NOAEL is not possible.

A 30-day study of rats exposed to doses of thallium at 1/20 the LD_{50} and below examined the effect on brain distribution and brain function in rats [33]. There was little difference between brain regions with respect of thallium concentrations. At the highest dose (1.6 mg/kg) rat brains showed lipid peroxidation, whereas at lower doses (0.8 mg/kg) there was no measureable effect.

Mammalian Oral Toxicity: Chronic

Manzo *et al.* [34] was the only chronic oral toxicity test located, which reported structural and functional changes in the peripheral nerves of rats.

Eighty female Sprague Dawley rats (180–200 g) were exposed to thallium as thallium sulfate in drinking water at 10 mg thallium/L (25 μm) and followed for 36 weeks. Overt symptoms included alopecia (hair loss), periorbital redness, and irritability. Alopecia occurred in about 20% of the treated animals. Overall mortality rate was 15% and 21% after 40 and 240–280 days of treatment, respectively.

At 240 days, following mean exposures of 1.4 mg/kg/day, a 44% decrease in motor action potential amplitude, a 30% decrease in sensory action potential amplitude, and a 25% increase in motor action potential latency were observed. In addition, axonal destruction and myelin sheath alterations were observed at this treatment level. Upon completion of the study, complete alopecia was observed in a number of subjects. Mortality (15–21%) was reported following 40 days of exposure to levels exceeding 1.4 mg/kg/day.

Mammalian Oral Toxicity: Other

According to the World Health Organization [24], seed-eating animals, farm animals, and predators exposed to thallium via the use of rodenticides exhibit adverse effects to their central nervous system and gastrointestinal tract. Additional adverse effects in mammals also include salivation from the mouth and nose in cattle and reduced growth in sheep [24].

Studies Relevant to Mammalian TRV Development: Thallium Ingestion Exposures

There are relatively few studies on the toxicity of thallium with most focusing on acute toxicity. Acute and subchronic thallium toxicity is characterized by anorexia, vomiting, diarrhea, skin changes including hair loss, dyspnea and nervous disorders, and respiratory failure.

Formigli *et al.* [31] and Downs *et al.* [25] evaluated the subchronic toxicity of thallium in rats. Downs *et al.* [25] showed that thallium affected growth and survival of rats, however, no statistical analyses were conducted, and the data presented were insufficient for further independent analysis. Although this study had shown apparent biologically significant effects of thallium at 0.002% thallium oxide (1.51 mg/kg/day), the lack of statistical analyses precludes the use of this study for TRV development. Consistent with those results are the findings of Manzo *et al.* [34], where alopecia, irritability, and periorbital redness were observed, as well as low chronic mortality rates (15–20% after 40 weeks).

Formigli *et al.* [31] found that reproduction was deleteriously affected by thallium when male Wistar rats were exposed to 0.3 mg thallium/kg/day (as thallium sulfate) in drinking water for 60 days. The LOAEL for sperm motility of thallium-exposed rats was $40 \pm 8.11\%$ at 0.3 mg thallium/kg/day compared to $66.5 \pm 4.47\%$ in controls. Chapin *et al.* [32] evaluated various reproductive endpoints in Swiss mice and determined a threshold for biological significance for each reproductive endpoint, including altered sperm motility. The study found that fertility significantly decreased when sperm motility was 37% or less. Thus, it is concluded that the thallium-induced decrease in sperm motility found in Formigli *et al.* [31] would likely cause a reduction in reproductive success, potentially affecting species at the population level. Moreover, the levels at which these effects occur could be protective of other adverse reported effects.

Mammalian Inhalation Toxicity

No studies on the inhalation toxicity of thallium were located.

Mammalian Dermal Toxicity

No studies on the dermal toxicity of thallium were located. However, it should be noted that thallium in the form of thallous ions can mimic potassium and can cross membranes [35]; this route of exposure is unaccounted for in studies.

SUMMARY OF AVIAN TOXICITY

Avian Oral Toxicity: Acute

Bean and Hudson [36] conducted an acute toxicity experiment on thallium in three Golden Eagles (*Aquila Chrysaetos*). Single oral doses of thallium administered as thallium sulfate in a gelatin capsule were given to each bird. One bird received a dose of 60 mg/kg and the remaining two birds received a dose of 120 mg/kg. Since the bird given the dose of 60 mg/kg survived and the remaining two birds died, the researchers concluded that the LD_{50} was between 60 and 120 mg/kg.

Avian Oral Toxicity: Other

A number of studies exist on the toxicity of thallium compounds to chicken embryos. Teratogenic effects [37], growth inhibition [38], and disturbances in the development of bones [39] were found to occur in chicken embryos after injection of thallium into the egg.

Karnofsky *et al.* [39] were among the first to describe the onset of achondroplasia when thallium sulfate was injected into the yolk or onto the chondrioallantoic membrane. Achondroplasia is the occurrence of short-limbed dwarfism marked by severe shortening of the long bones of the wing and leg, bending of the tibia and overgrowth of the upper jaw (parrot beak, 38). Karnofsky *et al.* [39] determined that LD_{50} values for thallium ranged from 1.3 and 2.0 mg/egg, depending on the timing of injection relative to developmental stage. The site of injection (yolk sac or chondroallantoic membrane) was also important [40] in the manifestation of toxicity.

Ford *et al.* [37] conducted a range-finding experiment to establish optimum conditions for inducing chondrodystrophy in developing chicken embryos. The researchers injected 0.2–1.6 mg of thallium sulfate into the yolk of white leghorn eggs and examined the developing embryos at different time intervals. An injection of 1.2 mg of thallium sulfate resulted in pronounced skeletal deformities that were characterized by alterations in cartilage formation, skeletal deformity, and defective maturation of the chondrocytes at embryonic day (ED) 11. In one of a series of research reports on this topic, Hall [41] injected 0.6 mg thallium sulfate on ED7 and studied the ensuing development at 2-day intervals up to ED18. In observing reduced growth of the long bones and the formation of necrotic cartilage, Hall [38] speculated that these consequences might be related to a thallium-induced depletion of acid mucopolysaccaride. Hall's experiments [38] defined the critical period of embryonic development as it relates to the point where tibias first deviate from normal after thallium injection. After thallium sulfate was injected into white leghorn eggs at ED5, the critical period was found to range from ED5 to 9.

Studies Relevant to Avian TRV Development: Thallium Ingestion Exposures

Although there are several studies on the toxicity of thallium to avian species, none of the data generated are appropriate for deriving an avian TRV. The egg injection studies were not used to derive an avian TRV since the studies do not accurately represent real-life thallium exposures to birds nor could the exposure regime be calculated back to an oral dose. Additionally, the study by Bean and Hudson [36] were not used since the study lacked statistical analysis, did not narrow the range in determining a more accurate LD_{50} value, and presented some design deficiencies.

SUMMARY OF AMPHIBIAN TOXICITY

No data are available.

SUMMARY OF REPTILIAN TOXICITY

No data are available.

RECOMMENDED TOXICITY REFERENCE VALUES

Toxicity Reference Value for Mammals

TRVs for Ingestion Exposures for the Class Mammalia

Formigli *et al.* [31] showed that male Wistar rats exposed to thallium sulfate in their drinking water for 60 days had a significant decrease in sperm motility compared to controls. Sperm motility was $40.5 \pm 8.11\%$ in rats at 0.3 mg Tl/kg/day, compared to $66.5 \pm 4.47\%$ in the control group. According to Chapin *et al.* [32], fertility in mice is reduced if sperm motility is 37% or less; therefore, the sperm effects seen in Formigili *et al.* [31] would likely cause a decrease in reproductive success, and a concomitant reduction in fitness.

The 0.3 mg Tl/kg/day dose of the male rat in this study was derived from a cohort of male rats exposed to 10 ppm of thallium sulfate in their drinking water for 60 days. The body weights in the study ranged from 0.350 kg to 0.380 kg. The average body weight, 0.365 kg, was used to calculate the average daily intake. According to the Toxicological Benchmarks for Wildlife [42], average body weights of the laboratory animals in the study should be used when calculating NOAELs and LOAELs [42]. Since data from the chronic toxicity test were not adequate to use the NOAEL/LOAEL approach, the approximation approach was used to derive the mammalian TRV [14]. An uncertainty factor of 20 was used to calculate the NOAEL-based TRV from the subchronic LOAEL (0.3 mg Tl/kg/day) and an uncertainty factor of 4 was used to calculate the LOAEL-based TRV from the study LOAEL (0.3 mg Tl/kg/day) [14]. Table 20.2 presents a summary of the relevant mammalian data for TRV derivation, and Table 20.3 presents the calculated mammalian TRVs. These TRVs were given a medium confidence rating. While study design was adequate, the lack of adequate chronic toxicity tests and the lack of data on other mammalian classes preclude a high confidence rating for these TRVs.

Toxicity Reference Value for Birds

Not available at this time.

Table 20.2 Summary of Relevant Mammalian Data for TRV Derivation

		Test Results			
Test Organism	**Test Duration**	**NOAEL (mg/kg/day)**	**LOAEL (mg/kg/day)**	**Effects Observed at the LOAEL**	**Study**
Rats (Wistar)	15 weeks	0*	1.51* (0.002%)	Alopecia, decreased kidney weight, decreased growth (males)	[25]
Rats	15 weeks	ND	0.01	Mild alopecia, lacrimation, trends in increased sodium, LDH, SGOT levels; decreased glucose	[30]
Rats	240 days	ND	1.4	Change in neuroconduction, lesions on myelin sheaths, 15–21% mortality after 40 d, alopecia	[34]
Rats (Wistar)	60 days	ND	0.3	Reproductive effects (decreased sperm motility)	[31]

NA = not applicable; ND = not determined.
*No statistical analyses performed on alopecia incidence and growth data.

Table 20.3 Selected Ingestion TRVs for the Class Mammalia

TRV	**Dose**	**Confidence Level**
NOAEL-based	0.015 mg/kg/day	Medium
LOAEL-based	0.075 mg/kg/day	Medium

Sources: [31,32].

Toxicity Reference Values for Amphibians

Not available at this time.

Toxicity Reference Values for Reptiles

Not available at this time.

IMPORTANT RESEARCH NEEDS

The limited availability of data on the toxicity of thallium to wildlife species precludes the development of a high-confidence TRV. Moreover, for many of the studies on thallium toxicity to wildlife, the experimental design did not incorporate realistic exposures and hence these studies were not useful for deriving TRVs. In general, more research on the toxicity of thallium to wildlife species is needed. Particularly warranted are long-term, chronic toxicity studies on mammals but there is also a need for studies on amphibians and reptiles, since thallous ions can readily cross membranes and could be absorbed without ingestion. There are no useful studies of thallium on birds, reptiles, or amphibians and therefore, it is impossible to obtain TRVs for these important wildlife receptors.

REFERENCES

[1] Kazantzis G. Thallium in the environment and health effects. Environ Geochem Health 2000;22:275–80.

[2] Hazardous Substance Data Bank (HSDB). Thallium [Internet]. Bethesda (MD): National Library of Medicine (US), National Institute of Health [cited 2012]. Available from: http://toxnet.njm.nih.gov/.

[3] U.S. GS. Thallium: statistics and information: U.S. Geological Survery. Washington (DC): Department of the Interior; 2014. Available at, http://minerals.usgs.gov/minerals/pubs/commodity/thallium/.

[4] Peter AL, Viraraghavan T. Thallium: a review of public health and environmental concerns. Environ Int 2005;31(4):493–501.

[5] Cotton FA, Wilkinson G. Advanced inorganic chemistry: a comprehensive text. 4th revised ed. New York (NY): John Wiley and Sons; 1980.

[6] Britten JS, Blank M. Thallium activation of the (Na + –K +)-activated ATPase of rabbit kidney. Biochim Biophys Acta 1968;159(1):160–6.

[7] Mulkey JP, Oehme FW. A review of thallium toxicity. Vet Hum Toxicol 1993;35 (5):445–53.

[8] Eskandari MR, Mashayekhi V, Aslani M, Hosseini MJ. Toxicity of thallium on isolated rat liver mitochondria: the role of oxidative stress and MPT pore opening. Environ Toxicol 2013 Aug 30;.

[9] Hanzel CE, Verstraeten SV. Thallium induces hydrogen peroxide generation by impairing mitochondrial function. Toxicol Appl Pharmacol 2006;216(3):485–92.

[10] Hoffman RS. Thallium toxicity and the role of Prussian blue in therapy. Toxicol Rev 2003;22(1):29–40.

[11] Agency for Toxic Substances and Disease Registry (ATSDR). Toxicological profile for thallium. Atlanta (GA): U.S. Department of Health and Human Services, Public Health Service; 1992.

[12] Merck. The Merck index: an encyclopedia of chemicals, drugs, and biologicals. 12th ed. Whitehouse Station (NJ): Merck and Company, Inc; 1996.

[13] Zitko V. Toxicity and pollution potential of thallium. Sci Total Environ 1975;4 (2):185–92.

[14] USACHPPM. Standard practice for wildlife toxicity reference values. Aberdeen Proving Ground (MD): U.S. Army Center for Health Promotion and Preventive Medicine; 2000, Technical Guide No. 254.

[15] Nuss P, Eckelman MJ. Life cycle assessment of metals: a scientific synthesis. PLoS One 2014;9(7):e101298.

[16] U.S. Environmental Protection Agency (U.S. EPA). Research and development: health and environmental effects document for thallium and compounds. Washington (DC): U.S. Environmental Protection Agency, Office of Solid Waste and Emergency Response; 1988.
[17] U.S. EPA. Ambient water quality criteria for thallium. Washington (DC): U.S. Environmental Protection Agency Office of Water Regulations and Standards; 1986.
[18] Ewers U. Environmental exposure to thallium. Sci Total Environ 1988;71(3):285–92.
[19] Sharma J, Sharma RL, Singh HB, Satake M. Hazards and analysis of thallium—a review. Toxicol Environ Chem 1986;11:93–116.
[20] Zitko V, Carson WV. Accumulation of thallium in clams and mussels. Bull Environ Contam Toxicol 1975;14(5):530–3.
[21] Zitko V, Carson WV, Carson WG. Thallium: occurrence in the environment and toxicity to fish. Bull Environ Contam Toxicol 1975;13(1):23–30.
[22] Callahan MA, Slimak MW, Gabel NW. Water-related environmental fate of 129 priority pollutants. Vol I. Introduction and technical background, metals and inorganics, pesticides and PCBs. Springfield (VA): Versar, Incorporated; 1979, Prepared for: U.S. Environmental Protection Agency, Office of Water Planning and Standards, Washington, DC. EPA-440/4-79-029a. NTIS No. PB80-204373.
[23] Frantz G, Carlson R. Effects of rubidium, cesium, and thallium on interlayer potassium release from transvall vermiculite. Soil Sci Soc Am J 1987;51(2):305–8.
[24] World Health Organization (WHO). Environmental health criteria, 182. Thallium. Geneva, Switzerland: WHO; 2002.
[25] Downs WL, Scott JK, Steadman LT, Maynard EA. Acute and sub-acute toxicity studies of thallium compounds. Am Ind Hyg Assoc J 1960;21:399–406.
[26] Achenbach C, Hauswirth O, Heindrichs C, Ziskoven R, Kohler F, Bahr U, et al. Quantitative measurement of time-dependent thallium distribution in organs of mice by field desorption mass spectrometry. J Toxicol Environ Health 1980;6(3):519–28.
[27] Aoyama H, Yoshida M, Yamamura Y. Induction of lipid peroxidation in tissues of thallous malonate-treated hamster. Toxicology 1988;53(1):11–8.
[28] Klaassen CD, Casarett LJ, Doull J. Casarett & Doull's toxicology: the basic science of poisons. 8th ed. New York (NY): McGraw-Hill Education; 2013.
[29] Leung KM, Ooi VE. Studies on thallium toxicity, its tissue distribution and histopathological effects in rats. Chemosphere 2000;41(1–2):155–9.
[30] Stoltz ML, Stedham MA, Brown LK. Subchronic (90-day) toxicity of thallium sulfate in Sprague-Dawley rats. Washington (DC): Dymanic Corporation; 1986, Report to the U.S. Enviornmental Protection Agency, Office of Solid Waste. Project No. 8702-L.
[31] Formigli L, Scelsi R, Poggi P, Gregotti C, Di Nucci A, Sabbioni E, et al. Thallium-induced testicular toxicity in the rat. Environ Res 1986;40(2):531–9.
[32] Chapin RE, Sloane RA, Haseman JK. The relationships among reproductive endpoints in Swiss mice, using the reproductive assessment by continuous breeding database. Fundam Appl Toxicol 1997;38(2):129–42.
[33] Galvan-Arzate S, Martinez A, Medina E, Santamaria A, Rios C. Subchronic administration of sublethal doses of thallium to rats: effects on distribution and lipid peroxidation in brain regions. Toxicol Lett 2000;116(1–2):37–43.
[34] Manzo LR, Scelsi R, Moglia A. Long-term toxicity of thallium in the rat. In: Chemical toxicology and clinical chemistry of metals. London (UK): Academic Press; 1983.
[35] Rodriguez-Mercado JJ, Altamirano-Lozano MA. Genetic toxicology of thallium: a review. Drug Chem Toxicol 2013;36(3):369–83.
[36] Bean JR, Hudson RH. Acute oral toxicity and tissue residues of thallium sulfate in golden eagles, Aquila chrysaetos. Bull Environ Contam Toxicol 1976;15(1):118–21.
[37] Ford JK, Eyring EJ, Anderson CE. Thallium chondrodystrophy in chick embryos. An histological and biochemical investigation. J Bone Joint Surg Am 1968;50(4):687–700.

[38] Hall BK. Thallium-induced achondroplasia in chicken embryos and the concept of critical periods during development. Teratology 1977;15(1):1–15.
[39] Karnofsky DA, Ridgway LP, Patterson PA. Production of achondroplasia in the chick embryo with thallium. Proc Soc Exp Biol Med 1950;73(2):255–9.
[40] Ridgway LP, Karnofsky DA. The effects of metals on the chick embryo: toxicity and production of abnormalities in development. Ann N Y Acad Sci 1952;55(2):203–15.
[41] Hall BK. Thallium-induced achondroplasia in the embryonic chick. Dev Biol 1972;28 (1):47–60.
[42] Sample BE, Suter GW. Toxicological benchmarks for wildlife: 1996 revision. Oak Ridge (TN): Oak Ridge National Laboratory; 1996.

PART V

Wildlife Toxicity Assessments of Pesticides

CHAPTER 21

Wildlife Toxicity Assessment for Aldrin and Dieldrin

Adam T. Deck, Emily N. Reinke, Wilfred C. McCain

Contents

INTRODUCTION

The cyclodiene pesticide aldrin (CAS No. 309-00-2) was widely used as a broad-spectrum pesticide between 1948 and 1974, in particular against soil insects that attack field and forage crops, vegetables, and fruits [1]. Like other organochlorine pesticides, use of aldrin was subsequently restricted due to concerns about potential threats to human health and its capacity to bioaccumulate in ecological food chains. While aldrin is no longer registered for use in the United States, its metabolic product, dieldrin, is frequently detected at environmental sites, indicative of its resistance to bio- or chemical degradation. Aldrin (1,2,3,4,10,10-hexachloro-1,4, 4a,5,8,8a-hexahydro-1,4:5,8-exo-dimethanonaphthalene) is structurally related and readily converted to dieldrin (1,2,3,4,10,10-hexachloro-6,7-epoxy-1,4,4a,5,6,7,8,8a-octahydro-endo,exo-1,4:5,8-dimethanonaphthalene), which can occur via biotic or abiotic epoxidation. Thus, epoxidation of aldrin occurs in the tissues of animals, in air by photooxidation, and in soil because of microbial action.

Wildlife Toxicity Assessments for Chemicals of Military Concern
http://dx.doi.org/10.1016/B978-0-12-800020-5.00021-1

However, while dieldrin is more stable in the environment, it is at least as toxic to insects and other animals as aldrin.

This Wildlife Toxicity Assessment summarizes the current knowledge of the toxicological impacts of aldrin on wildlife (mammals, birds, reptiles, and amphibians). Evaluating the toxicity of aldrin contributes to the derivation of Toxicity Reference Values (TRVs) that could serve as screening-level benchmarks for wildlife near contaminated sites. The protocol for the development of this assessment is documented in the U.S. Army Center for Health Promotion and Preventive Medicine (USACHPPM) Technical Guide (TG) 254, *Standard Practice for Wildlife Toxicity Reference Values* [2].

TOXICITY PROFILE

Environmental Fate and Transport

Although aldrin is readily degraded in the environment because of biotic and abiotic processes, its primary metabolic product, dieldrin, resists further transformation and is persistent in the environment. Even though the use of aldrin as a pesticide has been abolished, a primary breakdown product, dieldrin, can be persistent and can bioaccumulate. These compounds probably continue to represent a toxicological threat to wildlife.

As discussed in the Agency for Toxic Substances and Disease Registry (ATSDR) [3], when aldrin was used as a pesticide, it was readily detected in the air. For example, a mean concentration of 0.4 nanograms per cubic meter (ng/m^3) was reported in 13% of 2,479 samples taken from 16 states during the period of 1972 to 1974. This compares to a mean concentration of 1.6 ng/m^3 dieldrin (94% incidence), a difference probably attributable to the ease with which aldrin undergoes conversion to dieldrin. Dieldrin is formed from aldrin by photochemical epoxidation in the presence of hydroxyl radicals. It is then dispersed in the atmosphere, where it might undergo subsequent wet or dry deposition.

The physical-chemical characteristics of aldrin (Table 21.1) and dieldrin explain the observed persistence of the chemicals in soil. The behavior of dieldrin in soil is emphasized since epoxidation of residual aldrin occurs readily because of microbial interaction. The nonpolar nature of dieldrin and its low solubility in water enhance binding to soil particles, leaving little potential for leaching to groundwater. However, during and subsequent to the 1970s, aldrin/dieldrin was detected in groundwater at contaminated environmental sites. In addition, even though the vapor pressure of the compound is low, dieldrin can also be released from soil to the atmosphere by volatilization.

Table 21.1 Summary of the Physical-Chemical Properties of Aldrin

CAS No.	309-00-2
Molecular weight	364.93
Color	White (pure), tan (technical grade)
Physical state	Crystalline solid
Melting point	104 °C
Boiling point	145 °C
Odor	Mild
Solubility in water	0.2 mg/L at 25 °C: miscible in most organic solvents
Partition coefficients:	
Log K_{ow}	3.01
Log K_{oc}	4.69
Vapor pressure at 25 °C	1.4×10^{-4} mm Hg
Henry's Law constant at 25 °C	3.2×10^{-4} atm.m^3/mole
Conversion factors	1 ppm = 14.96 mg/m^3 1 mg/m^3 = 0.067 ppm

Source: ATSDR [7].

Summary of Mammalian Toxicity

Mammalian Oral Toxicity: Acute

Single oral median lethal dose (LD_{50}) values for aldrin were published by a number of investigators. For example, Gaines [4] surveyed the acute lethality of 98 pesticides and two metabolites of dichlorodiphenyltrichloroethane (DDT) in Sherman rats, deriving values for aldrin (grade and vehicle unstated) of 39 and 60 milligrams per kilogram (mg/kg) for males and females, respectively. These data show substantial agreement with LD_{50} values from a number of studies summarized by Hodge *et al.* [5]. Values ranging from 10 to 120 mg/kg were listed for nine laboratory and domestic animals and cattle.

In a more recent study, laboratory animals were acutely exposed to aldrin while endpoints other than lethality were monitored. Ali and Shakoori [6] provided aldrin in the diet for 48 hours to 8 female Sprague Dawley rats, half of which were euthanized after 24 hours, the remainder after 48 hours. The authors reported the ingested dose to be approximately 20 mg aldrin/kg (body weight)/day. Blood samples were collected to measure clinical chemistry and hematological parameters, and excised livers were processed for histopathological and biochemical studies. Compared to controls, marked fluctuations occurred in hematological parameters of aldrin-exposed rats. For example, statistically significant increases were observed in white blood cell count, mean corpuscular volume, and mean corpuscular hemoglobin; decreases occurred in hemoglobin concentration, red blood cell count, packed cell volume, and mean corpuscular hemoglobin concentration.

The 48-hour values closely resemble the intermediate data. Changes in the activity of some serum enzymes and in concentrations of other clinical chemistry parameters were evident after 24 hours, with further increases or decreases beyond the 24-hour levels after 48 hours. In a single example, an overall doubling of the serum activity of isocitrate dehydrogenase was observed after 48 hours (570.26 ± 44.04 SIU/mL in exposed rats versus 243.18 ± 14.8 in controls, the 24-hour activities being intermediate). When aldrin-exposed rats were compared to controls, far fewer fluctuations were observed in liver biochemistry parameters than in serum. In addition, the true nature of any apparent changes in liver biochemistry is uncertain since the results were expressed per gram of liver rather than per specific amount of supernatant protein. However, some hepatic changes caused by aldrin treatment are suggested by the parallel histological findings of hepatomegaly, necrosis, vacuolation, and fatty degeneration. In another phase of the research, the same authors used female Sprague Dawley rats dosed at 8 mg/kg/day to extend the in-life phase of the experiment to 15 days. Ali and Shakoori [6] reported a similar pattern of changes in hematology, clinical chemistry, liver biochemistry, and histopathology compared to those described in the 48-hour study.

Mammalian Oral Toxicity: Subacute

No data were found describing the toxicological effects in laboratory animals exposed to aldrin under subacute conditions. However, a study by Hutson [7] explored the putative metabolites of the compound by gavaging male CFE rats and male CF1 and LACG mice with radiolabeled dieldrin after a period in which subjects received either 20 parts per million (ppm) (rats) or 10 ppm (mice) dieldrin in the diet for 28 days. Animals were maintained in metabolic cages for an additional 8 days before euthanization. Among the key findings to emerge from this study was that most of the counts were recovered from the feces, with far fewer in the urine or deposited in tissue. In all cases, the metabolites of dieldrin appeared to be the same and included 12-hydroxydieldrin and its glucuronide, 4,5-trans-dihydroaldrindiol, an unspecified pentachloroketone metabolite, and an aldrin-derived dicarboxylic acid. Since most or all administered aldrin is rapidly converted to dieldrin, these compounds might also be expected to represent major products of aldrin metabolism.

Mammalian Oral Toxicity: Subchronic

An article by Pallade *et al.* [8] described a toxicological experiment in rats (strain and sex unstated) in which some of the dosing strategies were subchronic to chronic in duration. For example, groups of 10 to 20 rats were gavaged with aldrin for 14 months as follows: 3 mg/kg/day for the first 6 months, then 4.5 mg/kg/day for 8 months. Sequential euthanizations allowed for an evaluation of the 3 and 4.5 mg/kg/day treatments. Stage two involved rats exposed to six treatments at doses (mg/kg/day) of 0 for 20 days, 3 for 73 and 97 days, 4.5 for 75 days, 5 for 81 and 105 days, 6 for 24 days, and 9 for 20 days. While the predominant effects were of hepatic origin, the authors noted that a dose-dependent increase in mortality was obtained when only the two lowest dose groups had rats that survived the entire duration of exposure. Unfortunately, insufficient data were presented in the report with which to identify a no observed adverse effect level (NOAEL) for increases in relative liver weight; however, it was implicit that levels below 5 mg/kg/day did not permit identification of a NOAEL. However, increases in relative liver weights were associated with decreases in glycogen in exposures ≥5 mg/kg/day, and serum cholesterol was increased in the 9 mg/kg/day groups, relative to controls. Steatosis (fatty liver) was also seen in histological preparations of the liver from 6- and 9-mg/kg/day animals. Mortality rates were presented as follows: 0% (0/20) at 3 mg/kg/day, 10% (4/10) at 4.5 mg/kg/day, 40% (8/20) at 5 mg/kg/day, 50% (10/20) at 6 mg/kg/day, and 85% (17/20) at 9 mg/kg/day.

The appearance of dieldrin as a metabolite of aldrin was the focus of a study in which aldrin was administered to beagle dogs at 0.6 mg/kg in gelatin capsules, 5 days per week for 10 months [9]. While the primary focus of the study was the formation of dieldrin in important metabolic, transport, and storage sites such as liver, blood, and adipose tissue, the authors reported a number of neurological consequences of treatment such as excitability and tremors. This outcome suggests that a time-weighted average dose of 0.43 mg/kg/day might be a lowest observed adverse effect level (LOAEL) for the neurological effects of aldrin in this animal model.

Mammalian Oral Toxicity: Chronic

The record for aldrin in the Integrated Risk Information System (IRIS) database [10] cites a number of chronic experimental studies focused on the liver as the principal target organ and on the capacity for inducing hepatic neoplasms in mice. Although not relevant to the development of a toxicity reference value, hepatic neoplasm induction appears to be characteristic of organochlorine pesticides in general for susceptible species such as mice, in which the spontaneous tumor rate in aging animals is appreciable. Hepatocellular tumors due to aldrin administration have not been observed in less susceptible species such as rats although the liver remains the primary target organ. For example, a 2-year study of aldrin in Osborne-Mendel rats by Fitzhugh *et al.* [11] was chosen as the principal study for noncarcinogenic effects in the development of a reference dose. Twelve rats/sex/group received dietary levels of 0, 0.5, 2, 10, 50, or 150 ppm aldrin. Survival was significantly affected at 100 and 150 ppm aldrin, and liver lesions were observed at all dietary levels, suggesting a LOAEL of 0.5 ppm [11]. For the aldrin record in IRIS, an associated dose of 0.025 mg/kg/day was derived by using a rat-specific default food factor of 0.05. While liver effects appeared to be the most sensitive index of aldrin toxicity in rats, the authors described the incidence of hemorrhagic urinary bladders and nephritis in male rats exposed to the compound at dietary concentrations of 50 ppm and above. This observation was consistent with a NOAEL of 0.5 mg/kg/day (equivalent to dietary levels of 10 ppm aldrin) and a LOAEL of 2.5 mg/kg/day.

Fitzhugh *et al.* [11] administered aldrin (0.2, 1.0, 2.0, and 5.0 mg aldrin/kg/day) or dieldrin (0.2, 1.0, 2.0, 5.0, and 10 mg dieldrin/kg/day) to dogs for up to 25 months via the diet. A variety of histopathological changes in the liver, kidney, and bone marrow were observed. While there were too few animals tested to allow any statistically supported conclusions to be drawn from these studies, the authors assigned dose levels of 0.2 mg/kg/day and 0.5 mg/kg/day as the associated NOAEL and LOAEL, respectively.

The review by Hodge *et al.* [5] summarized NOAELs and LOAELs from earlier studies that were based on changes in relative liver weight in rats and dogs that resulted from dietary supplementation with aldrin. Assuming a dietary default food factor of 0.05 for rats, these doses ranged from 0.125–1.25 mg/kg/day for a NOAEL to 0.025–7.5 mg/kg/day for a LOAEL. The individual values are provided in Figure 21.1. If a default food factor of 0.025 is used, the equivalent NOAEL and LOAEL in dogs are 0.025 and 0.075 mg/kg/day, respectively.

The experimental studies of Pallade *et al.* [8] included a chronic exposure component in which rats (sex and strain unstated) were gavaged with 3 mg aldrin/kg/day for 6 months, then with 4.5 mg aldrin/kg/day for 8 months. To monitor gross and histopathological changes in the liver, serial euthanization of 10 exposed and 10 control animals took place. Statistically significant increases in relative liver weight were evident after the first

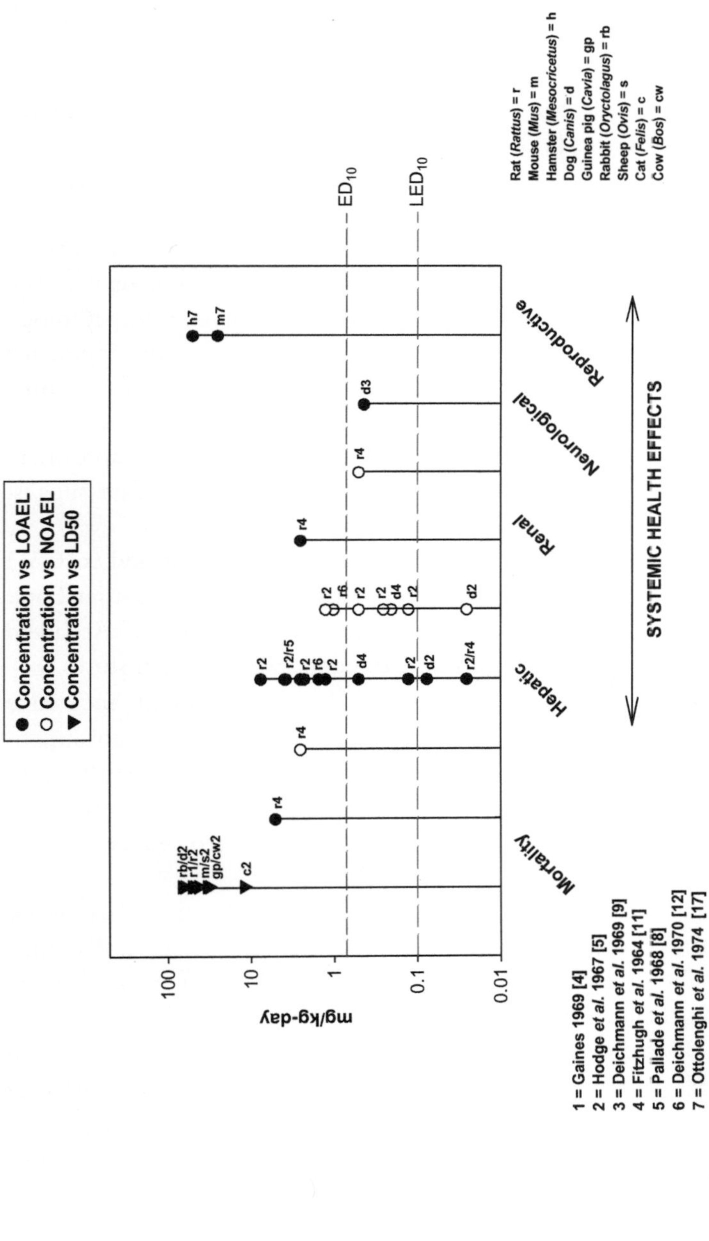

Figure 21.1 Aldrin: Health Effects to Mammals. Letters by symbols represent test species, and numbers represent study.

euthanization point at 31 days. Subsequently, the time weighted average (TWA) dose level of 3.85 mg/kg/day was determined to represent a LOAEL for the hepatotoxic effects described in this experiment.

The potential of aldrin to induce tumors was the primary focus of an experiment in which 50 Osborne-Mendel rats/sex/group were fed 0, 10, 15, or 25 ppm aldrin for the first 10 weeks followed by 0, 20, 30, or 50 ppm of the pesticide for the remainder of the study, which ran for approximately 20 months [12]. Animals were monitored regularly for clinical signs, after which they were subjected to a gross necropsy. Excised pieces of the liver, kidney, and lungs were processed for histopathological examination. A number of compound-related changes were observed; for example, reduced survival was seen among high-dose females. Changes in the gross appearance of some organs at necropsy and in liver histopathology were evident, but no apparent aldrin-induced increases in tumor incidence were evident. Among the important noncarcinogenic effects of the compound were increases in relative liver weight in male rats exposed to 30 and 50 ppm aldrin. Assuming a default food factor of 0.05 was appropriate to the Osborne-Mendel rats employed in this study, these data suggest 1.0 mg/kg/day as a NOAEL and 1.5 mg/kg/day as a LOAEL.

The occurrence of hepatic vein thrombosis was investigated in C3H mice that had been given organochlorine pesticides as part of earlier Food and Drug Administration (FDA) toxicological investigations [13]. Archived tissue blocks and/or sections from the Davis and Fitzhugh [14] study provided the experimental evidence for Reuber [13] to evaluate the incidence of hepatic thrombosis and update the hepatic tumor incidence according to the 1975 experimental pathology standards. An 8% incidence of hepatic vein thrombosis because of aldrin exposure was on the borderline of statistical significance and potentially implied a compound-related effect in this animal model.

The National Cancer Institute (NCI) sponsored a 2-year toxicological study of aldrin and dieldrin in Osborne-Mendel rats and B6C3F1 mice [15]. Fifty (10 for control) rats/sex/dose were fed 30 or 60 ppm aldrin for either 74 weeks (males) or 84 weeks (females). All survivors were observed for a period of 37 to 38 weeks postexposure and prior to termination. In the same study, 50 (10 for control) mice were exposed for 80 weeks to differing dietary concentrations of aldrin amounting to TWA concentrations of 4 and 8 ppm in males and 3 and 6 ppm in females. Survivors were removed from the exposure conditions for a 10- to 13-week postexposure observation period prior to termination. Incidence of tumors and of noncarcinogenic lesions was compared to that in a small number of controls [10–20] and a larger number of untreated animals from similar bioassays of other chemicals.

In addition to the increased incidence of hepatocellular carcinomas in male mice, a number of aldrin-related carcinogenic effects were described in rats, such as combined thyroid follicular cell adenomas and carcinomas in low-dose rats of both sexes, and cortical cell adenomas in low-dose females. However, the lack of a clear-cut dose-response effect suggested that these "effects" may not have been compound-related. As described by the authors, a number of noncancerous inflammatory, degenerative, or proliferative lesions were seen with broadly similar frequency in both aldrin-treated and control animals. Such findings did not allow compound-related NOAELs and LOAELs to be defined for these effects.

A review article by Deichmann *et al.* [16] described an experiment in which 50 female Sprague Dawley and Osborne-Mendel rats/group were chronically exposed to 0, 20, or 50 ppm aldrin in the diet (overall duration unspecified). There was a marked dose-dependent reduction in survival, especially in high-dose rats, suggesting systemic toxicity of aldrin. No other compound-related effects were mentioned in the report. However, tumors of the mammary gland and uterus, the primary focus of the study, were observed in controls as well as in treated animals. There were no compound-related increases in tumors of the liver.

Mammalian Oral Toxicity: Other

Ottolenghi *et al.* [17] examined reproductive and developmental effects of aldrin in pregnant Syrian golden hamsters and CD1 mice. The hamsters received doses of 50 mg/kg/day aldrin in corn oil via gavage on gestation days (GD) 7, 8, and 9, and the mice received doses of 25 mg/kg/day via gavage on GD 9 only. All hamsters were euthanized on GD 14, and all mice were euthanized on GD 18. The uteri of the euthanized specimens were examined for live versus dead fetuses, visceral and skeletal abnormalities, and so on. Both hamsters and mice showed an increased number of dead fetuses and a similar pattern of fetal anomalies such as open eye, cleft palate, and webbed feet. These data suggest that the doses of aldrin used in this study may be considered as LOAELs for reproductive and developmental effects.

In a later experiment, Gellert and Wilson [18] dosed pregnant Sprague Dawley rats with 3 mg/kg aldrin in sesame oil on GDs 14 to 20 and monitored the F_1 females for changes in ovary, uterine, or adrenal weights, while F_1 males were compared to untreated controls in mating trials. No aldrin-related effects were observed.

Studies Relevant for Mammalian TRV Development for Ingestion Exposures

Many of these toxicological studies suggest that the liver or the central nervous system is the target organ of aldrin exposure. In an attempt to forge the link between the observed toxicological consequences of the chemicals in different target species and the known effects of the compounds at the cellular and subcellular levels, Stevenson *et al.* [19] examined the toxicology and metabolism of aldrin/dieldrin. The authors expressed the opinion that many of the toxicological/carcinogenicity studies of aldrin and dieldrin do not conform to current experimental design and documentation. The experimental designs of many of the studies had been altered after the studies had begun, and exact measurements of exposure were often not possible, requiring the use of default food factors in many cases. Notwithstanding the inadequacies of many of the experiments that provided the data on which this Stevenson *et al.* [20] survey was built, there is adequate evidence that aldrin/dieldrin is one of a number of organochlorine compounds that promote spontaneous liver tumor formation in aged mice and induce a number of noncarcinogenic hepatotoxic effects in other research models.

Although the liver appears to be the primary target of aldrin toxicity, such effects are limited to cancer mechanisms or metabolism. Aldrin has been linked to deleteriously affecting the survival of adults [11] and the development of offspring [17]. Survival of adult dogs in the Fitzhugh and Deichmann studies was not a robust endpoint since few animals were used, and deaths were attributable to treatment by the authors.

Ecological and health-related relevance of some hepatic observations may be equivocal if linked to cell proliferation and not to adverse noncancer cellular effects. However, where liver mass data were corroborated with necrosis, fatty degeneration, and other adverse hepatic effects, and where increased exposures led to decreased survival and developmental abnormalities, the hepatic observations reported herein clearly have relevance. This logic provides the basis for inclusion of the studies listed in Table 21.2.

Neurological effects that are either marked (e.g., convulsion) or less obvious (e.g., wasting) are important adverse health effects [9,11]. The Ottolenghi *et al.* [17] study on the developmental effects of aldrin directly addresses an ecologically relevant parameter, but only one dose level was used, thus precluding a dose-response relationship. These data suggest that adverse effects to pups occur at levels higher than those reported to have caused neurological effects in other species [9,11]. Because data from a chronic study are not ideally the most relevant for deriving a TRV, the Fitzhugh *et al.* study on the effects of

Table 21.2 Summary of Relevant Mammalian Data for TRV Derivation

Test Organism	Test Duration	Test Results: NOAEL (mg/kg/day)	LOAEL (mg/kg/day)	Effects Observed at the LOAEL	Study
Dogs (Beagles)	10 months	NA	0.43	Neurological signs, excitation, tremors, etc.	[13]
Rats (Osborne-Mendel)	2 years	NA	0.025	Histopathological lesions in the liver	[14]
		2.5	5.0	Adult mortality	
Dogs (Mongrel)	25 months	0.2	0.5	Histopathological lesions of the liver, kidney, and bone marrow; mortality/convulsions at 2 mg/kg/day and higher	
Rats	Various	0.125–1.25	0.025–7.5	Increases in relative liver weight	[9]
Dogs	ND	0.025	0.075	Increases in relative liver weight	
Rats	14 months	3	4.5	Mortality; increases in relative liver weight (steatosis)	[12]
Rats (Osborne-Mendel)	Lifetime	1.0	1.5	Increases in relative liver weight	[15]
Golden hamster	GD-7	NA	50	Increased mortality and abnormalities in litters	[20]

GD = gestation day; NA = not applicable; ND = not determined.

aldrin in rats [11] is the most pertinent to TRV development among the literature reviewed for this WTA. In that 2-year study, rats exposed to aldrin showed increased mortality with increased doses of aldrin.

Mammalian Inhalation Toxicity

No data were available on the inhalation toxicity of aldrin.

Mammalian Dermal Toxicity

No data were available on the dermal toxicity of aldrin.

Summary of Avian Toxicology

Numerous studies have been conducted to survey the presence of pesticides, polychlorinated biphenyls (PCBs), polycyclic aromatic hydrocarbons (PAHs), dioxins, and dioxin-like residues in the bodily tissues, brains, or eggs of birds. Some of the studies included in the bibliographic database measured the dose-related impacts of organochlorine pesticides such as aldrin and dieldrin on birds in closely controlled experimental settings. An important concept to emerge from the research of Stickel and coworkers (Stickel *et al.* [20], for example) was that it is possible to estimate levels of toxicants or their primary metabolites in the brains of birds indicating (or correlating with) a state of ill health or failure to thrive for a particular species. A related range of mostly lower concentrations of organochlorine residues representing "no adverse impact" tissue levels may also be discernable.

In passerine species, Friend and Franson [21] collated mean sub- and suprathreshold brain concentration data for aldrin, dieldrin, and other organochlorine pesticides. Respective concentrations of less than 8 or greater than 18 ppm wet weight for dieldrin were documented; however, these data contrast with experimental observations in Pheasants given capsules containing 0.5, 1.0, or 1.5 mg aldrin/week for seven weeks [22]. Dieldrin concentrations in the brains of 50% (22 in total) of high-dose recipients that died within 48 hours of the start of the experiment ranged from 2.13 to 4.25 ppm wet weight; equivalent values among survivors were 0.12–0.20 ppm.

Elevated levels of dieldrin in the brain have also been observed in birds involved in a "die-off" in the field [23]. In 1972 and 1974, a number of dead or moribund Snow Geese were found in an area of Texas near rice fields that had been flooded with heavy rains soon after aldrin-treated seeds were planted. Brains of Snow Geese found moribund had an average concentration of 8.2 ppm dieldrin, while brains of a subset of dead birds averaged

14.1 ppm dieldrin. Flickinger cites an earlier monograph by Stickel *et al.* [24] to support a lower limit 4 to 5 ppm for the lethal level of dieldrin in brains of experimental and wild birds; this level agrees with the data presented by Hall *et al.* [22].

Kan [25] reviewed the ability of a number of organochlorine pesticides, including aldrin and dieldrin, to accumulate in birds by comparing the measured amounts of target substances in body fat or egg to the amount of these same substances in feed. When the parameters were expressed as a ratio, aldrin and dieldrin were shown to be among the compounds with high accumulation potential. By injecting fertile chicken eggs with either 0, 10, 20, 30, 40, or 50 ppm technical grade aldrin in corn oil, Kuzan and Prahlad [26] demonstrated that aldrin and dieldrin may be harmful to developing embryos. Hatchability was comparatively unaffected; however within 3 days of hatching, increased mortality occurred in chicks that received the higher concentrations of aldrin.

Studies Relevant for Avian TRV Development for Ingestion Exposures

Field studies have shown that aldrin can cause severe toxicity to exposed avian populations [23]. However, these data cannot be used to derive a TRV because exposure information was unavailable. The availability of suitable experimental data on the toxicity of aldrin to avian species was limited to two studies [22,26]. The study in which Kuzan and Prahlad injected chicken eggs with several concentrations of aldrin [26] showed no effects on development, hatchability, and histology but showed an increase in post-hatch chick mortality. However, this study cannot be used to derive a TRV because the experimental design does not permit a realistic exposure route/level.

Hall *et al.* [22] conducted an oral toxicity study in which pheasants received three levels of encapsulated aldrin on a weekly basis for 6 weeks. Observed effects included diminished growth and survival in pheasants exposed to the higher levels of aldrin. The NOAEL for growth was 0.5 mg aldrin/kg/week, and the corresponding LOAEL was 1.0 mg aldrin/kg/week. Since TRVs are based on daily exposures, the weekly exposures incorporated by this study slightly compromise the use of this data for an avian TRV. However, given the low availability of suitable data and the toxicokinetics of aldrin (rapid conversion to dieldrin with deposition in fat), the weekly exposure levels were divided by seven to estimate the daily exposures. Dividing the NOAEL and LOAEL values by seven yielded daily estimates of 0.07 mg/kg/day and 0.14 mg/kg/day, respectively.

Amphibian Toxicology

Vijay and Jayantha [27] reported aldrin LC_{50} values in *Rana hexadactyla* of 2.6 ppm (24 hours) and 2.4 ppm (48 hours) from waterborne exposures. The authors also reported oral LD_{50} values of 2.2 mg/kg after 24 hours and 2.5 mg/kg after 48 hours.

Reptilian Toxicology

Aldrin/dieldrin was one of a number of organochlorine pesticides surveyed in nonviable Morelet's crocodile eggs collected from two lagoons in northern Belize [28]. Although other residues such as dichlorodiphenyldichloroethelene (DDE) and methoxychlor were found, none of the eggs tested positive for aldrin. In one particular egg, the total concentration of organochlorines was 0.7 ppm. Lagoon sediments and nest material also tested positive for a range of organochlorine pesticides.

RECOMMENDED TOXICITY REFERENCE VALUES

Toxicity Reference Values for Mammals

TRVs for Ingestion Exposures for the Class Mammalia

Decreased survival was chosen as the endpoint on which the aldrin TRV is based because survival has clear and direct implications for wildlife populations. Rat mortality in the Fitzhugh *et al.* [11] study showed a clear dose-response. Other mortality, reported in dogs, was weak given few animals and other extraneous circumstances [9,11]. Other endpoints were ambiguous regarding adverse effects (e.g., hepatic changes). Using a default food factor of 0.05 for rats, a NOAEL of 2.5 mg/kg/day was determined for this study, based on original aldrin levels that had been converted from ppm. The Fitzhugh *et al.* [11] study was suitably designed and executed, and since it evaluated the chronic toxicity of aldrin, the data meet the minimum requirements as stated in Section 2.2 of the Standard Practice (TG 254) [2]. Therefore, no uncertainty factors are required in the derivation of the TRV. The TRV for aldrin was derived using the Benchmark Dose approach; the results are presented in Table 21.3. Although three orders of mammals are

Table 21.3 Selected Ingestion TRVs for the Class Mammalia

TRV	Dose	Confidence Level
LED_{10}	0.1 mg/kg/day	Medium
ED_{10}	0.7 mg/kg/day	Medium

Source: [11].

represented by the studies, a medium confidence rating is applied to the TRVs due to the required estimation of dose in mg/kg/day units and the poor experimental design of many of the studies. Comparison of these values with those of reported adverse effects in other studies and species strongly suggests that the derived TRVs are relevant to and protective of mammalian species.

TRVs for Inhalation and Dermal Contact Exposures for the Class Mammalia

Not data with which to derive these TRVs are available at this time.

Toxicity Reference Values for Birds

TRVs for Ingestion Exposures for the Class Aves

The endpoint chosen as the basis of the aldrin TRV for avian species was decreased growth seen in pheasants fed encapsulated aldrin [22]. A lower growth rate can have ecologically relevant effects on birds because fledglings with greater mass have a higher probability of recruiting into the breeding population [29]. Since only one study [22] was located that characterized aldrin oral toxicity in birds, and that study was of subchronic duration, the TRV was derived using the approximation approach (Table 21.4). An uncertainty factor of 10 was used to obtain the NOAEL-based TRV from the subchronic NOAEL (0.07 mg/kg/day), and an uncertainty factor of 4 was used to derive the LOAEL-based TRV from the subchronic LOAEL (0.14 mg/kg/day). A low confidence rating was assigned to this avian TRV because only one study was found, and the design of its experiments included weekly, not daily, administration of aldrin.

TRVs for Inhalation and Dermal Contact Exposures for the Class Aves

No data with which to derive these TRVs are available at this time.

Table 21.4 Selected Ingestion TRVs for the Class Aves

TRV	Dose	Confidence Level
NOAEL-based	0.007 mg/kg/day	Low
LOAEL-based	0.035 mg/kg/day	Low

Source: [22].

Toxicity Reference Values for Amphibians

TRVs for Ingestion Exposures for the Class Amphibia

Only very limited data on the effects of aldrin on amphibian species are available. Vijay and Jayantha [27] reported oral LD_{50} values for aldrin in *R. hexadactyla* of 2.2 mg/kg after 24 hours and 2.5 mg/kg after 48 hours. Since these data does not meet the minimum requirements outlined in the Standard Practice, Section 2.2 [2], uncertainty factors must be used for TRV derivation from the available acute toxicity data. The TRV is based on the 24-hour LD_{50} value with an uncertainty factor of 100 for the NOAEL-based TRV and 20 for the LOAEL-based TRV. The amphibian TRV received a low confidence rating due to the lack of data and of representative species (Table 21.5).

TRVs for Inhalation and Dermal Contact Exposures for the Class Amphibia

No data with which to derive these TRVs are available at this time.

Toxicity Reference Values for Reptiles

TRVs for Ingestion, Inhalation, and Dermal Contact Exposures for the Class Reptilia

No data with which to derive these TRVs are available at this time.

IMPORTANT RESEARCH NEEDS

The limited availability of data on the toxicity of aldrin to wildlife species precludes the development of a high-confidence TRV. Hence, more studies on the toxicity of aldrin to wildlife species are needed. Particularly warranted are long-term (subchronic) toxicity studies focusing on nonrodent species and exploring neurological endpoints. These studies should build on previous findings as well as refine consistent dose-response relationships.

Table 21.5 Selected Ingestion TRVs for the Class Amphibia

TRV	Dose	Confidence Level
NOAEL-based	0.02 mg/kg/day	Low
LOAEL-based	0.1 mg/kg/day	Low

Source: [27].

Reproductive studies investigating effects on F1 and F2 generations are also needed. Although a number of studies on the effects of alcrin on mammalian species have been undertaken, many of these are old and of poor quality. Further experimentation, with close attention to experimental design, would greatly increase the confidence levels of aldrin TRVs. Controlled laboratory data are also needed for reptiles and bird investigating reproduction.

REFERENCES

[1] Hazardous Substances Data Bank (HSDB). Aldrin [Internet]. Bethesda (MD): National Library of Medicine (US) [cited 2001]. Available from: http://toxnet.nlm.nih.gov/cgi-bin/sis/search2/f?./temp/~1EtSKz:1

[2] USACHPPM. Standard practice for wildlife toxicity reference values. Aberdeen Proving Ground (MD): U.S. Center for Health Promotion and Preventive Medicine; 2000, Technical Guide 254.

[3] ATSDR. Toxicological profile for aldrin/dieldrin. Atlanta (GA): U.S. Department of Health and Human Services, Public Health Service; 2002.

[4] Gaines TB. Acute toxicity of pesticides. Toxicol Appl Pharmacol 1969;14:515–34.

[5] Hodge HC, Boyce AM, Deichmann WB, Kraybill HF. Toxicology and no-effects levels of aldrin and dieldrin. Toxicol Appl Pharmacol 1967;10:613–75.

[6] Ali SS, Shakoori AR. Toxicology of aldrin in rats. Punjab Univ J Zool 1990;5:1–56.

[7] Hutson DH. Comparative metabolism of dieldrin in the rat (CFE) and in two strains of mouse (CF1 and LACG). Food Cosmet Toxicol 1976;14:577–91.

[8] Pallade S, Popovici C, Rotaru G, Gabrieliescu E, Dorobantu M. Quelques donnees concernant l'action de l'aldrin sur le foie. Med Lav 1968;59:346–56.

[9] Deichmann WB, Keplinger M, Dressler I, Sala F. Retention of dieldrin and DDT in the tissues of dogs fed aldrin and DDT individually and as a mixture. Toxicol Appl Pharmacol 1969;14:205–13.

[10] U.S. EPA. Integrated risk information system [Internet]. Cincinnati (OH): Office of Health and Environmental Assessment, National Center for Environmental Assessment (US) [cited 2014]. Available from: http://www.epa.gov/iris/subst/0130.htm.

[11] Fitzhugh OG, Nelson AA, Quaife ML. Chronic oral toxicity of aldrin and dieldrin in rats and dogs. Food Cosmet Toxicol 1964;2:551–62.

[12] Deichmann WB, MacDonald WE, Blum E, Bevilacqua M, Radomski J, Keplinger M, et al. Tumorigenicity of aldrin, dieldrin and endrin in the albino rat. Ind Med Surg 1970;39:426–34.

[13] Reuber MD. Hepatic vein thrombosis in mice ingesting chlorinated hydrocarbons. Arch Toxicol 1977;38:163–8.

[14] Davis KJ, Fitzhugh OG. Tumorigenic potential of aldrin and dieldrin for mice. Toxicol Appl Pharmacol 1962;4:187–9.

[15] National Cancer Institute (NCI) National Institutes of Health. Bioassays of aldrin and dieldrin for possible carcinogenicity. Bethesda (MD): National Cancer Institute; 1978.

[16] Deichmann WB, MacDonald WE, Lu FC. Effects of chronic aldrin feeding in two strains of female rats and a discussion of the risks of carcinogens in man. Toxicology and Occupational Medicine 1979;407–13.

[17] Ottolenghi AD, Haseman JK, Suggs F. Teratogenic effects of aldrin, dieldrin and endrin in hamsters and mice. Teratology 1974;9:11–6.

[18] Gellert RJ, Wilson C. Reproductive function in rats exposed prenatally to pesticides and polychlorinated biphenyls (PCB). Environ Res 1979;18:437–43.

[19] Stevenson DE, Walborg Jr EF, North DW, Sielken Jr RL, Ross CE, Wright AS, et al. Monograph: reassessment of human cancer risk of aldrin/dieldrin. Toxicol Lett 1999;109:123–86.
[20] Stickel LF, Stickel WH, Dyrland RA, Hughes DL. Oxychlordane, HCS-3260 and nonochlor in birds: Lethal residues and loss rates. J Toxicol Environ Health 1983;12:611–22.
[21] Friend M, Franson JC. Chlorinated hydrocarbon insecticides. In: Friend M, Franson JC, editors. Field manual of wildlife diseases: general field procedures and diseases of birds. Madison (WI): U.S. Geological Survey; 1999.
[22] Hall JE, Greichus YA, Severson KE. Effects of aldrin on young pen-reared pheasants. J Wildl Manage 1971;35:429–34.
[23] Flickinger EL. Effects of aldrin exposure on snow geese in Texas rice fields. J Wildl Manage 1979;43:94–101.
[24] Stickel WH, Stickel LF, Spann JW. Tissue residues of dieldrin in relation to mortality of birds and mammals. In: Proceedings of the first Rochester conference on Toxicity; 1969, Rochester, New York.
[25] Kan CA. Accumulation of organochlorine pesticides in poultry: a review. J Agric Food Chem 1978;26:1051–5.
[26] Kuzan FB, Prahlad KV. The effects of 1,2,3,4,10,10-hexachloro-1,4,4a,5,8,8a-hexahydro endo, exo-5,8-dimethionaphthalene (aldrin) and sodium ethylenebisdithiocarbamate (Nabam) on the chick. Poult Sci 1975;54:1054–64.
[27] Vijay JK, Jayantha RK. Toxic effects of aldrin on histopathology of intestine in the frog *Rana hexadactyla*. J Ecobiol 1990;2:161–5.
[28] Wu TH, Rainwater TR, Platt SG, McMurry ST, Anderson TA. Organochlorine contaminants in Morelet's crocodile (*Crocodylus moreletii*) eggs from Belize. Chemosphere 2000;40:671–8.
[29] Both C, Visser ME, Verboven N. Density-dependent recruitment rates in great tits: the importance of being heavier. Proc R Soc B 1999;266:465–9.

CHAPTER 22

Wildlife Toxicity Assessment for Chlordane

Emily N. Reinke, Adam T. Deck

Contents

INTRODUCTION

Chlordane (CAS No. 12789-03-6) is a mixture of structurally related cyclodienes (primarily octachloro-tetrahydro-methanoindanes) that was widely used as a broad-spectrum organochlorine-pesticide from 1947 to 1978 [1]. Subsequently, the use of chlordane was severely curtailed because of increasing concerns about its potential carcinogenicity and reports linking it and other organochlorine compounds to the onset of physiological deficits and reproductive failure in birds [2]. Thus, from 1978 on, chlordane's only approved use was as an underground termiticide, and even this application was cancelled after 1988 [3]. Since chlordane bioaccumulates, it is still readily detected in environmental media. Its heavy use in the 1950s and 1960s and its environmental persistence are evidenced by the estimate that 25 to 50% of the 70,000 tons of chlordane produced since 1946 remains unaltered in the environment to this day [3].

The CAS number given above refers to "technical" chlordane, a formulation manufactured by the Velsicol Corporation that contains *cis*-chlordane (α-chlordane) and *trans*-chlordane (γ-chlordane) in varying proportions,

Wildlife Toxicity Assessments for Chemicals of Military Concern
http://dx.doi.org/10.1016/B978-0-12-800020-5.00022-3

along with other ingredients. The individual isomers (CAS No. 5103-74-2 for *trans*-chlordane and 5103-71-9 for *cis*-chlordane) are also available commercially in highly purified preparations (>99% by weight). This Wildlife Toxicity Assessment summarizes current knowledge of the likely harmful impacts of chlordane on wildlife and reports Toxicity Reference Values (TRVs) for chlordane. The intended use of TRVs is to serve as protective exposure standards for screening-level ecological risk assessments for wildlife inhabiting contaminated sites. The protocol for the development of TRVs for this assessment is provided in the U.S. Army Center for Health Promotion and Preventive Medicine (USACHPPM) Technical Guide (TG) 254, *Standard Practice for Wildlife Toxicity Reference Values* [4].

TOXICITY PROFILE

Environmental Fate and Transport

As discussed in the Agency for Toxic Substances and Disease Registry (ATSDR) [3], many lines of evidence link the presence of chlordane in air today to the compound's former use as a pesticide. Its persistence is further supported by the fact that it has been detected in indoor air 15 years after a house had been treated with it for termites. Air monitoring data suggest that most of the airborne chlordane exists in the vapor phase.

Chlordane is present in the nation's water bodies as run-off from urban and agricultural uses, although subsequent partitioning to sediment is favored. Alternatively, vaporization can occur depending on temperature, water turbulence, and wind conditions, as well as the extent and composition of suspended matter. As might be expected for an agricultural pesticide and termiticide, most of the chlordane detected in the environment exists in soil in which the compound readily absorbs to organic matter and only volatilizes slowly over time.

Degradation of the compound in air occurs because of photolysis and oxidation, processes that apparently favor the *trans* isomer. In soils, chlordane persists for long periods in many different locations; this is perhaps its most striking feature. Mass-balance studies have accounted for the greater part of the originally applied compound to the soil 10 or more years after application, often with little evidence of chemical or biochemical transformations. However, loss of chlordane from light sandy soils is more likely to occur than from soil with high organic content [3]. Biological degradation of chlordane has been demonstrated in some microbial species, including the fungi *Aspergillus niger* and *Phanerochaete chrysosporium*. Nonetheless, overall rates of

Table 22.1 Summary of Physical-Chemical Properties of Chlordane

CAS No.	12789-03-6 (technical), 5103-71-9 (cis-), 5013-74-2 (trans)
Molecular weight	409.76
Color	None
Physical state	Viscous liquid
Melting point	106–107 °C (cis-), 104–105 °C (trans-)
Boiling point	175 °C (at 2 mm Hg)
Odor	None
Solubility in water	0.056–1.85 mg/L at 25 °C: miscible with Hydrocarbon solvents
Partition coefficients:	
Log K_{ow}	5.54
Log K_{oc}	3.49 – 4.64
Vapor pressure at 25 °C	(for cis-) 2.2×10^{-5} mm Hg (supercooled liquid), 3.0 H 10^{-6} (crystal); (for trans-) 2.9 H 10^{-5} mm Hg (supercooled liquid), 3.9 H 10^{-6} (crystal)
Henry's Law constant at 25 °C	4.86×10^{-5} atm.m^3/mole
Conversion factors	1 ppm = 16.76 mg/m^3 1 mg/m^3 = 0.06 ppm

Sources: U.S. EPA [1], ATSDR [3].

degradation appear to be very low. Table 22.1 summarizes the physical data relating to the environmental fate and transport of chlordane.

Summary of Mammalian Toxicity

Mammalian Oral Toxicity - Acute

A number of researchers have published single oral median lethal dose (LD_{50}) values for chlordane. For example, Gaines [5] surveyed the acute lethality of 98 pesticides and two metabolites of dichlorodiphenyltrichloroethane (DDT) in Sherman rats, deriving values for chlordane (grade and vehicle unstated) of 335 and 430 mg/kg for males and females, respectively. These values are consistent with that of 311 mg/kg in male Wistar rats receiving technical-grade chlordane in cottonseed oil [6]. However, Wistar rats receiving protein-deficient diets had significantly lower LD_{50} values in the latter experiment.

A meeting abstract by Harbison [7] addressing the acute lethality of several pesticides including chlordane in adult versus newborn Sprague Dawley rats reported respective LD_{50} values of 344 and 1121 mg/kg for the compound. Truhaut *et al.* [8] calculated an oral LD_{50} value of 350 mg/kg for Wistar rats (sex unstated) receiving a single oral dose of technical chlordane in olive oil. Golden hamsters (with an LD_{50} of 1720 mg/kg) appeared to be far less susceptible to the acute toxic effects of the compound. Berman *et al.* [9] used the simplified up-down procedure to derive an acute oral LD_{50} of 520 mg/kg for chlordane in female F344 rats.

In a number of studies, laboratory animals that had received acute exposures to chlordane were monitored for health effects other than lethality. For example, Den Tonkelaar and Van Esch [10] exposed male Wistar and SPF rats to 2, 5, 10, 20, 50, or 200 parts per million (ppm) chlordane in the diet for 2 weeks, then measured the specific activities of aniline hydroxylase, aminoantipyrine demethylase, and hexobarbital oxidase in hepatic 10,000 × g supernatants. Aminoantipyrine demethylase appeared to be the most sensitive index of chlordane toxicity, with an apparent dietary LOAEL of 10 ppm and a corresponding NOAEL of 5 ppm for the compound. These dietary levels equate to doses of 0.5 and 0.25 mg/kg/day, respectively, if a default food consumption factor of 0.05 is used. Ogata and Izushi [11] exposed male Donryu rats to 100 mg/kg "chlordane 2%" in olive oil via gavage once a day for 4 days to assess compound-related fluctuations in clinical chemistry parameters and changes in liver weight and histopathology. As determined 24 hours after the final dose, a number of clinical chemistry parameters were altered in the chlordane-receiving group compared to the control group, including elevations in triglycerides, cholesterol, and gamma-glutamyl transferase activity, as well as a reduction in serum glucose levels. Absolute and relative liver weights were increased due to chlordane, and there was evidence of altered histopathology in the livers of treated rats. The principal metabolites of chlordane detected in the blood were oxychlordane, heptachlor epoxide, and *cis*- and *trans*-nonachlor.

Chatterjee *et al.* [12] administered a single dose of 15 mg/kg chlordane (grade unstated) to male Wistar rats and observed greater than 40% mortality among the subjects during a 15-day follow-up period. This response was accompanied by increases in liver and kidney weights, histopathological lesions in the liver and kidney, and overall growth retardation among the survivors. However, the response was mitigated in similarly treated animals who received L-ascorbic acid as a dietary supplement. Moser *et al.* [13] gavaged female F344 rats with single or multiple doses of pesticides including chlordane at concentrations of up to 291 mg/kg; (e.g., 40% chlordane plus 60% other cyclodiene compounds). Survivors were assessed in a battery of neurobehavioral tests that scored such responses as excitability, neuromuscular coordination, motor activity, and grip strength on a semi-quantitative scale. The authors noted a cumulative toxicity that was evident when results from the single-dose studies were compared to those that incorporated a 14-day dosing regimen. The authors suggested NOAELs of 52 and 5 mg/kg for single versus multiple dose effects of chlordane on the neurobehavioral responses of F344 rats.

Barnett and Dorough [14] subjected both sexes of Sprague Dawley rats to various acute, subacute, and subchronic oral dosing regimens of ^{14}C-labeled HCS-3260, a 98% pure formulation of chlordane containing a 3:1 cis:trans-isomer ratio. Fecal elimination of the radioactivity was emphasized, although a number of cyclodiene products of chlordane metabolism were tentatively identified in various tissue extracts. For example, adipose tissue contained measurable amounts of oxychlordane, concentrations of which declined during the posttreatment recovery period.

Berman *et al.* [9] included chlordane in a screen of 10 chemicals surveyed for toxicological consequences in a 14-day experiment in which eight female F344 rats/group were dosed. Initially, doses were 0, 5, 16, 52, and 156 mg/kg, although data from the two highest groups were excluded from the reported findings because of early deaths. Survivors were bled to obtain plasma for clinical chemistry analyses; samples of the liver, kidney, spleen, thymus, and adrenals were examined histopathologically. Changes in relative liver weight and hepatic centrilobular cytomegaly, as well as reductions in such liver-related clinical chemistry parameters as plasma alkaline phosphatase activity, were observed. Corresponding NOAELs and LOAELs for the responses to chlordane were 5 and 16 mg/kg/day for liver weight increase, 0 and 5 mg/kg/day for histopathological change, and 0 and 5 mg/kg/day for plasma alkaline phosphatase reduction.

Another study having mechanistic implications for the hepatic impact of chlordane in mice was that of Whysner *et al.* [15], who used the ^{32}P-DNA postlabeling technique to search for DNA adduct formation. B6C3F1 mice received either a single oral gavage dose or a 2-week dietary exposure to 50- or 200-ppm chlordane, respectively. Benzidine and 2-acetylaminofluorene were used in parallel exposures as positive controls to demonstrate the effectiveness of the postlabeling technique, but only negative results were obtained for chlordane in the mice.

Mammalian Toxicity: Subchronic

Five Osborne-Mendel rats/sex/group were fed casein-sufficient or -deficient diets containing 0, 125, 250, or 500 ppm chlordane (grade unstated) for

28 days [16]. Clinical signs were monitored daily, and food consumption and body weights were monitored weekly. Excised livers were weighed and compared to the body weights at termination. Some fluctuations of tissue and serum enzyme activities were noted according to chlordane concentration and dietary status. However, the most marked effects of the pesticide on the rats were the dramatic increases in relative liver weights compared to controls. These were statistically significant at all concentrations of chlordane, irrespective of protein sufficiency/deficiency. Based on the body weight and food consumption data at the lowest dietary chlordane concentration (125 ppm), a subacute LOAEL within the range of 8 to 14 mg/kg/day is suggested for the compound in Osborne-Mendel rats.

Tryphonas *et al.* [17] investigated the effects on the immune system of Sprague Dawley rats following exposure to technical chlordane and to its technical-grade components *cis*-nonachlor or *trans*-nonachlor. Seven rats/sex/dose were exposed to 0, 0.25, 2.5, or 25 mg/kg/day for 28 days by oral gavage, and alterations in various immunologic endpoints were measured. Chlordane and the two isomers all had various effects on the immune system, with the *cis* and *trans* forms being the more immunotoxic, increasing immunoglobulin (IgG) levels in both male and female rats. The *trans* form also increased the white blood cell, B-lymphocyte, natural killer cell, T-suppressor/cytotoxic lymphocyte, and double positive T lymphocyte counts in females. Splenocytes from female rats exposed to both the *cis* and *trans* forms were less resistant to challenge by *S. typhimurium* and *L. monocytogenes*, respectively.

Al-Omar *et al.* [18] conducted a subchronic toxicity study on chlordane (with and without lead oxide) to investigate its effects on the testicular function and architecture of BALB/c mice. Twelve males/group received 0, 75, or 275 mg/kg commercial chlordane in corn oil by gavage for up to 35 days. Other groups received lead oxide in addition to chlordane or lead oxide alone. Body weights were recorded weekly and three mice/group were euthanized at 2, 3, 4, and 5 weeks for histopathological examination of the testes. A number of structural and functional deficits were observed in mice exposed to chlordane only, including a loss of testicular weight, damaged tubules, reductions in seminiferous tubule diameter, and reductions in the number of spermatogonia, primary and secondary spermatocytes, spermatids, and Sertoli cells. Lead enhanced the effect of chlordane but was unable to induce the same effects on its own. Given the statistically significant impact of chlordane on these parameters at the lowest concentration, 75 mg/kg/day chlordane represents the LOAEL for chlordane's effects on the male mouse reproductive system.

Mahon and Oloffs [19] described an experimental protocol whose primary aim was to determine the effect of chlordane within the carbon tetrachloride/rat liver cirrhosis model. The design featured a 10-week exposure to male Long-Evans BLU-LE rats (phase 1) followed by 10 weeks of recovery (phase 2). The authors identified the key group as those rats receiving carbon tetrachloride and chlordane during phase 1 followed by chlordane alone during phase 2. However, the group of animals having consumed 0.05 mg/kg/day of both α- and(γ-chlordane daily on a piece of carrot during both exposure phases (20 weeks in all) also provided relevant data. Some minor changes in relative liver weight were observed in those members of this group who were euthanized at 10 weeks. However, no such changes were observed in animals receiving chlordane for the full 20 weeks, and no signs of histopathological lesions were apparent at either time point. These data suggest 0.05 mg/kg/day as a subchronic NOAEL for chlordane in Long-Evans rats.

A subchronic dosing regimen has also been used to investigate the possible effects of chlordane on ATPase activities in the brain of male Sprague Dawley rats. As described in a meeting abstract, Drummond *et al.* [20] mixed 0, 25, 50, or 100 ppm chlordane with either iron-deficient or -sufficient lab chow, which was then made available to nine animals/group for 12 weeks. Three animals/group were euthanized at 4, 8, or 12 weeks after exposure; "P2" fractions of the brain were prepared and assayed for specific Na^+, K^+, oligomycin-sensitive, and oligomycin-insensitive Mg^{++} ATPase activities. The Na^+, K^+, and oligomycin-sensitive Mg^{++} ATPase activities were dose-dependently reduced at all the time intervals post-chlordane treatment, whereas oligomycin insensitive Mg^{++} ATPase activity was unaffected.

Oral exposure of Sprague Dawley rats to *trans*-nonachlor for 90 days found that the compound was not overtly toxic to the animals [21]. Rats exposed to 0, 5, 13, or 50 ppm *trans*-nonachlor exhibited an increase in oxidative stress markers and an increase in liver microsomal enzymes.

Other animal models have been used to demonstrate the toxicological effects of chlordane, including three Holstein cows that received the compound in gelatin capsules daily for 60 days [22]. The formulation employed was highly purified HCS 3260 containing 74% ∀-chlordane and 24% (-chlordane. The dosage was considered equivalent to 50 pounds (lbs) of feed/day containing 1, 10, or 100 ppm chlordane. Given an approximate weight of 1,400 lbs for each animal, these concentrations were equivalent to approximate doses of 0.036, 0.36, and 3.6 mg/kg/day. The authors did not address the issue of a subthreshold point of departure for overt toxicity but did demonstrate the appearance of chlordane and its metabolites in milk and in fat biopsy tissue. For example, the major product of chlordane in milk was oxychlordane. However, levels of this and other metabolites declined dramatically when exposure was terminated.

Mammalian Toxicity: Chronic

In a chronic study by Ingle [23], Osborne-Mendel rats (20/dose group) were fed laboratory chow spiked with 0, 5, 10, 30, 150, or 300 ppm chlordane for 2 years. Clinical signs, including tremors, hyperexcitability, convulsions, prostration, and other effects, were evident in those groups receiving 30, 150, or 300 ppm chlordane-dosed chow. Animals in the two highest dose groups also displayed reduced food consumption, decreased growth and survival, fetal toxicity, and increases in liver and kidney weights. Retardation in growth rate occurred at high doses (150 and 300 ppm). There was also an increase in the mortality rate at these doses, such that rats receiving 150 and 300 ppm chlordane showed significantly higher mortality rates than the lower dose groups. Newborn rats from the 150 and 300 ppm dose groups had definite symptoms of fetal toxicity, including death, when allowed to remain with their lactating mothers. It was also noted that the milk from chlordane-treated mothers contained a sufficient amount of chlordane to elicit effects in suckling newborns. These effects included death and impaired growth and development. Newborn rats born to chlordane-treated females, which were then removed from their mothers and placed with control females for lactation, showed no signs of toxicity. Histopathological impacts of chlordane were also evident in the liver and kidney, with cellular hypertrophy and bile duct proliferation among the hepatic effects. Some liver damage was evident in those exposed to 30 ppm chlordane, while those receiving 10 ppm were essentially unaffected. The authors considered the doses to be 0, 0.35, 0.65, 1.5, 8.0, and 17 mg/kg/day, which would justify designating 0.65 mg/kg/day as a NOAEL and 1.5 mg/kg/day as a LOAEL for the toxic effects of the compound.

A full-scale bioassay of chlordane for possible carcinogenicity has been published as a National Cancer Institute (NCI) report [24] and is in the open literature [25]. Fifty Osborne-Mendel rats and B6C3F1 mice/sex/group were exposed to the compound in laboratory chow for 80 weeks, and then observed for an additional 29 weeks prior to sacrifice. The researchers used a formulation of technical chlordane that contained 71.7% α- and 23.1%

γ- isomers and changed the concentrations of the compound in the chow at various points during the course of the experiment. These adjustments resulted in time-weighted average (TWA) chlordane concentrations in chow of 0, 203.5, and 407 ppm for the male rats; 0, 120.8, and 241.5 ppm for the female rats; 0, 29.9, and 56.2 ppm for the male mice; and 0, 30.1, and 63.8 ppm for the female mice. In common with a number of other NCI-sponsored bioassays of the period, the experimental design incorporated a comparatively small number of concurrent controls (10 rats/sex and 20 mice/sex) but obtained relevant data from pooled controls involved in other NCI-sponsored studies. All animals were observed twice daily for clinical signs, weighed at regular intervals, and palpated for masses at each weighing. Premature decedents and the animals that were euthanized at term were necropsied, and a full suite of organ and tissue samples was processed for histopathological examination. The U.S. EPA's IRIS summary of this study gave the average daily doses as 10.2 and 20.4 mg/kg/day in male rats; 6.0 and 12 mg/kg/day in female rats; 4.3 and 8 mg/kg/day in male mice; and 4.3 and 9.1 mg/kg/day in female mice [1]. These values were apparently derived using food consumption factors of 0.05 in rats and approximately 0.15 in mice. There were few signs of compound-related histopathological lesions among the rats. However, both sexes of high-dose rats displayed reduced body weight gain of 10% or greater compared to controls, suggesting a NOAEL for this effect of 6–10.2 mg/kg/day. In contrast to the results in rats, gains in body weight were not reduced in chlordane-treated mice. However, histopathological lesions were noted in the liver in which a dose-dependent increase in hepatocellular carcinomas was reported (2/18, 16/48, and 43/49 in males; and 0/19, 2/47, and 34/49 in females for the control, low-dose, and high-dose groups, respectively).

Velsicol Corporation, the manufacturer of chlordane, has sponsored a number of toxicity/carcinogenicity studies. For example, Khasawinah and Grutsch [26] exposed 80 ICR SPF mice/sex/group to 0, 1, 5, or 12.5 ppm technical chlordane in feed for 104 weeks and reported data on the resulting clinical signs, body weight changes, and food and water consumption. The authors also looked at hematological and clinical chemistry measurements and urinalyses, organ weights at term, gross pathology, and histopathology. No clinical signs were attributable to chlordane, and no dose-related body weight changes were observed. There were some effects on hematological and clinical chemistry parameters, but these were not consistent and occurred only in the high doses. Overwhelmingly, the primary target organ appeared to be the liver, in which absolute and relative increases in organ weights were evident. In addition, both neoplastic and nonneoplastic changes were observed in the liver of exposed animals, the former including a dose-dependent increase in hepatocellular adenomas and hepatic hemangiomas in male subjects, the latter including atrophy, necrosis, and fatty degeneration, among others. While the neoplasm data do not lend themselves to NOAEL/LOAEL designation, and while, in general, tumorigenic endpoints are inapplicable to TRVs for wildlife, the non-neoplastic hepatic responses to chlordane in this animal model suggest a NOAEL of 1 ppm. This is equivalent to a daily dose of 0.15 mg/kg/day [1].

Barrass *et al.* [27] examined the role of cellular proliferation and liver hyperplasia in the etiology of hepatic tumor development using a complex protocol in which a 2-year hepatocarcinogenic assay was paralleled by a 6-month study that examined serially liver histopathology and thyroid follicular cell proliferation. In the first part of the experiment, 50 ppm dietary levels of chlordane were made available to 100 male C57BL/10J mice, inducing 23/79 combined hepatocellular adenomas and carcinomas as compared to 10/400 in historical controls. Nontumorigenic responses included centrilobular hypertrophy and generalized liver enlargement. The second part of the study featured the administration of chlordane in the diet as before but for a 6-month duration in which groups of five male mice were euthanized on days 2, 3, 4, 5, 8, 15, 29, 99, and 190 after the start of dosing. Replicating cells were labeled by infusing 5-bromo-2-deoxyuridine (BrdU) at 15 milligrams per liter (mg/mL) via osmotic mini-pumps implanted subcutaneously 3 days before necropsy, on days 4, 5, 8, 18, 29, 190, and 247. Sections of liver, thyroid, and duodenum (as a positive control) were examined for signs of cellular proliferation using anti-BrdU immunostaining. In the liver, the peak hepatocyte labeling index (LI) was observed after 8 days of chlordane treatment with subsequent time-related decreases toward control levels. Treatment-then-withdrawal groups had LIs similar to those of concurrent controls. These data represent a direct demonstration of the ability of liver from C57BL/10J mice to proliferate when challenged with a non-genotoxic carcinogen such as chlordane. Individual values for discussed studies are provided in Figure 21.1.

A study involving sledge dogs (*Canis familiaris*) in Greenland exposed female breeding dogs to organohalogen-contaminated Mink whale blubber for 90 weeks, to include gestation. Additionally, their offspring were also exposed to organohalogen-contaminated Mink whale blubber for 51 weeks following weaning [28]. All exposed animals had an increase in liver lesions.

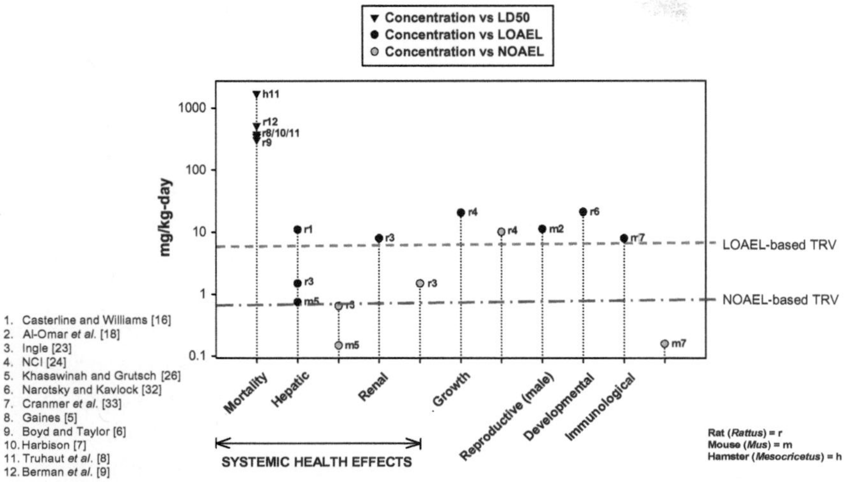

Figure 22.1 Chlordane: Health Effects to Mammals. Letters by symbols represent test species, and numbers represent study.

Mammalian Oral Toxicity: Other

A number of experimental studies have focused on prenatal exposure of experimental animals to chlordane, including [1] reproductive/developmental and teratogenicity studies in which compound-related impacts to reproductive performance and the incidence of fetal abnormalities were examined, [2] studies on endocrine dysfunction, and [3] experiments assessing chlordane's ability to alter the immune competence of offspring during development. In the first category, a review by Deichmann [29] describes a six-generation reproductive study in Swiss mice, reporting a NOAEL of 25 ppm (concentration of chlordane in feed equivalent to a dose of 3.25 mg/kg/day) for effects of the compound on fertility and viability.

More recently, Chernoff and Kavlock [30] included chlordane in a survey of the teratogenic potential of several compounds in CD-1 mice. Pregnant females were administered 50 mg/kg by gavage in corn oil on gestation days (GD) 8–12, a treatment that induced few if any reproductive or developmental effects in either the dams or their progeny.

Various numbers of pregnant female rats (strain unstated) were gavaged during GD 7–17 with either 0, 20, 40, or 80 mg/kg chlordane, the highest dose of which induced deficits in maternal body weight gain and 50% fatalities among the dams [31]. However, when fetuses were examined on GD

20, no statistically significant reproductive, developmental, or teratogenic impacts of chlordane were observed, thereby suggesting 80 mg/kg/day as a NOAEL for these effects.

Narotsky and Kavlock [32] included chlordane in a developmental toxicity screen in which the compound was administered by gavage at 0, 21, or 28 mg/kg/day to pregnant F344 rats during GD 6-19. The dams were allowed to deliver their progeny that were then examined on postpartum days (PD) 1, 3, and 6. Litter weights were monitored on PD 1 and 6, and the number of implants was counted to assess prenatal loss. In contrast to earlier reproductive and developmental experiments on chlordane, some developmental impacts of the compound were evident in the offspring, including dramatic reductions in postnatal viability and pup weight in the absence of any prenatal effects. A LOAEL of 21 mg/kg/day can be based on postnatal effects on pup viability. A summary of the health effects of chlordane to mammals is provided in Figure 22.1.

The ability of chlordane to affect the endocrine status of experimental animals was addressed in an experiment in which pregnant F2 hybrid mice received 0.16 or 8 mg/kg analytical-grade chlordane throughout gestation [33]. At birth, the pups were weighed, examined for viability, gross defects, and gender and were subsequently assessed for endocrine status (on PD 101). This evaluation included the measurement of corticosterone in plasma, *in vitro* assessments of corticosterone production in excised adrenals, and the capacity of excised liver pieces to metabolize corticosterone. Fifty-five percent of offspring born to high-dose dams died; thus, the findings of endocrine status are essentially confined to the low-dose group. Nonetheless, a complex pattern of results emerged in which the corticosterone concentration in the plasma of low-dose males was elevated although *in vitro* adrenal corticosterone production was unaffected. Total hepatic metabolism was comparable to control values.

Perturbation of steroid hormone levels due to chlordane was also examined by Cassidy *et al.* [34], who exposed pregnant Sprague Dawley rats to 0, 100, 500, or 5,000 micrograms per kilogram (μg/kg) chlordane between GD 4 and lactation day (LD) 21. Pups were similarly treated up to PD 80. At the completion of this exposure regimen, pups were subjected to a number of behavioral tests, including their ability to find their way out of a maze, their startle response, and the mating responses of males. Serum levels of testosterone and of chlordane metabolites such as heptachlor, heptachlor epoxide, and oxychlordane were measured at PD 85, the latter parameters displaying dose-related increases with treatment. In general, the chlordane-receiving offspring performed better in the behavioral tests, including the mating trials, than did the controls. These observations and the fluctuation of testosterone levels according to treatment caused the authors to speculate that cyclodienes such as chlordane may "masculinize" sexually dimorphic functions and behaviors by either mimicking steroid hormone activities or altering their levels.

A study reported by Spyker-Cranmer *et al.* [35] is typical of several that have sought to show an altered immune competency of offspring exposed to chlordane *in utero*. Pregnant female BALB/c mice were treated throughout gestation at 0, 0.16, or 8 mg/kg to enable contact mediated immune (CMI) responses to oxazolone to be monitored among the offspring. This was achieved by measuring the extent of ear-swelling 3 days after challenge with the latter compound. The authors concluded that chlordane appeared to reduce the extent of the immune response in the progeny of dams receiving the highest dose. The authors also carried out the Jerne hemolytic plaque-forming cell (PFC) assay in which 120-day-old subjects were immunized with sheep erythrocytes. Freshly isolated spleen cells were then challenged with sheep erythrocytes, and the released antibody (IgM) was detected by adding complement, which caused zones of lysis in the plated background of sheep erythrocytes surrounding the spleen cells. Once again, positive results (an increase in plaque formation compared to controls) were seen at the higher exposure level, suggesting a NOAEL of 0.16 mg/kg/day. It was thought that chlordane might reduce the activity or number of T-suppressor or T-effector cells. However, these results contrast with those of Johnson *et al.* [36], who measured some of the same responses in spleen cells isolated from adult female B6C3F1 mice exposed to 0.1, 1.0, 4.0, or 8.0 mg/kg/day of 99% chlordane by gavage in corn oil for 14 days. The authors carried out the Jerne plaque-forming assay and assessed the proliferation of spleen cells in response to concanavalin A, phytohemagglutinin, and other mitogens. In general, they concluded that chlordane might not be a serious threat to immunocompetence, at least because of its exposure to adults.

The contrast between the immunotoxic effects of chlordane in prenatally exposed subjects versus those exposed as adults was explored in a study by Barnett *et al.* [37]. These researchers exposed pregnant BALB/c mice to chlordane at 0, 4, or 8 mg/kg for a total of 18 days during pregnancy; the pups were born and nursed in the normal way. A group of nonpregnant adult female mice was also exposed to chlordane for a similar time. Subjects were monitored for bone marrow hematopoietic activity at 14, 100, and

200 days postexposure, by means of an *in vitro* granulocyte-macrophage colony forming unit (GM-CFU) assay, and an *in vivo* colony forming unit assay in the spleen. The latter approach measured the immunotoxic effects of chlordane on more primitive stem cell precursors. Animals treated prenatally with chlordane showed significant reductions in their bone marrow GM-CFU at 14 days of age and when they had reached adulthood (100 and 200 days). Prenatal treatment with chlordane also reduced the number of spleen colony-forming units in exposed subjects. However, direct treatment of 80-day-old adult mice with chlordane had no effect on these parameters, consistent with the results of Johnson *et al.* [36].

Blyler *et al.* [38] exposed developing BALB/c mice to chlordane *in utero* by dosing pregnant females with chlordane at either 0 or 8 mg/kg for 18 gestation days. Myeloid hematopoietic activity of bone marrow cells from 6-week-old offspring was measured *in vitro* through the colony-forming effects of exogenously added recombinant forms of the cytokines granulocyte/macrophage-colony stimulating factor, macrophage-CSF, and interleukin 3 (IL-3). There was a gender-specific reduction in the numbers of bone marrow colony-forming units in response to these challenges (females only), although prenatal treatment with chlordane did not significantly affect the number of viable bone marrow cells in either male or female mice.

The effects of exposure to an organochlorine mixture containing chlordane on preimplantation development of porcine embryos were investigated by Campagna *et al.* [39]. Oocytes were fertilized *in vitro* and then exposed to the mixture for 9 days. The rate of development was slowed, and there was a decrease in the mean number of blastomeres produced. Exposure of embryos to a metabolized organochlorine mixture did not alter developmental markers.

Bondy *et al.* [40] analyzed gender differences in tissue depletion of *trans*-nonachlor, *trans*-chlordane, and the metabolite oxychlordane in Sprague Dawley rats. Each dose group consisted of 30 males and 30 females; the animals were gavaged with 2.5 mg/kg/day of each of the test compounds. Oxychlordane accumulated in the liver of rats exposed to both *trans*-nonachlor and *trans*-chlordane. *Trans*-nonachlor and oxychlordane persisted in the abdominal fat of females, while *trans*-chlordane was eliminated by the end of the study. Males were able to clear approximately half of the oxychlordane.

As evidence of environmental contamination by chlordane and other organochlorine contaminants in the Arctic and elsewhere has developed, multiple studies have investigated the impact of organochlorine exposure on species found in the Arctic. In grizzly bears (*Ursus arctos horribilis*) of British Columbia, and in polar bears (*Ursus maritimus*), chlordane levels were found to increase during hibernation, as the fat levels became depleted [41,42]. Additionally, it has been found that chlordane levels correlate negatively with bone mineral density following analysis of skulls from 1892 to 2002 [43] and in the penile bone (or os penis) in subadults and adults [44]. The same study found an inverse correlation between ovary length and chlordane levels in female polar bears. Chlordane levels are also inversely correlated with testosterone levels in polar bears [45], and increased levels can lead to diffuse glomerular capillary wall thickening [46].

Studies Relevant for Mammalian TRV Development for Ingestion Exposures

Table 22.2 summarizes the studies that highlight the effects of chlordane. The liver appears to be the primary target organ of chlordane toxicity in experimental studies [9–12,19,23,24,27,47]; other target organs include the testis [18], kidney [12], and possibly the brain [13,20]. Bone mineral density is also a target of chlordane toxicity [43,44].

Although the liver seems to be the primary target of chlordane toxicity, it is difficult to relate confidently hepatic effects to ecologically relevant effects for deriving a TRV. Other chlordane-induced effects such as testicular effects [21], developmental toxicity, and reduced weight gains [26,35] are likely more suitable as a basis for a mammalian TRV.

Al-Omar *et al.* [21] showed that 75 and 275 mg/kg chlordane caused a significant reduction in testis weight, diameter of the seminiferous tubules, and in the number of spermatagonia and primary and secondary spermatocytes and spermatids in Swiss mice. These results indicate that chlordane is a likely reproductive toxicant, but since no functional assays were conducted, it is difficult to determine if the effects on male reproductive parameters would result in a diminished ability to reproduce successfully. Although a clear link to functional reproduction is lacking, the results from this study provide useful supporting evidence for reproductive effects of chlordane.

Narotsky and Kavlock [32] showed a significant decrease in the number of live pups from dams exposed to 21 mg/kg/day chlordane as well as decreased weight gain in chlordane-exposed dams. This study was chosen as a basis for the TRV since there was a clear link between exposures to chlordane and decreased reproductive performance. In addition, significant

Table 22.2 Summary of Relevant Mammalian Data for TRV Derivation

Test Organism	Test Duration	Test Results			Study
		NOAEL (mg/kg/d)	LOAEL (mg/kg/d)	Effects Observed at the LOAEL	
Rat (Osborne-Mendel)	28 days	NA	8–14	Increase in relative liver weight	[16]
Mouse (BALB/c)	2, 3, 4, and 5weeks	NA	75	Structural and functional perturbation of the male reproductive system	[18]
Rat (Osborne-Mendel)	2 years	0.65	1.5	Increased mortality, decreased growth at high concentrations, fetal toxicity, and hepatic hypertrophy	[23]
Rat (Osborne-Mendel)	2 years	10.2	20.4	Reduced body weight gain of more than 10%	[24]
Mouse (ICR SPF)	2 years	0.15	0.75	Hepatic necrosis, atrophy, and fatty degeneration. Absolute and relative increases in liver weights	[26]
Rat (F344)	GD 6–19	NA	21	Developmental toxicity (reduced postnatal survival and weight gain in pups)	[32]
Mouse (BALB/C)	Throughout gestation	0.16	8	Immunocompetence	[33]

NA = not applicable; GD = gestation day.

effects occurred at 21 mg/kg/day, which is lower than the effects seen in the Al-Omar *et al.* [18] study (75 mg/kg/day). Importantly, although the dose to the pups occurred via the mother, this is the only mechanism by which chlordane could reach the pups. Therefore, the dose of 21 mg/kg/day as received by the dams is suitable as a dose for the pups. Generally, decreased reproductive output (and lower growth rate) in natural populations is likely to result in deleterious population-level effects; hence, this endpoint is suitable for TRV development based on previously outlined criteria [4].

Mammalian Inhalation Toxicity: Acute

No data are available at this time.

Mammalian Inhalation Toxicity: Subacute

A detailed report on the inhalation toxicology of chlordane contains a subacute component in which 10 Wistar rats/sex/group were exposed to 0, 5.8, 28.2, 154, or 413 micrograms per liter (μg/L), 8 hours/day, 5 days/week for up to 28 days [48]. Exposure to chlordane occurred in 0.5 m^3 stainless steel and glass chambers. Body weights were monitored on days 0, 3, 7, and then weekly. Urinalysis and hematological and clinical chemistry parameters were measured in five animals/sex/group on day 28. Survivors were necropsied at term, major organs were weighed, and a full range of organs and tissues were processed for histopathological examination (control versus high-dose). Sections of the lung, liver, and kidney were examined in all groups. Premature deaths and instances of weight loss among the survivors suggest that 154 and 413 μg/L were above the maximum tolerated dose. However, a dose of 28.2 μg/L appeared to be at or near the toxicity threshold since clinical signs and a reduction in food consumption were observed in animals exposed to this chlordane concentration. There was no indication of toxicity in animals exposed to 5.8 μg/L, but there were some signs of fluctuation in clinical chemistry parameters in those exposed to 28.2 μg/L, including a reduction in blood glucose. Similarly, while there were no apparent histopathological effects of chlordane in rats exposed to 5.8 μg/L, those receiving 28.2 μg/L displayed hepatocellular hypertrophy. This demarcation between the effects of chlordane at the two exposure levels justifies the choice of 5.8 μg/L as a NOAEL and 28.2 μg/L as a LOAEL.

Mammalian Inhalation Toxicity: Subchronic

The report by Khasawinah *et al.* [48] also contains an account of subchronic inhalation experiments on chlordane in Wistar rats and cynomolgus monkeys. Varying numbers of Wistar rats and six monkeys/sex/group were exposed to chlordane concentrations of 0, 0.1, 1, or 10 μg/L for 90 days, followed by a 90-day recovery period for a subset of rats only. In addition to the suite of observations and analyses as described for their subacute inhalation study, Khasawinah *et al.* measured the levels of chlordane in blood, liver, and adipose tissue. They also measured the activity of hepatic cytochrome P450 in exposed subjects, examined the eye color of the animals before and in the final week of exposure, and carried out pulmonary function tests on the monkeys before exposure and once during week 12 following exposure. Rectal temperatures were measured in all monkeys and in five rats/sex in the control and high-dose groups prior to exposure and periodically throughout the study.

Although there were few compound-related effects of chlordane in monkeys at the concentrations employed, there was some evidence of chlordane-induced toxicity in the intermediate and high-dose rats. For example, there was a statistically significant reduction in the total and mean corpuscular hemoglobin concentration in males by week 5, which was enhanced by week 13 and apparent in females at the later time point. Other dose-dependent hematological changes were observed, as were a variety of impacts to the liver, including increased weight, enlargement, centrilobular hypertrophy, and an increase in the activity of cytochrome P450. These findings justify the choice of 0.1 μg/L as a NOAEL for the rats; however, an equivalent value for the monkeys would be >10 μg/L.

Mammalian Inhalation Toxicity: Chronic

No data are available.

Studies Relevant for Mammalian TRV Development for Inhalation Exposures

There are few inhalation toxicity studies on chlordane. The inhalation study of Khasawinah *et al.* [48] demonstrated that the hepatic impacts of chlordane occur regardless of the route of administration. Supporting this are [^{14}C]-labeled chlordane studies that have shown that once absorbed, inhaled chlordane has the same kinetics as orally administered chlordane [14,49]. Thus, the liver was shown to be the target organ in like manner to the oral exposure studies described elsewhere in this WTA. As discussed in Khasawinah [48], the hepatic toxicity manifested at exposure concentrations in excess of 0.1 μg/L (the suggested NOAEL) were accompanied by dose-dependent increases in the levels of "total" chlordane in adipose tissue, liver, and blood. For the most part, the components observed in rat tissues included oxychlordane, heptachlor epoxide, trans-nonachlor, and an unidentified component designated as "compound C." Heptachlor, *cis*-chlordane and *trans*-chlordane were much less heavily represented.

There is insufficient data on the inhalation toxicity of chlordane to derive a TRV for mammalian inhalation exposures. Given that the kinetics of chlordane is the same for both oral and inhalation exposures, it would be possible to use oral toxicity data as a basis for an inhalation TRV. However, there are no data comparing the relative absorption of chlordane from the inhalation and oral routes; hence, a TRV for inhalation exposures to chlordane for mammalian wildlife cannot be derived.

Mammalian Dermal Toxicity

Two studies featuring the topical administration of chlordane to experimental animals were identified [50,51]. In the first, Schop *et al.* [50] examined the capacity of the pesticide to induce micronuclei (MN) in femoral bone marrow and nuclear aberrations (NA) in the hair follicles of male CD-1 mice that had received a single application of the compound in dimethyl sulfoxide (DMSO) at 1/8, 1/16, or 1/32 the dermal LD_{50}. For chlordane, these exposure levels were 125, 250, and 500 mol/kg, each of which was associated with a dose-dependent incidence of NA formation. The results obtained for chlordane in the MN test were less clear-cut, although a positive effect was obtained at the highest dose, equivalent to a LOAEL of 205 mg/kg and a NOAEL of 102 mg/kg. A LOAEL of 51 mg/kg would apply to chlordane in the NA test. More recently, a meeting abstract by Blaylock and Mehendale [51] described the topical application of chlordane to female Balb/c mice and the extent of its effects on oxazolone-induced contact sensitivity. The pesticide appeared to have the ability to reduce oxazolone-induced ear swelling whether applied prenatally or at the time of sensitization. The dose-response component of the study suggested that 20 g of chlordane would represent a LOAEL for this effect.

Studies Relevant for Mammalian TRV Development for Dermal Exposures

The few studies on the dermal toxicity of chlordane do not provide sufficient information on which to base a dermal TRV.

Summary of Avian Toxicology

Many studies have surveyed an array of pesticides, polychlorinated biphenyls (PCB), polycyclic aromatic hydrocarbons (PAH), dioxins and dioxin-like residues in the bodily tissues, brains, or eggs of birds, with the underlying intent of correlating the levels of these environmentally important contaminants with symptoms of health impairment, impacts on species abundance, and reproductive deficits, among other effects. In one of a series of dose-response studies of pesticides in birds, Stickel *et al.* [52] determined the levels of chlordane metabolites in brain and carcass that resulted from feeding a number of different species of birds with measured amounts of chlordane and/or related compounds. Results included estimates of the levels of chlordane (and its metabolites) in the brain that are indicative of (or correlate with) a state of ill-health/failure-to-thrive for a particular species, as well as a range of (mostly lower) concentrations of chlordane residues representing "no adverse impact" tissue levels. Many subsequent field studies have used the Stickel criteria to calibrate their field measurements, thereby drawing inferential evidence of the degree of threat to which their subjects were exposed from the prevailing environmental contamination at the time the field samples were collected.

The Stickel *et al.* [53] experiments were conducted on a number of bird species that had been captured and kept in large outdoor flight cages equipped with swinging perches. Cage-conditioned birds were provided water *ad libitum*, and their food was spiked to predetermined levels with the chemical under investigation. In one series of experiments in the study, 10 female European Starlings (*Sturnus vulgaris*)/group were exposed to 200, 300, 400, or 500 ppm HCS-3260 in turkey starter crumbles. Other experimental groups receiving HCS-3260 included female Common Grackles (*Quiscalus quiscula*) (200 ppm), male Brown-Headed Cowbirds (*Molothrus ater*) (200 ppm), and male Red-Winged Blackbirds (10, 50, 100, and 200 ppm). Additional male Grackles received 12.5 ppm oxychlordane, whereas other groups of male cowbirds and female starlings received 12.5, 25, 50, or 100 ppm nonachlor. Whole brains and 20 g subsamples of homogenized carcasses were extracted in solvent and measured for oxychlordane, heptachlor epoxide, and *cis*- and *trans*-chlordane on a ppm wet weight basis. Some animals receiving the higher concentrations of HCS-3260 (200 ppm and above) showed a range of clinical signs including decreased activity, convulsions, loss of balance, and death. Importantly, those that died displayed a higher range of concentrations of oxychlordane in the brain than did survivors. For example, Cowbirds that died at 200 ppm HCS-3260 had oxychlordane brain concentrations ranging from 19.1 to 22.1 ppm wet weight, whereas those that survived to term had ranges of 2.3 to 4.8 ppm wet weight. Comparative ranges of oxychlordane in the brains of Grackles receiving HCS-3260 were 12.9 to 25 ppm for dead birds versus 1.3 to 2.1 ppm for surviving birds. In general, there appeared to be an inverse relationship between dose level and time-to-death, albeit with a considerable amount of interspecies variation because of the chemical insult. The authors demonstrated time- and dose-dependent oxychlordane accumulation, with residue levels slowly declining ($t_{1/2}=63$ days) when exposure was discontinued. HCS-3260 is also a precursor of tissue-borne oxychlordane, whereas technical chlordane is a precursor of oxychlordane and heptachlor epoxide.

Many studies have been conducted on the effects of exposure to various organohalogens and the impacts of these on breeding and reproductive behavior in the Glaucous gull (*Larus hyperboreus*). Oxychlordane levels in wild breeding birds were positively correlated with time away from the nest, and where the birds' diet consisted primarily of fish, the maternal levels negatively correlated with chick growth, an effect that was not seen where the diet primarily consisted of eggs and young from nearby seabird colonies [53]. The rate of return to the nesting site was also negatively correlated with oxychlordane blood levels. Increased chlordane levels also resulted in smaller eggs [54] and increased progesterone levels in male birds [55]. Basal metabolic rate of these birds was also negatively correlated with chlordane blood levels [56].

Although a number of studies have shown that chlordane is toxic to birds, the focus of laboratory experiments has been in correlating tissue burdens with toxicity. As a result, no suitable data are available with which to derive an avian TRV. Stickel *et al.* [52] is one of the few laboratory toxicity studies involving oral exposure of birds to chlordane. However, the focus of this effort was to correlate brain chlordane levels with mortality. Details of the exposure duration and effects were not presented, thus precluding the use of this study for TRV derivation.

Amphibian Toxicology

No data are available at this time.

Reptilian Toxicology

Feeding studies that report toxic effects of chlordane on reptiles are currently limited. The only study that has been conducted analyzed the effects of organochlorine pesticide mixtures in alligators (*Alligator mississippiensis*) and their eggs [57]. Female alligators were treated with a 0.2 mg/kg/day pesticide mixture including chlordane for 274 ± 8 days. Eggs were found to have an increase in organochlorine pesticides and decreased viability, indicating maternal transfer of the pesticides and an impact on survival. Other evidence of maternal transfer in reptiles is provided by Stewart *et al.* [58] and Van de Merwe *et al.* [59], both of which found a positive correlation between maternal blood chlordane levels and egg levels in the Leatherback turtle (*Dermochelys coriacea*) and the Green Sea Turtle (*Chelonia mydas*).

Studies by Willingham and Crews [60] and Willingham *et al.* [61] used an experimental design that featured the perturbation of the temperature-dependent sex determination of the red-eared slider turtle (*Trachemy scripta elagans*) by the challenge of estrogen or estrogen-like substances such as the organochlorine pesticides. Eggs in stage 17, the temperature-sensitive period, were spotted with 0.125 to 28 ng (dependent upon type of organochlorine) in a volume of 5 μL of organochlorines (including chlordane) in DMSO, with the solvent as a negative control and estradiol in DMSO as a positive control. The sex ratio among the ensuing progeny gave a measure of the degree of resulting "feminization," an excess of which is likely to be detrimental to the population. Greater than 35% of the eggs incubated at 28.2 °C hatched as female turtles, although eggs incubated at this temperature normally yield male offspring (0% females were obtained in control incubations). The ability of chlordane in this system to mimic estradiol and override the temperature-dependent sex determination of the red-eared slider turtle demonstrates the possible capacity of this and other organochlorine pesticides to disrupt developmental responses governed by the levels of estrogens.

Studies in loggerhead turtles (*Caretta caretta*) have found an impact of organocholorine exposure and immune function [62,63]. Analysis of blood levels for multiple different organochlorines, including chlordane, found that chlordane negatively correlated with red blood cells, hemoglobin, and hematocrit levels, indicating anemia; lymphocyte levels were also decreased. Lysozyme activity was also negatively affected by increasing chlordane levels.

Although these data strongly suggest that chlordane could potentially alter turtle sex ratios, the study designs are unsuitable for TRV derivation since the exposures included multiple organochlorine compounds and did not follow a dose-response design.

RECOMMENDED TOXICITY REFERENCE VALUES

TRVs for Ingestion Exposures for the Class Mammalia

Developmental effects in rats of pre- and postnatal maternal exposure to chlordane [32] were chosen as a basis for the TRV. Chlordane-exposed dams showed decreased weight gain and had fewer live litters and fewer live pups on postnatal day 6 compared to control dams [32]. Since the developmental effects were apparent during postnatal maternal exposure, it was hypothesized that exposure of pups occurred via the mother's milk. This result was also supported by a previous study [23]. Decreased reproductive output in natural populations is likely to result in deleterious population-level effects; hence, this endpoint is suitable for TRV development based

Table 22.3 Selected Ingestion TRVs for the Class Mammalia

TRV	Dose	Confidence
NOAEL-based	2.1 mg/kg/day	Medium
LOAEL-based	21 mg/kg/day	Medium

Source: [32].

on previously outlined criteria [4]. Although the developmental effects occurred once pups were born, the weanling stage can be considered a sensitive life stage since it is short and pups are entirely dependent on the mother for survival. Therefore, this exposure is a surrogate for chronic exposure. Because the dose-response data does not include a NOAEL, the TRV was derived using the NOAEL-LOAEL approach; the results are presented in Table 22.3. The LOAEL for the developmental and growth effects was 21 mg chlordane/kg/day, which is the value for the LOAEL-based TRV. The NOAEL-based TRV is 2.1 mg chlordane/kg/day obtained by dividing the LOAEL by an uncertainty factor of 10 [4]. This TRV is assigned a medium confidence rating due to a shortage of longer-term studies on other mammalian species.

IMPORTANT RESEARCH NEEDS

Data clearly indicates that chlordane is toxic to mammalian, avian, and reptilian wildlife. However, few studies are available that provide data suitable for TRV development. The major research need regarding chlordane toxicity is more studies focused on relating dose to toxic effect. Mainly, oral toxicity studies on avian, reptilian, and perhaps amphibian species are needed. An emphasis on chronic toxicity data would be most useful for the development of higher-confidence TRVs.

REFERENCES

[1] U.S. EPA. Integrated risk information system. Chlordane. In: U.S. Environmental Protection Agency NCfEA, editor. Integrated risk information system, chlordane. Cincinnati (OH): U.S. Environmental Protection Agency; 2001.

[2] Fry DM. Reproductive effects in birds exposed to pesticides and industrial chemicals. Environ Health Perspect 1995;103(Suppl. 7):165–71.

[3] ATSDR. Toxicological profile for chlordane. In: Registry AfTSaD, editor. Atlanta (GA): Centers for Disease Control and Prevention, U.S. Department of Health and Human Services, Public Health Service; 1994.

[4] USACHPPM. Standard practice for wildlife toxicity reference values. Aberdeen Proving Ground (MD): U.S. Army Center for Health Promotion and Preventive Medicine; 2000, Technical Guide No. 254.

[5] Gaines TB. Acute toxicity of pesticides. Toxicol Appl Pharmacol 1969;14(3):515–34.

[6] Boyd EM, Taylor FI. The acute oral toxicity of chlordane in albino rats fed for 28 days from weaning on a protein-deficient diet. IMS Ind Med Surg 1969;38 (12):434–41.

[7] Harbison RD. Heptachlor, chlordane, and parathion toxicity in adult, newborn, and phenobarbital-treated newborn rats. Toxicol Appl Pharmacol 1973;25:472–3.

[8] Truhaut R, Gak JC, Graillot C. Modality and mechanism of the toxic action of organochlorine insecticides. I. Comparative study of the acute toxic effects on the hamster and the rat. Eur J Toxicol Environ Hyg 1974;7(3):159–66.

[9] Berman E, Schlicht M, Moser VC, MacPhail RC. A multidisciplinary approach to toxicological screening: I. Systemic toxicity. J Toxicol Environ Health 1995;45(2):127–43.

[10] den Tonkelaar EM, van Esch GJ. No-effect levels of organochlorine pesticides based on induction of microsomal liver enzymes in short-term toxicity experiments. Toxicology 1974;2(4):371–80.

[11] Ogata M, Izushi F. Effects of chlordane on parameters of liver and muscle toxicity in man and experimental animals. Toxicol Lett 1991;56(3):327–37.

[12] Chatterjee K, Banerjee SK, Tiwari R, Mazumdar K, Bhattacharyya A, Chatterjee GC. Studies on the protective effects of l-ascorbic acid in chronic chlordane toxicity. Int J Vitam Nutr Res 1981;51(3):254–65.

[13] Moser VC, Cheek BM, MacPhail RC. A multidisciplinary approach to toxicological screening: III. Neurobehavioral toxicity. J Toxicol Environ Health 1995;45 (2):173–210.

[14] Barnett JR, Dorough HW. Metabolism of chlordane in rats. J Agric Food Chem 1974;22(4):612–9.

[15] Whysner J, Montandon F, McClain RM, Downing J, Verna LK, Steward 3rd RE, et al. Absence of DNA adduct formation by phenobarbital, polychlorinated biphenyls, and chlordane in mouse liver using the 32p-postlabeling assay. Toxicol Appl Pharmacol 1998;148(1):14–23.

[16] Casterline Jr JL, Williams CH. The effect of 28-day pesticide feeding on serum and tissue enzyme activities of rats fed diets of varying casein content. Toxicol Appl Pharmacol 1971;18(3):607–18.

[17] Tryphonas H, Bondy G, Hodgen M, Coady L, Parenteau M, Armstrong C, et al. Effects of cis-nonachlor, trans-nonachlor and chlordane on the immune system of Sprague-Dawley rats following a 28-day oral (gavage) treatment. Food Chem Toxicol 2003;41(1):107–18.

[18] Al-Omar MA, Abbas AK, Al-Obaidy SA. Combined effect of exposure to lead and chlordane on the testicular tissues of Swiss mice. Toxicol Lett 2000;115(1):1–8.

[19] Mahon DC, Oloffs PC. Effects of sub-chronic low-level dietary intake of chlordane on rats with cirrhosis of the liver. J Environ Sci Health B 1979;14(3):227–45.

[20] Drummond L, Chetty KN, Coll MI, Bailey M, Desaiah D. In vivo effect of chlordane on rat brain ATpase system. Fed Proc Fed Am Soc Exp Biol 1980;39(998).

[21] Bondy G, Curran I, Doucet J, Armstrong C, Coady L, Hierlihy L, et al. Toxicity of trans-nonachlor to Sprague-Dawley rats in a 90-day feeding study. Food Chem Toxicol 2004;42(6):1015–27.

[22] Dorough HW, Hemken RW. Chlordane residues in milk and fat of cows fed HCS 3260 (high purity chlordane) in the diet. Bull Environ Contam Toxicol 1973;10 (4):208–16.

[23] Ingle L. Chronic oral toxicity of chlordane to rats. Arch Ind Hyg Occup Med 1952;6 (4):357–67.

[24] NCI. Bioassay of chlordane for possible carcinogenicity. In: U.S. Department of Health Education, and Welfare, editor. Bioassay of chlordane for possible carcinogenicity. Bethesda, (MD): U.S. Department of Health, Education, and Welfare; 1977.
[25] Reuber MD, Ward JM. Histopathology of liver carcinomas in (C57BL/6 N X C3H/HeN)F1 mice ingesting chlordane. J Natl Cancer Inst 1979;63(1):89–92.
[26] Khasawinah AM, Grutsch JF. Chlordane: 24-month tumorigenicity and chronic toxicity test in mice. Regul Toxicol Pharmacol 1989;10(3):244–54.
[27] Barrass N, Stewart M, Warburton S, Aitchison J, Jackson D, Wadsworth P, et al. Cell proliferation in the liver and thyroid of C57BL/10 J mice after dietary administration of chlordane. Environ Health Perspect 1993;101(Suppl. 5):219–23.
[28] Sonne C, Leifsson PS, Dietz R, Kirkegaard M, Jensen AL, Shahmiri S, et al. Greenland sledge dogs (*canis familiaris*) develop liver lesions when exposed to a chronic and dietary low dose of an environmental organohalogen cocktail. Environ Res 2008;106(1):72–80.
[29] Deichmann WB. Toxicology of DDT and related chlorinated hydrocarbon pesticides. J Occup Med 1972;14(4):285–92.
[30] Chernoff N, Kavlock RJ. An *in vivo* teratology screen utilizing pregnant mice. J Toxicol Environ Health 1982;10(4–5):541–50.
[31] Usami M, Kawashima K, Nakaura S, Yamaguchi M, Tanaka S, Takanaka A, et al. Effect of chlordane on the prenatal development of rats. Eisei Shikenjo Hokoku 1986;(104):68–73.
[32] Narotsky MG, Kavlock RJ. A multidisciplinary approach to toxicological screening: II. Developmental toxicity. J Toxicol Environ Health 1995;45(2):145–71.
[33] Cranmer JS, Avery DL, Grady RR, Kitay JI. Postnatal endocrine dysfunction resulting from prenatal exposure to carbofuran, diazinon or chlordane. J Environ Pathol Toxicol 1978;2(2):357–69.
[34] Cassidy RA, Vorhees CV, Minnema DJ, Hastings L. The effects of chlordane exposure during pre- and postnatal periods at environmentally relevant levels on sex steroid-mediated behaviors and functions in the rat. Toxicol Appl Pharmacol 1994;126(2):326–37.
[35] Spyker-Cranmer JM, Barnett JB, Avery DL, Cranmer MF. Immunoteratology of chlordane: cell-mediated and humoral immune responses in adult mice exposed *in utero*. Toxicol Appl Pharmacol 1982;62(3):402–8.
[36] Johnson KW, Holsapple MP, Munson AE. An immunotoxicological evaluation of gamma-chlordane. Fundam Appl Toxicol 1986;6(2):317–26.
[37] Barnett JB, Blaylock BL, Gandy J, Menna JH, Denton R, Soderberg LS. Long-term alteration of adult bone marrow colony formation by prenatal chlordane exposure. Fundam Appl Toxicol 1990;14(4):688–95.
[38] Blyler G, Landreth KS, Barnett JB. Gender-specific effects of prenatal chlordane exposure on myeloid cell development. Fundam Appl Toxicol 1994;23(2):188–93.
[39] Campagna C, Ayotte P, Sirard MA, Bailey JL. An environmentally relevant mixture of organochlorines, their metabolites and effects on preimplantation development of porcine embryos. Reprod Toxicol 2008;25(3):361–6.
[40] Bondy GS, Coady L, Doucet J, Armstrong C, Kriz R, Liston V, et al. Enantioselective and gender-dependent depletion of chlordane compounds from rat tissues. J Toxicol Environ Health A 2005;68(22):1917–38.
[41] Polischuk SC, Norstrom RJ, Ramsay MA. Body burdens and tissue concentrations of organochlorines in polar bears (*Ursus maritimus*) vary during seasonal fasts. Environ Pollut 2002;118(1):29–39.
[42] Christensen JR, MacDuffee M, Macdonald RW, Whiticar M, Ross PS. Persistent organic pollutants in British Columbia grizzly bears: consequence of divergent diets. Environ Sci Technol 2005;39(18):6952–60.
[43] Sonne C, Dietz R, Born EW, Riget FF, Kirkegaard M, Hyldstrup L, et al. Is bone mineral composition disrupted by organochlorines in East Greenland polar bears (*Ursus maritimus*)? Environ Health Perspect 2004;112(17):1711–6.

[44] Sonne C, Leifsson PS, Dietz R, Born EW, Letcher RJ, Hyldstrup L, et al. Xenoendocrine pollutants may reduce size of sexual organs in East Greenland polar bears (*Ursus maritimus*). Environ Sci Technol 2006;40(18):5668–74.

[45] Oskam IC, Ropstad E, Dahl E, Lie E, Derocher AE, Wiig O, et al. Organochlorines affect the major androgenic hormone, testosterone, in male polar bears (*Ursus maritimus*) at Svalbard. J Toxicol Environ Health A 2003;66(22):2119–39.

[46] Sonne C, Dietz R, Leifsson PS, Born EW, Kirkegaard M, Letcher RJ, et al. Are organohalogen contaminants a cofactor in the development of renal lesions in East Greenland polar bears (*Ursus maritimus*)? Environ Toxicol Chem 2006;25(6):1551–7.

[47] Malarkey DE, Devereux TR, Dinse GE, Mann PC, Maronpot RR. Hepatocarcinogenicity of chlordane in B6C3F1 and B6D2F1 male mice: Evidence for regression in B6C3F1 mice and carcinogenesis independent of ras proto-oncogene activation. Carcinogenesis 1995;16(11):2617–25.

[48] Khasawinah AM, Hardy CJ, Clark GC. Comparative inhalation toxicity of technical chlordane in rats and monkeys. J Toxicol Environ Health 1989;28(3):327–47.

[49] Nye DE, Dorough HW. Fate of insecticides administered endotracheally to rats. Bull Environ Contam Toxicol 1976;15(3):291–6.

[50] Schop RN, Hardy MH, Goldberg MT. Comparison of the activity of topically applied pesticides and the herbicide 2,4-D in two short-term in vivo assays of genotoxicity in the mouse. Fundam Appl Toxicol 1990;15(4):666–75.

[51] Blaylock BL, Newsom KK, Holladay SD, Shipp BK, Bartow TA, Mehendale HM. Topical exposure to chlordane reduces the contact hypersensitivity response to oxazolone in BALB/c mice. Toxicol Lett 1995;81(2–3):205–11.

[52] Stickel LF, Stickel WH, Dyrland RA, Hughes DL. Oxychlordane, HCS-3260, and nonachlor in birds: lethal residues and loss rates. J Toxicol Environ Health 1983;12 (4–6):611–22.

[53] Bustnes JO, Miland O, Fjeld M, Erikstad KE, Skaare JU. Relationships between ecological variables and four organochlorine pollutants in an artic glaucous gull (*Larus hyperboreus*) population. Environ Pollut 2005;136(1):175–85.

[54] Verboven N, Verreault J, Letcher RJ, Gabrielsen GW, Evans NP. Differential investment in eggs by arctic-breeding glaucous gulls (*larus hyperboreus*) exposed to persistent organic pollutants (investissement différentiel dans les œufs par des *larus hyperboreus* nichant dans l'arctique et exposés à des polluants organiques persistants). The Auk 2009;126(1):123–33.

[55] Verreault J, Letcher RJ, Ropstad E, Dahl E, Gabrielsen GW. Organohalogen contaminants and reproductive hormones in incubating glaucous gulls (*Larus hyperboreus*) from the Norwegian Arctic. Environ Toxicol Chem 2006;25(11):2990–6.

[56] Verreault J, Shahmiri S, Gabrielsen GW, Letcher RJ. Organohalogen and metabolically-derived contaminants and associations with whole body constituents in Norwegian Arctic glaucous gulls. Environ Int 2007;33(6):823–30.

[57] Rauschenberger RH, Wiebe JJ, Buckland JE, Smith JT, Sepulveda MS, Gross TS. Achieving environmentally relevant organochlorine pesticide concentrations in eggs through maternal exposure in *Alligator mississippiensis*. Mar Environ Res 2004;58(2–5):851–6.

[58] Stewart KR, Keller JM, Templeton R, Kucklick JR, Johnson C. Monitoring persistent organic pollutants in leatherback turtles (*Dermochelys coriacea*) confirms maternal transfer. Mar Pollut Bull 2011;62(7):1396–409.

[59] van de Merwe JP, Hodge M, Whittier JM, Ibrahim K, Lee SY. Persistent organic pollutants in the green sea turtle chelonia mydas: nesting population variation, maternal transfer, and effects on development. Mar Ecol Prog Ser 2010;403:269–78.

[60] Willingham E, Crews D. Sex reversal effects of environmentally relevant xenobiotic concentrations on the red-eared slider turtle, a species with temperature-dependent sex determination. Gen Comp Endocrinol 1999;113(3):429–35.

[61] Willingham E, Rhen T, Sakata JT, Crews D. Embryonic treatment with xenobiotics disrupts steroid hormone profiles in hatchling red-eared slider turtles (*Trachemys scripta elegans*). Environ Health Perspect 2000;108(4):329–32.
[62] Keller JM, Kucklick JR, Stamper MA, Harms CA, McClellan-Green PD. Associations between organochlorine contaminant concentrations and clinical health parameters in loggerhead sea turtles from North Carolina, USA. Environ Health Perspect 2004;112 (10):1074–9.
[63] Keller JM, McClellan-Green PD, Kucklick JR, Keil DE, Peden-Adams MM. Effects of organochlorine contaminants on loggerhead sea turtle immunity: comparison of a correlative field study and in vitro exposure experiments. Environ Health Perspect 2006;114(1):70–6.

PART VI

Wildlife Toxicity Assessments of Industrial and Military-Related Chemicals

CHAPTER 23

Wildlife Toxicity Assessment for Acetylene

Marc A. Williams

INTRODUCTION

In its purest form, acetylene (or ethyne—systematic name; CAS No. 74-86-2) is a colorless, odorless gas; structurally, it is a simple alkyne hydrocarbon. However, depending on the manufacturing process (e.g., when generated by adding water to calcium carbide), acetylene might contain trace contaminates of phosphine and arsine and a characteristic garlic/onion aroma. In the pure form, acetylene is quite unstable, especially when pressurized. The danger is that on initiation by intense heat or a shockwave, acetylene decomposes quite readily and explosively, and for that reason, acetylene is frequently shipped and dissolved in liquid form in dimethylformamide or acetone at standard temperature and pressure [1].

Acetylene is used as the feedstock in the manufacture of vinyl chloride, vinylidene chloride, vinyl acetate, and acrylates, among others. It is also an intermediate in the synthesis of acrylic acid, tetrahydrofuran, and some chlorinated solvents, and it is used in oxyacetylene torches when mixed with air;

Wildlife Toxicity Assessments for Chemicals of Military Concern
http://dx.doi.org/10.1016/B978-0-12-800020-5.00023-5

for example, in welding, cutting, or soldering metals [1,2]. The key to its use in metallurgy is the intense heat with which acetylene/oxygen mixtures burn (>3,300 °C or 6,000 °F), making acetylene the third hottest natural chemical flame after cyanogen and dicyanoacetylene, which burn at 4,525 °C and 4,490 °C, respectively. As briefly mentioned above, gaseous acetylene is explosive when compressed or heated. Lewis pointed out that while acetylene does not constitute a serious toxic hazard *per se*, it is a highly dangerous fire hazard when exposed to heat, flame, or oxidizers [3]. Acetylene might become violently explosive when mixed with any of a substantial number of chemical substances with which it is incompatible. These agents include copper, halogens, molten potassium, silver, and mercury (Hg).

This chapter provides a Wildlife Toxicity Assessment (WTA) that summarizes current knowledge of the toxicological impacts of acetylene on wildlife. Evaluating the toxicity of acetylene is intended to contribute to the derivation of Toxicity Reference Values (TRVs) that could serve as screening-level benchmarks for wildlife near contaminated sites. The protocol for the performance of this assessment is documented in the U.S. Army Center for Health Promotion and Preventive Medicine (USACHPPM) Technical Guide (TG) 254, *Standard Practice for Wildlife Toxicity Reference Values* [4]. Table 23.1 presents a summary of the physical and chemical properties of acetylene.

TOXICITY PROFILE

Environmental Fate and Transport

Acetylene is released to the atmosphere from such sources as internal combustion engines, particularly the diesel engine, emissions from which reportedly constitute greater than 14% of all released hydrocarbons [2]. Consequently, acetylene is a ubiquitous component of urban air, with reported concentrations ranging to 300 parts per billion (ppb). The bioconcentration factor (the equilibrium ratio of the concentration of a chemical in an exposed organism to the concentration of the chemical in the surrounding water) of acetylene is 3.48 (estimated from its solubility in water), with a Henry's Law constant of 2.17×10^{-2} atm/m^3/mol at 25 °C [2].

When used as a feedstock in the manufacture of vinyl chloride and other substances, acetylene can be lost to the environment during chemical processing and waste disposal. Eleven chemical companies are listed in the Hazardous Substances Data Bank (HSDB) as manufacturers of the gas, some at multiple

Table 23.1 Summary of the Physical-Chemical Properties of Acetylene

CAS No.	74-86-2
Molecular weight	26.04 g mol^{-1}
Formula	C_2H_2 (H-C-C-H)
Color	Colorless
Physical state	Gas
Density/Specific Gravity	0.377 g/cm^3
Melting/freezing point	– 80.7 °C (–113.2 °F; 192.3 K) Triple point at 1.27 atm
Boiling point	–84 °C (–119.0 °F, product sublimes)
Autoignition temperature	335 °C (635 °F; 635 K)
Odor	Faint, ethereal, slightly sweet
Solubility in water	Slightly soluble at 1.2 g/L at 20–25 °C
Solubility in other solvents	Soluble in ethanol, diethyl ether, benzene, chloroform, acetone
Partition coefficients:	
Log K_{ow}	K_{ow} = 2691.53 (log = 3.44): estimated from water solubility
Log K_{oc}	K_{oc} = 4508.17 (log = 3.66): estimated from water solubility
Vapor pressure at 25 °C	5.24×10^3 mm Hg
Henry's Law constant at 25 °C	2.17×10^{-2} $atm/m^3/mol$ at 25 °C
Vapor density	0.91 (air = 1)
Conversion factors	1 ppm = 1.07 mg/m^3 1 mg/m^3 = 0.94 ppm

Source: HSDB (2): Data are provided under standard temperature and pressure (STP) conditions of 25°C (77 °F) and 100 kPa.

sites [2]. Volatilization of soil-borne acetylene is likely, based on acetylene's high vapor pressure of 5.24×10^3 mm Hg at 25 °C [2]. As listed in Table 23.1, these and other physical-chemical characteristics contribute to the volatilization of acetylene from the surface of marine and freshwater systems. The HSDB [2] also discusses the terrestrial fate of acetylene in the context of its possible degradation by abiotic or biotic processes. For example, many plants and bacteria metabolize acetylene to ethylene via nitrogen-fixing mechanisms. One bacterial species in particular, *Nocardia rhodochrous*, is capable of constitutively catabolizing acetylene via acetaldehyde to acetyl coenzyme A as its sole source of carbon and energy [6]. During combustion of ponderosa pine (*Pinus ponderosa*), 17% of the emitted hydrocarbons might contain acetylene, which is also present in the combustion emissions of diesel-engineered and other vehicular traffic and from petroleum refineries [7]. It is unsurprising that *N. rhodochrous* exhibits this metabolic capacity [6,7].

Summary of Mammalian Toxicity

Given the gaseous nature of acetylene at ambient temperatures and pressures, there are no data on the toxicological effects of the compound when administered via the oral route. Furthermore, no reports of studies on the toxicology of acetylene via the inhalation route were identified in biomedical and environmental literature, although Lewis [3] describes acetylene as a simple asphyxiant that, at sufficient concentrations, will dilute the available oxygen (partial O_2 pressure, or pO_2) in the air to a level that will not support life. An asphyxiant (e.g., Argon) may reach 33% of the air/gas mixture that is inhaled, reducing the pO_2 to approximately 75% of normal before appreciable symptoms develop [3]. Initial symptoms present as rapid respiration and air hunger, followed by impaired muscular coordination and nausea. Moreover, approximately 75% asphyxiant in air is fatal. Note, however, that impurities in commercial acetylene may cause symptoms before an asphyxiating concentration of C_2H_2 is reached, and approximately 10% in air may cause slight intoxication [2,3]. The National Institute of Occupational Safety and Health (NIOSH)-recommended exposure ceiling value for acetylene is 2,500 parts per million (ppm) (or 2,662 mg/m^3; 0.25%), a figure that could be valuable in assessing the significance of environmental concentrations [8].

Summary of Avian Toxicology

No toxicological data for the effects of acetylene on avian species were located.

Summary of Amphibian Toxicology

No toxicological data for the effects of acetylene on amphibian species were located.

Summary of Reptilian Toxicology

No toxicological data for the effects of acetylene on reptiles were located.

RECOMMENDED TOXICITY REFERENCE VALUES

Toxicity Reference Values for Mammals

TRVs for Ingestion Exposures for the Class Mammalia

At this time, it is not possible to derive a TRV for oral route of exposure for acetylene since there are no studies on the toxicity of this compound when administered orally. Due to the gaseous nature of the compound, it cannot be administered orally.

TRVs for Inhalation Exposures for the Class Mammalia

Not available at this time.

TRVs for Dermal Exposures for the Class Mammalia

Not available at this time.

TRVs for Amphibians

Not available at this time.

TRVs for Reptiles

Not available at this time.

IMPORTANT RESEARCH NEEDS

The severe lack of data on the toxicity of acetylene to wildlife species precludes the development of a TRV. Hence, additional and more detailed toxicological studies of the compound and its derivatives are recommended. Chronic toxicity studies on nonmammalian wildlife such as birds, reptiles, and amphibians are particularly warranted.

REFERENCES

[1] Downie NA. Industrial gases. London (UK): Blackie Academic and Professional; 1997, Chapter 2, Gas technology; p. 120–123.
[2] HSDB. Acetylene. National Library of Medicine [Internet] 2014. Available from: http://toxnet.nlm.nih.gov/cgi-bin/sis/search/a?dbs+hsdb:@term+@DOCNO+166.
[3] Lewis Sr RJ. Sax's dangerous properties of industrial materials. 8th ed. New York (NY): John Wiley & Sons, Inc.; 1992.
[4] USACHPPM. Standard practice for wildlife toxicity reference values. Aberdeen Proving Ground (MD): U.S. Army Center for Health Promotion and Preventive Medicine; 2000, Technical Guide No. 254. Revision forthcoming.
[5] O'Neil MJ, editor. The Merck index–an encyclopedia of chemicals, drugs, and biologicals. Cambridge (UK): Royal Society of Chemistry; 2013.
[6] Kanner D, Bartha R. Metabolism of acetylene by Nocardia rhodochrous. J Bacteriol 1982;150(2):989–92.
[7] Whitby RA, Altwicker ER. Acetylene in the atmosphere: sources, representative ambient concentrations, and ratios to other hydrocarbons. Atmos Environ 1978;12:1289–96.
[8] NIOSH. National Institute for Occupational Safety and Health pocket guide to chemical hazards: acetylene. NIOSH [Internet] 2011. Available from: http://www.cdc.gov/niosh/npg/npgd0008.html.

CHAPTER 24

Wildlife Toxicity Assessment for Benzo[a]Pyrene

Marc A. Williams, Christopher Salice, Gunda Reddy

Contents

INTRODUCTION

Benzo[a]pyrene (B[a]P) (CAS No. 50-32-8) is one of a large number of polycyclic aromatic hydrocarbons (PAHs) that are formed during the incomplete combustion of organic matter. Though structurally related, the compounds display considerable heterogeneity in the number and spatial arrangement of their fused aromatic rings. Mixtures of PAHs are ubiquitous in the air because of stack emissions, smoke from household wood and coal fires, cigarettes, barbecue grills, automobile exhausts, and so on. Fossil fuels are also a major source of PAHs, and this is of relevance to military sites. The compounds may be secondarily dispersed to other environmental media. Seventeen PAHs including B[a]P have been identified as being especially important for environmental monitoring and compliance purposes [1]. These can be further identified by whether or not they have been demonstrated or are thought likely to induce or promote tumor formation in experimental animals. Benzo[a]pyrene is the most well-studied PAH, a

Wildlife Toxicity Assessments for Chemicals of Military Concern
http://dx.doi.org/10.1016/B978-0-12-800020-5.00024-7

cancer-inducing substance for which toxicological data serve as quantitative benchmarks for the entire carcinogenic subgroup. Thus, although toxicological data on the carcinogenicity of other PAHs are poor, quantitative estimates of their carcinogenic potency can be expressed relative to that of B[a]P using toxicity equivalency factors (TEFs) [2]. The carcinogenic potency of B[a]P (defined as increased risk per mg/kg/day) was first estimated using data from several animal test systems. The geometric mean determination was estimated at 3.1 mg/kg/day. The cancer potency for a number of related PAHs was estimated and then each was rounded off to a simple multiple (1.0, 0.1, 0.01, or 0.001) of the B[a]P value. PAHs always occur in complex mixtures, and the TEF approach may be used to provide an aggregate assessment of cancer risk, rather than focusing on B[a]P alone.

Much of the experimental data on the toxicity of B[a]P is irrelevant to the development of wildlife Toxicity Reference Values (TRVs) for the compound, since their primary focus is on tumor formation. In general, TRVs are based on noncancer effects, of which there is paucity of studies for B[a]P. The TRVs presented are intended to serve as benchmarks for screening-level ecological risk assessments. This Wildlife Toxicity Assessment (WTA) summarizes available information on the likely effects of B[a]P on wildlife. The TRVs are documented in the U.S. Army Center for Health Promotion and Preventive Medicine Technical Guide 254, *Standard Practice for Wildlife Toxicity Reference Values* [3].

TOXICITY PROFILE

Environmental Fate and Transport

There is an extensive bibliography on the dispersion of PAHs such as B[a]P in environmental media, with HSDB [4] and ATSDR [1] providing summaries of the available information. The fact that B[a]P has been detected in all environmental media and in foodstuffs [4] attests to the compound's ability to become widely dispersed when released to the atmosphere, and to the comparative ease with which it can become taken up into the food chain.

The ATSDR [1] discusses the potential for B[a]P and other PAHs to become dispersed throughout environmental media following their release to the atmosphere in combustion emissions. Natural sources of PAH emissions include forest fires and volcanoes, while anthropogenic sources include the residential burning of wood; generation of industrial power; incineration of waste; production of coal tar, coke and asphalt; and exhaust emissions from use of fossil fuels, as in automobiles, among many others. Typically, a complex mixture of PAHs will be released by the processes of which B[a]P is but a single, and not necessarily the most heavily represented, component.

Airborne PAHs undergo widespread dispersion in the atmosphere but will ultimately deposit in surface water or soil because of wet or dry deposition. In surface water, the compounds can volatilize, photolyze, oxidize, biodegrade, bind to suspended particles or sediments, and/or bioaccumulate in aquatic organisms. However, the combined low water-solubility and vapor pressure, and high octanol-water partition coefficient (K_{ow}) of B[a]P suggest that the compound will partition to the soil compartment primarily, with sediment secondary, and air, water, and biota much less heavily favored [1]. Table 24.1 summarizes the physical and chemical properties of B[a]P.

The capacity for PAHs to undergo abiotic degradation in the atmosphere, for example, by photooxidation, is governed largely by the physical nature of the compound. Compounds absorbed to soot are more resistant to photochemical reactions than pure compounds; however, photolysis may be an important process by which B[a]P molecules in water can be degraded. A number of microbial genera are able to break down PAHs to related but not necessarily simplified structures. In addition, some algae and fungi are

Table 24.1 Summary of the Physical-Chemical Properties of Benzo[a]pyrene

CAS No.	50-32-8
Molecular weight	252.32
Color	Pale yellow
Physical state	Monoclinic crystals/plates
Melting point	179 °C
Boiling point	310–312 °C (at 10 mm Hg)
Odor	Faint aromatic
Solubility in water	1.6–2.3 μg/L at 25 °C: miscible with hydrocarbon solvents
Partition coefficients:	
Log K_{ow}	5.97–6.06
Log K_{oc}	6.74
Vapor pressure at 25 °C	$5.5 - 5.6 \times 10^{-9}$ mm Hg
Henry's Law constant at 25 °C	4.9×10^{-7} atm.m^3/mole
Conversion factors	1 ppm = 10.32 mg/m^3 1 mg/m^3 = 0.097 ppm

Sources: ATSDR [1], HSDB [4].

effective, with biodegradation being an important process by which PAHs can be removed from soils and sediments [1].

Summary of Mammalian Toxicity

Mammalian Oral Toxicity

By far, the majority of research on the toxicity of B[a]P has involved mammalian test species largely as a surrogate for human exposure and toxicity. The vast majority of these studies, therefore, are of little utility for wildlife TRVs. Other than a few key research articles that characterized effects of B[a]P on TRV-relevant endpoints (e.g., survival, growth, reproduction), most articles focus on various elements of the carcinogenic mechanism and manifestation of B[a]P. With the advent of the "omics" revolution for example, a wide array of studies have been produced that are more mechanistically oriented.

Mammalian Oral Toxicity: Acute

There are no listed values for the median lethal dose (LD_{50}) of B[a]P via the oral route in such secondary references as IARC [5], ATSDR [1], Lewis [6], and HSDB [4]. However, many studies have explored the acute toxicity of B[a]P to experimental animals. For example, Reddy *et al.* [7] studied the induction of nuclear anomalies in the gastrointestinal tract of B6C3F1 mice gavaged with a single dose 192 mg/kg of B[a]P (and other PAHs) in dimethyl sulfoxide. All experimental animals were sacrificed 24 to 48 hours after dosing, and the incidence of nuclear anomalies was monitored in the forestomach epithelium by quantitative histopathology/cytology. By using a range of PAHs of differing carcinogenic potential, the authors concluded that the induction of nuclear anomalies by the tested compounds was broadly aligned to their relative carcinogenic potency in the gastrointestinal tract. However, in general, cancer studies are not considered highly relevant to a WTA.

Saunders *et al.* [8] evaluated the effect of a single oral dose of B[a]P in male rats. The endpoints included neurobehavioral effects as well as antioxidant enzyme levels and lipid peroxidation in selected brain regions. Animals were given a single oral dose (25–2,000 mg/kg) of B[a]P and motor activity was determined at 2, 4, 6, 12, 24, 48, 72, and 96 hours posttreatment. Benzo [a]pyrene treatment markedly affected motor activity of males that were generally restricted to early observation periods and higher concentrations of the chemical. Treatment with B[a]P also affected brain enzyme levels that suggested the involvement of oxidative stress in neurobehavioral effects in a mechanism that was partly mediated by inhibited antioxidant scavenging systems of the brain [8]. Although the lowest observed effect level was 50 mg/kg, the effects were not observed beyond 48 hours, which suggested recovery from the initial exposure.

Knuckles *et al.* [9] evaluated the acute toxicity of B[a]P to F-344 rats. An equal number of rats/sex/group were dosed once by oral gavage with 0, 100, 600, or 1,000 mg benzo[a]pyrene/kg in peanut oil. Rats were sacrificed 14 days later, following which blood specimens were taken to quantify a number of hematological parameters, and organs (e.g., ovaries, liver, stomach, kidney, and testes) were weighed and evaluated for morphological and histopathological analysis. The liver-to-body-weight ratio was significantly lower in male and female treated animals at 100 mg/kg and 600 mg/kg, respectively, as compared to controls. The white blood cell counts and mean cell hemoglobin concentrations were also markedly lower in the males only at 600 mg/kg as compared to the controls.

The study by Reddy *et al.* [7] illustrated the propensity of PAHs like B[a]P to cause profound site-specific lesions when administered to experimental animals (e.g., at doses from 48 to 192 mg/kg/day), for which the development of skin papilloma in the nude mouse skin-painting protocol is the best documented example. However, several reports showed that at least a portion of the applied dose traverses the absorption barrier that could also be transported to remote sites. Kleisch *et al.* [10] studied chromosome damage of mouse bone marrow by B[a]P administered to (101 × C3H) F_1 mice in single oral doses ranging from 0 to 250 mg/kg. Chromosome analysis of bone marrow specimens containing 2,000 polychromatic erythrocytes at 12, 30, or 54 hours postdosing showed treatment-related increases in micronuclei and in a dose-dependent manner.

Benzo[a]pyrene can cause both target-specific lesions and effects in distant organs or tissues that result from migration of B[a]P from the site of exposure. Benzo[a]pyrene occupies the hydrocarbon hydroxylase receptor (AHH), and similar to observations with dioxins, can elicit various toxic metabolites. Simultaneously, this leads to the breakdown of PAHs, thereby limiting bioaccumulation and decreasing the overall toxic load.

Mammalian Oral Toxicity: Subacute

Anselstetter and Heimpel [11] studied the subacute effects of B[a]P and development of hematotoxicity in DBA/2 and BDF1 mice. Severe bone marrow depression, with near complete destruction of pluripotent hematopoietic stem cells, was seen in female DBA/2 mice after administration of oral B[a]P (125 mg/kg body weight (bw)/day) for 13 days. Extreme resistance to bone marrow toxicity was observed in BDF1 mice fed for 19 days [11].

Mammalian Oral Toxicity: Subchronic

Ramesh *et al.* [12] carried out a subchronic study in F-344 rats exposed to B[a]P in the diet at concentrations equivalent to doses of 0, 5, 50, or 100 mg/kg for 90 days. Five animals/sex/group were sacrificed after 30, 60, and 90 days, and liver microsomal preparations were monitored for aryl hydrocarbon hydroxylase (AHH) activity. Both dose- and time-dependent increases in AHH activity were obtained, potentially contributing to the formation of toxic reactive metabolites and disease symptoms in target organs such as the liver.

Knuckles *et al.* [9] evaluated the subchronic toxicity of B[a]P to F-344 rats for 90 days. Forty rats/sex/group were given 0, 5, 50, and 100 mg/kg/day of B[a]P administered via feed. The actual doses received (consumed) by the animals were approximately 10% of the nominal doses; that is, 0.5, 5, and 10 mg/kg bw/day, respectively. The effects of B[a]P on body weight, hematological parameters, and organ mass, gross alteration, and histopathology were also studied. The diet was a mixture of lab meal that was obtained from the Purina Ralston Company and the respective B[a]P concentration. The liver, kidney, stomach, prostate, testes, and ovaries were weighed and evaluated for gross and histopathological changes. Data obtained from hematological analyses and evaluation of the organs was analyzed by two-way ANOVA, Fisher's Exact Test, and the Cochran-Armitage test.

At 90 days, Knuckles *et al.* [9] found that the red blood cells (RBC) and hematocrit levels had significantly decreased at 50 mg/kg/day for males and 100 mg/kg/day for females. Hemoglobin levels had significantly decreased for males and females at 100 mg/kg/day.

Wolford *et al.* [13] established reference blood parameter data for mice and rats. The study presented typical blood parameter values observed in mice and rats. Although Knuckles *et al.* [9] states that RBC, hematocrit, and hemoglobin levels had significantly decreased at particular doses at 90 days, these values fall within the normal range [13], indicating that the findings were not biologically significant.

No significant differences between the control and experimental groups regarding body weight were reported. Tubular casts in the kidney tissues were reported to be abnormal from the controls in males at 50 and

100 mg/kg/day after 90 days. Eighty percent of the males were affected at the 50 mg/kg/day level and 100% of the males were affected at 100 mg/kg/day. Ten percent of the females at 50 mg/kg and 100 mg/kg were found to have abnormalities in the tubular casts in the kidneys [9].

DeJong *et al.* [14] studied the immunotoxicological effects of B[a]P to male Wistar SPF rats. Eight rats/group were dosed via gavage with 3, 10, 30, or 90 mg benzo[a]pyrene/kg bw in soybean oil. After 35 days of exposure, the rats were necropsied and body weight, blood, and organs were evaluated. The immunotoxicological parameters studied were bone marrow, thymus, spleen, and lymph nodes. The adrenals, brain, bone marrow, colon, caecum, jejunum, heart, kidney, liver, lung, lymph nodes, esophagus, pituitary, spleen, stomach, testis, and thymus were evaluated for gross and histopathological changes. All organs were weighed with the exception of the bone marrow, pituitary, and GI tract.

The immunotoxicological parameters and organ weights were assessed using one-way ANOVA. The critical values in the blood data were determined by using the protocol per Statistical Tables and Biometry and The Principles and Practice of Statistics in Biological Research [15,16]. The histopathological data were evaluated using Fisher's exact test.

The results in the study stated that there was a significant difference in body weight between the control and the 3 mg/kg dose groups. This was true in the 90 mg group but not in the intermediate 10 and 30 mg groups. There was a significant difference in RBC, hemoglobin, and hematocrit between the control and the 10 mg/kg dose groups. Although all values in the DeJong *et al.* [14] study were statistically significant, the findings are not biologically significant. The RBC levels fall within a normal range as defined previously [13]. The hemoglobin levels in the DeJong *et al.* study were slightly higher [14] than the values stated in Wolford *et al.* [13] yet are not considered biologically significant since the increase was not that much higher compared to the reference value. The hematocrit value was higher than the reference value in Wolford *et al.* [13], although it was not considered biologically significant since the other blood parameters, including the RBC and hemoglobin levels, fell within the normal range, and it was likely that since the RBC and hemoglobin levels increased, the hematocrit levels were also increased.

Mammalian Oral Toxicity: Chronic

No studies on the chronic toxicity of B[a]P were located.

Mammalian Oral Toxicity: Other

MacKenzie and Angevine [17] conducted an experiment to examine the reproductive and developmental effects of oral B[a]P exposure. The study assessed the effects of B[a]P to CD-1 mice upon exposure during the fetal stage. Thirty to 60 pregnant female mice/group were gavaged with 0, 10, 40, or 160 mg/kg bw in 0.2 mL corn oil on gestation days (GDs) 7 to 16, then allowed to deliver and nurse their young. After treatment and giving birth to pups, the litter size, pup weight, and percentage of viable litters were recorded. Litters (F_1) were allowed to remain with their mothers until postnatal day 20, at which point the dams were removed. At 7 to 8 weeks old, each F_1 male was housed with two untreated virgin females every 5 days for 25 days. Pregnant females from these matings were sacrificed on GD 19, and the number of implants, fetuses, and resorptions was recorded. The F_2 young were examined for gross abnormalities. At 7 to 8 weeks old, each F_1 female was housed with untreated proven breeder males for periods of up to 6 months. All resulting F_2 young were counted and examined for gross abnormalities on day 1 of life, then sexed and weighed at 4 days of age.

The data on litter size for F_1 and F_2 populations were analyzed using the least significant difference method. The data on pup weight and the fertility index in the F_1 males and females were analyzed using nonparametric statistics. There was no significant difference between the control and all experimental dose groups regarding the mean litter size and percentage of viable litters in the F_1 population. There was a statistically significant decrease in mean pup weight in all B[a]P exposed groups (F_1 population). A significant decrease in fertility index (females pregnant/females exposed to males $\times$ 100) had resulted in the F_1 males and females at 10, 40, and 160 mg B[a]P /kg/ day. F_2 mean litter size by F_1 females was significantly lower at 10 mg/kg/ day compared to controls. The results indicate that the effects of B[a]P exposure during development had manifested in adult reproduction for the F_1 population. These results indicate that a prenatal exposure of 10 mg/kg/ day is an appropriate lowest observable adverse effect level (LOAEL).

Kristensen *et al.* [18] reported that B[a]P might suppress development of primordial oocytes during fetal life. Benzo[a]pyrene was given to female NMRI mice, 10 mg/kg/body weight by oral intubation on days 7 to16 of gestation (F0 generation). The female pups (F1 generation), thus exposed prenatally to B[a]P, showed markedly reduced fertility with few ovarian follicles compared to controls.

Gao *et al.* [19] evaluated the effects of orally and interperitoneally administered B[a]P on female mouse survival, growth, and cervix morphology.

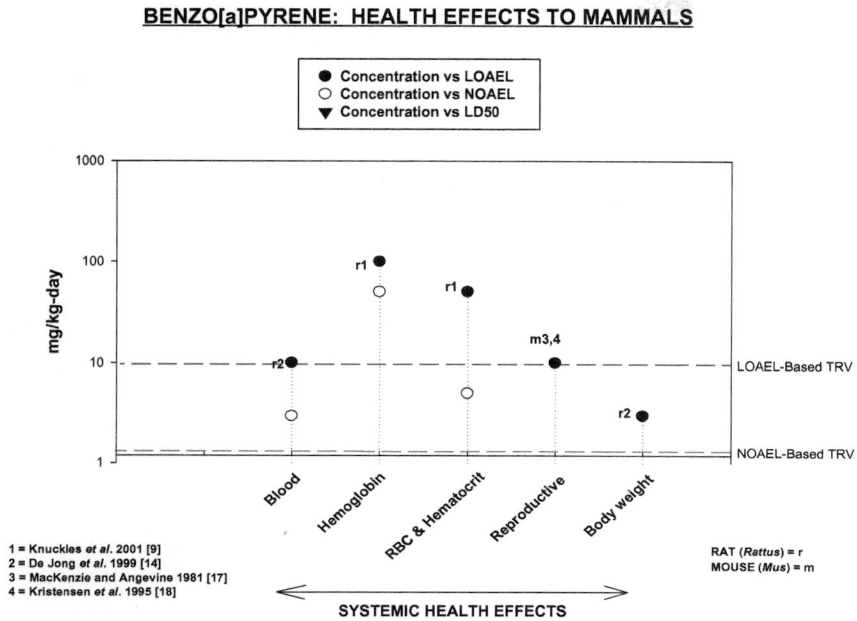

Figure 24.1 Benzo[a]Pyrene: Health Effects to Mammals. Letters by symbols represent test species, and numbers represent study.

Female mice received either an oral or interperitoneal dose of 0, 2.5, 5, or 10 mg B[a]P/kg body weight, twice a week for 14 weeks. B[a]P treatment caused a variety of effects including increased enzyme activity of CYP1A1, creatine kinase, and aspartate aminotransferase (AST), and decreased activity of glutathione-S-transferase in the cervix. Cervix morphology was altered and weight was reduced in B[a]P treated mice and 5.0 and 10 mg B[a]P/kg significantly decreased survival and body mass in female mice. Overall, IP injection of B[a]P caused more severe effects than did oral gavage. There were no significant effects at 2.5 mg B[a]P/kg body weight. Table 24.2 summarizes the relevant mammalian oral toxicity data for TRV derivation

Studies Relevant for Mammalian TRV Development for Ingestion Exposures

Although the majority of studies on B[a]P focus on carcinogenic effects, several studies are potentially suitable for deriving a mammalian TRV. Ramesh *et al.* [12] found that there was a dose-dependent and time-dependent increase in AHH activity. However, due to the lack of histopathological analyses to determine if increased levels of AHH had an adverse effect on the liver, this

Table 24.2 Summary of Relevant Mammalian Oral Toxicity Data for TRV Derivation

Test Organism	Test Duration	Test Results: NOAEL (mg/kg/day)	Test Results: LOAEL (mg/kg/day)	Test Results: Effects Observed at the LOAEL	Study
Mice (CD-1)	GDs 7–16	ND	10	Reproductive effects (fertility, decreased birth weight)	[17]
Mice (NMRI)	GDs 7–16	ND	10	Decreased ovarian follicles, litters, litter size	[18]
Rats (Wistar SPF; males)	35 days	ND	3	Decreased body weight, decreased growth	[14]
		3	10	Increased RBC, hematocrit, and hemoglobin	
Rats (F-344)	90 days	5	50	Increased RBC and hematocrit levels (males)	[9]
		50	100	Increased hemoglobin levels	

GD, gestation day; ND, not determined; RBC, red blood cell(s).

study will not be used to derive a mammalian TRV. Knuckles *et al.* [9] and DeJong *et al.* [14] assessed the toxic effects of B[a]P during subchronic exposures. Knuckles *et al.* [9] found that there was no significant effect of B[a]P on body weight whereas DeJong *et al.* [14] found a significant decrease in body weight in B[a]P exposed rats. The blood parameter data in both studies were not biologically significant as determined by comparisons with reference data in Wolford *et al.* [13]. The only endpoint suitable for TRV development in the aforementioned studies was the altered growth of B[a]P exposed rats (at the lowest level, 3 mg, and at the highest level, 90 mg/kg, but not at the 10 mg or 30 mg levels), as seen by DeJong *et al.* [14]. This study, however, was not chosen for TRV development because of the biological relevance of the statistical 3 mg/kg LOAEL finding and because the MacKenzie and Angevine [17] study was considered more suitable because it was a multigenerational reproductive study with clear effect-based endpoints identified.

MacKenzie and Angevine [17] conducted a developmental/reproductive study showing significant reproductive effects of B[a]P in mice. The study assessed the effects of B[a]P on the development of F_1 pups exposed during gestation, and when they matured, on the reproduction of the F_1 population. Importantly, although this study is not chronic in the classic sense (i.e., study duration), the exposure occurred during a sensitive life stage (development), so it was considered equivalent in weight of a chronic exposure regime [3]. Pregnant dams exposed to B[a]P at 10 mg/kg/day from GD 7 to 16 had given birth to F_1 pups with significantly lower birth weights. Subsequently, when the F_1 mice (exposed to 10 mg/kg/day during development only) reached adulthood, females produced significantly smaller, or no (at higher doses) litters and males and females had a significantly lower fertility index indicating that the effects of prenatal exposure to B[a]P were evident even in adulthood at the lowest experimental dose (10 mg/kg/day). As a result, the data regarding fertility index will be used to derive the mammalian TRVs. This result, with regard to female mice of the NMRI strain, is supported by the findings of Kristensen *et al.* [18], in which a LOAEL of 10 mg/kg/day was established for reduction in litters and litter size and in ovarian follicles. As outlined in TG254, endpoints used to derive a TRV should have a clear link to an ecologically relevant effect [3]. For B[a]P, the reduction in reproductive output and the apparent long-lasting effects of prenatal exposure shown by MacKenzie and Angevine [17] has clear implications for the sustainability of exposed mammalian populations.

Mammalian Inhalation Toxicity

As summarized in ATSDR [1], several inhalation studies exist that address the carcinogenicity of B[a]P via the inhalation route, especially those resulting in site-of-impact lesions to the respiratory tract. Greenwood *et al.* [20] conducted an acute inhalation study pertaining to noncarcinogenic effects of B[a]P emphasizing the neurotoxicological effects at the molecular level in Sprague Dawley rats. The maternal rats were exposed to a dose of B[a]P: carbon black aerosol (100 $\mu g/m^3$) via inhalation on GD 15 for 4 hours. An Sp1 transcription factor consensus sequence was examined by electrophoretic mobility shift analysis of nuclear extracts from various brain regions in pups on postnatal days 3, 5, 7, 10, and 15. The results showed changes in the developmental expression of Sp1 abundance. The data obtained on the temporal and spatial regulation of gene expression in the brain indicate that an effect of the transplacental deposition of desorbed B[a]P to the fetus is in-utero neurotoxicity.

Archibong *et al.* [21] exposed pregnant F-344 rats to B[a]P via the inhalation route at 25, 75, and 100 $\mu g/m^3$ for 4 hours a day for 10 days (GDs 11–20) and evaluated effects on their offspring. Fetal survival decreased in a dose-dependent manner from 78.3 to 33.8% from lowest to highest exposure levels compared to nearly 100% survival in controls. There were also significant decreases in plasma progesterone, estrogen, and prolactin in exposed females associated with B[a]P exposure.

Archibong *et al.* [21] also evaluated effects of inhaled B[a]P on intratesticular function in F-344 rats. Rats were exposed to 75 $\mu g/m^3$ for 4 hours a day for 60 days. At day 60 and then the following 24, 48, and 72 hours, plasma testosterone and luteinizing hormone were measured as indicators of testicular function. Testis weight, total tubule mass, and total tubular length decreased 33, 27, and 39% compared to controls and the number of homogenization-resistant spermatids was markedly reduced in B[a]P exposed rats; plasma testosterone was significantly lower and LH significantly higher in exposed rats as well. Taken together, the data suggest 60-day exposures to B[a]P contributed to reduced endocrine and spermatogenic function.

Wormley *et al.* [22] evaluated the neurotoxic effects in the F1 generation when parental generation were exposed to B[a]P via inhalation. Dams were exposed on GDs 11–21 for 4 hours per day and maintained to term and pups weaned on post-partum day (PD) 30. Electrophysiological studies were conducted PD 60–70 and revealed impacts to long-term potentiation, a cellular correlate of memory and learning.

Ramesh *et al.* [23] evaluated the effect of inhaled B[a]P on fertility endpoints in male F-344 rats. Rats were exposed to 75 μg B[a]P/m^3 for 4 hours a day for 60 days while controls were unexposed. Blood samples were collected after study cessation and again at 24, 48, and 72 hours. Inhaled B[a]P caused reductions in testis weight, components of spermatogenic and steroidogenic compartments of the testis, and progressive motility and mean density of stored spermatozoa were also significantly reduced. There was also a decrease in plasma testosterone and an increase in luteinizing hormone in B[a]P exposed rats. The authors concluded that subchronic exposure to B[a]P via inhalation reduced testicular and epididymal function in rats.

Mammalian Dermal Toxicity

No studies were found.

Mammalian Toxicity from Uncommon Exposure Pathways

There were a variety of studies on B[a]P toxicity in which the exposure route evaluated could be considered uncommonly studied. With growing apparent concerns about the effects of PAHs on humans, a number of studies that evaluated effects of parental exposure on fetuses and newborns were identified. Although in most cases, the endpoints evaluated were generally subtle and not likely relevant for TRV development. Nonetheless, several are highlighted to provide perspective on current environmental toxicity of B[a]P. Bouayed *et al.* [24] evaluated the effects of lactational exposure to B[a]P on postnatal neurodevelopment, gene expression, and behavior. Overall, they found changes in neurodevelopment, gene expression, and behavior (assessed using different mazes) in mice whose mothers were given oral doses of either 2 or 20 mg B[a]P/kg for 14 days and there were significant effects of B[a]P on neonatal mice at both exposure levels.

Summary of Avian Toxicology

Hoffman and Gay [25] found that B[a]P causes embryotoxicity in avian species. Their study design consisted of surface applications of a mixture containing B [a]P to Mallard Duck eggs during a critical period of development (72 hours of development). Proportions of B[a]P (i.e., 0.02, 0.10, or 0.50%) were combined with a petroleum hydrocarbon mixture that was applied to the egg surface [25]. The B[a]P mixture of 0.02% resulted in a significant decrease in embryonic growth and increased occurrence of abnormal survivors. Levels of 0.1% B[a]P resulted in a significant decrease in survival and more abnormal survivors than the 0.05% B[a]P group. Levels of 0.5% B[a]P resulted in mortality of all animals [25]. Abnormal survivors were characterized as having bill defects, gastroschisis, incomplete ossification, stunted embryos with incomplete feather formation, eye and brain defects, anophthalmia, and exencephaly.

Others have assessed the embryotoxicity of B[a]P in Chicken Hens *Gallus domesticus*, and determined mean weight change, food intake, egg production, eggshell thickness, liver weight, and hepatic microsomal proteins after a 5-day treatment with 5 mg of B[a]P that was dissolved in 0.25 mL corn oil and placed in gelatin capsules. The hens were orally administered B[a]P capsules without any marked differences in the measured endpoints [26]. A related study tested B[a]P for embryotoxicity in Chicken Hen *Gallus domesticus* embryos [27]. The study design consisted of injecting 2.0 mg/kg B[a]P into egg yolks that were preincubated for 4 days, following which embryonic mortality was measured 2 weeks later [27]. The vehicle was peanut oil, lecithin, and water. Thirty percent embryonic mortality had resulted, a significant difference from the control. None of the data in the avian studies will be used to derive avian TRVs. Chen *et al.* [25] did not find any significant differences, and the egg injection study by Brunstrom *et al.* [27] and an egg surface application study by Hoffman and Gay [25] do not accurately reflect real-life exposures for birds.

Amphibian Toxicology

Studies designed to obtain more information on the carcinogenic effects of B[a]P have been conducted [28–32]. To date, no amphibian studies have evaluated the noncarcinogenic health effects of B[a]P.

Reptilian Toxicology

There are little to no data on the toxicity of B[a]P to reptiles. In fact, the only studies identified were focused more on polycyclic aromatic hydrocarbons as a whole. One study [33] was designed to improve understanding of the potential effects of polycyclic aromatic hydrocarbons (PAHs) on the development of snapping turtles (*Chelydra serpentina*) residing in a PAH contaminated wildlife refuge. Turtle eggs from clean reference sites were exposed to a crude oil mixture but also included a B[a]P only exposure. There were limited effects of B[a]P on development except for turtles from one reference site where there was a linear relationship between deformity severity and B[a]P exposure concentration. Exposure levels are difficult to determine because exposure mixtures were administered to the eggshell. Nonetheless, these are among the only data that indicate a potential adverse effect of B[a]P on turtle hatchlings. Marsili *et al.* [34] also demonstrated that Italian wall lizards (*Podarcis sicula*) could prove useful as bioindicators of adverse ecological effects for habitat near oil fields. They linked several biomarkers (enzymic activity of P450 1A1; PAH metabolites) with exposures to a suite of chemical stressors, which included PAHs. While neither study was suitable for TRV development, the studies suggest potential for adverse ecological effects to reptiles [33] and that exposures in the field may occur [34].

RECOMMENDED TOXICITY REFERENCE VALUES

Toxicity Reference Values for Mammals

TRVs for Ingestion Exposures for the Class Mammalia

Results from a reproductive study on mice [17] showed significant effects of B[a]P on the fertility index in male and female F_1 mice and on mean litter size in female F_1. These reproductive effects have clear implications for the sustainability of exposed populations; thus, this study is ideal for the derivation of a mammalian TRV. The LOAEL for effects on the fertility index was 10 mg/kg/day, which was the lowest dose tested. The MacKenzie and Angevine [17] study was suitably designed and executed, and was equivalent to chronic exposure data, since exposures occurred during a sensitive life

Table 24.3 Selected Ingestion TRVs for the Class Mammalia

TRV	Dose (mg/kg/day)	Confidence Level
NOAEL-based	1.0	Medium
LOAEL-based	10	Medium

Source: [17].

stage [3]. Since the dose-response data does not include a no observable adverse effect level (NOAEL), the TRV was derived using the NOAEL-LOAEL approach, and the results are presented in Table 24.3. An uncertainty factor of 10 was used to derive the NOAEL-based TRV from the LOAEL [3]. The TRVs are the same for both males and females since the exposure effects on fertility were similar. A medium confidence rating was applied to these TRVs since the study was appropriately designed and representative of a chronic study. Too few mammalian orders were represented, which precluded a high confidence TRV.

TRVs for Inhalation Exposures for the Class Mammalia

Not available at this time.

TRVs for Birds

Not available at this time.

TRVs for Amphibians

Not available at this time.

IMPORTANT RESEARCH NEEDS

The limited availability of data on the toxicity of B[a]P to wildlife species precludes the development of a high-confidence TRV. Hence, more studies on the toxicity of B[a]P to wildlife species are needed. Particularly warranted are long-term, chronic toxicity studies on mammals and additional studies on nonmammalian wildlife such as birds, reptiles, and amphibians. Although there are a number of studies on the effects of B[a]P on mammalian species, many of these are of poor quality and cannot be used to generate TRVs. Further experimentation, with close attention to experimental design, would greatly increase confidence in B[a]P TRVs.

REFERENCES

[1] Agency for Toxic Substances and Disease Registry (ATSDR). Toxicological profile for polycyclic aromatic hydrocarbons. Atlanta (GA): U.S. Department of Health and Human Services, Public Health Service; 1995.

[2] United States Environmental Protection Agency (U.S. EPA). Provisional guidance for quantitative risk assessment of polycyclic aromatic hydrocarbons. Cincinnati (OH): Environmental Criteria and Assessment Office, Office of Health and Environmental Assessment; 1993, EPA/600/R-93/089.

[3] US Army Center for Health Promotion and Preventive Medicine (USACHPPM). Standard practice for wildlife toxicity reference values. Aberdeen Proving Ground (MD): U.S. Army Center for Health Promotion and Preventive Medicine; 2000, Technical Guide No. 254.

[4] Hazardous Substance Data Bank (HSDB). Benzo(a)pyrene [Internet]. Bethesda (MD): National Library of Medicine (US), National Institutes of Health [cited 2001]. Available from: http://toxnet.njm.nih.gov/.

[5] International Agency for Research on Cancer (IARC). Benzo(a)pyrene. Monographs on the evaluation of carcinogenic risk of chemicals to man, vol. 32. Lyon, France: International Agency for Research on Cancer; 1983, p. 211–24.

[6] Lewis Sr RJ, Sax NI. Sax's dangerous properties of industrial chemicals. 8th ed. New York (NY): Van Norstrand Reinhold; 1992.

[7] Reddy TV, Stober JA, Olson GR, Daniel FB. Induction of nuclear anomalies in the gastrointestinal tract by polycyclic aromatic hydrocarbons. Cancer Lett 1991;56 (3):215–24.

[8] Saunders CR, Das S, Ramesh A, Shockley D, Mukherjee S. Benzo(a)pyrene-induced acute neurotoxicity in the F-344 rat: role of oxidative stress. J Appl Toxicol 2006;26 (5):427–38.

[9] Knuckles ME, Inyang F, Ramesh A. Acute and subchronic oral toxicities of benzo(a) pyrene in F-344 rats. Toxicol Sci 2001;61(2):382–8.

[10] Kliesch U, Roupova I, Adler ID. Induction of chromosome damage in mouse bone marrow by benzo(a)pyrene. Mutat Res 1982;102(3):265–73.

[11] Anselstetter V, Heimpel H. Acute hematotoxicity of oral benzo(a)pyrene: the role of the Ah locus. Acta Haematol 1986;76(4):217–23.

[12] Ramesh A, Inyang F, Hood DB, Knuckles ME. Aryl hydrocarbon hydroxylase activity in F-344 rats subchronically exposed to benzo(a)pyrene and fluoranthene through diet. J Biochem Mol Toxicol 2000;14(3):155–61.

[13] Wolford ST, Schroer A, Gohs FX, Gallo PP, Brodeck M, Falk HB, et al. Reference range data base for serum chemistry and hematology values in laboratory animals. J Toxicol Environ Health 1986;18(2):161–88.

[14] De Jong WH, Kroese ED, Vos JG, Van Loveren H. Detection of immunotoxicity of benzo(a)pyrene in a subacute toxicity study after oral exposure in rats. Toxicol Sci 1999;50(2):214–20.

[15] Sokal RR, Rohlf FJ. Statistical tables. 2nd ed. New York (NY): W.H. Freeman and Company; 1981, p. 156–62.

[16] Sokal RR, Rohlf FJ. Biometry, the principles and practice of statistics in biological research. 2nd ed. New York (NY): W. H. Freeman and Company; 1981.

[17] MacKenzie KM, Angevine DM. Infertility in mice exposed in utero to benzo(a)pyrene. Biol Reprod 1981;24(1):183–91.

[18] Kristensen P, Eilertsen E, Einarsdottir E, Haugen A, Skaug V, Ovrebo S. Fertility in mice after prenatal exposure to benzo(a)pyrene and inorganic lead. Environ Health Perspect 1995;103(6):588–90.

[19] Gao M, Li Y, Sun Y, Shah W, Yang S, Wang Y, et al. Benzo(a)pyrene exposure increases toxic biomarkers and morphological disorders in mouse cervix. Basic Clin Pharmacol Toxicol 2011;109(5):398–406.
[20] Greenwood M, Hood DB, Inyang F, Nayyar T, Ramesh A. Modulation in the developmental expression profile of Sp1 subsequent to transplacental exposure of fetal rats to desorbed benzo(a)pyrene following maternal inhalation. Inhal Toxicol 2000;12 (6):511–35.
[21] Archibong AE, Inyang F, Ramesh A, Greenwood M, Nayyar T, Kopsombut P, et al. Alteration of pregnancy related hormones and fetal survival in F-344 rats exposed by inhalation to benzo(a)pyrene. Reprod Toxicol 2002;16(6):801–8.
[22] Wormley DD, Chirwa S, Nayyar T, Wu J, Johnson S, Brown LA, et al. Inhaled benzo (a)pyrene impairs long-term potentiation in the F1 generation rat Dentate gyrus. Cell Mol Biol (Noisy-le-Grand) 2004;50(6):715–21.
[23] Ramesh A, Inyand F, Lunstra DD, Niaz MS, Kopsombut P, Jones KM, et al. Alteration of fertility endpoints in adult male F-344 rats by subchronic exposure to inhaled benzo (a)pyrene. Exp Toxicol Pathol 2008;60(4–5):269–80.
[24] Bouayed J, Desor F, Rammal H, Kiemer A, Tybl E, Schroeder H, et al. Effects of lactational exposure to benzo(a)pyrene (B(alpha)P) on postnatal neurodevelopment, neuronal receptor gene expression and behavior in mice. Toxicology 2009;259 (3):97–106.
[25] Hoffman DJ, Gay ML. Embryotoxic effects of benzo(a)pyrene, chrysene, and 7,12-dimethylbenz(a)anthracene in petroleum hydrocarbon mixtures in mallard ducks. J Toxicol Environ Health 1981;7(5):775–87.
[26] Chen SW, Dziuk PJ, Francis BM. Effect of four environmental toxicants on plasma Ca and estradiol 17β and hepatic P450 in laying hens. Environ Toxicol Chem 1994;13(5):789–96.
[27] Brunstrom B, Broman D, Naf C. Embryotoxicity of polycyclic aromatic hydrocarbons (PAHs) in three domestic avian species, and of PAHs and coplanar polychlorinated biphenyls (PCBs) in the common eider. Environ Pollut 1990;67(2):133–43.
[28] Sadinski WJ, Levay G, Watson MC, Hoffman JR, Bodell WJ, Anderson AL. Relationships among DNA adducts, micronuclei, and fitness parameters in Xenopus laevis exposed to benzo(a)pyrene. Aquat Toxicol 1995;32(4):333–52.
[29] Fernandez M, Jaylet A. An antioxidant protects against the clastogenic effects of benzo (a)pyrene in the newt in vivo. Mutagenesis 1987;2(4):293–6.
[30] Fernandez M, Gauthier L, Jaylet A. Use of the newt larvae for in vivo genotoxicity testing of water: results on 19 compounds evaluated by the micronucleus test. Mutagenesis 1989;4(1):17–26.
[31] Marty J, Lesca P, Jaylet A, Ardourel C, Riviere JL. In vivo and in vitro metabolism of benzo(a)pyrene by the larva of the newt, Pleurodeles walti. Comp Biochem Physiol 1989;93(2):213–9.
[32] Morse HR, Jones NJ, Peltonen K, Harvey RG, Waters R. The identification and repair of DNA adducts induced by waterborne benzo(a)pyrene in developing Xenopus laevis larvae. Mutagenesis 1996;11(1):101–9.
[33] Van Meter RJ, Spotila JR, Avery HW. Polycyclic aromatic hydrocarbons affect survival and development of common snapping turtle (*Chelydra serpentina*) embryos and hatchlings. Environ Pollut 2006;142(3):466–75.
[34] Marsili L, Casini S, Mori G, Ancora S, Bianchi N, D'Agostino A, et al. The Italian wall lizard (*Podarcis sicula*) as a bioindicator of oil field activity. Sci Total Environ 2009;407 (11):3597–604.

CHAPTER 25

Wildlife Toxicity Assessment for Diphenylamine

Christine Ann Arenal, Bradley E. Sample

Contents

INTRODUCTION

Diphenylamine (DPA), chemical formula $(C_6H_5)_2NH$, is a dimer of aniline manufactured by heating the parent monomer in the presence of aniline hydrochloride. DPA is a colorless-to-brown crystalline solid that finds a number of industrial/commercial applications, for example, in the manufacture of dyes and the protection of apples during storage. The compound is also used as an antioxidant, fungicide and antihelminthic agent, and as a stabilizer of nitrocellulose explosives and solid-fuel rocket propellants [1].

Though DPA is used in the food industry for storage and pest control, the compound is thought to have toxic effects in persons exposed in the workplace. Therefore, the National Institute for Occupational Safety and Health (NIOSH) and the American Conference of Governmental Industrial Hygienists (ACGIH) have set time-weighted average threshold limit values of 10 mg/m^3 over an 8-hour (ACGIH) or 10-hour (NIOSH) exposure period [1]. Furthermore, the U.S. Environmental Protection Agency's

Wildlife Toxicity Assessments for Chemicals of Military Concern
http://dx.doi.org/10.1016/B978-0-12-800020-5.00025-9

(U.S. EPA) Office of Pesticides has promulgated a 10 ppm tolerance for residues of DPA in or on pears and apples. To protect environmentally exposed humans, the U.S. EPA's Integrated Risk Information System (IRIS) carries a chronic oral reference dose (RfD) of 0.025 milligram compound per kilogram body weight per day (mg/kg/day) for the compound [2]. The same value was specified in the Health Effects Assessment Summary Tables (HEAST) for subchronic exposure to DPA [3].

This Wildlife Toxicity Assessment (WTA) summarizes available information on the potential harmful effects of DPA to wildlife, emphasizing threshold doses for the onset of noncancer effects, as described in reports of experimental studies of the compound. Available dose-response data for DPA may support the establishment of Toxicity Reference Values (TRVs) that could serve as protective thresholds for wildlife exposed to DPA. This WTA is consistent with protocols outlined in the U.S. Army Center for Health Promotion and Preventive Medicine Technical Guide 254, *Standard Practice for Wildlife Toxicity Reference Values* [4].

TOXICITY PROFILE

Environmental Fate and Transport

No data were found on the fate and transport of DPA in the environment, although some characteristics can be inferred from the compound's physical-chemical properties (Table 25.1). For example, the low solubility of DPA in water suggests that the compound will not readily partition to, or be dispersed in, aqueous media. Additionally, a low log K_{ow} (3.50) suggests low mobility in soil, though DPA would be expected to volatilize from moist soil given the estimated Henry's Law constant of 2.69×10^{-6} atm-m^3/mole [1].

DPA has been shown to biodegrade in media seeded with bacteria of an appropriate species, assuming favorable conditions are maintained for the organism in question. Christodoulatos *et al.* [5] demonstrated the ability of some Pseudomonades, whether as pure strains or in activated sludge, to degrade DPA in bench-top (shaker flasks) or batch reactor experiments. This finding suggests that environmental DPA could be biodegraded in the presence of soil-borne microflora. More recently, Shin and Spain [6] isolated two aerobic bacteria strains (*Burkholderia* sp strain JS667 and *Ralstonia* sp strain JS668) from DPA-contaminated sediment that can grow using DPA as the sole source for carbon, nitrogen, and energy. The DPA degradation pathway was determined and the authors speculate that this pathway developed relatively recently in response to DPA contamination.

Table 25.1 Summary of the Physical-Chemical Properties of Diphenylamine

CAS No.	122-39-4
Molecular weight	169.22
Color	Colorless through brown
Physical state	Crystals
Melting point	53–54 °C
Boiling point	302 °C
Odor	Floral
Solubility in water	53 mg/L @ 20 °C; freely soluble in organic solvents
Partition coefficients:	
Log K_{ow}	3.50
Log K_{oc}	ND
Vapor pressure	6.70×10^{-4} mm Hg @ 25 °C
Henry's Law constant	2.69×10^{-6} atm-m^3/mole @ 25 °C (estimated)
Conversion factors	1 ppm = 6.92 mg/m^3 1 mg/m^3 = 0.14 ppm

Sources: HSDB [1], International Program on Chemical Safety (IPCS) [11].

Summary of Mammalian Toxicity

Mammalian Oral Toxicity: Acute

Although there are few data on the acute toxicity of DPA in laboratory animals, Lewis [7] reported LD_{50} values for the compound of 3,000 mg/kg in rats and 300 mg/kg in mice. Some researchers have examined DPA for acute toxicological effects other than lethality. For example, Kronevi and Holmberg [8] administered the compound to 150 female NMRI mice in a single gavage dose of 600 mg/kg (vehicle not stated). Forty-six mice died within 5 days and another nine within 10 days. A total of 95 animals were reported to be symptom-free and survived for 1 month after dosing. The authors described a number of pathological effects of DPA, although it is unclear from the report whether the observations were in mice that had died prematurely or in those that had survived the full duration of the experiment. Principal effects were acute nephritis characterized by swelling and vacuolation of the proximal tubule epithelium, and dilation of some tubules with the presence of luminal cellular debris [8].

Acute toxicity of DPA was illustrated by the dose-dependent induction of nephropathy in 10 male Syrian hamsters/group that were gavaged on 3 consecutive days with 0, 400, 600, or 800 mg DPA/kg/day in peanut oil [9]. Pathological lesions in identically treated Sprague Dawley rats were observed in only 2/10 high-dose subjects. The rats displayed apex-limited necrosis of the medullary interstitial cells and vasa recta, with degeneration of the renal interstitial matrix. By contrast, no gross or microscopic renal lesions were observed in Mongolian gerbils receiving the same treatment as the hamsters and rats.

In an attempt to determine the biochemical mechanism of DPA-induced nephropathy in Syrian hamsters, Lenz [10] used an exposure regimen similar to that described by Lenz and Carlton [9]. He excised the kidneys and teased apart the papilla, outer medulla, and cortex, then analyzed each for reduced glutathione-stimulating hormone (GSH). A dose-dependent decrease in GSH levels of the renal cortex was observed within an hour after a single oral dose of DPA at 200, 400, or 600 mg/kg. However, no significant changes in outer medullary or papillary GSH were seen with the same treatment. In some experiments, the authors reduced the concentration of renal papillary GSH by 81% with a prior intraperitoneal injection of 500 mg L-buthionine sulfoxime/kg. In administering a non-toxic dose (400 mg/kg) of DPA via the oral route to L-buthionine sulfoxime pretreated hamsters, the authors concluded that DPA-induced renal papillary necrosis in the Syrian hamster is probably not associated with a fluctuation in renal papillary or outer medullary GSH levels nor mediated through oxidative cellular injury.

A detailed review of the toxicology of DPA [11] referenced numerous unpublished reports, including a study on the toxicokinetics of ^{14}C-labeled DPA in Sprague Dawley rats [12]. As described in IPCS [11], five rats/sex/group received a single oral dose of ^{14}C-labeled DPA in corn oil at 5 mg/kg, or an identical dose preceded by 14 days administration of a nonradioactive DPA at the same concentration. A third group of rats received a single oral dose of 750 mg/kg. In all cases, animals were kept in metabolism cages for a 168-hour period during which 94 to 105% of the ^{14}C-labeled-DPA dose was recovered in urine, feces, or carcass. The majority of DPA (68 to 89%) was recovered in the urine, regardless of the dosing regimen employed in the study. Separation and analysis of the products revealed the parent compound and its sulfate and glucuronide conjugates 4, 4-dihydroxydiphenylamine, 4-hydroxydiphenylamine, indophenol, 3-hydroxydiphenylamine, and 2-hydroxydiphenylamine [12].

Mammalian Oral Toxicity: Subacute

One subacute mammalian toxicity study was identified. In this study, an unspecified number of female Sprague Dawley rats were exposed to DPA as a 2.5% dietary supplement [13]. Exposure duration was 3 to 6 weeks at a dose equivalent to 1,250 mg/kg/day (assuming U.S. EPA standard defaults of 350 g body weight and a 0.05 food factor) [14]. The primary observation in these studies was an inability to concentrate urine in DPA-treated rats after a minimum exposure period of 2 weeks. This was seen as a 50% reduction in maximal urine osmolality (U_{max}) from more than 3,000 to approximately 1,500 milliosmoles/kilogram water (mOsm/kg H_2O) in all treated animals compared to controls. Pathological examination revealed that only about 10% of the rats had gross defects in the corticomedullary region of their kidneys. Other manifestations of nephropathy included cystic dilation of the collecting ducts and the appearance of intraluminal proteinaceous cysts. By contrast, the glomeruli, proximal loops, and interstitium appeared to be normal. The authors instituted a number of physiological experiments to explore the early functional deficits of DPA-induced nephropathy, including measurements of inulin clearance, solute-free water reabsorption, free-water clearance, glucose and bicarbonate reabsorption, and ultrastructural tracer and micropuncture studies. Together, data from these studies identify the collecting duct as the location of physiological lesions associated with a failure to maximally concentrate urine.

Mammalian Oral Toxicity: Subchronic

The review by IPCS [11] describes a number of unpublished studies in which DPA was administered to experimental animals for 90 days. Botta [15] exposed 15 CD-1 mice/sex/group to diets containing 0, 10, 520, 2,600, or 5,200 parts per million (ppm) DPA, equivalent to doses of 0, 1.7, 94, 440, and 920 mg/kg/day in males and 0, 2.1, 110, 560, and 1,100 mg/kg/day in females, respectively. A suite of toxicological endpoints was monitored, including clinical signs, body weight changes, food consumption, ophthalmological parameters, hematology and clinical chemistry, organ weight, and gross/histopathology. No compound-related changes were noted in body weight, food consumption, or ophthalmological parameters. However, one characteristic clinical sign in high- and mid-dose animals was the appearance of a greenish stain on the fur in the urogenital area. In the two highest dose groups in both sexes, statistically significant reductions in the red blood cell (RBC) count and hematocrit along with statistically significant and dose-dependent increases in mean corpuscular hemoglobin (MCH), mean corpuscular volume (MCV), and mean corpuscular hemoglobin concentration (MCHC) were noted when compared to the controls.

There were also increases compared to controls in the relative weights of the liver and spleen of male mice receiving DPA at the two highest doses, whereas the kidney and heart were significantly different only in the high-dose group. Similar increases were seen in the relative spleen weights of females at the two highest doses while increases in liver and kidney weights were confined to the high-dose group. Spleen, liver, and kidney in both male and female mice had a darkened appearance at necropsy; a feature that was also apparent in the spleen of female mice receiving DPA at the three highest doses. Histopathological lesions in the kidney were characterized by the deposition of pigment and the formation of cysts in kidney tubules in 9/15 high-dose males. Additionally, increased cellularity of the bone marrow was evident in animals in the mid- and high-dose groups. The authors of the study identified the dose level of 1.7 mg/kg/day in males as a no observed adverse effect level (NOAEL) for DPA, based on the lesions that were apparent in the spleen at the three highest doses. However, because this effect was more evident in female than in male mice, a value of 2.1 mg/kg/day was selected as a more appropriate NOAEL by IPCS [11].

IPCS [11] describes an experiment analogous to the mouse study in which 10 Sprague Dawley rats/sex/group were exposed for 90 days to 0, 150, 1,500, 7,500, or 15,000 ppm DPA in the diet [16]. These levels were equivalent to doses of 0, 9.6, 96, 550, and 1,200 mg DPA/kg/day in males and 0, 12, 110, 650, and 1,300 mg/kg/day in females. An increase in the incidence of green staining of the anogenital area was apparent in females exposed to 1,500 ppm DPA or more. Body weight gain was reduced in animals exposed at the two highest doses, with the urine of high-dose animals darker than that of other groups. Fluctuations in hematological parameters were similar to those described in the mouse study [15]. Statistically significant, but not necessarily biologically significant, changes were seen in some clinical chemistry parameters. For example, there was a dose-dependent increase in the concentration of plasma cholesterol that achieved statistical significance at the three highest dose levels in female rats. Similar to the mouse study, organ weight, necropsy, and histopathology data indicated the spleen, liver, and kidney to be the primary target organs of DPA toxicity. A NOAEL of 12 mg/kg/day was identified based on an increase in liver weight in females receiving DPA at 1,500 ppm or more, and based on a statistically significant and dose-dependent increase in plasma cholesterol levels. NOAEL values of 96 (males) and110 (females) mg/kg/day for spleen and kidney effects in DPA-exposed Sprague Dawley rats were also identified.

IPCS [11] also summarizes a 90-day study by the same author [17] in which four beagle dogs/group received 0, 10, 25, or 50 mg DPA/kg/day in gelatin capsules. Although a full suite of toxicological parameters was evaluated, there were few clear-cut effects of the compound in exposed groups. Consequently, the high dose of 50 mg/kg/day was identified as the NOAEL for the beagle dog.

A Toxic Substances Control Act Submission (TSCATS) to the U.S. EPA gives a brief summary of a 90-day dietary DPA feeding study in rats in which 10 rats/sex/group (strain unstated) received food containing 0, 0.01, 0.03, 0.1, 0.3, or 1.0% DPA [18]. Clinical signs were monitored daily, while body weight was monitored twice weekly for the first 28 days and then once/week thereafter. Food consumption was determined for the first month and hematological parameters were measured at the end of the study in control rats and those exposed to 0.3 and 1.0% DPA. All animals were necropsied and samples of lung, heart, liver, kidney, spleen, testis, pancreas, and adrenals were evaluated for histopathological effects. Clinical signs of toxicity, including increased mortality, were evident in some rats receiving the highest dose. The most sensitive changes associated with the compound were those observed in relative organ weights. For example, there was a statistically significant, dose-dependent increase in relative liver weight in female rats compared to controls at the lowest dose. This suggests that a dose equivalent to 0.01% DPA in feed would result in the lowest observed adverse effects level (LOAEL) in this animal receptor. Using a standard default food factor of 0.05 mg food/kg body weight [14], this would approximate 5 mg/kg/day.

Mammalian Oral Toxicity: Chronic

One of the early studies to describe the chronic effects of DPA in experimental animals was Safouh *et al.* [19] who dosed Sprague Dawley rats with the compound in the diet for periods of up to a year. Laboratory chow contained 1.5 or 2.5% DPA, with and without additional sulfur-containing amino acids, d-l-methionine and 1-cysteine (0.25 and 0.5%). Body weight changes were monitored during the in-life phase of the study and periodically animals were placed in metabolism cages for the collection of urine. Blood samples were taken before periodic sacrifices at 0.5, 1, 2, 5, 8, and 10 months. Kidneys were excised to assess renal cyst formation and to permit the microdissection of dilated and undilated nephrons. Compared to controls, rats receiving DPA showed diminished body weight gain and reduced hemoglobin levels. The urine of DPA-exposed animals displayed a marked reduction in osmolality compared to controls, a functional deficit that appeared to precede the onset of renal cyst formation. Mean maximal urinary concentrating capacity was significantly decreased in rats exposed to

2.5% DPA for 2 months compared to control values. Morphological changes in the kidney (tubular dilation and cyst formation) were most evident in the outer medullary zone, but some cortical collecting tubes were also affected. Occasionally, dilated tubules were filled with cellular debris and pigment. There was no evidence of cystic alteration or degenerative changes in the proximal tubules. Microdissection studies delineated the anatomical site of the dilation as the juncture of the collecting ducts. Because renal cyst formation was evident in rats exposed to 1.5 and 2.5% DPA in feed, a dose equivalent to the lower concentration represents the LOAEL for DPA. Assuming a default food factor of 0.05 mg food/kg body weight for the rat [14], the 1.5% DPA feed concentration approximates a dose of 75 mg/kg/day.

Male Sprague Dawley rats (number unstated) were fed a flavored laboratory chow into which 1% wet weight (w/w) DPA had been mixed [20]. The period of treatment was from 5 to 20 months, during which various microsurgical techniques were used to compare kidney function with untreated controls receiving only the vehicle (methyl cellulose). The parameters evaluated included tubular diameters, intratubular hydrostatic pressures, single-nephron glomerular filtration rates, tubular fluid-to-plasma inulin ratios, time of recovery of [^{3}H]inulin after single nephron microinjection, and transit times within the loop of Henle. Although very few nephrons were dilated, transit times were longer through the loop of Henle despite the similarity in single glomerular filtration rates between dilated and undilated tubules. The authors considered DPA to be responsible for occlusion of affected tubules by the formation of intraluminal debris. The consequent dilation of affected tubules was accompanied by compression of adjacent (noncystic) tubules, further contributing to the functional deficit.

In another series of experiments by the same research group, kidney disease development in response to 1% DPA in the diet demonstrated that the ability to concentrate urine in male Sprague Dawley rats was diminished compared to controls as early as 5 weeks after the onset of treatment [21]. The progressive degeneration of the kidneys was monitored by light, transmission, and scanning electron microscopy and by autoradiography after an intraperitoneal injection of radioactive thymidine. The authors described a suite of morphological changes, including a multilayering of tubular epithelial cells and increases in the number of nuclei and ^{3}H-thymidine-labeled nuclei. Cellular necrosis was evident with the appearance of intraluminal debris after 10 weeks. Frank cysts formation was apparent after 24 weeks.

The U.S. EPA's IRIS record for DPA [2] uses a 2-year feeding study in dogs as its principal study [22]. Two beagles/sex/group were exposed to laboratory chow containing 0, 0.01, 0.1, or 1.0% DPA. Although food consumption did not differ among the groups, body weight gain was reduced throughout the study in both sexes exposed to 1% DPA, and in those exposed to 0.1% DPA at later time points. Hematological monitoring indicated that mid- and high-dose animals were anemic, and pathological examination suggested that liver, spleen, kidney, and bone marrow were the principal target organs. Lesions observed at necropsy included fatty change in the liver and hemosiderosis of the spleen, kidney, and bone marrow in dogs exposed to DPA at the highest dose. Based primarily on decreased body-weight gain, the IRIS reviewers selected the dietary concentration of 0.01% as the subthreshold point-of-departure for toxic effects from which they calculated a NOAEL of 2.5 mg/kg/day. Application of an uncertainty factor (UF) of 100 resulted in a verified RfD of 0.025 mg/kg/day [2].

The same researchers exposed 20 Slonaker-Addis rats/sex/group to 0, 0.01. 0.1, 0.5, or 1.0% DPA in the diet for 2 years while monitoring toxicological parameters such as clinical signs, body-weight gain, food consumption, hematology, gross pathology, and histopathology [23]. The number of survivors at 240 days was broadly similar among the treatment groups. However, body-weight gain was reduced in males, possibly associated with decreased food consumption. Thomas *et al.* [23] reported the findings of a number of separate but related experiments within the same research report. For example, groups of 12 female and 3 male rats/group were exposed to dietary concentrations of DPA of 0, 0.1, 0.25, and 0.5% DPA from 5 to 14 weeks of age. A mated subgroup formed the basis for a two-generational reproductive study, and exhibited few differences in reproductive parameters between the controls and treated groups.

In another experiment, urine samples were collected from five male rats/group that had received 0, 0.1, or 0.5% DPA for 175 and 250 days. No differences in urinary protein content were observed between the groups. Together, the experiments of Thomas *et al.* [23] in rats confirm the primary effect of DPA toxicity in this animal model to be renal histopathological lesions. The lesions were characterized by cystic dilatation of renal tubules, interstitial inflammation, and the accumulation of proteinaceous fluid and pigment. By contrast, there appeared to be no similar abnormalities in the proximal convoluted tubules or glomeruli. A dietary level of 0.01% DPA was the subthreshold point-of-departure for these toxic effects, approximating a NOAEL of 5 mg/kg/day when using a default food factor of 0.05 mg food/kg body weight for rats [14].

A more recent series of experiments by Rohrbach *et al.* [24] demonstrated renal cyst formation in response to dietary administration of 2% (w/w) DPA in C57BL/6 mice for 10 months. Exposed animals developed tubular cysts in both the cortical and medullary areas of the kidney. The authors used immunohistological techniques to demonstrate the loss of the specific glomerular basement membrane glycoprotein, bamin, in DPA-treated mice. While this effect was not evident in controls, the relationship of the impairment of bamin synthesis in basement membranes to tubular cyst formation is unclear. A notable finding in these mice was that while the DPA-treated group appeared in good health, they did not gain weight during the treatment period, whereas the controls did gain weight.

A number of unpublished chronic toxicity studies of DPA have been extensively summarized in IPCS [11], including one by Botta [25] in which 60 CD-1 mice/sex/group were exposed to 0, 520, 2,600, or 5,200 ppm DPA in their diet for up to 78 weeks. These concentrations were equivalent to doses of 0, 73, 370, and 760 mg/kg/day in male mice and 0, 90, 460, and 940 mg/kg/day in females. Among the clinical signs reported was a progressive green staining of the fur in the urogenital region. The two highest dose groups showed a general failure to thrive and displayed increased mortality by 52 weeks. Body weight gain was reduced in a dose-dependent manner in both sexes as was food consumption in males. Interim blood samples were taken from a subset of animals that were sacrificed at 52 weeks. Hemoglobin and the RBC counts were lower in the two highest dose groups compared to controls, while MCV, MCH, and MCHC were higher than controls in the same two groups. These effects were essentially unchanged at the termination of the experiment. Dose-related increases in the absolute and relative weights of the spleen, liver, and heart were seen in animals of each sex at the interim and final sacrifices. These changes were accompanied by a darkening in appearance of the liver and spleen.

Histopathological findings included hematopoiesis and pigment deposition in the liver in the two highest dose groups. Splenic congestion was apparent in all groups, while the level of bone marrow cellularity was increased in a dose-dependent manner. IPCS [11] described amyloiditis in the adrenals, thyroid, kidney, stomach, small intestine, and ovaries of the high-dose females. No signs of compound-related tumor formation were found in any organ or tissue. A dose equivalent to the dietary concentration of 520 ppm DPA (73 mg/kg/day) was identified as the NOAEL based on decreased body weight gain, decreased survival, and significant hematological and gross and microscopic pathological changes. However, because abnormalities in the spleen were evident at all dose levels, the dose of 73 mg/kg/day is the LOAEL for the effects of DPA in this organ.

Botta [26] also exposed 60 Sprague Dawley rats/sex/group to doses of 0, 200, 750, 3,750, or 7,500 ppm DPA (males) or 0, 150, 500, 2,500, or 5,000 ppm DPA (females) in the diet for up to 2 years [11]. These doses were equivalent to 0, 8.1, 29, 150, and 300 mg/kg/day in male and 0, 7.5, 25, 140, and 290 mg/kg/day in female rats. An unexpected finding described in IPCS [11] was that the incidence of mortality was markedly higher in controls and the low-dose group (up to 78%) when compared to the high-dose subjects (about 45%). A dose-dependent reduction in body weight gain was seen with increasing dose, despite the lack of a corresponding decrease in food consumption. Similar hematological changes to those seen in the previous study in CD-1 mice [25] were seen in the rats, along with a dose-dependent increase in absolute and relative spleen weight in the two highest dosed female groups and the high-dosed males. Necropsy revealed a darkened spleen in all groups of exposed males, with histopathological changes in the kidney, liver, spleen, and bone marrow.

The lowest concentration of DPA in the diet (150 to 200 ppm) was determined to be the NOAEL for toxicity based on hematological changes and histopathology of the spleen, kidney, and liver. These DPA dietary concentrations are approximately 7.5 to 8.1 mg/kg/day. More severe splenic changes (darkened and/or enlarged) in males at $\geq$750 ppm and in females at 500 ppm (29 and 25 mg/kg/day, respectively) were considered LOAEL values for this response.

A third unpublished study examining the chronic toxicity of DPA administered 0, 10, 25, or 100 mg DPA/kg/day in gelatin capsules to four dogs/sex/group for 52 weeks [27,11]. Although a full suite of toxicological parameters was monitored, no compound-related effects were seen in mortality, body-weight changes, food consumption, or ophthalmic parameters. However, high-dose males did show decreased hemoglobin, RBC count and hematocrit, while mid- and high-dose males had increased platelet counts when compared to controls. There were inconsistent, and not necessarily compound-related, changes in plasma cholesterol and blood urea nitrogen levels. There were no gross or histopathological differences among the treatment groups. As with the mouse and rat studies [25,26], there was no indication of tumor formation in the dogs [27]. A NOAEL of 10 mg/kg/day was identified based on hematological changes [11].

Mammalian Oral Toxicity: Other

A number of unpublished toxicological studies have addressed the potential for DPA to induce reproductive and developmental effects in laboratory animals. For example, Elf Atochem [28] reported a study in which 18 pregnant female New Zealand white rabbits/group were gavaged on gestation days (GDs) 7 to 19 with 0, 33, 100, or 300 mg/kg/day DPA in 1% methylcellulose. This study was also cited in IPCS [11]. The experiment conformed to the Organization for Economic Cooperation and Development (OECD) guideline No. 414 for studies on the teratogenicity of potentially toxic compounds. The study emphasized and assessed the potential reproductive and developmental toxicity of DPA on parameters such as changes in body weight/food consumption; mortality and/or clinical signs in the dams; number of corpora lutea; litter size and viability; implantation losses; fetal weights; and the incidence of fetal malformations, anomalies, and skeletal variations. Clinical signs evident in exposed dams were a decrease in food consumption and body-weight gain in high-dose rabbits, along with a greenish discoloration of the urine in high- and mid-dose animals. Despite these observations, there were no significant necropsy findings, and all evaluated reproductive and developmental parameters were unchanged between the exposed and control groups.

IPCS [11] cited an unpublished study by Rodwell [29] that examined the reproductive and developmental toxicity of DPA in two generations of rats. Twenty-eight Sprague Dawley rats/sex/group were exposed in the diet to 0, 500, 1,500, or 5,000 ppm DPA concentrations approximating doses of 0, 40, 120, and 400 mg/kg/day in males and 0, 46, 130, and 450 mg/kg/day in females. Test diets were made available to the F_0 generation for 70 days prior to mating, and then repeated during the F_1 generation. It is unclear from the summary in IPCS [11] whether dams and sires were treated with DPA after mating, (i.e., through gestation, lactation, and weaning), although that is the expected method for a two-generational study. The F_0 generation was observed for deaths, clinical signs, body weight/food consumption, mating and fertility indices, reproductive performance, organ weights, and gross and histological appearance. F_1 offspring were observed from birth through lactation for clinical signs, deaths, weights, and sex ratio distribution. All underwent necropsy. F_0 and F_1 individuals receiving the highest dose of DPA developed a bluish tinge to their fur, and a bluish liquid began to accumulate in the cages. Swelling of the mammary glands was seen in females, and palpable masses in the lateral and ventral abdominal regions appeared transiently. Reduced body weights and body-weight gains were significant in the two highest dose groups. The relative weights of the kidney, liver, and spleen were increased at these same dose levels. At necropsy, animals at all doses displayed an enlarged and discolored spleen (e.g., in 9/28 F_1 males and 6/28 F_1 females receiving DPA at the lowest dose). Splenic changes were more common at the higher doses along with a suite of pathological responses in the liver and kidney (e.g., hepatocytic hypertrophy and the appearance of pigment in kidney tubules). Reproductive responses to DPA were seen as reduced litter size and mean pup weights in the high-dose progeny. IPCS [11] recognized the difficulty of deriving a reasonable NOAEL where toxic responses (spleen and liver) were evident at the lowest dose. However, for reproductive toxicity, a dose of 120 mg/kg/day (1,500 ppm DPA in diet) was identified as the NOAEL based on reduced litter size in both generations. The low dose of 46 mg/kg/day (500 ppm DPA in diet) was identified as a developmental NOAEL based on a reduced mean body weight in F_2 pups.

Springborn Laboratories also reported two developmental studies, the first gavaging groups of six pregnant female Sprague Dawley rats on GDs 6 to 15 with 0, 10, 50, 100, 200, 300, or 400 mg/kg/day DPA in corn oil, with termination on GD 20 [30,11]. This first study represented a range-finding experiment for the main study in which groups of 25 pregnant female Sprague Dawley rats were given 0, 10, 50, and 100 mg/kg/day on GDs 6 to 15 with termination on GD 20 [31,11]. Two deaths occurred in the high-dose group in the range-finding study, with dose-related reductions in food consumption and body-weight gain at a dose of 100 mg/kg/day or greater. Necropsy revealed a purple-black discolored spleen in high- and mid-dose dams ($\geq$100 mg/kg/day), with early resorptions and declining uterine and fetal weights in the 200 and 300 mg/kg/day dose groups [30].

The effects of DPA on the spleen in the range-finding study were also observed in 5/25 high-dose animals in the main study. However, no treatment-related effects were observed on gross necropsy or for the reproductive and developmental parameters evaluated (e.g., uterine weights, number of viable fetuses, number of early and late resorptions, number of corpora lutea, number of fetal malformations and variations). The NOAEL for maternal toxicity was 50 mg/kg/day, based on changes in the spleen at the high dose. A NOAEL for developmental toxicity was 100 mg/kg/day, the highest dose tested [31,11].

To date, evidence from chronic exposure studies indicates that DPA is not likely a carcinogen [25,26]. Findings consistent with this observation had been supported by gene reversion and other genotoxicity tests conducted prior to 1995, which have provided uniformly negative results for DPA [11,32–34]. However, Santovito *et al.* [35] recently found cytogenetic effects (increased micronucleus formation) in human lymphocytes exposed to DPA at concentrations of 1.25 μg/mL and higher. The authors indicated that further evaluation to delineate actual genotoxicity is needed.

Studies Relevant for Mammalian TRV Development for Ingestion Exposures

A substantial number of toxicological and physiological studies have been identified that demonstrate the toxicity of DPA in experimental animals. Many of these studies provide data for the identification of NOAELs, making them highly relevant for the development of mammalian TRVs. As shown in Table 25.2, available acute, subacute, subchronic, and chronic toxicological information indicate that the kidney is the primary target organ in laboratory rats. The occurrence of kidney cysts in DPA-treated mice suggests that the response of the kidney to DPA occurs in laboratory mice as well. DPA-induced polycystic disease in rats has been used as a model to study similar renal disease in humans. While much is known about the onset of DPA-induced kidney disease and its associated functional deficits, the biochemical mechanisms and associated pathology remain elusive.

Although the kidney is the primary target organ of DPA toxicity, a detailed summary of unpublished toxicological studies compiled in an addendum to the joint meeting of the Food and Agricultural Organization/World Health Organization (FAO/WHO) panel of experts on pesticide residues has shown that organs other than the kidney are affected by this compound [11]. Principal findings are that other affected organs (e.g., liver, spleen, and bone marrow, etc.) are highly sensitive to DPA exposure. Carcinogenic and genotoxic results are not used in TRV derivation.

Table 25.2 and Figure 25.1 provide the available DPA mammalian NOAEL and LOAEL values from the experimental studies described in this survey.

Mammalian Inhalation Toxicity

No relevant studies were located that examined the inhalation toxicity of DPA.

Mammalian Dermal Toxicity

A study by Siglin [36] monitored a full suite of toxicological parameters at the completion of a 3-week dermal patch test in New Zealand rabbits [11]. Animals were exposed for 6 hours/day, for 21 consecutive days to DPA under an occlusive skin patch, with dose levels approximating 0, 100, 500, or 1,000 mg/kg/day. No adverse effects of DPA on the skin were

Table 25.2 Summary of Relevant Mammalian Data for TRV Derivation

		Test Results			
Test Organism	**Test Duration**	**NOAEL (mg/kg/day)**	**LOAEL (mg/kg/day)**	**Effects Observed at the LOAEL**	**Study**
Mice (CD-1)	90 days	2.1	110	Darkening of the spleen, as well as hemosiderosis and congestion	[15]
		110	560	Decreased RBC count and hematocrit, increased MCH, MCV, and MCHC (hematological parameters); increased pigmentation and slight hematopoiesis in liver	
		560	1,100	Darkened kidneys, pigment deposition, and cystitis	
Rats (Sprague Dawley)	90 days	12	96	Increases in relative liver weight in females, as well as hematopoiesis and pigment	[16]
		110	650	Increased relative weight of kidney and spleen; hematopoiesis, hemosiderosis, and congestion in spleen and pigmentation in kidneys; decreased RBC count and hematocrits	
Dogs (Beagles)	90 days	50	NA	NA	[17]
Rats (ND)	90 days	NA	5	Increases in relative liver weight in females	[18]
Rats (Sprague Dawley)	1 year	NA	75	Renal cyst formation	[19]
Dogs (Beagles)	2 years	2.5	25	Decreased body weight gain and increases in relative weight of liver and kidney; anemia	[22]
		25	250	Hemosiderosis (accumulation of abnormal amounts of granular ferric oxide and protein) in kidney and spleen; fatty change in liver	
Rats (Slonaker-Addis)	2 years	5	50	Renal cyst formation	[23]

Continued

Table 25.2 Summary of Relevant Mammalian Data for TRV Derivation—cont'd

		Test Results			
Test Organism	**Test Duration**	**NOAEL (mg/kg/day)**	**LOAEL (mg/kg/day)**	**Effects Observed at the LOAEL**	**Study**
Mice (CD-1)	78 weeks	NA	73	Increase in relative weight of spleen, with histopathological changes	[25]
		90	460	Increased mortality; decreased growth and food consumption; decreased RBC count and hematocrit, increased MCH, MCV, and MCHC (hematological parameters); increased pigmentation and slight hematopoiesis in liver; increased pigmentation in kidneys	
Rats (Sprague Dawley)	2 years	NA	8.1	Darkened spleen in males	[26]
		7.5	25	Pigment deposition in kidney; decreased RBC count and hematocrit, increased MCH, MCV, and MCHC (hematological parameters); hematopoiesis and pigment deposition in liver	
Dogs (Beagles)	1 year	10	25	Increased number of platelets in males	[27]
Rabbits (New Zealand)	GD 7–19	100	300	Decreased food consumption and body weight gain	[28]
Rats (Sprague Dawley)	2 generations	NA	46	Enlarged and discolored spleen, as well as congestion	[29]
		46	130	Reduced litter size; reduced mean body weight of F_1 pups; pigment deposition in kidneys; liver hypertrophy	
Rats (Sprague Dawley)	GD 6–15	100	200	Increased fetal resorptions and decreased uterine and fetal weights	[30]
Rats (Sprague Dawley)	GD 6–15	100	NA	NA	[31]

NA, not applicable; GD, gestational day; TRV, toxicity reference value; mg/kg/day, milligrams per kilogram per day; NOAEL, no observed adverse effect level; LOAEL, lowest observed adverse effect level.

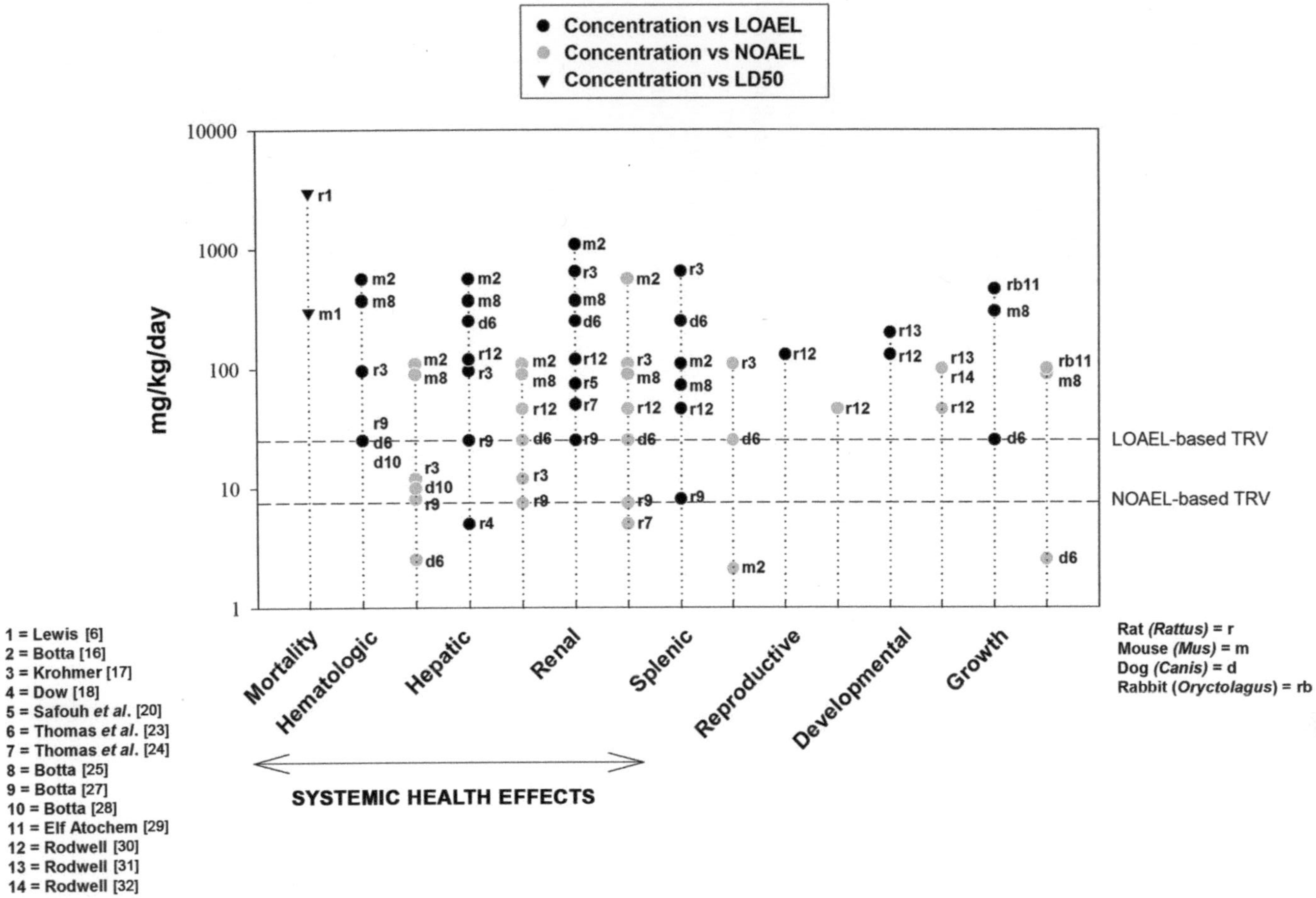

Figure 25.1 Health effects of Diphenylamine in mammals. Letters by symbols represent test species; numbers represent study.

noted at the applied levels. However, dark-red foci in the stomach were observed at necropsy in 4/5 high-dose males, 2/5 high-dose females, and 1/5 mid-dose animals of either sex. The author of the study identified 100 mg/kg/day as the NOAEL for changes observed in the stomach. The NOAEL for dermal effects was 1,000 mg/kg/day, the highest dose tested [36].

Mobil Corporation [37] also reported a 13-week dermal study using 10 Sprague Dawley rats/sex/group, with an exposure regime of 5 days/week at 0, 500, or 2,000 mg/kg/day. DPA was applied to shaved skin on the dorsal surface of rats, which were then fitted with an Elizabethan collar to prevent interfering with the site. Most of the monitored toxicological parameters showed no compound-related effects, including skin irritation, sperm evaluation, urinalysis, hematology, and clinical chemistry. However, absolute and relative liver weights were found at necropsy to be increased in high-dose females and in males at all doses. Similarly, the relative weights of kidneys of high-dose males were increased compared to controls. Notable histopathological changes included vacuolization of the adrenal cortex in 4/10 high-dose males. The authors recommended a subchronic dermal NOAEL for DPA of 500 mg/kg/day, while an increase in absolute and relative liver weight indicated the 500 mg/kg/day dose level as a suitable LOAEL.

Summary of Avian Toxicology

Avian Oral Toxicity: Acute

No relevant studies were identified for acute exposures.

Avian Oral Toxicity: Subchronic

No relevant studies were identified for subchronic exposures.

Avian Oral Toxicity: Chronic

A 2% mixture of DPA in feed was made available to 12 white leghorn chickens for a maximum of 120 days [38]. Using a default body weight of 1.5 kg [39] and a calculated food consumption of 0.106 kg/day (calculated using the White-Leghorn Chicken allometric equation in U.S. EPA [40], a dosage of 2% in the feed is equivalent to approximately 1,413 mg/kg/day. Nine of 12 Chickens (75%) died within 90 to 120 days of the onset of the study. Body-weight gain in surviving Chickens was 20% lower than controls. At necropsy, all organs were normal in appearance, except the kidneys, which were edematous. Histologically, the periphery of affected kidneys exhibited detachment of the epithelium from the basal membrane. Tubules appeared full of amorphous periodate-Schiff's-staining cellular debris. However, no overt cysts were observed.

Avian Oral Toxicity: Other

No other relevant avian studies describing DPA toxicity were identified.

Relevance for Avian TRV Development

One chronic avian toxicity study representing one species (Chicken) and one order was available [38]. Severe toxicity (i.e., death) was observed at the only dose level administered. As with mammals, evidence of kidney damage was observed in birds. This study was not sufficient to develop reliable TRVs; however, the results of this study are discussed below.

Summary of Amphibian Toxicology

Amphibian Toxicity: Other

Fort *et al.* [40] conducted 4-day Frog Embryo Teratogenesis Assay Xenopus (FETAX), and 30-day limb development tests using the South African clawed frog (*Xenopus laevis*). FETAX tests were conducted using both pond water and sediment extracts from a site with observed amphibian malformations and mortality. Additionally, sediment extracts were used in the 30-day limb development tests. Both control water and purified sand extract (standard FETAX) and reference water and sediment extracts were compared to the affected pond water and sediment extracts. Exposure to either pond water or sediment extracts from the contaminated site for 4 days resulted in larval malformations (e.g., craniofacial, mouth, and eye deformities), and 30-day exposure to sediment extracts induced limb maldevelopment.

Through the use of several toxicity characterization methods, compounds associated with developmentally toxic extracts were identified. DPA, estimated at a concentration of 42.3 micrograms per liter (μg/L), was among these compounds. Therefore, DPA was further studied as a single compound in supplemental 4-day FETAX and 30-day limb development tests. Larval effects in the 4-day test included maldeveloped gut, visceral hemorrhage, and muscular kinking, with a 96-hour median teratogenic concentration (EC_{50}) of 39.6 mg DPA/L in FETAX solution and 21.2 mg DPA/L in reference pond water. Conversely, no adverse effects on limb development were observed in *X. laevis* exposed to DPA for 30 days. The authors also determined a median lethal concentration (LC_{50}) of greater than 100 mg DPA/L.

Relevance for Amphibian TRV Development

One study of the effects of DPA on amphibians was located [40]. This study represents one species (*Xenopus*), though this is a standard test species for evaluating effects in amphibians. Effects observed during embryo-larval development at a concentration of 21.2 mg/L included maldeveloped gut, visceral hemorrhage, and muscular kinking. This study is relevant for TRV development.

Summary of Reptilian Toxicology

No relevant reptilian DPA toxicity studies were identified.

RECOMMENDED TOXICITY REFERENCE VALUES

Toxicity Reference Values for Mammals

TRVs for Ingestion Exposures for the Class Mammalia

The toxicological database for DPA includes chronic as well as reproductive and developmental data. More than three species of two orders are represented, as well as chronic NOAEL and LOAEL values. As seen in all species tested, the primary target organ for DPA is the kidney. Damage to this organ has been observed at the cellular level in most studies evaluated [19–24,26], and is expressed physiologically as a reduced ability to concentrate urine [13,19]. Kidney damage may result in hematological (e.g., anemia), neurological (e.g., depression, lethargy, seizures, coma, trembling, weakness) and/or gastrointestinal signs (e.g., nausea, anorexia, stomatitis, vomiting, diarrhea). In particular, animals that are unable to adequately concentrate urine may become dehydrated, while anemic animals may exhibit excessive fatigue. These effects are highly relevant to wildlife at the individual and possibly the population level. Individuals exhibiting clinical signs of kidney dysfunction are potentially weaker and thus more susceptible to death from disease, as well as more vulnerable to predation.

Reproductive and developmental effects of DPA were also described [29–31,34]; however, the thresholds for these effects (LOAELs $\geq$120 mg/kg/day) were greater than those for kidney damage (LOAELs $\geq$25 mg/kg/day). It should be noted that 25 mg/kg/day was also the LOAEL for reduced body weight gain in beagle dogs exposed to DPA for 2 years [22]. The liver, spleen, and bone marrow are also affected by DPA; however, the ecological relevance for endpoints such as enlargement and discoloration, as observed in the liver and spleen, are unclear. Therefore, reproductive and developmental effects, as well as liver, spleen, and bone marrow changes, were not used to develop the mammalian TRV for the oral ingestion of DPA.

Kidney damage was observed in all species and is considered an ecologically relevant endpoint. Therefore, the lowest chronic LOAEL and highest chronic NOAEL (lower than the lowest LOAEL) for kidney effects were selected as the high and low TRVs (Figure 25.1). In this case, the LOAEL and NOAEL values are from the same study [26]. The minimum data requirements as outlined in Section 2.2 of USACHPPM [4] were met; therefore, UFs are not required for TRV development. Table 25.3 presents the selected TRVs. Given the consistency of effects within the studies and among the animal models (rat, mouse, dog), a high level of confidence is given to the derivation of a mammalian TRV for DPA.

TRVs for Ingestion Exposures for Mammalian Foraging Guilds

TRVs specific to particular guild associations (e.g., small herbivorous mammals) have not yet been derived. However, the class-specific TRVs shown in Table 25.3 may be considered to apply to herbivorous small mammals because both rats and mice are members of this guild. TRVs may also apply to carnivorous mammals, which are represented in the data for dogs. As with the class-specific TRVs, three species and two orders are represented so confidence in the TRVs is high. Data to derive TRVs for other guild associations (e.g., strict insectivorous mammals) were not identified.

TRVs for Inhalation Exposures for the Class Mammalia

No data were identified permitting derivation of a mammalian inhalation TRV for DPA.

TRVs for Dermal Exposures for the Class Mammalia

Two studies that evaluated the dermal effects of DPA in Class Mammalia were available. Although two species are represented (rabbits and rats), the effects differed between the species and are of uncertain ecological relevance. For example, rabbits exposed to DPA over a 3-week period exhibited dark-red foci of the stomach [36]. In contrast, rats dermally exposed to DPA for 13 weeks had increased liver (all male dose groups and the female high-dose group) and kidney weights (male high-dose group). In addition,

Table 25.3 Selected Ingestion TRVs for the Class Mammalia

TRV	Dose (mg/kg/day)	Confidence Level
NOAEL-based	7.5	High
LOAEL-based	25	High

Source: [26].

40% of the high dose males exhibited vacuolization of the adrenal cortex. Although the significance of these effects is unknown, both studies indicate potentially adverse effects at 500 mg/kg/day or greater. Because damage to the stomach occurred at 100 mg/kg/day, it is possible that potentially relevant effects could occur at levels less than 100 mg/kg/day. However, the current data are insufficient for TRV development.

Toxicity Reference Values for Birds

TRVs for Ingestion Exposures for the Class Aves

One study that evaluated the effects of oral ingestion of DPA in birds was available. However, this study used just one dose level (2% in feed or 1,413 mg/kg/day) [38]. The effects at this dose were severe, with nine of 12 (75%) individuals dying by the end of the study. Of those that survived, body-weight was 20% lower than controls. As with mammals, damage to the kidneys was observed. This suggests that the kidney is a primary target organ for DPA in birds as well as mammals. Although avian TRVs cannot be developed from these sparse data, applying an UF of 100 (UF of 10 for interspecies variability and UF of 10 for lethality), a LOAEL of 14 mg/kg/day could be estimated. To determine a NOAEL, an UF of 5 was used to approximate a NOAEL of 2.8 mg/kg/day. These thresholds are relatively close to the values observed for mammals (NOAEL-based TRV of 7.5 and LOAEL-based TRV of 25 mg/kg/day). Although not reliable for risk assessment, these TRV estimates should be considered when developing dosing regimens in future toxicity studies.

TRVs for Inhalation Exposures for the Class Aves

No data were identified permitting derivation of an avian inhalation TRV for DPA.

TRVs for Dermal Exposures for the Class Aves

No data were identified permitting derivation of an avian dermal TRV for DPA.

Toxicity Reference Values for Amphibians

One study of the effects of DPA on amphibians was located. This study demonstrated clear toxicological effects on the developing embryo (e.g., maldeveloped gut, visceral hemorrhage, and muscular kinking) [40]. Since these changes adversely affect the health or ecology of the whole organism, this

Table 25.4 Selected Aquatic TRVs for Amphibians

TRV	Concentration (mg/L)	Confidence Level
NOAEC-based	2.12	Low
LOAEC-based	0.21	Low

Source: [40].

study was relevant for TRV derivation. However, the minimum data requirements as described in Section 2.2 of USACHPPM [4] were not met (only a single species represented). Therefore, amphibian DPA TRVs were developed by the approximation approach.

The 4-day FETAX study was conducted during a critical life-stage (embryo-larval development); therefore, the lowest EC_{50} derived from this study (21.2 mg/L) was considered to represent a chronic lowest observed adverse effects concentration (LOAEC). An UF of 10 was applied to account for interspecies variability, resulting in a chronic LOAEC of 2.12 mg/L. From the LOAEC a chronic no observed adverse effects concentration (NOAEC) of 0.21 mg/L was derived by applying an UF of 10.

Table 25.4 presents the derived TRVs. Given the paucity of available data, the reliance on the EC_{50}, and toxicity information for only a single species, a low level of confidence was given to amphibian TRVs.

Toxicity Reference Values for Reptiles

No data were identified permitting derivation of a TRV for DPA toxicity to reptiles.

IMPORTANT RESEARCH NEEDS

Mammalian TRVs derived for DPA have high confidence due to the consistency in effects across studies and species. However, additional species and taxonomic orders should be evaluated to provide a greater breadth of interspecific data. In addition, toxicity studies that examine demographic factors such as birth, death, and recruitment would have much greater ecological significance. The additional data would increase confidence in mammalian TRVs and enable the development of TRVs for specific foraging guilds. Inhalation and dermal studies on mammals were either lacking or limited for DPA. The paucity of toxicity data for birds and the absence of toxicity data for reptiles preclude TRV derivation. Additionally, only one study was available to derive TRVs for amphibians, resulting in low confidence in the

TRVs. Appropriate acute, subacute, subchronic, and especially chronic DPA toxicity data derived through biologically relevant exposure routes are needed before reliable avian, amphibian, and reptilian TRVs can be developed. Research studies should include experimental models of species genetically, biologically, and behaviorally similar to wildlife exhibiting the greatest propensity for toxicant exposure. For birds, the limited data suggest that dosages under 1,400 mg/kg/day need to be evaluated. Experimental design should attempt to mimic both exposure type and duration, and include assessments of long-term effects.

REFERENCES

[1] Hazardous Substance Data Bank (HSDB). Diphenylamine [Internet]. Bethesda (MD): National Library of Medicine (US), National Institute of Health [cited 2012]. Available from: http://toxnet.njm.nih.gov/.
[2] U.S. Environmental Protection Agency (U.S. EPA). Integrated risk information system, [Internet]. Cincinnati (OH): Office of Health and Environmental Assessment, National Center for Environmental Assessment; 2014. Available from, http://www.epa.gov/IRIS/.
[3] U.S. EPA. Health effects assessment summary tables. FY-1997 Update. Washington (DC): Office of Research and Development, Office of Emergency and Remedial Response; 1997, EPA 540-R-97-036. PB97-921199.
[4] USACHPPM. Standard practice for wildlife toxicity reference values. Aberdeen Proving Ground (MD): U.S. Army Center for Health Promotion and Preventive Medicine; 2000, Technical Guide No. 254.
[5] Christodoulatos C, Koutsospyros A, Brodman BW, Korfiatis GP. Biodegradation of diphenylamine by selected microbial cultures. J Environ Sci Health 1997 Jan;A32:15–30.
[6] Shin KA, Spain JC. Pathway and evolutionary implications of diphenylamine biodegradation by *Burkolderia* sp. strain JS667. Appl Environ Microbiol 2009;75 (9):2694–704.
[7] Lewis Sr RJ. Sax's dangerous properties of industrial materials. 8th ed. New York (NY): John Wiley & Sons, Inc.; 1992.
[8] Kronevi T, Holmberg B. Acute and subchronic kidney injuries in mice induced by diphenylamine (DPA). Exp Pathol 1979;17(2):77–81.
[9] Lenz SD, Carlton WW. Diphenylamine-induced renal papillary necrosis and necrosis of the pars recta in laboratory rodents. Vet Pathol 1990;27(3):171–8.
[10] Lenz SD. Investigation of regional glutathione levels in a model of chemically induced renal papillary necrosis. Food Chem Toxicol 1996;34(5):489–94.
[11] International Program on Chemical Safety (IPCS). Diphenylamine [Internet]. Addendum to the Joint Meeting of the FAO/WHO Expert Group on Pesticide Residues [cited 1998]. Available from: http://www.inchem.org/documents/jmpr/jmpmono/v098pr07.htm.
[12] Wu D. Diphenylamine: rat metabolism study. Plainsboro (NJ): Xenobiotic Laboratories, Inc; 1993, Submitted to WHO by the DPA Task Force (as cited in IPCS, 1998). XBL No. 92081, Report No. RPT00131. Unpublished study.
[13] Eknoyan G, Weinman EJ, Tsaparas A, Tisher CC, Yarger WE, Suki WN, et al. Renal function in experimental cystic disease of the rat. J Lab Clin Med 1976;88(3):402–11.

[14] U.S. EPA. Appendix C - guidelines and methodology used in the preparation of health effect assessment chapters of the consent decree water criteria documents. Fed Regist 1980;45(231):79347.
[15] Botta JA. 90-day subchronic toxicity evaluation of diphenylamine in the mouse. Mount Vernon (IN): T.P.S., Inc; 1992, Submitted to WHO by the DPA Task Force (as cited in IPCS, 1998). Unpublished Report No. 426E-001-034-91.
[16] Krohmer RW. 90-day subchronic toxicity evaluation of diphenylamine in rats. Mount Vernon (IN): T.P.S., Inc; 1992, Submitted to WHO by the DPA Task Force (as cited in IPCS, 1998). Unpublished Study No. 426C-10-034-91.
[17] Krohmer RW. 90-day evaluation of diphenylamine in the dog. Mount Vernon (IN): T.P.S., Inc; 1992, Submitted to the WHO by the DPA Task Force (as cited in IPCS, 1998). Unpublished Study No. 426C-501-034-91.
[18] Dow Chemical Company. Results of 90-day dietary feeding of diphenylamine to rats with cover letter dated 05/10/94. Cincinnati (OH): U.S. Environmental Protection Agency; 1994, EPAOTS0557241.
[19] Safouh M, Crocker JF, Vernier RL. Experimental cystic disease of the kidney: sequential, functional and morphologic studies. Lab Invest 1970;23(4):392–400.
[20] Gardner KD, Solomon S, Fitzgerrel WW, Evan AP. Function and structure in the diphenylamine-exposed kidney. J Clin Invest 1976;57(3):796–806.
[21] Evan AP, Hong SK, Gardner K, Park YS, Itagaki R. Evolution of the collecting tubular lesion in diphenylamine-induced renal disease. Lab Invest 1978;38(3):244–52.
[22] Thomas JO, Ribelin WE, Woodward JR, DeEds F. The chronic toxicity of diphenylamine for dogs. Toxicol Appl Pharmacol 1967;11(1):184–94.
[23] Thomas JO, Ribelin WE, Wilson RH, Keppler DC, DeEds F. Chronic toxicity of diphenylamine to albino rats. Toxicol Appl Pharmacol 1967;10(2):362–74.
[24] Rohrbach DH, Robinson LK, Murrah VA. Loss of the basement membrane matrix molecule, bamin, in diphenylamine-treated mice. Matrix 1993;13(5):341–50.
[25] Botta JA. 18-month oncogenicity evaluation of diphenylamine in the mouse. Mount Vernon (IN): T.P.S., Inc; 1994, Submitted to the WHO by the DPA Task Force (as cited in IPCS, 1998). Unpublished Report No. 426H-002-646-91.
[26] Botta JA. 24-month combined oncogenicity/toxicity evaluation of diphenylamine in rats. Mount Vernon (IN): T.P.S., Inc; 1994, Submitted to the WHO by the DPA Task Force (as cited in IPCS, 1998). Unpublished report No. 426D-102-048-91.
[27] Botta JA. One year chronic study of diphenylamine in dogs. Mount Vernon (IN): T.P. S., Inc; 1994, Submitted to the WHO by the DPA Task Force (as cited in IPCS, 1998). Unpublished Report No. 426B-502-044-91.
[28] Elf Atochem North America, Inc. Effect of diphenylamine on pregnancy of the New Zealand white rabbit with cover letter dated 04/13/94. Cincinnati (OH): U.S. Environmental Protection Agency; 1994, Microfiche Number EPAOTS0557206.
[29] Rodwell DE. Two-generation reproductive study in rats with diphenylamine (DPA). Spencerville (OH): Springborn Laboratories, Inc., Life Sciences Division; 1993, Submitted to WHO by the DPA Task Force (as cited in IPCS, 1998). Unpublished Study No. 3255.2.
[30] Rodwell DE. Range finding teratology study in rats with diphenylamine (DPA). Spencerville (OH): Springborn Laboratories, Inc., Life Sciences Division; 1992, Submitted to the WHO by the DPA Task Force (as cited in IPCS, 1998). Unpublished Study No. 3255.2.
[31] Rodwell DE. Teratology study in rats with diphenylamine (DPA). Spencerville (OH): Springborn Laboratories Inc., Life Sciences Division; 1992, Submitted to the WHO by the DPA Task Force (as cited in IPCS, 1998). Unpublished Study No. 3255.3.
[32] McGregor DB, Riach CG, Hastwell RM, Dacre JC. Genotoxic activity in microorganisms of tetryl, 1,3-dinitrobenzene and 1,3,5-trinitrobenzene. Environ Mutagen 1980;2 (4):531–41.

[33] Mobil Corporation. An Ames Salmonella/Mammalian microsome mutagenesis pre-incubation screening assay for determination of potential mutagenicity with diphenylamine with cover letter dated 04/26/94. Cincinnati (OH): U.S. Environmental Protection Agency; 1994, EPAOTS0572471.
[34] Mobil Corporation. An Ames Salmonella/Mammalian microsome mutagenesis pre-incubation screening assay for determination of potential mutagenicity with diphenylamine with cover letter dated 05/03/94. Cincinnati (OH): U.S. Environmental Protection Agency; 1994, EPAOTS0557284.
[35] Santovito A, Cervella P, Delpero M. Micronucleus frequency in human lymphocytes after exposure to diphenylamine in vitro. Mutat Res 2012;747(1):135–7.
[36] Siglin JC. Repeated dose dermal toxicity: 21-day study. Spencerville (OH): Springborn Laboratories, Inc., Life Sciences Division; 1999, Submitted to the WHO by the DPA Task Force. Unpublished study No. 3255.1.
[37] Mobil Corporation. Thirteen-week dermal assay of diphenylamine in rats with cover letter dated 04/26/94. Cincinnati (OH): U.S. Environmental Protection Agency; 1994, EPAOTS0572473.
[38] Sorrentino F, Fella A, Pota A. Diphenylamine-induced renal lesions in the chicken. Urolog Res 1978;6(2):71–5.
[39] U.S. EPA. Recommendations for and documentation of biological values for use in risk assessment. Cincinnati (OH): U. S. Environmental Protection Agency Environmental Criteria and Assessment Office; 1988, EPA/600/6-87/008.
[40] Fort DJ, Rogers RL, Copley HF, Bruning LA, Stover EL, Helgen JC, et al. Progress toward identifying causes of maldevelopment induced in Xenopus by pond water and sediment extracts from Minnesota, USA. Environ Toxicol Chem 1999;18:2316–24.

CHAPTER 26

Wildlife Toxicity Assessment for Ethylene

Rachel M. Hebert, Allison M. Jackovitz

Contents

INTRODUCTION

Ethylene is a colorless gas produced in vast quantities by the petrochemical industry. While widely used as an industrial intermediate in the production of plastics and polymers, ethylene is also used to stimulate ripening in fruits and vegetables. Polyethylene, ethylene oxide (EO), ethylene dichloride, and ethylbenzene are all derived from the compound. In several countries, ethylene is classified as an asphyxiant due to its ability to lower available air concentrations when present at high concentrations [1]. It also acts as a central nervous system depressant; however, more common hazards are flammability and explosion [2]. Ethylene is released into the environment as a product of burning vegetation, a by-product of petroleum refinement, via incomplete combustion of fossil fuels, in automobile and diesel exhausts, and by sewage treatment plants [3]. The compound is also produced and emitted by all plants, humans, and other mammals and, hence, is present naturally in the environment. A summary of the physical and chemical characteristics of ethylene are shown in Table 26.1.

Wildlife Toxicity Assessments for Chemicals of Military Concern
http://dx.doi.org/10.1016/B978-0-12-800020-5.00026-0

Table 26.1 Summary of the Physical-Chemical Properties of Ethylene

CAS No.	74-85-1
Molecular weight	28.05
Color	Colorless
Physical state	Gas
Melting point	−169 °C
Boiling point	−103.7 °C
Odor	Sweet
Solubility	131 mg/L in water at 20–25 °C slightly soluble in benzene, ethanol, acetone; soluble in diethyl ether
Partition coefficients:	
Log K_{ow}	1.13
Log K_{oc}	2.0–2.5
Vapor pressure at 25 °C	5.2×10^4 mm Hg
Vapor density	0.978 (air = 1)
Henry's Law constant at 25 °C	2.3×10^{-1} atm.m^3/mole
Conversion factors	1 ppm = 1.15 mg/m^3 1 mg/m^3 = 0.87 ppm

Sources: HSDB [3], IARC [1].

This chapter summarizes current knowledge of the toxicological impacts of ethylene on wildlife. Evaluating the toxicity of ethylene is intended to contribute to the derivation of Toxicity Reference Values (TRVs), which could serve as protective exposure standards for wildlife near contaminated sites.

TOXICITY PROFILE

Environmental Fate and Transport

Ethylene is a ubiquitous component of the atmosphere, produced both naturally and during manufacturing. Ethylene is listed as a High Production Volume chemical; 36 million pounds and more than 1 billion pounds were produced or imported into the U.S. in 1990 and 2002, respectively. Based on the compound's high vapor pressure of 5.2×10^4 and a Henry's Law constant of 2.3×10^{-1} atm-m^3/mole at 25 °C, volatilization of soil-borne ethylene is likely. As listed in Table 26.1, several physical-chemical characteristics also favor volatilization of ethylene from the surface of marine and freshwater systems. Model rivers and lakes estimate a volatization half-life of 30 minutes and 2 days, respectively. Although sparingly soluble in water, the compound has been detected in oceans, lakes, and rivers in and around the U.S. For example, concentrations of up to 35 nanoliters per liter (nL/L) were measured in water samples taken in the Gulf of Mexico [3].

Researchers have suggested several mechanisms by which ethylene is degraded in the environment. For example, it reacts with photochemically produced hydroxyl radicals, a process that has a half-life of about 1.9 days. Nitrate radicals and ozone also cause ethylene degradation at a lower rate. Additionally, pure culture research shows that the compound can be broken down by microbial action [3].

Given the volatility of ethylene, it is unlikely to persist in the environment long enough for terrestrial wildlife exposure to occur. In the event that it would occur, the most likely routes would be inhalation or dermal. Notably, ethylene is readily converted to ethylene oxide (EO), a somewhat more stable compound that also poses some risks to wildlife, is carcinogenic to rats and mice, and is a known carcinogen (IARC Group 1) to humans. Ethylene readily undergoes oxidation to EO *in vivo* and *in vitro* [4]. Considerable evidence suggests that ethylene is metabolized to EO by cytochrome P450 [3,4].

Summary of Mammalian Toxicology

Mammalian Toxicity: Oral

Given the gaseous nature of ethylene at ambient temperatures and pressures, there are no data on the toxicological effects of ethylene when administered via the oral route.

Studies Relevant for Mammalian TRV Development for Ingestion Exposures

No relevant studies are available at this time for mammalian TRV development for oral exposures.

Mammalian Inhalation Toxicity: Acute

Rats exposed to ethylene at concentrations of up to 500,000 parts per million (ppm) for 5 hours suffered no long-term adverse effects [3]. The apparent tolerance to high concentrations of ethylene in animals precludes the identification of a reliable compound-specific median lethal concentration (LC_{50}). However, it appears that concentrations of ethylene necessary to induce anesthesia nearly induce hypoxia when the proportion of oxygen in the mixture is reduced.

Inhibition of cytochrome P450 is correlated to the concentration of EO in the blood, as shown in an acute inhalation study [4]. Fennell *et al.* exposed male F344 rats to 300, 600, and 1,000 ppm ethylene for up to 6 hours.

Exposure was through the nose only, and blood was drawn throughout. During the first hour of exposure, blood EO concentrations increased to 600 ppm ethylene and then decreased to approximately half of the peak blood concentration. Effects were less pronounced in the 300 and 1,000 ppm groups. Additionally, livers were collected at various stages from four male F344 rats/dose, and in each case, liver microsomal cytochrome P450 activity was decreased.

Although there are few, if any, toxicological consequences of ethylene exposure via inhalation, the ability of the compound to react with biological macromolecules has been demonstrated. *N*-(2-hydroxyalkyl)valine adducts of hemoglobin and 7-alkylguanine adducts of liver and lymphocytes were detected consistently in male Sprague Dawley rats exposed to 1-alkenes [5]. Researchers exposed animals to 1-alkenes, including ethylene, at 300 ppm for 12 hours per day (hr/day) on three separate days. Exposure took place in a conically shaped steel chamber with a glass door and walls. At termination, aliquots of blood and tissue samples were measured for 1-alkenes. The formation of deoxyribonucleic acid (DNA) adducts in liver and lymphocytes were monitored using the ^{32}P-postlabeling technique; hemoglobin adducts were detected using gas chromatography/mass spectrometry. While the concentration of 1-alkenes in blood and different tissues increased with the increasing number of carbon atoms, levels of hemoglobin and DNA adducts decreased.

Mammalian Inhalation Toxicity: Subchronic

An experiment by Vergnes and Pritts [6] used a subacute dosing regimen to examine ethylene's ability to induce the formation of micronuclei in the bone marrow of male F344 rats and B6C3F1 mice. Ten animals per group were exposed to 0, 40, 1,000, and 3,000 ppm ethylene, 6 hr/day, 5 days per week (days/wk) for 4 weeks, with bone marrow collected 24 hours after the last exposure. Exposure took place in a steel inhalation chamber with glass windows. Examination of cell smears revealed little, if any, formation of micronuclei or polychromatic nuclei in the ethylene-exposed groups. However, the incidence of these features was significantly increased in the cells of control animals, who received 200 ppm EO. Although ethylene is not classifiable as to its carcinogenicity in humans, EO is carcinogenic in rats and mice and is rated as a known carcinogen in humans [1].

Walker *et al.* [7] used the same biomarkers as Eide *et al.* [5] to quantitate the conversion of ethylene to EO: *N*-(2-hydroxyethyl)valine in hemoglobin; *N*7-(2-hydroxyethyl)guanine in DNA; and *Hrpt* mutants in splenic T cells. Male F344 rats and B6C3F1 mice were exposed to A) 0 or 3,000 ppm ethylene for 1, 2, or 4 weeks; or B) 0, 40, 1,000, or 3,000 ppm ethylene for 4 weeks in an effort to generate dosimetry data to extrapolate for human risk assessment. The authors showed that cytochrome P4502E1-mediated conversion of ethylene to EO saturates at levels that are insufficient to trigger the toxic responses typical of EO exposure [7].

The Chemical Industry Institute of Toxicology (CIIT) completed two full-scale studies on the toxicology of ethylene [8,9]. Rhudy *et al.* [8] exposed 15 albino rats/sex/group to 0, 300, 1,000, 3,000, or 10,000 ppm ethylene, 6 hr/day, 5 days/wk for 13 weeks. Ethylene was delivered in unspecified inhalation chambers. During the study, clinical signs, mortality, body weights, and food consumption were monitored daily. Clinical chemistry, hematological, and urinalysis parameters were monitored in controls and high dose groups on days 6, 45, and 83. Full necropsies and histopathological evaluations were carried out on all survivors at termination. As there were no compound-related differences noted from ethylene exposure, the authors determined that rats can tolerate exposure of up to 10,000 ppm for 90 days.

Mammalian Inhalation Toxicity: Chronic

The second CIIT study [9] exposed 120 F344 rats/sex/group to 0, 300, 1,000, or 3,000 ppm ethylene 6 hr/day, 5 days/wk, for up to 24 months. Exposure took place in four glass and stainless steel chambers. Animals were euthanized on an interim basis (after 6, 12, and 18 months). Subjects and all survivors were monitored for survival, clinical signs, ophthalmologic characteristics, hematology, clinical chemistry, and urinalysis. Again, no compound-related effects were seen in control and high-dose groups that underwent histopathological examinations.

Studies Relevant for Mammalian TRV Development for Inhalation Exposures

Full recovery of faculties after cessation of exposure to ethylene reinforces the challenge in deriving a TRV for the compound [10]. However, the formation of 2-hydroxyethyl derivatives of hemoglobin and guanine shows that ethylene has biochemical activity, which is most likely mediated through its EO intermediate. Others have addressed the disparity between the benign effects of ethylene and the more severe effects of EO in toxicological tests [6,7,10]. Bolt estimated that 1,000 ppm ethylene was equivalent to about 7.5 ppm EO, a concentration that is likely below the threshold at which any toxicological impacts would become apparent [10].

Mammalian Inhalation Toxicity: Other

Aveyard and Collins [11] used Organisation for Economic Cooperation and Development (OECD) guidelines to test the effects of ethylene on the fertility, pregnancy, maternal and suckling behavior, and F1 growth and development of ethylene-exposed rats (strain unstated). Ten parental animals/group were exposed head-only to 0, 200, 1,000, or 5,000 ppm ethylene, 6 hr/day from 2 weeks prior to mating until gestation day (GD) 20 (males) or postpartum day 4 (females). No animals died during treatment, and no compound-related effects were evident in weight gain, food consumption, fertility, fecundity, litter characteristics, or in pathology or histopathology in either generation.

Mammalian Dermal Toxicity

No relevant studies regarding ethylene dermal toxicity to mammals are available at this time.

Summary of Avian Toxicology

No toxicological data for the effects of ethylene on avian species were located. Ecotoxicological research on the effects of this compound on birds is recommended.

Summary of Amphibian Toxicology

No toxicological data for the effects of ethylene on amphibian species were located. Ecotoxicological research on the effects of this compound on amphibians is recommended.

Summary of Reptilian Toxicology

No toxicological data for the effects of ethylene on reptiles were located. Ecotoxicological research on the effects of this compound on reptiles is recommended.

RECOMMENDED TOXICITY REFERENCE VALUES FOR MAMMALS

TRVs for Ingestion Exposures for the Class Mammalia

At this time, it is not possible to derive a TRV for the oral route of exposure to ethylene due to the lack of toxicity data and the gaseous nature of the compound, which renders an oral exposure unlikely.

TRVs for Inhalation Exposures for the Class Mammalia

There is no evidence of toxicological effects occurring at ethylene levels that would be likely in the environment. However, there are toxicological effects through its EO intermediate. Such effects are molecular in nature and were not linked to more toxicologically relevant endpoints such as tumor formation, morbidity, or death. As a result, a TRV for ethylene cannot be derived at this time.

IMPORTANT RESEARCH NEEDS

The limited availability of data on the toxicity of ethylene to wildlife species precludes the development of a TRV. Hence, additional studies of the compound and its derivatives are recommended. The toxicity and biochemical activity of ethylene and EO warrant research to explain the relationship between these compounds. Chronic toxicity studies on nonmammalian wildlife such as birds, reptiles, and amphibians are particularly necessary.

REFERENCES

[1] International Agency for Research on Cancer (IARC). Evaluation of the carcinogenic risk of chemicals to humans. Ethylene 1994;60:45–72.
[2] American Conference of Governmental Industrial Hygienists. Documentation of the threshold limit values for physical agents. 7th ed. Cincinnati (OH): American Conference of Governmental Industrial Hygienists, Incorporated; 2009.
[3] Hazardous Substances Data Bank (HSDB). Ethylene [Internet]. Bethesda (MD): National Library of Medicine (US), National Institutes of Health [cited 2001]. Available from: http://toxnet.njm.nih.gov/.
[4] Fennel TR, Snyder RW, Parkinson C, Murphy J, James RA. The effect of ethylene exposure on ethylene oxide in blood and on hepatic cytochrome p450 in Fischer rats. Toxicol Sci 2004;81(1):7–13.
[5] Eide I, Hagemann R, Zahlsen K, Tareke E, Torngvist M, Kumar R, et al. Uptake, distribution and formation of hemoglobin and DNA adducts after inhalation of C2-C8 1-alkenes (olefins) in the rat. Carcinogenesis 1995;16:1603–9.
[6] Vergnes JS, Pritts IM. Effects of ethylene on micronucleus formation in the bone marrow of rats and mice following four weeks of inhalation exposure. Mutat Res 1994;324:87–91.
[7] Walker VE, Wu KY, Upton PB, Ranasinghe A, Scheller N, Cho MH, et al. Biomarkers of exposure and effect as indicators of potential carcinogenic risk arising from in vivo metabolism of ethylene to ethylene oxide. Carcinogenesis 2000;21:1661–9.
[8] Rhudy RL, Lindberg DC, Goode JW, Sullivan DJ, Gralla EJ. Ninety-day subacute inhalation study with ethylene in albino rats. Toxicol Appl Pharmacol 1978;45:285.
[9] Hamm Jr TE, Guest D, Dent JG. Chronic toxicity and oncogenicity bioassay of inhaled ethylene in Fischer-344 rats. Fundam Appl Toxicol 1984;4:473–8.
[10] Bolt HM. The carcinogenic risk of ethene (ethylene). Toxicol Pathol 1998;26:454–6.
[11] Aveyard L, Collins CJ. OECD 421-reproduction/developmental toxicity screening study by head-only exposure: experience with ethylene. Teratology 1997;55:47.

CHAPTER 27

Wildlife Toxicity Assessment for Hexachlorocyclohexane (HCH)

Allison M. Jackovitz, Rachel M. Hebert

INTRODUCTION

Hexachlorocyclohexane (HCH) is a mixture of eight or more stereoisomers used as an insecticide to protect crops. Technical-grade HCH (tech-HCH) is manufactured by the photochlorination of benzene, a process yielding a product of variable composition, typically containing 60 to 70% α-HCH, 5 to 12% β-HCH, 10 to 15% γ-HCH, 6 to 10% δ-HCH, and some other minor components. Because of its synthetic relationship to benzene, tech-HCH has also been known historically as benzene hexachloride, or BHC. Structurally, the stereoisomers differ only in the arrangement of the chlorine atoms on the six-carbon cyclohexane ring. The α-HCH isomer exists as a pair of enantiomers, while the other isomers are not optically active. Production of tech-HCH in the U.S. was banned in 1976, but the compound may still be used in other countries [1]. Hexachlorocyclohexane and its component isomers are referred to by a number of scientific, trivial, and proprietary names, many of which are listed in the Hazardous Substances Data Bank record for the compound [2]. The acronym HCH will be used

Wildlife Toxicity Assessments for Chemicals of Military Concern
http://dx.doi.org/10.1016/B978-0-12-800020-5.00027-2

throughout this chapter to imply any formulation of hexachlorocyclohexane, without regard to the isomeric composition. Where specific isomers are identified, the information pertains only to that isomer.

Lindane (CAS No. 58-89-9) is a purified preparation of γ-HCH, and it contains more than 99% of that specific isomer. The terms "lindane" and "tech-HCH" should not be used interchangeably, as the physical, chemical, and biological properties of the isomers differ. Only γ-HCH (1α, 2α, 3β, 4α, 5α, 6β-hexachlorocyclohexane) has any appreciable insecticidal activity [3]. Production of lindane is highly inefficient, contributing to a pollution problem in the form of "HCH muck." Dumpsites, which have been reported in Asia, South America, and Europe, are another source of HCH contamination [4].

The isomer γ-HCH is a seed treatment for barley, corn, oats, rye, sorghum, and wheat. Previously, γ-HCH was used in veterinary products to control mites, lice, and other pests; however, no products are currently registered in the United States for this use. All isomers are toxic to animals to varying degrees and are shown to be persistent in the environment [1,5]. The isomers α-, β-, and γ-HCH were designated as Persistent Organic Pollutants (POPs) at the Stockholm Convention [6]. As such, these isomers need to be phased out and their dumpsites decontaminated in an environmentally responsible manner.

Differential degradation and metabolism of HCH compounds result in a nonuniform uptake by animal systems. The highest concentrations of both α- and γ-HCH are found in soil, air, and water. With respect to acute exposure for mammals, γ-HCH is the most toxic, followed by the α-, δ-, and β-isomers. For chronic exposure, β-HCH demonstrates the greatest toxicity followed by the α-, γ-, and δ-HCH isomers [1]. Most research on various HCH isomers focuses on γ-HCH; however, it is appropriate to discuss each, because all isomers are found in the environment.

This Wildlife Toxicity Assessment summarizes current knowledge of the likely harmful impacts of HCH on wildlife and identifies levels for the onset of toxicological effects as described in reports of experimental studies of the compound. Evaluating the toxicity of HCH is aimed at establishing Toxicity Reference Values (TRVs), which are values that can serve as protective exposure standards for wildlife ranging near contaminated sites. The protocol for the performance of this assessment is documented in the U.S. Army Center for Health Promotion and Preventive Medicine (USACHPPM) Technical Guide (TG) 254, *Standard Practice for Wildlife Toxicity Reference Values* [7].

TOXICITY PROFILE

Environmental Fate and Transport

All the isomers of hexachlorocyclohexane have been detected in environmental media throughout the world. Aerial dispersion of HCH is prohibited; however, HCH volatizes easily, and is thus widely found in airborne vapors and particulates. Environmentally collected samples of HCH tend not to match the typical isomer distribution found in tech-HCH [1]. Numerous studies have found that degradation of HCH isomers via microbial metabolism or catalytic activity of soils is possible [8–12]. The physical-chemical properties of the HCH isomers are summarized in Table 27.1.

Table 27.1 Summary of the Physical-Chemical Properties of HCH Isomers

	α-HCH	β-HCH	γ-HCH	δ-HCH
CAS No.	319-84-6	319-85-7	58-89-9	319-86-8
Molecular weight	290.83			
Color	Brown to white	No data	White	No data
Physical state at 25 °C	Crystalline solid	Crystalline solid; monoclinic prisms	Crystalline solid; monoclinic prisms	Fine plates
Melting point	159–160 °C	314–315 °C	112.5 °C	141–142 °C
Boiling point at 760 mmHg	288 °C	60 °C (at 0.5 mmHg)	323.4 °C	60 °C (at 0.36 mmHg)
Odor	Phosgene-like	No data	Slightly musty	No data
Solubility in water	69.5 mg/L at 28 °C; 10 ppm	5 ppm	17 ppm	10 ppm
Partition coefficients:				
Log K_{ow}	3.8	3.78	3.72	4.14
Log K_{oc}	3.57	3.57	3.0; 3.57	3.8
Vapor pressure (mmHg) at 25 °C	4.5×10^{-5}	3.6×10^{-7} (20 °C)	4.2×10^{-5} (20 °C)	3.5×10^{-5}
Henry's Law constant at 25 °C	6.86×10^{-6}	4.5×10^{-7}	3.5×10^{-6}	2.1×10^{-7}
Conversion factors	1 ppm = 4.96 mg/m^3; 1 mg/m^3 = 0.2 ppm			

Source: ATSDR [1].

Research on the selective metabolism of the enantiomers of α-HCH showed that the (+) enantiomer is preferentially metabolized in the Arctic Ocean, arctic lakes and watersheds, the North American Great Lakes, and the Baltic Sea [13]. In contrast, in some marine regions—the Bering Sea, Chukchi Seas, and parts of the North Sea—the preference is reversed, and the (-) enantiomer of α-HCH is depleted. Apparent pseudo-first-order rate constants in the eastern Arctic Ocean are $0.12\ y^{-1}$ for (+) α-HCH and $0.030\ y^{-1}$ for (-) α-HCH. These rate constants are 3 to 10 times higher than those observed for basic hydrolysis in seawater [13].

Based on the movement of α-HCH in cold, arctic air and of γ-HCH in warm air from the lower United States, ambient levels of HCH might be temperature-dependent. Concentrations of up to $50\ ng/m^3$ HCH in air have been reported [13]. Photodegradation makes a comparatively insignificant contribution to the removal of HCH from the air; the primary removal process is deposition to soil and surface water during rainout [1].

As might be expected for a pesticide intended primarily for the protection of crops, the primary repository for HCH is soil, from which the compound does not readily migrate to other media [1]. HCH is capable of leaching to groundwater; low mobility and wind erosion can facilitate its volatilization. Biodegradation is also a viable mechanism for HCH removal from the soil because a number of bacteria, algae, and fungi use HCH as a carbon source. Metabolic products of these processes include benzene compounds and penta-, tetra-, and tri-chlorinated cyclohexane [1].

Biodegradation of HCH also occurs in surface water, given a suitable microbial environment. Many surface water bodies in the United States have detectable levels of HCH; concentrations of up to 300 ng/L have been reported [1]. Comparable HCH concentrations have also been found in groundwater samples in the United States although current thinking is that leaching from the soil is a comparatively slow process. Concentrations of up to 100 ng/L have been monitored in drinking water [1].

Summary of Mammalian Toxicology

Mammalian Oral Toxicity: Acute

A number of researchers have published oral median lethal dose (LD_{50}) values for isomers of HCH or for one or more of its technical-grade formulations. LD_{50} values of 184 and 157 milligrams per kilogram (mg/kg) γ-HCH were established in male albino Wistar rats fed either a normal protein diet or standard laboratory chow [14]. These values contrasted with an

LD_{50} of 95 mg/kg in rats receiving a protein-deficient diet [14]. When investigating the acute lethality of 98 pesticides and two metabolites of dichlorodiphenyltrichloroethane (DDT) in Sherman rats, Gaines [15] derived LD_{50} values for γ-HCH of 88 and 91 mg/kg in male and female rats, respectively. An oral LD_{50} value of 100 mg/kg was calculated for Wistar rats (gender not defined, sex unstated) receiving a single oral dose of γ-HCH in olive oil [16]. Golden hamsters appeared to be less sensitive to the acute toxic effects of γ-HCH as an LD_{50} of 360 mg/kg was reported [16]. In a range-finding experiment prior to studying the immunomodulatory effects of γ-HCH in female Swiss albino and BALB/c mice, Meera *et al.* [17] reported an acute oral LD_{50} of 120 mg/kg for γ-HCH in female mice, while Lahiri and Sircar [18] obtained an LD_{50} for the γ-isomer of 86 mg/kg in female Swiss mice. These data show a reasonable consensus on the range of LD_{50} values for γ-HCH in experimental animals.

The acute oral lethality of tech-HCH is strikingly different from that of γ-HCH. In a study conducted by Dikshith *et al.* [19], Swiss albino mice, Wistar albino rats, and albino rabbits (strain unstated) were gavaged with tech-HCH in peanut oil, and LD_{50} values of 1,434.6, 1,752.9, and 1,362.4 mg/kg, respectively, were obtained [19]. The content of the γ-isomer in the authors' technical-grade formulation was only 10.6%, with relative proportions of the α- and β-isomers of 72.4 and 11.9%, respectively. In Krishnakumari *et al.* (1982) [20], an LD_{50} value of 4,174 mg/kg in male and 5,673 mg/kg in female CFT-Wistar rats was reported for "technical X-factor," a mixture derived from tech-HCH by removing most of the β-isomer. The same group reported oral LD_{50} values of 2,624 mg/kg and 5,636 mg/kg for adult female rats exposed to tech-HCH and technical X-factor, respectively [21]. Collectively, these data suggest either enhanced toxicity of γ-HCH as compared to that of the other isomers or antagonistic effects upon combination with other isomers.

Mammalian Oral Toxicity: Subacute

In a study investigating endpoints other than lethality, male rabbits (strain unstated) were gavaged with daily doses of 7 mg/kg (0.1 LD_{50}) γ-HCH [22]. Three dosing regimens were used: A) a total of 0.5 LD_{50} γ-HCH (equal doses) during one week; B) a total of 0.5 LD_{50} γ-HCH (equal doses) over 1 week with a subsequent 1-week recovery period; or C) a total of 2.0 LD_{50} γ-HCH (equal doses) during 4 weeks (0.5 LD_{50} after 1 week, 1.0 LD_{50} after 2 weeks, 1.5 LD_{50} after 3 weeks, and 2.0 LD_{50} after 4 weeks). The levels of hemoglobin and hematocrit were unaltered in each of the tested rabbits;

however, a comparative 25% increase in the reticulocyte count was observed in rabbits that received 2.0 LD_{50} γ-HCH. A statistically significant decrease in leucocyte count and an increase in the number of micronuclei-bearing lymphocytes were found.

A number of studies on the subacute toxicity of HCH to laboratory animals have been conducted, as well. For example, Srinivasan *et al.* [23] exposed male albino Wistar rats to dietary β- or γ-HCH at concentrations of 800 ppm for 2 weeks and observed glucosaria in treated animals. Kidney dysfunction, marked by increased levels of urea and creatine in response to γ-HCH, was observed. In contrast, urinary protein levels in rats exposed to γ-HCH were lower than controls. Glucosaria was generally higher in animals supplemented with γ-HCH than β-HCH feed, suggesting the enhanced toxicity of the γ-isomer [23].

The toxicity of tech-HCH and γ-HCH was compared when male Swiss albino mice were exposed to dietary supplements of A) 400- or 800-ppm tech-HCH or B) 200- or 400-ppm γ-HCH for 2 weeks [24]. In a preliminary trial, doses of 600- and 800-ppm γ-HCH produced 75 and 100% mortality of mice within 1 week of administering the compound. The striking dose-dependent increases in relative liver and kidney weights were accompanied by dramatic increases in serum lipids, especially triglycerides. Moreover, the overall extent of these changes was greater in mice receiving tech-HCH than in those receiving γ-HCH [24]. Histopathological examination of pieces of excised liver revealed instances of hepatomegaly in animals exposed to either dietary level of tech-HCH. Taking the changes in relative organ weights as the most sensitive toxicological parameter under investigation in this study, lowest observed adverse effect levels (LOAELs) of 60 mg/kg/day tech-HCH (400 ppm) and 30 mg/kg/day γ-HCH (200 ppm) are suggested from the data. Derivation of the LOAELs assumed a food consumption rate of 0.15 mg food/kg body weight/day by default for the mice employed.

Mammalian Oral Toxicity: Subchronic

A substantial number of studies on the toxicology of HCH have focused on histopathological lesions in the liver and their potential for subsequent progression to form tumors. For example, hepatic tumors were reported in male Wistar rats exposed to diets containing up to 1,500 ppm of various HCH isomers for periods of up to 72 weeks [25]. When animals were necropsied, the livers were weighed, and the animals were examined macroscopically and processed for histopathology. Most of the toxicological consequences of treatment were confined to rats that received α-HCH, with histopathological lesions in the liver observed at all exposure levels and durations [25]. Moreover, rats exposed to 1,000-ppm α-HCH or greater for 48 weeks or more displayed increased "nodular hyperplasia" and hepatocellular carcinoma.

Others have attempted to assess the universality of the α-HCH/liver tumor effect by exposing other species of experimental animals to dietary supplements of 500-ppm α-HCH [26]. The compound induced nodular hyperplasia and hepatocellular carcinomas in the livers of mice, but not in male Wistar rats or Syrian golden hamsters. Among various strains of mice tested, DDY mice showed the highest incidence of liver nodule formation, while C57BL/6 mice showed the lowest [26]. Liver effects were also seen in randomly bred black male (HPBC57BL) mice exposed to diets supplemented with 500 ppm α-HCH for various periods of up to 50 weeks [27]. When animals were necropsied and examined for gross lesions in the liver and lungs, hepatocellular neoplasia was seen in mice after 21 weeks' exposure and with increasing incidence. Centrolobular megalocytosis was seen in all treated mice after 1 week of treatment.

The γ-isomer of HCH has been the subject of intense toxicological scrutiny. For example, Dikshith *et al.* gavaged 16 male rats/group (strain unstated) with 0 or 17.6 mg/kg/day in peanut oil for 90 days. Toxicological effects of γ-HCH were marked by the onset of necrotic changes to the liver and the formation of testicular lesions, implicating 17.6 mg/kg/day as an unbounded LOAEL [28]. Based on enzymic and histological changes, the authors posited the testes as more sensitive than the liver to γ-HCH.

Shivanandappa and Krishnakumari mixed tech-HCH (72% α-, 5% β-, 13.6% γ-, and 8% δ-HCH) with laboratory chow to administer doses of 0, 100, 250, 750, 1,500, and 3,000 ppm (equivalent to 0, 5, 12.5, 37.5, 75, and 150 mg/kg/day). These preparations were given to male CFT-Wistar rats for 90 days, at which point tissues were weighed and processed for histopathology. While all the members of the high-dose group died, survivors in the lower groups displayed dose-dependent hypertrophy of kidneys, adrenals, and especially liver, suggesting a LOAEL of 5 mg/kg/day [29]. That being said, no histopathological, histochemical, or biochemical changes were seen at the 5 mg/kg/day level.

As mentioned earlier, "technical X-factor" is a β-HCH depleted preparation that has also been tested for toxicological activity in experimental animals. For example, male CFT-Wistar rats were fed dietary concentrations of 0, 10, 50, 250, 750, 1,500, or 3,000 ppm (approximately equivalent to doses of 0, 0.4, 2, 11, 32, 64, and 135 mg/kg/day) for 90 days. The content in these experiments was 16% α–, less than 1% β–, 16% γ–, 50% δ-HCH, and 16% other components such as ortho- and trichlorobenzenes. Dose-dependent hypertrophy of liver, kidney, and adrenals was seen only beyond 750 ppm [30]. No vital organs showed abnormal histopathology at lower points. Collectively, these data suggest a LOAEL of 750 ppm (32 mg/kg/day) and a no observed adverse effect level (NOAEL) of 250 ppm (11 mg/kg/day) for technical X-factor in CFT-Wistar rats [30].

Clinical signs indicative of CNS-depression led to early sacrifice of 5/10 high-dose males and 6/10 high-dose females in a feeding study with β-HCH [31]. SPF-derived Wistar Riv: TOX rats were given either 0, 2, 10, 50, or 250 mg/kg feed β-HCH for 13 weeks. Animals that received lower doses exhibited a number of sublethal responses, including liver effects and changes in hematological parameters. The authors concluded that β-HCH caused liver enzyme induction at doses of 2 mg/kg/day and greater, thus establishing 2/mg/kg/day as the unbounded LOAEL [31].

Changes in calcium metabolism and kidney effects were the focus of a study in which male Long-Evans hooded rats were gavaged with 0, 10, or 20 mg γ-HCH/day for 10 weeks. The primary toxicological effects were alterations to kidney structure and function [32]. Along with an increase in kidney weight and kidney-to-body-weight ratio, there was a decrease of calcium in the urine. Since the responses were observed in animals receiving either dose level, 10 mg/kg/day may be considered an unbounded LOAEL for the kidney effects observed [32].

Mammalian Oral Toxicity: Chronic

One of the earliest definitive reports of the toxicological effects of HCH in experimental animals was that of Fitzhugh *et al.*, who exposed 10 Wistar rats/sex/group to diets supplemented with α-HCH, β-HCH, γ-HCH, or tech-HCH. The α-isomer was administered at 0.5, 2.5, 5, and 40 mg/kg/day; the β-isomer was administered at 0.5, 5, and 40 mg/kg/day; the γ-isomer was administered at 0.25, 0.5, 2.5, 2, 20, 40, and 80 mg/kg/day; and tech-HCH was administered at 0.5, 2.5, 5, and 40 mg/kg/day. There were two control groups. The tech-HCH employed in the experiment consisted of 64% α–, 10% β–, 13% γ–, 9% δ–, and 1.3% γ-HCH. Body weights and food consumption were monitored weekly, and most animals were allowed to live out their natural life span. Although no differences in food consumption rates were observed, body weight gain was significantly reduced at 40 mg/kg/day of α-HCH, 5 mg/kg/day of β-HCH in females, 80 mg/kg/day of γ-HCH, and 40 mg/kg/day of tech-HCH, making these the respective LOAELs [33]. Corresponding NOAELs for body weight gain were 5 mg/kg/day of α-HCH, 0.5 mg/kg/day of β-HCH, 40 mg/kg/day of γ-HCH, and 5 mg/kg/day of tech-HCH. Absolute and relative liver weights were monitored at termination, and the threshold exposure rates showed good agreement when putative NOAELs and LOAELs for changes in relative liver weight were compared to those associated with the onset of histopathological lesions. The LOAELs for liver effects were 5 mg/kg/day for α-HCH, 0.5 mg/kg/day for β-HCH, 5 mg/kg/day for γ-HCH, and 5 mg/kg/day for tech-HCH. Corresponding NOAELs were 2.5 mg/kg/day for α-HCH, γ-HCH, and tech-HCH. The histopathological effects of HCH were less profound in the kidney than in the liver, and higher doses of HCH were required to induce testicular atrophy among male Wistar rats [33]. The histopathological effects of HCH did not explicitly address the potential of the compound to induce tumors in the major organs of Wistar rat models. Many long-term studies of HCH have focused on tumor induction, which is inappropriate for TRV derivation when taken alone but can assist in a weight-of-evidence approach when combined with other endpoints [7].

An additional long-term exposure study in experimental animals involved the administration of γ-HCH to four beagles/sex/group at dietary levels of 0, 25, 50, or 100 ppm for 104 weeks. The dietary concentrations were equivalent to doses of 0, 0.625, 1.25, and 2.5 mg/kg/day. Blood samples were obtained prior to exposure and periodically to measure hematological and clinical chemistry parameters. Animals were also subjected to periodic urinalysis, ophthalmoscopy, electroencephalogram readings, complete necropsy, and comprehensive histopathology of major organs. No compound-related effects on body weight gain, food consumption, ophthalmology, hematology, urinalysis, or histopathology were observed [34]. Animals exposed to 2.5 mg/kg/day γ-HCH showed a slightly darkened liver and a statistically significant increase in serum alkaline phosphatase activity after 52 and 102 weeks. The authors concluded that their negative findings at 1.25 mg/kg/day over 2 years suggest a NOAEL, which is comparable with a NOAEL of 1.25 mg/kg in rats [35]. The respective LOAEL is 2.5 mg/kg/day.

The carcinogenic action of tech-HCH was investigated in a study that exposed inbred Swiss mice to the compound orally and via skin painting. In the study, researchers exposed 30 Swiss mice/sex/group to 100 ppm tech-HCH (68.7% α-, 6.5% β-, and 13.5% γ-HCH) in their diet for 80 weeks, gavaged mice at a dose of 10 mg/kg/day for 80 weeks, and applied 0.25 mg tech-HCH to the dorsal skin of mice in 0.1 milliliter (mL) olive oil twice a week for 80 weeks. Although not very prominent, convulsions and circling in one direction with drooping ears were observed [36]. Additionally, corneal opacity was seen in animals dosed orally in the latter half of the study. Tumors that developed as a result of exposure were marked morphologically by the development of a rough surface and the appearance of yellow nodules. Histopathologically, a sequence of lesions was evident, starting with the onset of oval cells and progressing to proliferation of the bile duct, hypertrophy of the parenchymal cells, development of hepatoma, and formation of hepatocellular carcinomas. While increased tumor occurrence was seen in animals treated orally and via skin painting, the increase was only statistically significant in the former.

There was a 100% incidence of hepatocarcinomas at 12 to 14 months of age when Swiss and BALB/c mice were exposed to 500 ppm tech-HCH (13% γ-HCH) in the diet [37]. Exposure began at 8 to 10 weeks of age. Liver tumors were not observed in Wistar rats and Syrian golden hamsters under similar experimental conditions.

Mammalian Oral Toxicity: Other

A substantial number of toxicological studies of HCH have examined the compound's capacity to disrupt reproductive and developmental characteristics in experimental animals or to induce teratogenic effects. Although the exposure durations in these studies were generally short, the studies were conducted during a sensitive life stage (i.e., gestation) with related endpoints (e.g., reproductive and developmental parameters); therefore, data collected from these studies are as valuable as those collected from chronic studies for the purpose of TRV derivation [7]. Among those studies in which HCH was administered to pregnant animals, gavage doses of 5 and 15 mg/kg/day γ-HCH administered to New Zealand white rabbits on gestation days (GDs) 6 to 18 and to CFY rats on GDs 6 to 16 had no effect on any reproductive, developmental, or teratogenic parameter investigated [38]. Likewise, gavage doses of 0, 6.25, 12.5, or 25 mg/kg/day "Benesan," a pesticidal formulation containing 50% γ-HCH, had no obvious reproductive, developmental, or teratogenic effects when administered to Wistar rats on GDs 6 to 15 [39].

Exposure to γ-HCH from conception decreased the mean number of kits born by 60% in a three-generation study with mink. While 1 mg/kg/day of γ-HCH administered via feed did not reduce the number of mated mink, it did reduce the proportion of mink that were subsequently born ($P < 0.1$) and the litter size ($P < 0.05$) [40]. Additionally, testes size was reduced in the γ-HCH-treated third generation males ($P < 0.05$), which may be related to the negative effects of the compound on steroid synthesis [40,41]. Data from what appears to be the same study found a decrease in the percentage of females accepting a second mating although mean litter size was unaffected [42]. Collectively, the data suggest the overall negative impact on prolificacy was greater in animals treated with γ-HCH from conception than in those treated from just before mating, thereby reinforcing the idea of enhanced toxicity during sensitive life stages [40,42].

To determine the effect of chronic exposure to HCH on female animals and their reproductive performance, pregnant rats were exposed to dietary β-HCH at concentrations of 0, 50, 200, 400, and 800 ppm or γ-HCH at concentrations of 0, 200, and 400 ppm. Concentrations of 0, 50, 200, 400, and 800 are equivalent to doses of 0, 2.5, 10, 20, and 40 mg/kg/day, assuming a default food consumption rate of 0.05 kg food/kg bw/day. Maternal mortality was observed in animals from the 40 mg/kg/day β-HCH dose group [43]. Dietary exposure at 10 and 20 mg/kg/day β-HCH had no significant effect during gestation and did not cause changes in the body weights of pups. However, all pups from the 20 mg/kg/day group died after 5 days, an effect that was partially evident (48% survival) in the pups of dams receiving 10 mg/kg/day β-HCH. Observed pup mortality suggests a LOAEL of 10 mg/kg/day and a NOAEL of 5 mg/kg/day β-HCH. Treatment with γ-HCH had no impact of female fertility in the 10 and 20 mg/kg/day groups [43]. Additionally, all pups born to mothers exposed to dietary γ-HCH survived, making 20 mg/kg/day γ-HCH a NOAEL.

Lindane did not appear to act as an endocrine-disrupting chemical when extra endpoints, including anogenital distance, nipple development, sexual maturation, estrous cycle, spermatogenesis, and so on, were measured. In the two-generation study, Crj:CD(SD)IGS rats (SPF) received 0, 10, 60, or 300 ppm of 99.5% pure γ-HCH (corresponding to a dose of 0, 0.5, 3.0, and 15 mg/kg/day, respectively). General toxicological effects (i.e., reduced gains in body weight, decreased food consumption, and deaths of females from the 15 mg/kg/day group) were seen in parental animals [44]. Convulsions and irritability were also seen in females receiving this dose. Centrilobular hepatocellular hypertrophy was seen upon pathological examination of males and females from the F_0 generations dosed as low as 0.5 mg/kg/day. Reproductive function was impacted in F_1 dams dosed at 15 mg/kg/day and thus their pups. Dams receiving 15 mg/kg/day demonstrated a lack of nursing or pup retrieval, which resulted in total litter loss and a decreased survival rate of F_2 offspring. Males and females that did survive demonstrated decreased weight gain, decreased thymus weight, and a delay in sexual maturation, implicating 15 mg/kg/day as the LOAEL and 3.0 mg/kg/day as the NOAEL [44].

As observed with previously described endpoints, mice are more susceptible than rats or rabbits to the reproductive, developmental, and teratogenic effects of HCH and its isomers. For example, Sircar and Lahiri gavaged pregnant Swiss mice with γ-HCH in either early (GDs 1–4), mid- (GDs 6–12), or late (GDs 14–19) pregnancy and monitored the reproductive and developmental deficits in formed fetuses (the first two groups) or pups (the last group). Administration of γ-HCH in early pregnancy prevented implantation; this effect was reversed by the administration of estrogen [45]. Administration of lindane in mid-pregnancy was associated with 100% resorption despite a normal number of implantation sites. Pups born to dams exposed to γ-HCH in late pregnancy died within 12 hours (high-dose group) or 5 days (low-dose group) of parturition [45]. The amount of γ-HCH administered to each animal was 50% of the LD_{50} (86 mg/kg) over the experimental period.

No reproductive, developmental or teratogenic effects were observed in Swiss mice fetuses from dams that had received a single gavage dose of 0, 5, 25, 50, 100, or 200 mg tech-HCH/kg in peanut oil on GD 9 [46]. Maternal weight gain during gestation was significantly reduced at the two highest dose levels, which appears to be due to total resorption of fetuses. Resorptions and reduced weight gain were also seen in the 25 and 50 mg/kg dose groups (although to a lesser degree), demonstrating a dose-response [46]. These findings were amplified in a subsequent full-scale reproductive/developmental study in which pregnant albino Swiss mice were gavaged with 0, 5, 25, or 50 mg/kg/day tech-HCH (72.43% α-, 11.95% β-, and 10.62% γ-HCH; and 2.5% other compounds) in peanut oil preimplantation (GDs 2–6) or postimplantation (GDs 6–12). All fetuses examined were free from teratogenic effects, although some appeared to have subcutaneous hemorrhages [41]. A toxicologically significant reduction in the number of fetuses/dam was observed in those mice exposed to high-dose tech-HCH during the postimplantation phase (GDs 6–12), reinforcing the dose-related effect of HCH on fetal resorption. Additionally, maternal weight gain during gestation was significantly reduced in the two highest dose levels, suggesting a LOAEL of 25 mg/kg/day and a corresponding NOAEL of 5 mg/kg/day for this effect [41].

A report by Lakkad *et al.*, is one of several in which impaired reproductive performance was observed in male experimental animals exposed to HCH. In this case, male Swiss mice were exposed to 500 ppm tech-HCH in the diet for 4, 6, or 8 months, and then mated with untreated females. The latter were necropsied mid-pregnancy to monitor the number of implants, dead embryos, percent fertility, and other reproductive parameters. Among the key results were significant decreases in the number of live embryos/female among the females mated with treated males compared to those females mated with untreated males [47]. In another approach, 10 male Wistar rats/group were exposed via gavage to either 6 mg γ-HCH/kg in peanut oil for 5 days (equivalent to 1.2 mg/kg/day) or a single dose of 30 mg γ-HCH/kg. Two weeks after treatment, reduced numbers of spermatids were seen in both groups [48]. The same research group reported relative reductions in testicular weight and the number of sperm and spermatid in the male offspring of female Bor: spf, TNO rats exposed to lindane [49]. Dams were treated with 6 mg/kg γ-HCH while lactating (postnatal days [PND] 9 or 14) or 1 mg/kg γ-HCH on PNDs 9 through 14. The researchers reported some alterations of the structural architecture of the testes of the male offspring and some subsequent changes in reproductive performance and sexual behavior. Additionally, testosterone levels were reduced to approximately 50% at puberty and adulthood in animals from the 6 mg/kg groups. Testosterone levels were also reduced in animals from the 1 mg/kg groups; however, the reduction was only significant at puberty [49].

Reproductive performance and mating behavior were also examined when male Wistar rats were exposed to HCH through their mothers who received a single oral dose of 30 mg γ-HCH/kg on GD 15 [50]. A key observation was the apparent reduction in sexual performance/activity in treated offspring, a finding apparently correlated with a decline in serum testosterone concentration, as was seen in the researchers' prior study [50,49]. When mating did occur, reproductive performance was the same as in controls (male progeny of unexposed dams) [50].

A two-generation study investigating the oral toxicity of lindane, as well as other selected compounds, was conducted in Sprague Dawley rats. Rats were exposed to 0, 10, 60, or 300 ppm lindane-supplemented feed at either 5 weeks of age (for the F_0 parents) or 3 weeks of age (for the F_1 parents). Rats were similarly exposed for 10 weeks prior to mating, during the mating period, through gestation, on delivery days, and during lactation. Observations included clinical signs (e.g., body mass and feed consumption), sperm characteristics, hormone levels, estrous cycles, vaginal opening, and prenuptial separation of the penis. No obvious effects were found in any of the parameters in the F_0, F_1, or F_2 offspring [51]. Few other details were reported.

Others have shown that tech-HCH alters the architecture of the male reproductive system following exposure of albino rats to 0, 3, or 6 mg/kg for 180 days [52,53]. Animals receiving either dose displayed histopathological degeneration of the muscular layer of the vas deferens [52]. Degeneration of the seminiferous tubules and decreased testicular weight were seen in animals dosed with 6 mg/kg/day [53].

Two studies examined the reproductive and developmental toxicology of HCH after male and female rats were dosed for an extended period of time prior to mating. In a limited three-generation reproductive study in Druckrey rats, tech-HCH did not affect reproductive and developmental parameters (e.g., pregnancy rate, gestation duration, pup weights, malformations and variations) when administered via feed at 125 and 250 ppm [54]. In a one-generation study in which both sexes of rat (strain unstated) were exposed to γ-HCH by gavage at 0, 10, or 20 mg/kg/day, male reproduction was unaffected [55]. Females exposed to 20 mg/kg/day had irregular estrous cycles but produced normal litters. However, all the high-dose and many of the low-dose progeny died shortly after birth. The authors speculated that the deaths were due to a maternal factor related to either lactation or a lack of maternal care [55].

Mammalian Inhalation Toxicity

No data on the toxicology of HCH via inhalation are available.

Mammalian Dermal Toxicity

Dikshith *et al.* examined the LD_{50} for tech-HCH when applied dermally and obtained values of more than 8,000 mg/kg in Wistar rats and 1,786.3 mg/kg in white rabbits with acetone [19]. A long-term skin painting study of tech-HCH in Swiss mice did not detect a statistically significant increase in tumor incidence compared to solvent-treated controls [36].

Summary of Avian Toxicology

Avian Oral Toxicity: Acute

Krishnakumari and Muralidhara found acute oral LD_{50} values to be similar when tech-HCH and technical X-factor were investigated in the domestic hen [56]. Single doses of the two formulations were administered via gelatin capsules at 150, 300, 600, 900, or 1,200 mg/kg. Two control groups were used; one received no treatment, and the other received empty capsules. Euthanized animals were necropsied and examined histopathologically after a three-week observation period. For tech-HCH (71% α-, 6.5% β-, 13% γ-, 7.0% δ-, and 1.5% ε-HCH) and technical X-factor, acute oral LD_{50} values were 596.8 mg/kg and 589.5 mg/kg, respectively. Based on the computed oral LD_{50} values, the authors posit that birds are more susceptible than albino mice, which were previously found to have oral LD_{50} values of 2,624 mg/kg and 5,636 mg/kg for tech-HCH and technical X-factor, respectively [21].

Avian Oral Toxicity: Subchronic

No data are available at this time.

Avian Oral Toxicity: Chronic

No data are available at this time.

Avian Oral Toxicity: Other

In a longer-term study, Whitehead *et al.* [57] exposed 12 Japanese Quail/group to either a control diet or one to which 200 mg/kg γ-HCH had been added. No differences were observed between the exposed and control groups. Egg production was identical in both groups, and similarly, eggshell thickness gradually declined in both groups, further demonstrating that there was no significant treatment effect. While there was no effect in Japanese Quail treated with 200 mg/kg γ-HCH, the authors previously saw a reduced rate of egg production (by 25–30%) when hens were exposed at 100 to 200 mg/kg γ-HCH [58].

The effect of γ-HCH on egg laying and eggshell characteristics was investigated in Indian Runner Ducks dosed for 8 weeks at 20 mg/kg either daily, 3 times per week, or twice per week. The researchers marked reductions in laying frequency; in hepatic, plasma, and vitellogenin levels; and in m-RNA production in the daily and 3 times per week dosing regimens, suggesting that lindane may be exerting a developmental effect in birds by inhibiting the synthesis or release of estrogens [59]. Effects were reversed in animals that also received an intramuscular injection of stilbestrol.

The γ-HCH isomer induced anemia marked by a reduction of hemoglobin and hematocrit with corresponding changes in mean cell volume and mean cellular hemoglobin in all birds exposed to a single dose at 5 mg/kg. Effects were seen across the six species used: House Sparrows (*Passer domesticus*), Baya Weavers (*Ploceus philippinus*), Common Mynas

(*Acridotheres tristis*), Rose-Ringed Parakeets (*Psittacula krameri*), Blue-Rock Pigeons (*Columba livia*), and Domestic Ducks (*Anas platyrhynchos*) [60].

Helberg *et al.* studied the impact of organochlorine contaminants on breeding wild, Great Black-Backed Gulls (*Larus marinus*). While the data were not conclusive because of the presence of other pesticides and an indication of possible nutritional issues, the authors noted a positive correlation between blood levels of β-HCH in female birds and the probability that a nesting site would be predated upon by ravens or hooded crows, resulting in smaller clutch size [61].

Avian Inhalation Toxicity

No data are available at this time.

Avian Dermal Toxicity

No data are available at this time.

Summary of Amphibian Toxicology

Amphibian toxicity data for HCH effects were limited to only three studies. Marchal-Segault and Ramade carried out a dose response experiment in which spawn of *Xenopus laevis* (African clawed frog) were incubated in water, water plus 0.1% acetone, or 0.5, 1.0, or 2.0 ppm γ-HCH ppm in water. The researchers monitored the lethality of γ-HCH to eggs and tadpoles over a period of a week and recorded decreased hatching rates at 2 ppm, although most of the aborted eggs contained well-developed embryos [62]. This suggested that the pesticide's toxicity might be associated with a disturbance in the hatching mechanism. The mortality of the tadpoles was relatively unaffected by the pesticide during the first week post-hatch, but by 6 weeks, 100% mortality was observed in the two highest exposure groups. Additionally, a 4 week delay in completion of metamorphosis was observed at the 0.5 ppm concentration, and those reaching metamorphosis weighed less than control animals [62]. Given these effects in the lowest exposure concentration, 0.5 ppm (0.5 mg/L) may be considered an unbounded LOAEL for growth and development.

Pawar and Katdare studied the effect of tech-HCH (13% γ-HCH) on the embryonic development of the frog, *Microhyla ornata* (ornate pygmy frog) and developed 96-hour LC_{50}s of 23.37 mg/L for embryos and 7.27 mg/L for tadpoles [63,64]. As evidenced by the LD_{50} values, the tadpoles were more sensitive than the embryos and developed a number of morphological and physiological deficits. The authors also observed clear dose-response relationships for percent hatching after 48 hours, percent abnormalities after 96 hours, and percent mortality after 96 hours when they exposed fertilized eggs (embryos) and 8-day-old tadpoles to various concentrations of tech-HCH [63]. Although the authors do not indicate at which dose these effects become statistically significant, effects become apparent at 10 mg/L and a dramatic decrease in hatching success and increase in abnormalities (e.g., retarded growth and upward and lateral curvature of the tail) and mortality occurred at 20 mg/L [63]. These results suggest a NOAEL of 1 mg/L and a LOAEL of 10 mg/L for developing embryos. Eight-day-old tadpoles exhibited abnormalities similar to those observed in embryos, including retarded growth, upward curved tail, poor pigmentation, and circular movement and loss of balance. Both studies indicate that the effects on tadpoles were observed at 10 mg/L, though it is suggested that they occur in all treated tadpoles. Additionally, the LC_{50} for tadpoles is stated to be 7.27 mg/L; therefore, it is presumed that sublethal effects were observed at concentrations lower than 7.27 mg/L. Consideration of the authors' comments and the LC_{50} suggests that 1 mg/L may represent an unbounded LOAEL for tadpoles.

Summary of Reptilian Toxicology

No toxicological data are available for the effects of HCH on reptiles.

RECOMMENDED TOXICITY REFERENCE VALUES

TRVs for Mammals

TRVs for Ingestion Exposures for the Class Mammalia

NOAELs and LOAELs from a variety of studies are presented in Table 27.2. Two chronic studies [33,34] and two reproductive/developmental studies [41,43] were identified as relevant to TRV derivation.

In the first chronic study, NOAELs and LOAELs for both reduced body weight gain and liver effects were identified for α-, β-, γ-, and tech-HCH. Increased relative liver weights and histopathological effects were present at lower doses than was reduced body weight gain; however, the LOAELs for liver effects could not be used for calculating the Benchmark Dose (BMD) because the number of deaths—and in which dose group they occurred—were not reported. The LOAELs for liver effects are 5 mg/kg/day for α-HCH, 0.5 mg/kg/day for β-HCH, 5 mg/kg/day for γ-HCH, and 5 mg/kg/day for tech-HCH. Corresponding NOAELs are 2.5 mg/kg/day for α-HCH, γ-HCH, and tech-HCH [33]. A beagle study conducted with

Table 27.2 Summary of Relevant Mammalian Data for TRV Derivation

Test Organism	Test Duration	Test Results: NOAEL (mg/kg/day)	Test Results: LOAEL (mg/kg/day)	Test Results: Effects Observed at the LOAEL	Study
Mouse – Swiss Albino	14 days	NA (tech)	60 (tech)	Increase in relative organ weights	[24]
		NA (γ)	30 (γ)	Increase in relative organ weights	
Rat – Unstated	90 days	NA (γ)	17.6 (γ)	Liver necrosis and testicular lesions	[28]
Rat – CFT Wistar	90 days	NA (tech)	5 (tech)	Hypertrophy of liver, kidneys, and adrenals	[29]
Rat – CFT Wistar	90 days	11 (X-factor)	32 (X-factor)	Hypertrophy of liver, kidneys, and adrenals	[30]
Rat – Wistar	13 weeks	NA (β)	2 (β)	Liver enzyme induction	[31]
Rat – Long-Evans Hooded	10 weeks	NA (γ)	10 (γ)	Kidney histopathology and functional deficits; decreased calcium in urine	[32]
Rat – Wistar	Lifetime	0.5 (β)	5 (β)	Reduced body weight gain	[33]
		NA (β)	0.5 (β)	Increased relative liver weight and histopathological effects	
		5 (α)	40 (α)	Reduced body weight gain	
		2.5 (α)	5 (α)	Increased relative liver weight and histopathological effects	
		40 (γ)	80 (γ)	Reduced body weight gain	
		2.5 (γ)	5 (γ)	Increased relative liver weight and histopathological effects	
		5 (tech)	40 (tech)	Reduced body weight gain	
		2.5 (tech)	5 (tech)	Increased relative liver weight and histopathological effects	
Dog – Beagle	104 weeks	1.25 (γ)	2.5 (γ)	Liver changes and altered serum alkaline phosphatase activity	[34]
Mouse – Swiss Albino	GD 6–12	5 (tech)	25 (tech)	Reduced maternal body weight gain	[41]
Rat – Wistar	1-Gen	5 (β)	10 (β)	Mortality of pups within 28 days of parturition	[43]
		20 (γ)	NA (γ)	Female fertility	
Rat – Crj:CD (SD)IGS-SPF	2-Gen	3 (γ)	15 (γ)	Decreased weight gain and thymus weight; delay in sexual maturation	[44]

NA = not applicable.

γ-HCH and lasting 104 weeks also seemed promising for TRV derivation; however, standard deviation values were not listed, and the number of animals dosed for the results pertaining to 50 weeks was not stated [34]. The authors did conclude, however, that negative findings at 1.25 mg/kg/day are consistent with a NOAEL of 1.25 mg/kg/day in rats [35].

Two reproductive/developmental studies were also evaluated for use in TRV derivation. In one study, Srinivasan *et al.* found a NOAEL and LOAEL of 5 mg/kg/day and 10 mg/kg/day for β-HCH, respectively [43]. They also found a NOAEL of 20 mg/kg/day for γ-HCH. There were no adequate dose-response data in the study with which to perform BMD analysis.

In the other reproductive/developmental study in which Swiss albino mice were exposed to tech-HCH, there was sufficient information to use reduced maternal weight gain as an endpoint to run BMD [41]. In order to do so, standard errors presented in the paper were converted to standard deviations. The BMD and BMDL (corresponding lower confidence limit) were based on one standard deviation from the control. A nonhomogenous variance model was used to extrapolate the data and produced a BMD of 7.89 and BMDL of 5.20, as presented in Table 27.3. The LOAEL and NOAEL for the same endpoints were 25 and 5 mg/kg/day, respectively.

A clear difference in toxicity among γ-HCH, tech-HCH, and technical X-Factor was observed in the available LD_{50} data (i.e., γ-HCH was the most toxic). However, trends in toxicity among HCH isomers or formulations within the other studied endpoints were not apparent (see Table 27.2).

TRVs for Ingestion Exposures for Mammalian Foraging Guilds

The TRVs specific to particular guild associations (e.g., small herbivorous mammals) have not yet been derived. However, the class-specific TRVs shown in Table 27.3 may be considered as applicable to small herbivorous mammals because mice are members of this guild. As with the class-specific TRVs, only two species among the mammalian foraging guilds are

Table 27.3 Selected Ingestion TRVs for Technical-Hexachlorocyclohexane for the Class Mammalia

TRV	Dose (mg/kg/day)	Confidence Level
BMDL	5.20	High
BMD	7.89	High

Source: [41].

represented, so confidence in the TRVs is low to medium. Data with which to derive TRVs for other guild associations (e.g., carnivorous mammals) are not available at this time.

TRVs for Inhalation Exposures for the Class Mammalia

No data were identified permitting the derivation of a mammalian inhalation TRV for HCH.

TRVs for Dermal Exposures for the Class Mammalia

Sufficient data required for development of dermal TRVs for the Class *Mammalia* were not available. However, acute LD_{50} values of greater than 8,000 mg/kg of tech-HCH for rats and 1,786 mg/kg for white rabbits have been identified [19].

TRVs for Birds

TRVs for Ingestion Exposures for the Class Aves

None of the studies located provided sufficient data from which to derive ingestion exposure TRVs for birds.

TRVs for Ingestion Exposures for Avian Foraging Guilds

The TRVs specific to particular guild associations (e.g., herbivorous birds) have not yet been derived.

TRVs for Inhalation Exposures for the Class Aves

No data from which to derive an avian inhalation TRV for HCH are available at this time.

TRVs for Dermal Exposures for the Class Aves

No data on the dermal toxicity of HCH to birds are available at this time.

TRVs for Amphibians

The effects of both γ-HCH and tech-HCH on amphibians (e.g., reduced hatching success, retarded growth in tadpoles) were consistent across the three studies and two species (*Xenopus laevis* and *Microhyla ornate*) evaluated. These two species represent one amphibian Order and two Families: Anura: Pipidae and Microhylidae. Marchal-Segault and Ramade [62] observed 100% mortality in the two highest dose groups by 6 weeks post-hatch, pointing to a LOAEL of 0.5 mg/L γ-HCH. Similarly, Pawar and Katdare [63,64] reported retarded growth and other abnormalities in 8-day-old tadpoles treated with tech-HCH. Although not explicitly stated, the results

Table 27.4 Selected Ingestion TRVs for Technical-Hexachlorocyclohexane for Amphibians

TRV	Exposure (mg/L)	Confidence Level
NOAEL-based	0.005	Low to Medium
LOAEL-based	0.05	Low to Medium

Source: [62].

suggest a LOAEL of 10 mg/L and a NOAEL of 1 mg/L. As these studies were conducted during sensitive life stages, they are considered chronic. Effects on reproduction, growth, and survival are directly relevant to the health and ecology of the whole organism and are considered appropriate endpoints for TRV derivation. As the lowest effect level, the unbounded LOAEL (0.5 mg/L) from Marchal-Segault and Ramade [62] was selected for development of the TRVs.

Since only one Order and two species were evaluated in the relevant studies, the minimum data set requirements of the Standard Practice, Section 2.2 [7] were not met. Therefore, TRVs based on an approximation of the NOAEL and LOAEL were developed for amphibians using uncertainty factors (UFs). An UF of 10 was applied to the LOAEL (0.5 mg/L) from Marchal-Segault and Ramade [62] to account for interspecific variability. A chronic NOAEL was estimated by applying an UF of 10 to the chronic LOAEL. Table 27.4 presents the selected TRVs. A low-to-medium level of confidence has been assigned to these TRVs because effects were highly consistent across two species, but a NOAEL for these endpoints (as determined within the experimental design of the available studies) was unavailable.

TRVs for Reptiles

No data are available at this time.

IMPORTANT RESEARCH NEEDS

The mammalian TRVs derived for HCH have received a high confidence level because many studies have been conducted utilizing several species (including two wildlife species) and representing a variety of toxicological endpoints. Any additional studies conducted should evaluate the reproductive, developmental, or growth effects of HCH on additional wildlife species and should employ multiple dose levels to achieve NOAELs and LOAELs for the evaluated endpoints. Inhalation and dermal studies on mammals were either limited or lacking for HCH. The TRV derivations for birds, amphibians, and reptiles were more uncertain than for mammals due to the paucity of high-quality toxicity data for birds, the lack of studies with sufficiently low dose levels to enable a NOAEL determination for amphibians, and the absence of toxicity data for reptiles. Before reliable avian, amphibian, and reptilian TRVs can be derived, HCH toxicity to these classes of wildlife need to be adequately characterized. Appropriate acute, subacute, subchronic, and specifically chronic HCH toxicity data for birds, amphibians, and reptiles, by all exposure routes, are needed. Research should be conducted on species that are genetically, biologically, and behaviorally similar to the wildlife most likely to be exposed. Additionally, the experimental design should mimic both exposure type and duration and should include assessments of long-term effects.

REFERENCES

[1] Agency for Toxic Substances and Disease Registry (ATSDR). Toxicological profile for alpha-, beta-, gamma-, and delta-hexachlorocyclohexane. Atlanta (GA): U.S. Department of Health and Human Services, Public Health Service; 2005.

[2] Hazardous Substances Data Bank (HSDB). Hexachlorocyclohexane [Internet]. Bethesda (MD): National Library of Medicine (US), National Institutes of Health [cited 2001]. Available from: http://toxnet.njm.nih.gov/.

[3] International Agency for Research on Cancer (IARC). Hexachlorocyclohexane (Technical HCH and Lindane). IARC Monogr Eval Carcinog Risk Chem Hum 1979;20:195–239.

[4] Nayyar N, Sangwan N, Kohli P, Verma H, Kumar R, Negi V, et al. Hexachlorocyclohexane: persistence, toxicity, and decontamination. Rev Environ Health 2014;29 (1–2):49–52.

[5] Hegeman WJ, Laane RW. Enantiomeric enrichment of chiral pesticides in the environment. Rev Environ Contam Toxicol 2002;173:85–116.

[6] United National Environment Program (UNEP). Report of the Conference of Parties of the Stockholm Convention on Persistent Organic Pollutants on the work of its fourth meeting; 2009 May 11; Geneva, Switzerland. UNEP/POPS/COP. 4/38.

[7] USACHPPM. Standard practice for wildlife toxicity reference values. Aberdeen Proving Ground (MD): U.S. Army Center for Health Promotion and Preventive Medicine; 2000, Technical Guide 254.

[8] Nagata Y, Endo R, Ito M, Ohtsubo Y, Tsuda M. Aerobic degradation of lindane (gamma-hexachlorocyclohexane) in bacteria and its biochemical and molecular basis. Appl Microbiol Biotechnol 2007;76:741–52.

[9] Phillips TM, Seech AG, Lee H, Trevors JT. Biodegradation of hexachlorocyclohexane (HCH) by microorganisms. Biodegradation 2005;16:363–92.

[10] Deo PG, Karanth NG, Karanth NG. Biodegradation of hexachlorocyclohexane isomers in soil and food environment. Crit Rev Microbiol 1994;20:57–78.

[11] Straube G. Microbial transformation of hexachlorocyclohexane. Zentralbl Mikrobiol 1991;146:327–38.

[12] Singh BK, Kuhad RC, Singh A, Tripathi KK, Ghosh PK. Microbial degradation of the pesticide lindane (γ-Hexachlorocyclohexane). Adv Appl Microbiol 2000;47:269–98.

[13] Bidleman TF, Leone AD, Falconer RL, Harner T, Jantunen LM, Wiberg K, et al. Chiral pesticides in soil and water and exchange with the atmosphere. Scientific World Journal 2002;2:357–73.
[14] Boyd EM, Chen CP. Lindane toxicity and protein-deficient diet. Arch Environ Health 1968;17:156–63.
[15] Gaines TB. Acute toxicity of pesticides. Toxicol Appl Pharmacol 1969;14:515–34.
[16] Truhaut R, Gak JC, Graillot C. Recherches sur les modalites et les mecanismes d'action toxique des insecticides organochlores. I - Etude comparative des effets de toxicite aigue chez le hamster et chez le rat. J Eur Toxicol 1974;7:159–66.
[17] Meera P, Rao PR, Shanker R, Tripahi O. Immunomodulatory effects of γ-HCH (Lindane) in mice. Immunopharmacol Immunotoxicol 1992;14:261–82.
[18] Lahiri P, Sircar S. Suppression of adrenocortical function in female mice by lindane (γ-HCH). Toxicology 1991;69:75–9.
[19] Dikshith TS, Raizada RB, Singh RP, Kumar SN, Gupta KP, Kaushal RA. Acute toxicity of hexachlorocyclohexane (HCH) in mice, rats, rabbits, pigeons and freshwater fish. Vet Hum Toxicol 1989;31:113–6.
[20] Krishnakumari MK, Muralidhara S, Majumder SK. Mammalian toxicity of technical X-factor I. An acute oral study. J Environ Sci Health B 1982;17:241–52.
[21] Krishnakumari MK, Muralidhara S, Majumder K. CFTRI Annu Rep India 1975;1–2.
[22] Grabarczyk M, Kopec-Szlezak J, Szczepanska I, Wozniak J, Podstawka U. The effect of gamma-hexachlorocyclohexane (Lindane) on blood cells, kidney and liver tissues in rabbits. Haematologia 1990;23:171–9.
[23] Srinivasan K, Ramesh HP, Radhakrishnamurty R. Renal tubular dysfunction caused by dietary hexachlorocyclohexane (HCH) isomers. J Environ Sci Health B 1984;19:453–66.
[24] Ravinder P, Srinivasan K, Radhakrishnamurty R. Biochemical toxicity of hexachlorocyclohexane and its γ-isomer in albino mice. Ind J Exp Biol 1989;27:248–51.
[25] Ito N, Nagasaki H, Aoe H, Sugihara S, Miyata Y. Development of hepatocellular carcinomas in rats treated with benzene hexachloride. J Natl Cancer Inst 1975;54:801–5.
[26] Nagasaki H, Kawabata H, Miyata Y, Inoue K, Hirao K, Aoe H, et al. Effect of various factors on induction of liver tumors in animals by the α-isomer of benzene hexachloride. Gan 1975;66:185–91.
[27] Tryphonas L, Iverson F. Sequential histopathological analysis of α-hexachlorocyclohexane-induced hepatic megalocytosis and adenoma formation in the HPB mouse. J Natl Cancer Inst 1983;71:1307–18.
[28] Dikshith TS, Tandon SK, Datta KK, Gupta PK, Behari JR. Comparative response of male rats to parathion and lindane: histopathological and biochemical studies. Environ Res 1978;17:1–9.
[29] Shivanandappa T, Krishnakumari MK. Histochemical and biochemical changes in rats fed dietary benzene hexachloride. J Exp Biol 1981;19:1163–8.
[30] Muralidhara S, Krishnakumari MK. Toxicity of technical X-factor to weanling rats. II. A subacute oral study. Environ Sci Health B 1983;18:283–99.
[31] Van Velsen FL, Danse LH, Van Leeuwen FX, Dormans JA, Van Logten MJ. The subchronic oral toxicity of the β-isomer of hexachlorocyclohexane in rats. Fund Appl Toxicol 1986;6:697–712.
[32] Andrews JE, Gray LE. The effects of lindane and linuron on calcium metabolism, bone morphometry and the kidney in rats. Toxicology 1990;60:99–107.
[33] Fitzhugh OG, Nelson AA, Frawley JP. The chronic toxicities of technical benzene hexachloride and its alpha, beta and gamma isomers. J Pharmacol Exp Therap 1950;100:59–66.
[34] Rivett KF, Chesterman H, Kellett DN, Newman AJ, Worden AN. Effects of feeding lindane to dogs for periods of up to 2 years. Toxicology 1978;9:273–89.

[35] World Health Organization. Technical report No. 545: pesticide residues in food. Report of the 1973 Joint Meeting of the FAO Working Party of Experts on Pesticide Residues and the WHO Expert Committee on Pesticide Residues; 1973 Nov 26-Dec 5; Geneva, Switzerland.
[36] Kashyap SK, Nigam SK, Gupta RC, Karnik AB, Chatterjee SK. Carcinogenicity of hexachlorocyclohexane (BHC) in pure inbred Swiss mice. Environ Sci Health B 1979;14:305–18.
[37] Munir KM, Soman CS, Bhide SV. Hexachlorocyclohexane-induced tumorigenicity in mice under different experimental conditions. Tumori 1983;69:383–6.
[38] Palmer AK, Bottomley AM, Worden AN, Frohberg H, Bauer A. Effect of lindane on pregnancy in the rabbit and rat. Toxicology 1978;9:239–47.
[39] Khera KS, Whalen C, Trivett G, Anders G. Teratogenicity studies on pesticidal formations of dimethoate, diuron and lindane in rats. Bull Environ Contam Toxicol 1979;22:522–9.
[40] Beard AP, Rawlings NC. Reproductive effects in mink (*Mustela vison*) exposed to the pesticides lindane, carbofuran, and pentachlorophenol in a multigeneration study. J Reprod Fertil 1998;113:95–104.
[41] Srivastava MK, Raizada RB. Prenatal effects of technical hexachlorocyclohexane in mice. J Toxicol Environ Health 1993;40:105–15.
[42] Beard AP, McRae AC, Rawlings NC. Reproductive efficiency in mink (*Mustela vison*) treated with the lindane, carbofuran, and pentachlorophenol. J Reprod Fertil 1997;111:21–8.
[43] Srinivasan K, Mahadevappa KL, Radhakrishnamurty R. Effect of maternal dietary hexachlorocyclohexane exposure on pup survival and growth in albino rats. Environ Sci Health B 1991;26:339–49.
[44] Matsuura I, Saitoh T, Tani E, Wako Y, Iwata H, Toyota N, et al. Evaluation of a two-generation reproduction toxicity study adding endpoints to detect endocrine disrupting activity using lindane. J Toxicol Sci 2005;((30) special issue):135–61.
[45] Sircar S, Lahiri P. Lindane (γ-HCH) causes reproductive failure and fetotoxicity in mice. Toxicology 1989;59:171–7.
[46] Dikshith TSS, Srivastava MK, Raizada RB. Fetotoxicity of hexachlorocyclohexane (HCH) in mice: morphological, biochemical and residue evaluations. Vet Hum Toxicol 1990;32:524–7.
[47] Lakkad BC, Nigam SK, Karnik AB, Thakore KN, Babu KA, Bhatt DK, et al. Dominant-lethal study of technical-grade hexachlorocyclohexane in Swiss mice. Mutat Res 1982;101:315–20.
[48] Dalsenter PR, Faqi AS, Webb J, Merker HJ, Chahoud I. Reproductive toxicity and tissue concentrations of lindane in adult male rats. Human Exp Toxicol 1996;15:406–10.
[49] Dalsenter PR, Faqi AS, Webb J, Merker HJ, Chahoud I. Reproductive toxicity and toxicokinetics of lindane in the male offspring of rats exposed during lactation. Human Exp Toxicol 1997;16:146–53.
[50] Dalsenter PR, Faqi AS, Chahoud I. Serum testosterone and sexual behavior in rats after prenatal exposure to lindane. Bull Environ Contam Toxicol 1997;59:360–6.
[51] Yamasaki K, Takahashi M, Yasuda M. Two-generation reproductive toxicity studies in rats with extra parameters for detecting endocrine disrupting activity: introductory overview of results for nine chemicals. J Toxicol Sci 2005;30:1–4.
[52] Gautam AK, Gandhi DN, Jani JP, Venkatakrishna-Bhatt H, Chowdhury AR. Histological and pharmacological changes in vas deferens of rats exposed to hexachlorocyclohexane. Res Commun Chem Pathol Pharmacol 1989;63:463–6.
[53] Chowdhury AR, Gautam AK. BHC-induced testicular impairments in rats. Ind J Physiol Pharmacol 1990;34:215–7.

[54] Srivastava MK, Raizada RB. A limited three-generation study of hexachlorocyclohexane (HCH) in rats. Food Chem Toxicol 2000;38:195–201.
[55] Gray Jr LE, Ostby J, Sigmon R, Ferrell J, Rehnberg G, Linder R, et al. The development of a protocol to assess reproductive effects of toxicants in the rat. Reprod Toxicol 1988;2:281–7.
[56] Krishnakumari MK, Muralidhara S. Acute oral toxicity of hexachlorocyclohexane and X-factor to poultry *Gallus domesticus*. Poultry Sci 1981;60:2599–602.
[57] Whitehead CC, Downie JN, Phillips JA. Some characteristics of the egg shells of quail fed gamma-BHC. Pesticide Sci 1974;5:275–9.
[58] Whitehead CC, Downie JN, Phillips JA. BHC not found to reduce the shell quality of hens' eggs. Nature 1972;239:411–2.
[59] Chakravarty S, Mandal A, Lahiri P. Effect of lindane on clutch size and level of egg yolk protein in domestic duck. Toxicology 1986;39:93–103.
[60] Lahiri P, Mandal A, Sircar S, Datta S, Chakraborty J, Chakraborty S. Insecticides and impaired reproductive success: studies on lindene (sic) toxication. In: Manna GK, Jana BB, editors. Impacts of environment on animals and aquaculture. West Bengal (India): University of Kalyani; 1990.
[61] Helberg M, Bustnes JO, Erikstad RE, Kristiansen KO, Skaare JU. Relationships between reproductive performance and organochlorine contaminants in great black-backed gulls (*Larus marinus*). Environ Pollut 2005;134:475–83.
[62] Marchal-Segault D, Ramade F. The effects of lindane, an insecticide, on hatching and postembryonic development of *Xenopus laevis* (Daudin) anuran amphibian. Environ Res 1981;24:250–8.
[63] Pawar KR, Katdare M. Toxic and teratogenic effects of fenitrothion, BHC and carbofuran on embryonic development of the frog *Microhyla ornata*. Toxicol Lett 1984;22:7–13.
[64] Pawar KR, Katdare M. Toxicity of sumothion, BHC and furadan to the tadpoles of the frog *Microhyla ornata*. Ind J Environ Health 1983;25:214–7.

CHAPTER 28

Wildlife Toxicity Assessment for Perchlorate

William S. Eck

Contents

INTRODUCTION

This Wildlife Toxicity Assessment is based on a thorough review of the scientific literature regarding the toxicological characteristics of perchlorates that may pertain to the health of wildlife (mammals, birds, reptiles, and amphibians) exposed to these substances. Perchlorate salts are used as oxidizers in solid rocket propellants, road flares, and munitions and in pyrotechnic formulations. They have been discovered in groundwater around military installations, fireworks factories, and munitions manufacturing facilities. Although the military uses large quantities of perchlorate salts, their use is not strictly related to munitions and solid rocket propellants. Perchlorate can be used as an etching and engraving agent and can also be used in paper matches, fireworks, road flares, and automobile air bags. Until the 1970s, perchlorate at very high dose levels was used to treat specific types of hyperthyroid conditions in the United States, but this

Wildlife Toxicity Assessments for Chemicals of Military Concern
http://dx.doi.org/10.1016/B978-0-12-800020-5.00028-4

practice was discontinued due to the occurrence of aplastic anemia [1]. Perchlorate is still used in other parts of the world, notably Europe, where it is used under the trade name Irenate® to treat hyperthyroidism resulting from Graves' Disease or treatment of cardiac arrhythmia and tachycardia with amiodarone [2].

This chapter describes the current knowledge of the toxic effects of perchlorate ions (ClO_4^-) to wildlife species. Effects of low levels of perchlorate ions to humans are beyond the scope of this chapter. The protocol for the development of this assessment is documented in the U.S. Army Center for Health Promotion and Preventive Medicine (USACHPPM) Technical Guide (TG) 254, *Standard Practice for Wildlife Toxicity Reference Values* [3].

TOXICITY PROFILE

Environmental Fate and Transport

A perchlorate anion consists of a chlorine atom (oxidation number of +7) surrounded by four oxygen atoms (oxidation number of −2) to form an oxychlorine anion (−1 charge). Perchlorate is a very stable anion that forms salts with cations such as sodium, potassium, and ammonium. Salts of perchlorates have high solubility in aqueous environments [4]. Although no specific information on degradation rates (half-lives) was located, ionic perchlorate can reportedly persist in surface and groundwater for more than a decade. This is due to the high energy of activation required for perchlorate to react with other constituents in the environment [5]. Once combustion is achieved, perchlorate salts result in strongly exothermic chemical reactions. It is for this reason that ammonium and other perchlorate salts are used in solid rocket propellants, fireworks, and munitions.

A major pathway for perchlorates to enter the environment is during the manufacture and recharge of munitions and solid rocket motors. These activities have resulted in perchlorate contamination of groundwater at many military installations and rocket manufacturing facilities [5,6]. Other anthropogenic sources of perchlorate include fireworks, safety flares, blasting explosives, electrochemically-prepared (ECP) chlorine products, and Chilean nitrate fertilizers [7]. It has been discovered fairly recently perchlorates are also formed naturally by the action of lightning on chlorine species in the atmosphere [8].

Perchlorates have been found in biological tissue at sites contaminated with perchlorates in soil and surface water. Smith *et al.* [9] found elevated concentrations in vegetation but rarely in rodent tissue. Cows drinking from a perchlorate-contaminated stream were found to have very low levels in plasma but exhibited no change in circulating thyroid hormone levels [10]. Results of a market basket survey in the Colorado River region found traces of perchlorate in leafy vegetables but at levels that would result in less than 10% of the reference dose recommended by the National Academy of Sciences [11,12]. A controlled laboratory study of food chain transfer found that perchlorate has a limited ability to bioaccumulate in aquatic systems [13].

As part of a series of studies on ammonium perchlorate (AP) in wildlife, Kendall [14] reported that raccoons on Longhorn Army Ammunition Plant (LHAAP) were potentially exposed to up to 5,021 micrograms (μg) perchlorate per day in their diets. Only trace amounts of perchlorate were detected in a total of 44 different animals; one sample contained perchlorate at less than the quantitation limit, while all other samples were below the detection limit of 1 part per billion (ppb). No impact on thyroid function was detected. Previous collection of small mammal tissue (e.g., in harvest mice [15]) found perchlorate concentrations of 589 to 2,170 ppb.

Key physical-chemical properties of the three most common perchlorate salts are provided in Table 28.1. Because the three perchlorate salts are ionic compounds, partitioning to air will be limited to suspension of particulates. All three are also highly soluble in water (1.5×10^5 to 2.1×10^5 milligrams per liter (mg/L) at 25 °C) and have been identified in both surface and groundwater.

Summary of Mammalian Toxicity

Published toxicological studies on perchlorates have focused primarily on laboratory mammals such as rats and mice. This is partly due to the medical interest in perchlorate's ability to block iodine uptake and thus prevent hyperthyroidism (Grave's Disease). Most toxicity evaluations of perchlorate focus on the levels of the thyroid hormones triiodothyronine (T3) and thyroxine (T4) because it is believed that deficiencies of these hormones affect growth, development, and metabolism in animals. As part of the homeostatic mechanism, thyroid-stimulating hormone (TSH) secreted by the pituitary is increased with decreased levels of T3 and T4 (T3, T4, and TSH are often measured concurrently). Virtually all vertebrates have thyroid glands [5]. However, few studies were found that included testing of thyroid hormone levels in wildlife species. The evaluation of perchlorate toxicity relies mostly on studies in common laboratory animals (mice, rats, and rabbits).

Due to the high solubility of perchlorate salts and the ubiquitous environmental distribution of their associated cations (Na^+, K^+, $NH_4{}^+$), this toxicological review assumes that the contribution from the cations to

Table 28.1 Summary of the Physical-Chemical Properties of Perchlorate Compounds

	Cations Associated with Perchlorate Anion		
Physical Property	$NH_4(ClO_4)$	$Na(ClO_4)$	$K(ClO_4)$
CAS No.	7790-98-9	7601-89-0	7778-74-7
Percent perchlorate ion by weight	84.67	81.22	71.8
Molecular weight[a]	117.50	122.44	138.56
Color	White	White	White
Physical state	Crystalline solid	Crystalline solid	Powder
Melting point[a]	dec 240 °C	482 °C	dec 440 °C
Boiling point[a]	N/A	N/A	N/A
Odor	No data	Odorless	Odorless
Solubility in water at 25 °C (mg/L)[b]	2.1×10^5	2.0×10^5	1.5×10^5
Partition coefficients[a]:			
Log K_{ow}	–5.84	–7.18	–7.18
Log K_{oc}	1.985	1.687	1.687
Henry's Law constant at 25 °C (atm-m^3/mole)[a]	2.8×10^{-11}	9.15×10^{-19}	2.17×10^{-19}
Vapor pressure at 25 °C (mm Hg)[a]	4.34×10^{-18}	1.47×10^{-25}	3.96×10^{-26}
Conversion factors (% is ClO_4^- ion by weight)	1 ppm = 4.80 mg/m^3 1 mg/m^3 = 0.208 ppm (84.7%)	1 ppm = 5.0 mg/m^3 1 mg/m^3 = 0.20 ppm (81.2%)	1 ppm = 5.66 mg/m^3 1 mg/m^3 = 0.177 ppm (71.8%)

[a]Dean, J.A. Lange's Handbook of Chemistry, 14th Ed. New York: McGraw-Hill, 1992. [80].
[b][81] dec = decomposes.

overall toxicity is negligible. The only exception to this assumption is for aquatic (or semiaquatic, e.g., amphibian) species due to the acidic nature of the ammonium ion. All observed effects are assumed to primarily be associated with the perchlorate anion (ClO_4^-). Additionally, all dose levels presented in the text and Tables 28.2 and 28.3 have been converted and are expressed in terms of perchlorate anion concentrations, rather than concentrations of the parent salt compound.

Mammalian Oral Toxicity: Acute/Subacute

In a brief review of perchlorate toxicity [16], it was noted that acute mammalian toxicity data were limited. The oral LD_{50} of ammonium perchlorate in white rats was reported to be 3556 milligrams per kilogram (mg/kg)

Table 28.2 Summary of Median Lethal Dose (LD_{50}) Data for Perchlorate Salts

Species	Route of Administration	ChemicalForm	Perchlorate Ion Concentration[a](mg/kg)
Rabbit	Oral	NH_4ClO_4	635–1,610
Mouse	i.p.	$NaClO_4$	934
Mouse	Oral	NH_4ClO_4	1,610–1,690
Guinea Pig	Oral	NH_4ClO_4	2,800
Rat	Oral	NH_4ClO_4	2,960–3,560

i.p. = intraperitoneal.
[a]Perchlorate ion is determined by multiplying the concentration of perchlorate salt by percent perchlorate ion by weight (Table 28.1).
Source: [16].

perchlorate [17,18].[1] LD_{50} values for several species (Table 28.2) approach the concentration, known as the "limit test" or "limit dose," where the compound is considered to be essentially non-toxic. It appears that rabbits are more sensitive to the acute effects of perchlorate than rats, mice, or guinea pigs.

In addition to the derived LD_{50}, Mannisto *et al.* [19] dosed Sprague Dawley rats through drinking water for four days at 0, 7.6, 15.3, and 76.3 mg/kg/day perchlorate ion. At the end of the fourth day, concentrations of T3 and T4 in blood serum were found to be depressed, while TSH was elevated at 15.3 mg/kg/day perchlorate ion. Nervous system effects were reported in rabbits exposed for 3 months at 190 mg/kg/day, yet no other details were provided. No direct references were found.

Khan *et al.* [20] investigated the effects of short-term repetitive exposures of perchlorate and chlorate in male Fischer 344 rats. One-hundred and sixty rats were evenly distributed into one of 16 treatments and were exposed via drinking water to various combinations of AP and/or sodium chlorate (SC) or a control via the drinking water for 7 days. Analytical data pertaining to salt concentrations and drinking rates were measured. Hypertrophy and colloid depletion of the thyroid were observed in the high-dose AP treatment (8.7 mg/L, 1.2 mg/kg/day), in most treatments containing sodium chlorate (> 0.5 mg/L or 0.69 mg/kg/day), and in all groups containing mixtures of these compounds [20]. Lower serum T4 concentrations were found in all mixtures exceeding 0.36 mg/L AP and 57 mg/L SC; however, T3 levels were similar across all 16 treatments. Of concern for this study is that SC levels of 0.5 mg/L were detected in the control water. There were no

[1]Note that the study may report additional clinical symptoms. However, the text is in Russian, and additional interpretation was not possible.

Table 28.3 Summary of Relevant Mammalian Data for TRV Derivation

			Test Results			
Test Type	**Test Organism**	**Test Duration**	**NOAEL (mg/kg/day)**	**LOAEL (mg/kg/day)**	**Effects Observed at the LOAEL**	**Study**
Chronic/ other	Rat (Sprague Dawley)	>70 days	2.54	25.4	Reduced ossification sites per litter for sternal centers and forelimb phalanges (transient effect).	[39]
	Rat (Sprague Dawley)	>70 days	0.254 (♂ + ♀)	2.54 (♂ + ♀)	Histopathologic changes to the thyroid: hypertrophy and hyperplasia.	[34]
	Rabbit (New Zealand white)	23 days (GD)	0.762 (♀)	8.81 (♀)	Hypertrophy of the thyroid gland. Authors noted no adverse effects on fetal development.	[40]
	Rat (Sprague Dawley)	2 generations	0.847 (F1)	2.54 (F1)	Histopathological changes to the thyroid: hyperplasia.	[42]
			2.54 (F1)	8.47 (F1)	Increased thickness of the corpus callosum; potential indication of neurotoxicity.	
			8.47 (P1)		No observed effects on reproduction or thyroid in adults. Unbounded NOAEL.	
	Rat (Long-Evans)	GD 6–21, PND 1–25	0.34	3.4	Changes in ovarian follicular morphology and number.	[45]

Chronic/ other	Prairie vole	180 days	0.79	NA	No observed effects when compared oxygen metabolism with controls from 51-day exposure; no effects in T4 or other signs.	[26]
	Rat (Sprague Dawley)	Precohabitation – 2 weeks, GD 0- lactation D10 (PPD 22)	0.85	25.4	Dams GD21: Hypertrophy of the thyroid, including mass, changes in plasma thyroid hormone levels.	
	Rat (Sprague Dawley)	Precohabitation – 2 weeks, GD 0- lactation D10 (PPD 22)	0.85	25.4	Dams LD10/22 – Changes in thyroid mass, hypertrophy, colloid depletion. Thyroid hormone changes at LD22. PPD22 male pup thyroid mass and colloid depletion; female pups thyroid mass, colloid, and T3.	[43]
			0.0085	0.085	Male pups PPD 22: Changes in thyroid hormone levels.	
			25.4	Not determined	No changes in found various reproductive indices measuring success. Pup weights larger at NOAEL compared with controls (not considered treatment-related).	

Continued

Table 28.3 Summary of Relevant Mammalian Data for TRV Derivation—cont'd

Test Type	Test Organism	Test Duration	Test Results NOAEL (mg/kg/day)	LOAEL (mg/kg/day)	Effects Observed at the LOAEL	Study
Chronic/ other	Rat (Sprague Dawley)	Precohabitation – 2 weeks, GD 0-lactation D10 (PPD 22)	25.4	Not determined	No dose-related changes in brain morphometry, histology, or motor behavior of pups at various early growth stages.	[44]
	Prairie vole Deer mouse	21 days + mating to PND 21	0.7 (voles) 1.1 (mice)	NA	Data based on reproductive success (see text). Results in thyroid hormone levels variable between perchlorate-water and perchlorate-food treatments.	[51]
	Rat (Wistar)	31 days PND to puberty	500	NA	No changes in delay of puberty in males; no changes in reproductive organ histology, weights. Testosterone increased to 250 mg/kg/day; equal to controls at 500 mg/kg/day; thyroid and TSH/T4 changes at 62.5 and 125 mg/kg/day, respectively.	[50]
Subchronic	Rat (Sprague Dawley)	14 days	0.847 (♂ + ♀)	8.47 (♂ + ♀)	Decrease in T4 thyroid hormone level. This study is a reanalysis of Springborn (1998) data by EPA analyzing raw data by gender, time, and treatment levels. The 30-day recovery observation after the 90-day exposure indicated an unbounded LOAEL at 0.05 mg/kg/day.	[4]

Subchronic			0.00847 (♀)	0.0423 (♀)	Increase in TSH thyroid hormone level. This study is a reanalysis of Springborn (1998) data by EPA analyzing raw data by gender, time, and treatment levels.	[4]
	Rat (Sprague Dawley)	15 days	0.0423 (♂)	0.169 (♂)	The 30-day recovery observation after the 90-day exposure indicated an unbounded LOAEL at 0.05 mg/kg/day.	
		90 days	0.00847 (♂ + ♀)	0.0423 (♂ + ♀)	Decrease in T4 thyroid hormone level. This study is a reanalysis of Springborn (1998) data by EPA analyzing raw data by gender, time, and treatment levels.	[22]
	Rat (Sprague Dawley)		0.00847 (♂)	0.0423 (♂)	The 30-day recovery observation after the 90-day exposure indicated an unbounded LOAEL at 0.05 mg/kg/day.	
		90 days	0.0423 (♂ + ♀)	0.169 (♂ + ♀)	Increase in TSH thyroid hormone level. This study is a reanalysis of Springborn (1998) data by EPA analyzing raw data by gender, time, and treatment levels. The 30-day recovery observation after the 90-day exposure indicated an unbounded LOAEL at 0.05 mg/kg/day.	

Continued

Table 28.3 Summary of Relevant Mammalian Data for TRV Derivation—cont'd

Test Type	Test Organism	Test Duration	Test Results: NOAEL (mg/kg/day)	Test Results: LOAEL (mg/kg/day)	Test Results: Effects Observed at the LOAEL	Study
Subchronic	Prairie vole	51 days	1.13	9.89	Lower T4 levels/thyroid mass, lower kidney mass.	[26]
	Rabbit	3 months	NA	190	Nervous system effects (not specified).	[16]
	Rat(Sprague Dawley)	14 days	3.66 (♂ + ♀)	9.69 (♂ + ♀)	Increase in thyroid-to-body weight ratio; author observed that sex does not statistically affect treatment results.	[24][a]
		14 days	0.0931 (♂)	0.375 (♂)	Decrease in T3 thyroid hormone level; author concludes that sex is unrelated to treatment results.	
		14 days	0.105 (♀) 0.375 (♂)	0.395 (♀) 0.942 (♂)	TSH increases with increase in dose; sex responds to dose differently.	
Acute	Rabbit	LD_{50} LD_{50}	635 – 1,610 1,610 – 1,690		Mortality. Mortality.	[16]

[a]Inferred from the statistical table; the author did not explicitly indicate the LOAEL or NOAEL.

changes in body weight or water consumption, nor were there any other clinical signs of toxicity.

Keil *et al.* [21] exposed female B6C3F1 mice to calculated doses of AP equivalent to receiving 0, 0.1, 1.0, 3.0, or 30 mg/kg/day via drinking water for 14 or 90 days. Doses were within 10% of targets. Thyroid histology, systemic thyroid hormone levels, organ and body weights, and several immunological indices were endpoints of concern. Immunological indices included cellularity of lymphoid organs, CD4/CD8 thymic and splenic subpopulations, stem cell synthesis, natural killer cell activity, cytotoxic T-cell activity, nitrite production of peritoneal macrophages, phagocytosis, IgM and IgG antibody response to T-dependent antigen (sheep red blood cell), delayed-type hypersensitivity, melanoma tumor challenge model, and an antinuclear antibody screening assay. Changes in T4 levels, but not in T3 or TSH, were found respective to treatment (3 and 30 mg/kg/day after 14 days; 1, 3, and 30 after 90 days). Effects in the immunological assays produced mixed results with few significant or dose-related outcomes. However, changes in delayed-type hypersensitivities were observed at the 30 mg/kg/day dose for mice at durations of both 14 and 90 days.

Mammalian Oral Toxicity: Subchronic

The U.S. Environmental Protection Agency (U.S. EPA) reanalyzed subchronic 14-day and 90-day exposure data [22] supplied by the U.S. Air Force Research Laboratory/Operational Toxicology Branch (AFRL/HEST) and originally analyzed by Springborn Laboratories [23]. In the reanalysis, a comparison was made of the interaction between gender, time, and dose levels. Sprague Dawley rats in groups of nine to10 per treatment per sex were exposed to different doses of ammonium perchlorate via drinking water over a period of 90 days, followed by a 30-day recovery period (of no perchlorate exposure). Thyroid hormone levels were measured at 14 and 90 days of treatment and measured again after the 30-day recovery period. Using data for the 14-day subchronic AP exposure (perchlorate ion levels of 0, 0.00847, 0.0423, 0.169, 0.847, and 8.47 mg/kg/day), the U.S. EPA reviewer, Crofton [22], reported no observed adverse effect levels (NOAELs) of 0.847 mg/kg/day and 0.00847 mg/kg/day perchlorate ions for T4 and TSH effects, respectively, in both males and females. The effect of perchlorate on T3 serum levels in males differed statistically from that found in females. The lowest perchlorate ion dose (0.00847 mg/kg/day) reduced the T3 levels in male rats. This dose was considered an unbounded lowest observed adverse effect level (LOAEL) (i.e., no NOAEL could be determined). In contrast, T3 levels in female rats were statistically unchanged at all dose levels. Consequently, the 8.47 mg/kg/day dose (as perchlorate ion) was considered an unbounded NOAEL.

The data from the 90-day exposure study were evaluated for interactions between gender, time, and perchlorate dose levels [22]. Sprague Dawley rats in groups of nine to10 per treatment per sex were exposed to different doses of AP in drinking water over a period of 90 days. Thyroid hormone levels were measured at 90 days and measured again after a 30-day recovery period in which there was no perchlorate exposure. An unbounded LOAEL of 0.00847 mg/kg/day was reported for both male and female rats based on decreases in T3 and T4. Although Crofton could not determine a NOAEL for T3 and T4, a LOAEL of 0.0423 mg/kg/day was calculated, based on increased TSH levels in both sexes.

In another subchronic study, King [24] reported significant gender differences in sensitivity to AP. Eight groups of six each male and female Sprague Dawley rats were administered AP via drinking water at eight different target doses ranging from 0 to 25.0 mg/kg/day perchlorate ion for a period of 14 days. At the conclusion of the study, the following endpoints were measured: [1] body weight; [2] thyroid/body weight ratio; [3] TSH level; [4] water ingestion rate; [5] thyroglobulin (Tg), T3 and T4, and [6] the T3 level. The study findings are described below.

1. Body weight. There were no statistical differences between dose groups, indicating that AP consumption did not affect appetite; however, there were gender-related differences. Male rats gained more weight (~95 g) than did female rats (~20 g).
2. Thyroid/body weight ratio. ANOVA analysis indicated there were statistical differences in thyroid/body weight ratios between male and female rats. Animals receiving doses of 11.4 mg/kg/day or greater experienced an increase in thyroid/body weight ratios when compared to controls.
3. TSH level. Among female rats, the NOAEL for elevated TSH levels was 0.124 mg/kg/day; for male rats, the NOAEL for elevated TSH was 0.443 mg/kg/day. A clear relationship between perchlorate dose and TSH levels was observed, with TSH concentrations increasing with increasing perchlorate dose.
4. Water ingestion rate. Water consumption was not influenced by the concentration of perchlorate administered, however there was a statistical difference in water consumption between males and females. Male Sprague Dawley rats had higher water consumption rates (~38 mL/day) compared to females (~28 mL/day).

5. Tg, rT3, and T4 levels. Iodized Tg is required to synthesize monoiodo-L-thyronine (MIT) and 3,5-diiodo-L-thyronine (DIT), which are combined to produce T3 and T4. rT3 is formed from T4 in peripheral tissues; rT3 increased consistently with dose. Although the levels of Tg in the blood serum of Sprague Dawley rats were determined after exposure to AP for 14 days [22], this study did not perform further analysis on the threshold dose. This was because Capen [25] demonstrated that the lowest dose that lowers T3 and/or T4, and simultaneously increased TSH, could be considered a threshold dose. Therefore, the statistical analyses focused on T3 and TSH levels. Moreover, the T4 hormone level, which was considered to be as important as T3 did not exhibit a significant dose-response relationship and, therefore, was not discussed in further detail.
6. T3 level. Circulating T3 levels were measured in eight groups of 12 Sprague Dawley rats (6 males, 6 females). Although T3 levels decreased with increased AP doses, the T3 hormone level decreased at a lower dose in female rats than in males. No NOAEL could be determined for female rats; in male rats, the NOAEL was 0.11 mg/kg/day.

Isanhart *et al.* [26] exposed five prairie voles (*Microtus ochrogaster*) to target AP exposures of 0, 1, and 10 mg/kg/day for 51 and 180 days. Voles were exposed to amounts of AP in the drinking water that resulted in average perchlorate ion exposures of 0, 1.13, and 9.89 mg/kg/day. The purpose of this study was to measure changes in energy metabolism by measuring oxygen consumption under normal resting and cold stress conditions (12 °C for 20 min followed by a drop to 5 °C until the voles became hypothermic). Perchlorate exposure did not affect metabolism (oxygen consumption relative to subchronic AP treatments and controls). No differences were found in oxygen consumption (metabolic rates) between treatments and durations in voles during the resting phase (room temperature) or in cold-challenged voles. No differences were found in plasma T4 or T3 concentrations between treatment groups; however, T4 concentrations per thyroid weight differed between controls and high-dose treatment groups. Mean kidney weights were lower for the high-dose group compared with the low-dose and control groups; however, the clinical significance of that finding is uncertain. There were no changes in liver weights relative to controls, and no other clinical signs of toxicity were observed.

Zhang *et al.* [27] studied the effects of AP exposure on male rat testes. Male rats were exposed for a period of 80 days to 0, 130, 260, or 520 mg/kg/day. Levels of thyroid hormones, testosterone in serum, and sperm motility were measured, and histological change in the testes was observed. The increase in body weight was significantly lower in the AP-exposed group than in the control group. There were no differences of serum free T3 levels between all AP-exposed groups and the control group, while serum TSH levels in middle- and high-AP-exposed groups decreased significantly when compared to controls. In terms of sperm motility, the percentages of Grade A and B sperm in the middle and high groups were 12.3 ± 2.52 and 14.8 ± 5.93, respectively, and 17.7 ± 4.63 and 15.8 ± 2.28, respectively; all of which were significantly lower than those of the control group. The percentages of Grade D sperm in middle and high groups were 38.0 ± 3.61 and 40.0 ± 8.99, respectively; these were significantly higher than those of the control group. No differences in serum testosterone levels were observed between the AP-exposed and control groups.

Mammalian Oral Toxicity: Chronic

Chronic exposure to rats was studied by Kessler and Kruskemper [28, original text in German, and cited in 18]. Groups of 40 rats were provided drinking water containing perchlorate at 1% weight/volume (w/v), approximately 513 mg/kg/day. Body and thyroid gland weights were the measurement endpoints. At the end of 0, 40, 120, 220, and 730 days, groups of 7 to 8 rats were euthanized, and body weight and thyroid gland measurements were made. Although no statistically significant differences were found in body weight for any of the exposure periods, statistically significant changes in thyroid histology were observed starting at day 40. Observed thyroid pathologies developed into fibroses and follicular adenomas as the experiment progressed.

In the second part of a two-part study, Isanhart *et al.* [26] exposed five prairie voles (*M. ochrogaster*) to AP in drinking water for 180 days. This treatment resulted in average perchlorate ion exposures of 0.79 mg/kg/day and measured oxygen consumption under cold stress conditions (12 °C for 20 min followed by a drop to 5 °C until the voles became hypothermic). Perchlorate exposure did not affect metabolism (oxygen consumption relative to subchronic AP treatments and controls). No differences were found in oxygen consumption (metabolic rates) between treatments and durations in the voles during the resting phase (room temperature) or in cold-challenged voles. No difference was found between plasma T4 concentrations from voles exposed to 0.79 mg/kg/day for 180 days and those exposed to 1.12 mg/kg/day for 51 days. No changes were found in kidney or liver weights relative to controls, and no other clinical signs of toxicity were observed.

Mammalian Toxicity: Other

Intratracheal instillation of AP was evaluated for its ability to induce pulmonary fibrosis in rabbits [29]. Twenty-five male rabbits were randomly allocated into five groups to receive AP, bleomycin, or a saline control. All rabbits were euthanized, and lung tissues were evaluated by light and electron microscopy. AP-instilled lung tissue showed inflammatory infiltrates, alveolar collapse, subpleural thickening, and lymphocyte proliferation. Electron microscopic evaluation of lung tissue showed massive fibroblast accumulation, collagen fiber hyperplasia, and dense collagen deposition. The histopathological changes were considered an effective indicator leading to the development of pulmonary fibrosis.

Another study observed thyroid hypertrophy in rabbits following perchlorate exposure [30, in German, and cited in 18]. Rabbits were administered a dose of 71.8 mg perchlorate anion/kg/day via the diet from conception through gestation day (GD) 21 or 28. Fetal thyroid weights in the experimental group were approximately four times higher than the control weights, while the maternal thyroid weights in the experimental group were three times higher than were found in controls [18].

Perchlorate effects on thyroid weight and deciduoma formation (uterine deciduoma are formed in pseudopregnant dams, similar to decidua in normal pregnancy) were investigated by inducing false pregnancy in female Wistar rats [31]. Female rats were administered 0.25% and 1.0% potassium perchlorate in drinking water for 7 days (GD 2–8). Since exposure occurred during gestation, these data were considered representative of a chronic exposure. At the end of the 7 day exposure period, thyroid weights were significantly increased at exposures of 1.0% w/v (599 mg/kg/day perchlorate ion); deciduoma formation was unaffected by either exposure level.

In a related study, Brown-Grant and Sherwood [32] performed a two-generation study of potassium perchlorate in rats. The dams were dosed with 1,740 mg/kg/day perchlorate ion at the termination of pregnancy with sacrifice of the pups on postpartum day (PPD) 9. Relative thyroid weight for both the dams and pups was significantly different from that of the controls. Due to the single treatment group, only the unbounded LOAEL (1,740 mg/kg/day perchlorate ion) was determined.

Postel administered potassium perchlorate to female guinea pigs at 528.5 mg/kg/day (converted from 1% w/v) during the final 21 to 48 days of gestation for a total of 27 days [33, cited in 6]. This investigation is considered a chronic exposure because the perchlorate was administered during a sensitive life stage of the species, and fetal development was potentially affected. Female guinea pigs received perchlorate via drinking water and subcutaneous injections of 0, 8, 18, or 32 μg T3 supplement per day. The control groups received 257 mg/kg/day perchlorate ion in drinking water, and saline solution via subcutaneous injection. Although maternal thyroid weight and histology were not affected, fetal thyroid weight averaged 15 times greater than that of controls. The authors reported that the T3 supplements at all levels did not appear to have mitigated the effects of perchlorate and were suspected to have intensified the observed effects.

York *et al.* [34] presented a toxicology evaluation of perchlorate based on a report to the Perchlorate Study Group by Argus Research Laboratories [38]. In this study, Sprague Dawley rats were administered AP at levels of 0.254, 2.54, and 25.4 mg/kg/day perchlorate ion over two generations. The P1 generation male rats were exposed to perchlorate over 140 days, and the females were exposed for at least 126 days. The F1 generation male rats were exposed to perchlorate for 124 days, and the females were exposed for at least 90 days. Due to the extended duration of perchlorate exposure, the P1 and F1 generation studies qualified as chronic. The study evaluated the effect of perchlorate on the male and female rats' reproductive systems, growth and development of F1 offspring, neonatal morbidity, mortality, and potential prenatal developmental toxicity. The study found 2.54 mg/kg/day perchlorate ion to be the LOAEL, the lowest dose that resulted in a statistically significant increase in thyroid hypertrophy/hyperplasia in the male and female F2 generation rats, the female F1 generation pups, the male F1 generation adults, and the male and female P generation rats. Consequently, the next lowest level, 0.254 mg/kg/day perchlorate ion, was considered the NOAEL.

York *et al.* [34] concluded that AP is not a reproductive toxicant in rats at doses as high as 25.4 mg/kg/day; no deaths, abortions, or premature deliveries could be attributed to this level of perchlorate exposure in rats. However, Argus [39] reported the appearance of developmental effects, including significantly reduced ossification sites per litter for sternal centers and forelimb phalanges at 25.4 mg/kg/day perchlorate ion. Moreover, a subset of these data was published in York *et al.* [34]. The USEPA suggests 25.4 mg/kg/day perchlorate ion as a LOAEL with a corresponding NOAEL of 2.54 mg/kg/day for developmental toxicity, and acknowledges that a definitive assessment cannot be made on the available data.

The National Research Council reviewed available data for establishing a Reference Dose (RfD) for human exposure and determined on the basis of a report by Greer *et al.* [35] to establish the human reference dose at 0.007 mg/kg/day from all sources (i.e., food and water) [36]. For human exposure, the EPA accepted this value, but a comparable value for wildlife was not established. There is currently pressure to reduce the human RfD due to findings that decreased levels of thyroid hormones can have adverse developmental effects on human sensitive subpopulations, such as persons with thyroid disorders, pregnant women, fetuses, and infants [37]. Under the guidelines of wildlife toxicity assessments, these outcomes are not relevant and can be measured only with extraordinary difficulty. Hence, the level at which a reproductive effect to wildlife from perchlorate exposure becomes significant is equivocal.

In a companion study on rabbits, York *et al.* [40] presented a toxicological evaluation of perchlorate by the Argus Research Laboratory [41] to the Perchlorate Study Group. York *et al.* [40] evaluated the maternal and embryo-fetal toxicity potential of AP via drinking water. Female rabbits were exposed to AP during gestation for a total of 23 days (GD 6–28). Considering the exposure occurred during a critical life stage of the species, this test is considered equivalent in value to a chronic study. Adult female rabbits were dosed at 0, 0.0847, 0.762, 8.81, 25.7, and 86.7 mg/kg/day perchlorate ion. For hypertrophy of the thyroid gland, the NOAEL for adult female rabbits was 0.762 mg/kg/day perchlorate ion; the LOAEL was 8.81 mg/kg/day perchlorate ion. York *et al.* [40] reported no adverse effects of AP on fetal development. Subsequent to the 2001 studies, York *et al.* [42] investigated the neurodevelopmental effects of AP in exposed adults and offspring of Sprague Dawley rats. In this study, mated rats were exposed to 0, 0.1, 1.0, 3.0 and 10 mg AP/kg/day (0, 0.0847, 0.847, 2.54, and 8.47 mg perchlorate ion/kg/day) in drinking water from GD 0 (mating) through lactation (postpartum) day (LD) 10. Because this represents a critical life stage of the species and the affected endpoint is potentially relevant to fetal development, the data from this study are considered equal in weight to data of chronic duration. Neurodevelopmental effects in offspring (evaluated by passive avoidance, water maze, motor activity, and auditory startle testing) were not observed at any dose group. In addition, there were no observed differences in brain weights, morphometry, neuropathology, body weights, feed consumption, clinical signs, or sexual maturation of pups. In dams, no effects were observed at parturition, in clinical signs, in body weights, or in feed and water consumption rates.

As with previous studies, thyroid toxicity was observed in the rats' offspring, but there were no observations of follicular hypertrophy or hyperplasia in dams [42]. The F1 generation pups culled at 5 days postpartum exhibited significantly different T3, T4, and TSH levels in the 0.847,

2.54, and 8.47 mg/kg/day perchlorate ion dose groups compared to controls. There was a nonsignificant increase in the incidence of follicular cell hypertrophy at all dose levels, with pronounced hypertrophy at the highest dose in male pups and at the two highest doses in female pups. Other thyroid effects included a dose-related decrease in thyroid follicular lumen diameter that was significant in the 0.847, 2.54, and 8.47 mg/kg/day perchlorate ion dose groups for males and in the 2.54 and 8.47 mg/kg/day perchlorate ion dose groups for females. In addition, lumen area was decreased in the 2.54 and 8.47 mg/kg/day perchlorate ion dose groups for males and the 8.47 mg/kg/day perchlorate ion dose group for females. Based on these thyroid effects, a NOAEL of 0.0847 for pups was identified by York *et al.* [42]. Thyroid hyperplasia in pups occurred in the two highest dose groups. As opposed to thyroid follicular hypertrophy (not considered a reliable biomarker for adverse effects of perchlorate), hyperplasia at a perchlorate ion dose of 2.54 mg/kg/day was considered the LOAEL for adverse thyroid effects, with a corresponding NOAEL of 0.847 mg/kg/day perchlorate ion. In addition, histological evaluation of the brain indicated increased thickness of the corpus callosum at the 8.47 mg/kg/day perchlorate ion exposure level in female pups culled 12 days postpartum (LOAEL). However, the method by which the corpus was measured may be considered subjective. The authors identified a NOAEL for this endpoint in pups at 2.54 mg/kg/day perchlorate ion.

In another developmental study conducted by York *et al.* [43], Sprague Dawley rats were exposed for periods longer than those in the recommended guidelines. Female rats were exposed *ad libitum* to AP via drinking water 2 weeks prior to cohabitation and continuing through pregnancy to PPD 22. Sequential terminations of subgroups occurred throughout the exposure period. Water consumption of each animal was monitored during each stage. The rats were exposed to 0, 0.01, 0.1, 1.0, or 30.0 mg AP/kg/day (0, 0.0085, 0.085, 0.85, or 25.4 mg perchlorate ion/kg/day, respectively). There were no deaths, adverse clinical symptoms, or necropsy findings considered to be exposure-related during the precohabitation, gestation, and/or lactation periods. There were no changes in the reproductive and developmental parameters (i.e., litter size, number of litters, corpora lutea, implantations, fetuses, resorptions, or postnatal pup deaths). Pups in the 0.01 mg/kg/day and higher dose groups were heavier at PPD 14 compared to controls, but pups in the 30.0 mg/kg/day dose were larger only at PPD 22. The authors consider this to be due to a decrease in weights for controls and not necessarily treatment related. Thyroid hormone levels

(TSH and T4) were altered for dams at GD 21 at exposures of 0.01 and greater, but T3 levels were affected only at exposures of 30.0 mg/kg/day. Thyroid weight and colloid depletion were also affected at 30.0 mg/kg/day. These effects were less marked at lactation day 10. Male and female PPD 22-pups had thyroid changes similar to those of GD 21-dams. The TSH levels varied between intervals of exposure and were not consistent between life stages or gender.

In a companion study, York *et al.* [44] investigated brain morphometry, histology, and the motor behavior of pups from the previous experiment [43]. Twenty linear morphometric measurements were made of each rat brain from pups at PPD 10 and 22. No treatment-related changes were found in any of the nine measurements of motor activity. Microscopic analysis of multiple sections of rat pup brains did not indicate any evidence of exposure-related neuropathologic changes. Although some differences were found between treatments in some of the brain measurements, they were inconsistent and not considered dose-related due to the lack of a dose-response relationship.

Baldridge *et al.* [45] exposed 24 pregnant Long-Evans rats to either 0, 5, or 50 mg/L AP (corresponding to 0, 0.4, or 4.0 mg/kg/day or 0, 0.34, and 3.39 mg perchlorate ion/kg/day, respectively) from GD 7 to 21 to postnatal days (PND) 24 and 25 through maternal exposure and lactation. Some of these rats were also coadministered T4 supplements along with perchlorate in the drinking water to simulate levels suggested for normal development in humans. On PND 24 and 25, female pups were euthanized, followed by histological examination of the ovaries. The number of ovarian (preantral and antral) follicles were reduced in the 4 mg/kg/day groups but not in the 0.4 mg/kg/day groups. Rats treated with 4 mg/kg/day and receiving T4 were no different from the controls, suggesting that T4 ameliorated the effects of the perchlorate ion. Many of these follicles counted in the high-dose group were also atretic (i.e., abnormally closed). No changes in the mean ovarian area were observed between treatments.

Only studies by Thuett *et al.* [46–48] and Roots *et al.* [49] were identified as having examined the effects of perchlorate in wild mammalian species. In these studies, captive-born, wild-type breeding pairs of deer mice (*Peromyscus maniculatus*) were exposed to 0, 1 nM,1 μM, and 1 mM AP (0, 1.59×10^{-5}, 1.60×10^{-2}, and 15.78 mg/kg/day perchlorate ion; 10 pairs per dose group) in drinking water. The Thuett *et al.* [46] and Roots *et al.* [49] studies represent preliminary data that were reported only in abstracts. As such, this information is presented here but not used for

TRV derivation. These preliminary data suggest that AP may adversely affect growth (i.e., significant decrease in pup body weights at the highest dose, and development (i.e., significant differences in kidney, liver, and heart weights in pups) of deer mice when exposed from the time of cohabitation until post lactation; however, the data were largely equivocal. Subsequent studies using the same dose groups extended the exposure time to the weaning of the third litter. Effects were evaluated in the second litter pups at PND 21 and were reported in Thuett *et al.* [47,48].

In Thuett *et al.* [47,48], reproductive and growth parameters of deer mice were evaluated. Although there was a dose-related decrease in litter size, the decrease was an insignificant effect at any dose level. Additionally, there were no significant differences in body or organ weights when the data were analyzed using individual litters as experimental units. When individual pups were considered as the experimental unit, a significant decrease in body weight was observed at the 1 μM dose level measured at PPD 5 and PPD 20. This effect was not observed in the highest (1 mM) dose group, however, so this effect was not dose-dependent, and its significance was uncertain. The only statistically significant dose-related effect was a decrease in the heart weights (using body weight as a covariate) of male pups in the 1 μM and 1 mM dose groups on PND 21. A similar trend was observed in female pups, but the decrease was not statistically significant. As described by the authors, effects on the heart have previously been related to improper functioning of the thyroid gland. Therefore, the decreased heart weights may have resulted from perchlorate's effects on the thyroid, as observed during the study.

When thyroid hormone concentrations were evaluated in these deer mice [48] concentrations of T4 were significantly greater than controls in the low- and medium-dose groups. Concentrations of T3 did not differ among dose groups, and testosterone (except for one case in the high-dose group) and estradiol were undetectable. Histological changes in the thyroid gland (i.e., decreased follicle number per unit area) were observed in the low- and high-dose groups, but not in the medium-dose group. These studies indicate that exposure to AP in deer mice resulted in altered thyroid hormone levels and histopathological changes to the thyroid gland that may be related to developmental effects observed in the heart. However, the results are equivocal and difficult to interpret given that the thyroid effects were not dose-dependent, and the biological significance of the reduced heart weight is uncertain. Using decreased heart weight as the endpoint, the 1 nM dose level (1.59×10^{-5} mg/kg/day perchlorate ion) may be considered the

NOAEL, and the 1 μM dose level (1.60×10^{-2} mg/kg/day perchlorate ion) may be considered the LOAEL.

One component of the AFRL/HEST perchlorate study was an evaluation of recovery of thyroid hormones (T3/T4) and TSH production following the cessation of perchlorate exposure [23]. Provision of clean water to rats for 30 days followed a 90-day perchlorate exposure. The result of the thyroid level measurement taken at the end of the 30-day period revealed differential responses of male and female rats during the recovery period. For males, T3 and TSH returned to the control level after 30 days of no perchlorate exposure, while T4 remained depressed at all dose levels. For female rats, no statistically significant levels were observed for T4 at any dose, while T3 remained depressed at 8.47 mg/kg/day, and TSH increased at a dose of 0.042 mg/kg/day. It is a widely accepted belief that development of an adverse impact on the thyroid requires a simultaneous depressed T3/T4 and an increased TSH; none of the administered doses produced this combined effect. The unbounded NOAEL for combined thyroid and pituitary hormone effects of perchlorate in rats in this study is 10 mg/kg/day. Since significant depression in T4 levels were not observed in females at any dose, a higher dose would need to be administered to rats before reliable a chronic NOAEL and LOAEL could be established.

A recent study was conducted to test for potential endocrine-disrupting effects on male reproductive development because of perchlorate exposure. Offspring from 23 Wistar rats were exposed by gavage to 0, 62.5, 125, 250, or 500 mg perchlorate/kg/day in a corn oil vehicle from PND 23 to 53 (31 days) [50]. Onset of puberty was assessed through preputial separation of the penis of males (an indication of delay in male reproductive maturity). Also assessed were the histology of testes, epididymides, prostate, seminal vesicles, and other relevant organs; organ and body weights; and serum testosterone, TSH, T3, and T4 levels were determined. No dose-related significant effects were found in any of the reproduction-related parameters, however testosterone levels increased to 250 mg/kg/day and decreased to control levels at exposures of 500 mg/kg/day. Only the 250 mg/kg/day treatment was statistically higher relative to controls. Changes in thyroid histology (e.g., follicular height and colloid area) occurred at 62.5 mg/kg/day. Serum TSH and T4 levels differed following an exposure of 125 mg/kg/day.

The effects of perchlorate-contaminated food versus water were studied in prairie voles (*M. ochrogaster*) and deer mice (*Peromyscus maniculatus*) for 22 days [51], followed by evaluations of reproductive success and subsequent exposure for nonmating individuals. Pups were necropsied and evaluated at

PND 21. Perchlorate-contaminated food was prepared by incorporating soybean plant matter from plants grown with perchlorate-irrigated water. During the 21-day exposure period, voles in the water-exposed and feed-exposed groups received 0.6 and 0.7 mg/kg/day, and mice received 0.9 and 1.1 mg/kg/day, respectively. During cohabitation (following the initial 21-day exposures), voles received 0.4 and 0.3 mg/kg/day, and mice received 1.0 and 1.1 mg/kg/day, respectively. Although some mortality occurred, it was not attributed to treatment. No statistical differences were reported in reproductive success between treatments for either species (e.g., number pups/litter, pup survival, duration to birth). Male voles in the perchlorate-food treatment had lower plasma levels of T4 than did the perchlorate-water or control groups; no other differences in either T3 or between sexes or treatments were found. Plasma T3 concentrations were lower for mice in the perchlorate-food group compared with controls; no other differences in T3 or T4 levels were found. No differences in thyroid hormone content were found. Measurements of thyroid cell height and mean colloidal area in voles exposed to perchlorate-water were different from those of voles exposed to perchlorate-food and those of voles in the control group.

Studies Relevant for Mammalian TRV Development for Ingestion Exposures

The primary target organ for perchlorate toxicity is the thyroid gland, with histopathological changes and alterations in thyroid hormone production the most common effects of toxicant exposure (Table 28.3 and Figure 28.1). Changes in thyroid configuration and hormone levels were found in New Zealand white rabbits [40], Sprague Dawley and Fischer 344 rats [20,22,24,34,42], and deer mice [46]. Given that the reserve capacity of thyroid hormone is less in rats and other mammals than in humans, the competition of iodine and the perchlorate ion at the sodium-iodide symporter (NIS), and the lack of marked adverse health effects, these data are of uncertain significance and are limited for use in determining a health-based threshold level.

In deer mice, thyroid hormone and organ effects did not show a clear dose-response relationship, and reliable NOAEL and LOAEL values could not be identified. Reduced heart weights, possibly related to altered thyroid function, were also reported for deer mice [47] but were very small in magnitude. Although a NOAEL and LOAEL for this effect could be developed, the reliability of these values is uncertain given the discontinuous

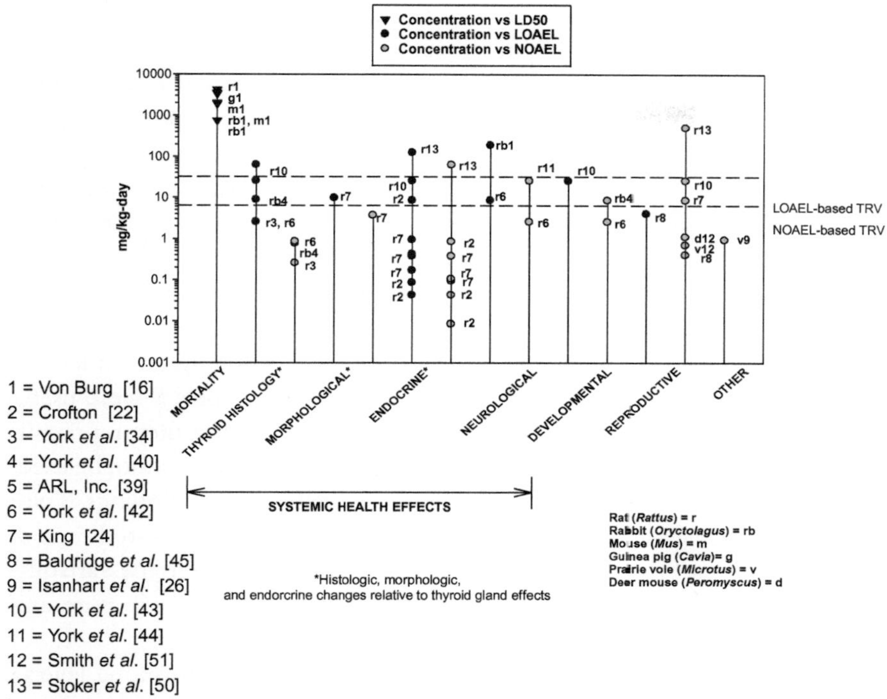

Figure 28.1 Perchlorate: Health Effects to Mammals.

dose-response relationships, minimal reduction in weights, and a lack of corroborative data for adverse effects observed for other studies. Other contributing factors included a three-order-of-magnitude difference between the NOAEL and LOAEL values from the studies of Thuett *et al.* [47] and the limited data that determined NOAEL, LOAEL, and LD_{50} values for rabbits, mice, guinea pigs, and rats [16]. Isanhart *et al.* [26] conducted a study of oxygen consumption under cold-stressed conditions in prairie voles for 51 and 180 days at target doses of 0, 1, or 10 mg/kg/day (0, 0.85, and 8.5 mg perchlorate ion/kg/day, respectively). No changes were found in oxygen consumption or clinical toxicity, however a trend was apparent for decreased T4 concentration relative to thyroid mass for 51-day exposures only. Other effects were not observed, and as such, might represent a LOAEL/NOAEL for another wildlife species. Care must be taken in evaluating these data, as small animal numbers coupled with suspect statistical comparisons (e.g., the comparison of data from 51-day controls with that from 180-day treatment) were reported.

Smith *et al.* [51] compared perchlorate exposures in prairie voles and deer mice using either perchlorate-incorporated food or perchlorate-containing water. Although levels of thyroid hormone levels were variable between treatments, they were not necessarily corroborated by histology, making interpretation difficult. However, this is the only study found that evaluated reproductive success in two wildlife species in a controlled laboratory environment using biologically incorporated food. In this respect, no differences were found in measures of reproductive success between treatments including controls (i.e., time to birth, number of pups/litter, and pup survival rates).

Several studies were found that were current, well-designed, and appropriate in the consideration of TRVs for mammals. The findings of York *et al.* [34,40,42] are of particular value because two different species were used. The studies by Thuett *et al.* [47,48] are important because they represent the only available data for a wildlife species. Despite this, they cannot be used in the derivation of mammalian TRVs because of inconsistent findings. Two orders and two families of *Mammalia* are represented in these studies and include *Rodentia* (*Muridae*) and *Lagomorpha* (*Leporidae*). Despite the variability with the deer mice studies, the effects from perchlorate exposure are consistent across the three species for which data are available.

Although the U.S. EPA reanalysis [22] measured only hormone levels, York *et al.* [34,40,42] conducted a broader toxicological evaluation in which NOAEL and LOAEL values were identified for various endpoints and included ecologically relevant organ systems (e.g., the reproductive system). No adverse health effects were found although Baldridge *et al.* [45] reported changes in ovarian follicles in the female pups of exposed dams. However, rigorous reproductive studies have not found any adverse effects, suggesting these findings are of uncertain biological significance.

The studies of York *et al.* [34,40,42–44] and Bekkedal *et al.* [52] are important in that no adverse health effects or changes in various measures of reproductive success, including behavior, were found in several long-term reproductive studies. The work evaluating male reproductive maturity with respect to perchlorate exposure found no evidence of adverse effects [50]. Thyroid histology (hyperplasia, hypertrophy) and changes in systemic thyroid levels were found in all of these studies, but these changes are of uncertain biological significance and do not differentiate between adaptive mechanisms and adverse effects. Changes in thyroid cell number or structure may be important as a precursor for cancer; however, cancer generally occurs in senescent individuals long after primary reproductive events and therefore is unlikely to have population-level consequences. Changes in

T3 or T4 levels have not been parameterized and thus provide no context from which to determine the levels at which adverse effects occur.

Data from acute studies where gavage methods were employed were considered only when subchronic or chronic studies were not available for a species (e.g., mice and guinea pigs, as presented in Von Burg [16]). However, given the lack of marked adverse health effects, mortality is an important endpoint.

Mammalian Inhalation Toxicity

Only one study that investigated the inhalation toxicity of perchlorate in mammals was found. However, no nominal or measured perchlorate concentration was reported. Moreover, other gases (e.g., H_2S, SO_2, HCl) were measured as vapor products of rocket propellants [53]. Von Burg [16] stated that because of the low vapor pressure of perchlorate salts and acids, exposure to fumes and vapors by wildlife would be negligible. Acute exposure to perchlorate fumes would be expected to affect the upper respiratory tract, producing signs and symptoms such as sneezing, coughing, chest pain, and pulmonary edema. No specific information on the dose required to produce these signs/symptoms was provided [16].

Using intratracheal instillation, Wu *et al.* [29] evaluated the potential for pulmonary fibrosis in rabbits. Twenty-five male rabbits were randomly allocated into five groups to receive AP, bleiomycin, or saline via intratracheal instillation. All rabbits were sacrificed and the lung tissue removed for histopathologic and electron-microscopic evaluation. The AP-instilled lung tissue showed inflammatory infiltrates alveolar collapse, subpleural thickening, and lymphocyte proliferation. Electron microscopic evaluation of lung tissue showed massive fibroblast accumulation, collagen fiber hyperplasia, and dense collagen deposition. The histopathologic results were indicative of development of pulmonary fibrosis.

In another study conducted in rabbits, Wu *et al.* [54] evaluated biomarkers for pulmonary fibrosis in rabbits. Random allocation of 25 male rabbits into five groups received AP, bleomycin, or saline by intratracheal instillation. All rabbits were euthanized, and total RNA was extracted from the lungs. Expression of types I and III collagens, transforming growth factor-β1 (TGF-β1), and tumor necrosis factor-α (TNF-α) were measured by reverse transcription-polymerase chain reaction (RT-PCR). The expression of type I and III collagen messenger RNA in low-, moderate-, and high-dose AP groups was significantly higher ($p < 0.01$ or $p < 0.05$) than that in the saline group. Significantly increased TGF-β(1) and TNF-α RNA levels were observed in the three AP groups as compared with the saline control group ($p<0.01$ or $p<0.05$).

Mammalian Dermal Toxicity

Mammalian toxicity data for the effects of perchlorate via dermal exposure were not located.

Summary of Avian Toxicology

Avian Oral Toxicity: Acute

No data were available for acute exposures of birds to perchlorate.

Avian Oral Toxicity: Subchronic

McNabb *et al.* [55] administered AP to 3- to 4-day post-hatch Northern Bobwhite quail (*Colinus virginianus*) chicks via drinking water. One study group was exposed to AP at concentrations of 0, 0.05, 0.5, 5, 50, and 250 mg/L for 2 weeks to measure thyroid hormone levels. A second group was exposed to ammonium perchlorate at concentrations of 0, 250, 500, 1,000, 2,000, and 4,000 mg/L to measure thyroid hormone levels and growth parameters for an exposure period of 8 weeks. The authors reported that plasma T3 and T4 thyroid hormones and thyroid weight were less sensitive indicators of thyroid function than thyroidal T4 concentrations. Thyroidal T4 was significantly reduced in dose groups of 0.5 mg/L or greater for 2 weeks, whereas a significant increase in thyroid weight occurred at doses exceeding 1,000 mg/L in the 8-week study. The NOAEL and LOAEL for decreased T4 hormone were determined using the 2-week exposure because doses in the 8-week exposure were too high for NOAEL determination. At the end of the 2-week exposure period, the 0.5 mg/L dose group (0.0326 mg/kg/day) was the LOAEL, and the 0.05 mg/L dose group (0.00326 mg/kg/day) was the NOAEL for decreased T4 hormone levels. The NOAEL and LOAEL for increased thyroid weight were 33 and 65 mg/kg/day perchlorate ion, respectively, based on thyroid weights at 8 weeks. Tibia growth was also significantly decreased at the end of the 8-week exposure period; the NOAEL and LOAEL were 130 and 261 mg/kg/day perchlorate ion, respectively (see Figure 28.2).

McNabb *et al.* [56] reports the results of a series of experiments (including those reported above [55]) with varying AP exposures (range includes 0.013, 0.025, 0.05, 0.5, 5, 25, 50, 250, 500, 1,000, 2,000, and 4,000 mg/L) and exposure durations (2, 4, or 8 weeks). Calculated doses as perchlorate ion were 0.00085, 0.0016, 0.0033, 0.033, 0.326, 1.6, 3.3, 16.3, 33, 65, 130,

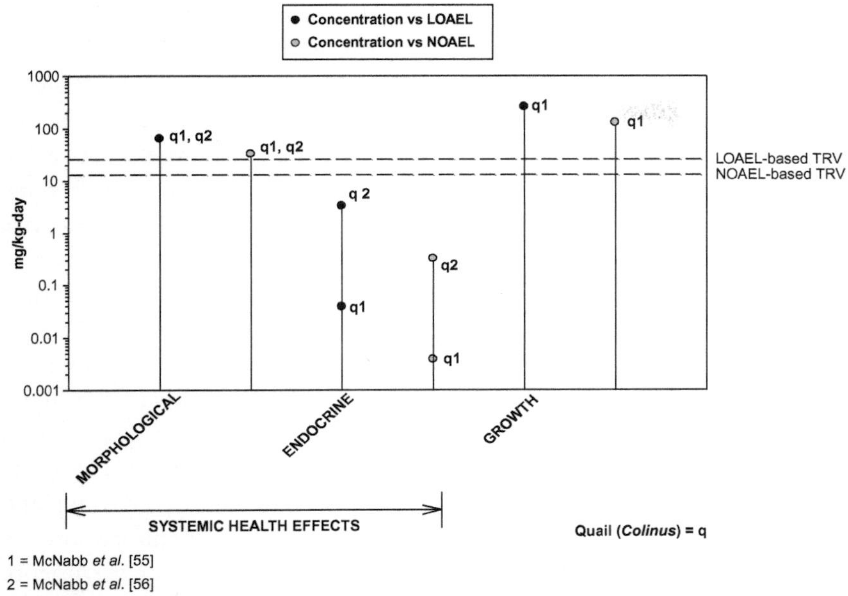

Figure 28.2 Perchlorate: Health Effects to Birds.

and 261 mg/kg/day. Significant reductions in T4 concentrations observed at 2 weeks in 0.5 and 5 mg/L dose groups were not apparent by 4 and 8 weeks (i.e., there was full compensation of T4 levels). Partial compensation was observed at 50 mg/L, but there was no evidence of compensation in the high-dose group (i.e., T4 levels were reduced at 8 weeks to a degree similar to the 2-week measurements). Despite the observed compensation at 50 mg/L, T4 concentrations remained significantly decreased compared to controls. Thyroid weight at 2 weeks was significantly increased at 500 mg/L or greater dose levels, whereas thyroid weight at 8 weeks was significantly increased at 1000 mg/L or greater. This study suggests that developing young quail have a limited ability to compensate T4 concentrations during early exposure to AP. Doses estimated in this WTA adopted the authors' conservative use of a daily water ingestion rate of 7.7% of adult quail body weight of 165 g. Using the results at 8 weeks, the NOAEL and LOAEL for T4 hormone levels were 0.326 and 3.3 mg/kg/day perchlorate ion, respectively. (Note: The 5 mg/L dose level was the highest no-effect dose level for the T4 endpoint; data for T4 values were not reported for the 25 mg/L dose level.) For increased thyroid weight, the NOAEL and LOAEL were 33 and 65 mg/kg/day perchlorate ion, respectively [55].

Smith and coworkers initiated a metabolism study of House Sparrows (*Passer domesticus*) [57], similar to the prairie vole study described above. Wild-caught, apparently disease-free Sparrows were acclimated for 2 weeks prior to exposure. Birds were exposed daily to 0, 12.5, or 25.0 mg perchlorate, administered in 100 μL of water via pipette for 2 to 3 weeks. After 2 to 3 weeks of perchlorate exposure, several birds became ill with a bacterial infection (House Finch Disease, *Mycoplasma gallisepticum*); most birds with the illness died within 2 to 3 days. Due to the fatalities, the study was restarted with fresh birds. The second group of birds received Tylan antibiotic, but four of these developed symptoms of House Finch Disease 2 weeks after the treatment, and this portion of the study was terminated. The investigators determined the perchlorate concentration in the livers, with the 12.5 mg/kg dose group having concentration in the liver of 57.41 ± 68.77 μg/g dry weight. The liver concentration of perchlorate in the 25.0 mg/kg treatment group was 234.48 ± 258.26 μg/g dry weight. The investigators found it noteworthy that none of the test animals showed symptoms of respiratory disease prior to the perchlorate dosing experiments.

Avian Oral Toxicity: Chronic

Zebra Finches (*Taeniopygia guttata*) were used by Rainwater *et al.* [58] as a model to examine the effects of environmentally relevant perchlorate concentrations on growth and behavior on growth and behavior of a granivorous passerine. On post-hatch days (PHD) 3 to 14, finch chicks were orally dosed once daily at 0, 10, 100, or 1,000 μg/g body weight. Multiple endpoints of growth and behavior were examined in birds up to 72 days of age. Mean body mass and tibiotarsal length were significantly reduced in the 100 and 1,000 μg/g dose groups from PHDs 3 to 14, and this reduction persisted until PHD 40. Mean liver mass (PHD 15) in the 100 and 1,000 μg/g dose groups was significantly greater than in the controls. Mean brain mass and mean nuclear volume of the song control nucleus area X in brains of 72-day-old males did not differ among dose groups. All endpoints of Finch behavior were significantly influenced by perchlorate exposure at the highest dose level. Compared to controls, birds in this dose group exhibited a greater begging intensity, decreased motivation for spontaneous movement (e.g., attempts to fly), and reduced capacity to wean themselves from parental care. On PHD 15, flight attempts were also significantly reduced in the 10 and 100 μg/g dose groups, and the proportion of perchlorate-dosed birds attempting flight was less than half that of the control group. These data can be construed to imply a LOAEL at 10 μg/g based upon begging behavior and flight attempts. A NOAEL was not observed in this study.

Avian Oral Toxicity: Other

Another avian study was available [18], but the secondary literature that described it did not indicate its duration of exposure (original study in German). Doses of 0, 14.4, 21.5, and 28.7 mg/kg/day perchlorate ion were administered to Chickens (n = 3). Thyroid and body weights were reduced beginning at the 21.5 mg/kg/day dose level. Other signs indicative of perchlorate toxicity were noted at all dose levels and included reduced feather exfoliation and sexual development, and failure of the bursa of Fabricius, the organ responsible for B-lymphocyte maturation. However the doses and toxicological endpoints obtained from the original paper noted that exposure duration prior to animal termination was not evident. Due to the lack of exposure duration information, a NOAEL could not be derived from the secondary literature.

Gentles *et al.* [59] exposed 23 adult female Northern Bobwhite to one of four exposures (0.12, 0.012, 0.0012, or 0 mg/L) of AP in the drinking water for 30 days (equivalent to 1, 0.1, 0.01, or 0 nM solutions, respectively). The liver, gizzard, heart, kidney, and thyroid were weighed at the end of the study, and eggs were collected daily during the study. The AP did not affect body weight or organ weight. Birds exposed in the 1 nM group had a decrease in mean colloidal volume and an increase in the height of the follicle cells of the thyroid. No change in the number of eggs was found. The concentrations of perchlorate found in the eggs were proportional to the exposure concentrations.

Avian Developmental Exposures

Gentles *et al.* [59] exposed 23 adult female Northern Bobwhite (*Colinus virginianus*) quail to 0, 0.01, 0.1, or 1 mM AP in drinking water for 30 days. Eggs laid by birds in all treatment groups, including the controls, were collected, dated, given an identification number, and weighed. On day 30 of exposure, 10 birds were euthanized by carbon dioxide asphyxiation, after which gross toxicological endpoints and thyroid histology were evaluated. Egg production and the accumulation of perchlorate in the eggs (n = 10) and liver (n = 5) were determined. Perchlorate did not affect body or organ weights significantly; however, at 1 mM, AP caused alteration of thyroid gland morphology. Perchlorate did not affect egg production, but significant accumulation was observed in the eggs and livers of exposed birds.

Studies Relevant for Avian TRV Development for Ingestion Exposures

Of the avian studies located, only three were suitable for TRV derivation [55,56,58]. The ERM study [18] provided insufficient information regarding exposure duration, and the Gentles *et al.* study [59] neither measured the dose (i.e., drinking rate) nor reported any biologically relevant effects. The McNabb *et al.* studies [55,56] represented a single species (the Northern Bobwhite) with effects observed for three endpoints (i.e., thyroid hormone levels, thyroid weight, and tibia growth). The NOAELs and LOAELs for these endpoints are outlined in Table 28.4. Although these studies are useful in developing an avian TRV for perchlorate, their confidence is reduced because only a single species is represented.

Avian Inhalation Toxicity

No data are available.

Avian Dermal Toxicity

No data are available.

Summary of Amphibian Toxicology

Amphibian Toxicity: Acute

Thyroid hormone function in *X. laevis* is affected by exposure to perchlorate in the 50 to 100 ppm exposure range [60]. Further, the metamorphosis of tadpoles was completely blocked at an exposure of 1,000 ppm perchlorate [61]. In addition, a concentration of 340 ppm potassium perchlorate blocked metamorphosis and affected thyroid function in the South American toad *Bufo arenarum* [61]. Dumont and Bantle [62] found an LC_{50} value of 496 ppm for ammonium perchlorate in the FETAX *Xenopus* test, wherein the EC_{50} value for malformations was 396 ppm. It was shown that ammonium perchlorate was not teratogenic but was an embryonic toxicant. Goleman *et al.* [60] found 5- and 70-day LC_{50} values of 510 ± 36 mg/L and 223 ± 13 mg/L, respectively, in *X. laevis* larvae exposed to AP.

As part of a series of studies on the effects of perchlorate on amphibian development, embryonic *X. laevis* were exposed to a single concentration of sodium perchlorate, ammonium chloride, or plain FETAX medium for 5 days, beginning fewer than 24 hours after fertilization. Ammonium chloride was more lethal than sodium perchlorate, with LD_{50} values of 118 and over 1,220 ppm, respectively. Thus, it is concluded that the ammonium ion contributes significantly to the toxicity of ammonium perchlorate [14].

Table 28.4 Summary of Relevant Avian Data for TRV Derivation

Test Type	Test Organism	Test Duration	Test Results: NOAEL (mg/kg/day)	LOAEL (mg/kg/day)	Effects Observed at the LOAEL	Study
Subchronic	Bobwhite Quail (*Colinus virginianus*)*	2 weeks	0.00326	0.0326	Decrease in thyroidal T_4 hormone level.	[55]
		8 weeks	33	65	Increased thyroid weight.	
		8 weeks	130	261	Decrease in tibia length. The authors note that tibia growth was decreased significantly, relative to increased ammonium perchlorate concentration.	
	Bobwhite Quail (*Colinus virginianus*)*	8 weeks	0.326	3.26	Decrease in thyroidal T_4 hormone level.	[56]
		8 weeks	33	65	Increased thyroid weight.	
	Zebra Finch (*Taeniopygia guttata*)	12 days	None	0.010	Greater begging intensity, decreased attempts to fly, reduced capacity to wean from parental care.	[58]

*= 3–4 week old chicks; sex unknown.

Kendall [14] also reported that when larval bullfrogs (*Rana catesbiana*) in Taylor-Krollos stages XV–XVII were exposed to 117 mg/L ammonium perchlorate for 96 hours, whole-body perchlorate concentrations increased linearly during the exposure period. The whole-body perchlorate content remained significantly elevated 48 hours after transfer of the bullfrogs to plain water. After 48 hours, their whole-body perchlorate content was not statistically different from that of the controls. The results suggest the elimination of perchlorate in tadpoles occurs relatively slowly, yet the location of perchlorate sequestration in the tadpoles is unknown.

Dean *et al.* [64] exposed green frog tadpoles (*Rana clamitans*) to sodium perchlorate under measured flow-through test conditions for 96 hours. Although other, shorter-exposure tests were also conducted, 96 hours is the standard exposure duration in water quality criteria development. Water hardness as $CaCO_3$ was 146 mg/L at the beginning of the test, and the pH ranged from 7.89 to 8.20 in all solutions throughout the duration of the test. The LC_{50} was 5,500 mg/L. The only observed sublethal effect was a loss of equilibrium in the test organism (i.e., the tadpole could not maintain proper orientation in the water). The EC_{50} value for this behavioral effect was 5100 mg/L.

Goleman and Carr [65] exposed *Xenopus laevis* tadpoles to ammonium chloride (AC) or sodium perchlorate (SP) to improve understanding of the relative toxicity of the ammonium cation. Exposure to AC was minimally more lethal than exposure to either AP or SP ($LC_{50}=83$ mg/L, 510 mg/L, and 2,780 mg/L for AC, AP, and SP, respectively). Incidence of edema was increased in SP although skeletal abnormalities were seen in high-dose groups from the SC and AC treatments (> 159 mg/L). Sparling and Harvey [66] also studied the relative toxicological influence of ammonium using various concentrations of ammonium perchlorate and ammonium bicarbonate. Northern leopard frog tadpoles (*Rana pipiens*) at Gosner stage 25 were exposed to concentrations of 0, 0.41, 1.13, 3.08, 8.4, or 61.3 mg/L perchlorate ion or at the relative same ammonium concentrations of ammonium bicarbonate for 7 days. Given the derivation of inconsistent results from the 7-day exposures, LC_{50} values were developed from the 96-hour data. The 7-day LD_{50} of 170 mg/L and 96-hour LC_{50} of 329 mg/L for perchlorate ion were calculated relative to the 7-day LD_{50} of 29.9 mg/L and 96-hour LC_{50} of 57.9 mg/L for ammonium ion. Based on similar values for other ammonium-containing compounds, and relationships between sublethal effects (e.g., growth measurements), the authors suggested that the ammonium cation was most responsible for the observed toxicity. This statement was supported by providing LC_{50} values for ammonium nitrate (25–32 mg/L) and by citing data from Sparling *et al.* [67], in which concentrations of up to the maximum amount (50 mg/L) of potassium perchlorate tested caused no adverse effects on growth.

As part of an effort to develop a more rapid method of evaluating thyroid-related endpoints, Tietge *et al.* [68] evaluated the early responses of various endpoints to perchlorate and two other thyroid hormone (TH) synthesis inhibitors. *Xenopus laevis* were exposed to 4 mg/L perchlorate for a period of 8 days. Endpoints included thyroid gland histology and cell numbers, circulating TH concentrations, and thyroidal TH and associated iodo-compounds. Thyroidal 3,5-diiodo-L-tyrosine (DIT), 3-monoiodo-L-tyrosine (MIT), and thyroxine (T4) were significantly reduced from day 2 onward. These reductions were the earliest indicators of TH synthesis inhibition. Histological effects were apparent on day 4 and became more exaggerated through day 8. Reductions in circulating T4 and increases in thyroid gland cell numbers were not apparent until day 6. Changes in thyroid histology and cell number are believed to represent compensatory effects that are modulated by circulating TSH.

Brausch *et al.* [69] reported possible mitigating factors for the effects of perchlorate in spadefoot toads (*Spea multiplicata*) and *X. laevis* exposed to environmental levels of perchlorate. In these studies, *S. multiplicata* were exposed to 60, 110, or 1,000 μg perchlorate per liter dissolved in natural surface water to determine the risks associated with perchlorate exposure in desert-adapted anurans. Hind- and forelimb development and tail resorption were measured to identify the effects of perchlorate exposure. No perchlorate-related effects on snout-vent length, hindlimb length, and percentage that were metamorphosed were observed in the highest treatment group (1,000 μg/L). These observations suggested that either *S. multiplicata* were not sensitive to the effects of perchlorate at the concentrations tested or that unidentified constituents of natural surface water mitigate perchlorate toxicity. To explore this issue further, *X. laevis* were exposed to 20 or 60 μg perchlorate/L in either natural surface water or synthetically prepared water (i.e., FETAX media). In this study, *X. laevis* exhibited no perchlorate-related antimetamorphic effects when exposed to perchlorate dissolved in surface water. By contrast, *X. laevis* experienced changes in percent metamorphosis ($p < 0.001$), time to metamorphosis ($p < 0.001$), snout-vent length ($p < 0.001$), and hindlimb length ($p < 0.001$) when compared to FETAX controls on exposure to perchlorate in FETAX media.

Weltje *et al.* [70] evaluated the relative acute and chronic sensitivity of amphibians and fish to toxicants in the water. Amphibians were more than

100 times more sensitive to perchlorate than were fish. The FETAX tests performed in *Xenopus* by Dumont and Bantle in 1998 [62] found an LC_{50} value of 496 ppm for AP. The EC_{50} for malformations was about 396 ppm. The authors concluded that AP was not teratogenic but was an embryonic toxicant. Moreover, Kinney *et al.* [61] found that an AP concentration of 1,000 ppm completely blocked metamorphosis of tadpoles. Others have investigated molecular mechanisms regulating amphibian metamorphosis by characterizing gene expression profiles in thyroids of *Xenopus laevis* tadpoles during spontaneous metamorphosis [71]. Using real-time PCR, elevated expression of slc5a5, tpo, tshr, and sarla mRNAs was detected at late prometamorphic and climax stages. For dio2 and dio3, but not dio1, developmental regulation of thyroidal expression was evident from a strong up-regulation at late stages. Conversely, expression of the DNA replication markers mcm2 and pcna declined in climax stages. The presence of functional feedback mechanisms at premetamorphic stages was examined in two experiments. Stage 52 tadpoles were exposed to 1.0 μg/L thyroxine (T4) for 72 hours. This treatment caused reduced mRNA expression of slc5a5, tpo, and dio2, while no significant detectable changes were found for tshr expression in thyroids and tshb expression in the pituitary. In another experiment, stage 46 tadpoles were treated with 20 mg/L sodium perchlorate for 5 and 10 days. Within this time, control tadpoles developed to stages 50 and 52, respectively. SP treatment resulted in up-regulation of slc5a5, tpo, and tshr mRNAs at both time points and increased dio2 mRNA expression at day 10. Effects of SP on thyroid histology were apparent on day 10 only.

Amphibian Toxicity: Subchronic

In the first of a two-part experiment, Tietge *et al.* [72] exposed two developmental stages of *Xenopus laevis* larvae (Nieuwkoop and Faber (NF) [73], stages 51 and 54) to five concentrations of sodium perchlorate (18, 62, 247, 972, and 4,000 μg/L as perchlorate ion) and a control for 14 days. Tadpoles exposed to concentrations of 247 μg/L (analytical concentration) and higher exhibited inhibited metamorphic development. Colloid depletion and follicular cell hypertrophy (i.e., larger cell size) was observed in all treatments (>17 μg/L); the severity was concentration-dependent. No treatment-related mortality occurred, and there was a positive relationship between dose and body weight in both developmental groups of tadpoles.

In the second of a two-part experiment, Goleman *et al.* [74] conducted a 14-day Endocrine Disruptor Screening and Testing Committee (EDSTAC)

Tier 1 frog metamorphosis assay with an exposure concentration of 14.04 mg/L AP or 11.89 mg/L perchlorate ion. The percentage of tadpoles completing tail resorption was reduced, and mean tail length was significantly greater in the treatment group (11.89 mg/L perchlorate ion) compared to controls. The authors note that tail length at the end of the experiment was less than at day 0; therefore, tail resorption was not completely inhibited by perchlorate ion at this exposure level. The authors also discuss the possibility that observed effects were due to the ammonium ion rather than the perchlorate ion. Because perchlorate is known to cause thyroid toxicity and specifically affects thyroid-hormone-level-sensitive aspects of growth and metamorphosis, the authors believed it was unlikely that the ammonium ion contributed to the observed effects.

In *X. laevis*, Goleman *et al.* [74] found that environmental concentrations of AP-derived perchlorate as low as 59 parts per billion (ppb) caused hypertrophy of thyroid follicles after 10 weeks of exposure. In the absence of perchlorate, these changes were reversible after a 28- to 30-day recovery period.

Amphibian Toxicity: Chronic

Four chronic studies of the effects of perchlorate exposures to amphibians were located. In a study reported by Sterner and Mattie [6], both male and female African clawed frogs (*Xenopus laevis*) were exposed to sodium perchlorate under static-renewal test conditions for 5 months following hatch to the end of the metamorphosis. At a single exposure level (1% or 8,120 mg/L perchlorate ion), the authors observed significant differences in blood chemistry measures (e.g., reduced number of circulating red blood cells) and retarded growth (e.g., fewer than half) relative to controls [6]. Additionally, maturation of the adult-specific immune system was affected at this exposure level, as evidenced by reduced numbers of adult-specific T-cells and reduced splenic B-cells (15% of controls). Therefore, an unbounded LOAEL for this study is 8,120 mg/L. This study was considered chronic because of its duration (5 months) and because it was conducted during a critical life stage (i.e., metamorphosis).

Dumont [75] found that concentrations of 1 mg perchlorate/L were sufficient to inhibit or delay metamorphosis in amphibians. Iodine is capable of blocking or overcoming the effects of AP within certain concentration ranges, but there appear to be differences in species response. Long-term exposure to AP concentrations above 100 mg/L is toxic, and embryos at these concentrations, both with and without iodine supplementation, died after 80 days of exposure.

Goleman *et al.* [63,74] conducted two ammonium perchlorate exposure experiments using *X. laevis*. The first experiment [63] included a range finding study in which approximately 50 fertilized eggs were exposed to one of seven concentrations of AP (0.001175, 0.01175, 0.1175, 1.175, 11.75, 117.5, and 1,175 mg/L) for 70 days and run in duplicate. Based on the results of the range finding study, two additional concentration groups were added. Measured exposure concentrations for these nine dose groups were 0.005, 0.018, 0.147, 1.412, 14.4, 133, 425, 585, and 1,175 mg/L AP (0.0042, 0.0152, 0.125, 1.196, 12.2, 113, 360, 495, and 995 mg/L perchlorate ion). LC_{50} values of 510 mg/L (432 mg/L perchlorate ion) for 5-day exposures and 223 mg/L (189 mg/L perchlorate ion) for 70-day exposures were determined. Hatching success was significantly reduced at exposures above 1000 mg/L ammonium perchlorate, and survival of larvae was significantly reduced to only 6 to 7% in the 425 mg/L exposure group. Larvae that survived the 425 mg/L exposure group had reduced snout-vent length (SVL). Other effects observed in the first experiment included concentration-dependent reductions in hind limb length, percent forelimb emergence (thyroid hormone-dependent process that indicates the beginning of metamorphic climax), and percent completing tail resorption (process that indicates the completion of metamorphosis). Hind limb length was significantly reduced at perchlorate ion exposure concentrations of 0.125 mg/L and greater. Forelimb emergence was the most sensitive endpoint with reductions starting at the lowest exposure level (0.0042 mg/L perchlorate ion), whereas reductions in percent completing tail resorption occurred at concentrationsof 0.0152 mg/L or more perchlorate ion.

In a second study, Goleman *et al.* [74] evaluated the effects of ammonium perchlorate on *X. laevis* using two environmentally relevant concentrations. Approximately 250 embryos were exposed to either 0.038 or 14.04 mg/L AP (actual concentrations were 0.06 and 14.1 mg/L AP or 0.06 and 11.9 mg/L perchlorate ion) for 70 days beginning less than 24 hours after oviposition, followed by a 28-day nontreatment recovery period. Effects similar to those observed in the first study [63] were observed, including significant reductions in hind limb length, percent forelimb emergence, and percent completing tail resorption at both dose levels. Additionally, whole-body T4 concentrations were reduced at the highest exposure level, and significant hypertrophy of the thyroid follicular epithelium occurred at both treatment concentrations. Moreover, the percentage of males at metamorphosis was decreased in both exposure groups compared to controls, suggesting that AP disrupts thyroid activity and impairs testes differentiation in developing *Xenopus*. The effects on metamorphosis and thyroid function were reversed after 28 days of nontreatment. Based on the results of this study, the low-dose group (0.06 mg/L perchlorate ion) represents an unbounded LOAEL for endpoints relating to metamorphosis, thyroid function, and gonadal differentiation.

Larval bullfrogs (*Rana catesbiana*) at Taylor-Krollos stage XV to XVII were exposed to 117 mg/L AP for 96 hours [14]. Whole-body perchlorate concentrations increased linearly during the exposure period. Whole-body perchlorate content was still significantly elevated 48 hours after transfer to plain water. After 48 hours, whole-body perchlorate content was not statistically different from that of controls. The results suggest elimination of perchlorate in tadpoles occurs relatively slowly, but the location of perchlorate sequestration in the tadpoles is unknown.

In the second part of a two-part study, Tietge *et al.* [72] exposed *X. laevis* at NF Stage 51 to target concentrations of sodium perchlorate at 0, 8, 16, 32, 63, or 125 μg/L (or analytical concentrations of 0, 9, 17, 34, 69, and 137 μg/L as perchlorate ion) for 44 days (until NF stage 66) and evaluated them histologically for thyroid effects and other gross effects. The authors reported that mean time to complete metamorphic development was increased at an exposure of 125 μg/L (137 μg/L analytical concentration). Thyroid hypertrophy was observable in frogs exposed to 125 μg/L (137 μg/L analytical concentration) and higher. Total thyroid area, as an indicator of size, was increased in frogs exposed to 60 μg/L (69 μg/L analytical concentration) and higher. Growth was not affected by perchlorate at these tested concentrations. The results suggest that while the 14-day exposures may have been more sensitive in the development of these measures, longer exposures may be important in the development of tolerance.

In an effort to discriminate the effects of the ammonium ion component, Goleman and Carr [65] conducted paired exposures to ammonium chloride (AC) or sodium perchlorate (SP) using *X. laevis*. Acute (5-day) and chronic (70-day) experiments were conducted. Ammonium perchlorate was used as an additional treatment in the chronic studies: *X. laevis* embryos were exposed to one of two concentrations (38 or 14,000 μg/L of AC, SP, or AP for 70 days through metamorphosis. Both of the AP concentrations and the high SP concentration inhibited hind limb length and development. A similar relationship was observed in colloid depletion and hyperplasia of the thyroid. Both AP and SP affected sex ratios, resulting in a greater percentage of females exposed to the high concentrations (10,645 and 10,672 μg/L perchlorate ion concentrations).

Sparling *et al.* [67] conducted two tests to determine [1] the concentration of perchlorate that would inhibit metamorphosis, and [2] if metamorphosis could be induced by the addition of iodide to the perchlorate treatment. Early larval *Hyla versicolor* tadpoles (Gosner [76] stage 24 or 25) were exposed to either 0, 2.2, 4.8, 10.5, 22.9, 33.8, or 50 ppm potassium perchlorate in the water using biweekly static renewal design until metamorphosis (70–100 days). In addition, two other treatments were added: a 0 ppm perchlorate with 0.10 ppm iodide (as KI); and a 30 ppm perchlorate with 0.10 ppm iodide. Survival was high throughout all treatments (>89%). No tadpoles in either the 22.9 or 33.8 ppm treatments completed metamorphosis, and only one tadpole in the 50 ppm treatment completed metamorphosis. No statistical differences were found between controls and the iodide treatments. However, the frequency of tadpoles that did complete their metamorphosis within the first 70 days differed between the tadpoles in these three treatments and all others that received perchlorate ($p < 0.0001$). Few significant differences between treatments were found in the time to complete metamorphosis, but there was a significant difference in the time to complete metamorphosis once a fore limb emerged. Though there were no significant differences among the treatment groups, tadpoles exposed to perchlorate without iodide took 0.5 to 4 days longer than those in any other group, on average, to complete their metamorphosis. An approximate EC_{50} of 3.63 was calculated; however, neither a 95% confidence limit nor a LOAEL could be determined. When 0.01 ppm iodide was added, 90 to 100% of tadpoles entered the metamorphic climax stage (regardless of perchlorate concentration); however, only 75% of those exposed at 2 ppm perchlorate, 82% at 4.8 ppm perchlorate, and 70% at 22.9 ppm perchlorate completed metamorphosis. These data suggest that *H. versicolor* do not preferentially transport perchlorate over iodide, and in environments where iodide is present, effects from perchlorate exposure may be ameliorated [67].

In order to evaluate the mode of action by which perchlorates interfere with the normal expansion of T and B lymphocytes and the development of an adult T-cell population, Rollins-Smith *et al.* [77] exposed postmetamorphic frogs to perchlorate. After 5 months, the treated animals, but not the untreated controls, had larger thyroid goiters, weighed significantly less and had significantly fewer erythrocytes, thymocytes, and splenocytes. However, the number of IgM-histocompatibility complex class II+ within this subset was not significantly different from that of controls. Also, splenocytes from perchlorate-treated frogs could respond normally by 3H-TdR incorporation to the T-cell mitogens, phyohemagglutinin-P and concanavalin-A.

Therefore, unlike perchlorate-treated larvae, perchlorate-treated juveniles appear to be able to develop T-cells with an adult phenotype competent to respond to activation and proliferation signals.

Amphibian Toxicity: Other

Theodorakis *et al.* [78] collected and evaluated thyroid histology from cricket frogs (*Acris crepitans*) collected from several perchlorate-contaminated streams in central Texas. There was no evidence of colloid depletion or follicle cell hyperplasia in any of the frogs studied ($n = 86$); however, moderate follicle cell hypertrophy was found in frogs collected from two sites with the highest perchlorate concentrations (~26 and 6 µg/L, estimated from bar graph). It is of note that data from water collected for perchlorate analyses at these sites were variable.

Studies Relevant for Amphibian TRV Development

In the acute study of perchlorate exposure to amphibians, the toxicant concentrations associated with effects included an LC_{50} of 5,500 mg/L and an EC_{50} for loss of equilibrium of 5100 mg/L [64]. Goleman and Carr [65] reported LC_{50} values of 83, 510, and 2,780 mg/L for AC, AP, and SP, respectively. Sparling and Harvey [66] attributed toxicity of acute AP studies to the ammonium, rather than the perchlorate ion (96-hr $LC_{50} = 329$; 7-day $LC_{50} = 170$ for perchlorate ion). The other studies [6,63,65,74] were conducted over an extended period (5 months or 70 days) and during a critical life stage (metamorphosis), and are thus considered relevant as chronic data. These studies represent multiple exposure concentrations as well as multiple endpoints, despite their having evaluated only a single amphibian species (*X. laevis*). In the Sterner and Mattie study [6], an unbounded LOAEL of 8,120 mg/L perchlorate ion was derived for aquatic amphibians. However, this exposure level exceeded the LC_{50} and EC_{50} values reported by Dean *et al.* [64], Goleman *et al.* [63,74], Tietge *et al.* [72], and Goleman and Carr [65], all of which collectively suggest that the effects of AP on aquatic amphibians occur at much lower levels.

For endpoints related to metamorphosis, forelimb emergence was the most sensitive endpoint with an unbounded LOAEL of 0.004 mg/L perchlorate ion [63]. However, this is not consistent with Tietge *et al.* [72] who observed no adverse effects on development from exposures of up to 0.069 mg/L to the same species within a comparable exposure duration.

The LOAEL finding is also not consistent with Goleman *et al.* [74], where forelimb emergence was delayed at 11.9, but not at 0.06 mg/L; or with Goleman and Carr [65], who reported that exposures of 10.6 mg/L perchlorate ion caused a depression in growth rates and in metamorphosis. Exposures of AP, not SP, at 23 mg/L (as perchlorate ion) caused a mild depression in growth rates and metamorphosis (Figure 28.3).

Other discrepancies exist. After controlling for perchlorate ion concentration, levels that were reported to cause altered metamorphosis and development (measured by hind limb development and NF stage) occurred in sodium perchlorate treatments at perchlorate ion concentrations of 10.7 mg/L. Changes were reported for frogs exposed to AP as perchlorate ion concentrations of 0.024 mg/L. Endpoints evaluating thyroid function represented by significant hypertrophy of the thyroid follicular epithelium and gonadal differentiation (reduced percentage of males) had an unbounded LOAEL of 0.06 mg/L perchlorate ion, yet these endpoints may be adaptive, and their biological relevance to the health of amphibian populations is difficult to interpret. Data from Theodorakis *et al.* [78] provide information from field exposures to another species; however, the importance of the presence of thyroid hypertrophy alone, combined with variable perchlorate concentrations measured in the streams, would not allow for a definitive interpretation of an adverse effect level. These data are instructional and do provide an additional line of evidence of a threshold whereby frogs are responding to environmental concentrations of perchlorate. The body of evidence from these chronic studies is insufficient for TRV development.

Summary of Reptilian Toxicology

Reptilian Oral Toxicity: Acute

No data relating to reptilians were found.

Reptilian Oral Toxicity: Subchronic

Eggs of red-eared sliders (*Trachemys scripta*) and snapping turtles (*Chelydra serpentine*) were topically exposed to either 200 or 177 μg/g as sodium perchlorate to determine impacts on glandular thyroxine concentrations, embryonic growth and development, and metabolic rates of hatchlings for a period of 2 months post-hatching. In red-eared sliders, *in ovo* perchlorate exposure delayed hatching, increased external yolk size at hatching, increased hatchling mortality, and reduced total glandular thyroxine concentrations in hatchlings. In snapping turtles, hatching success and standard metabolic rates

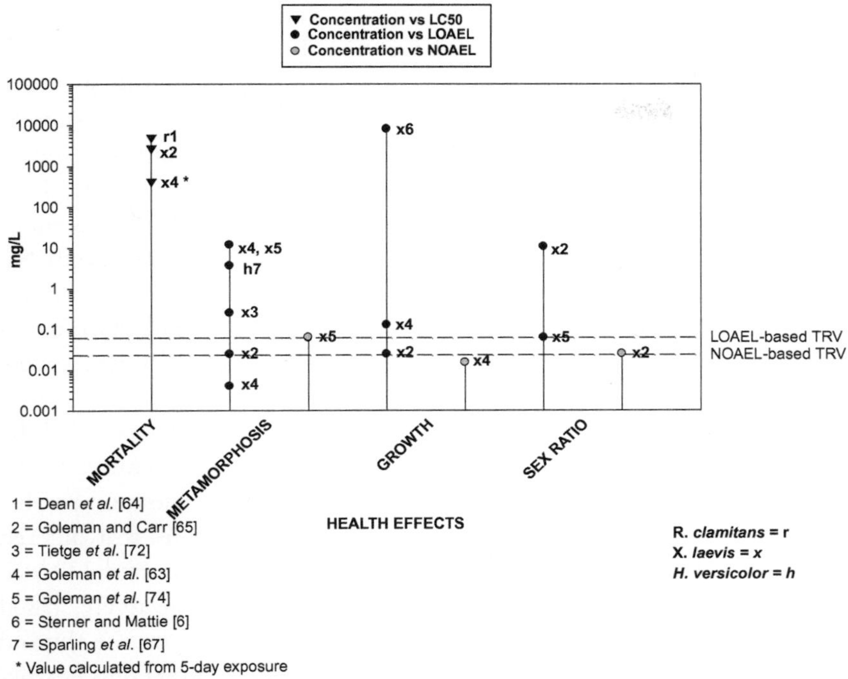

Figure 28.3 Perchlorate: Health Effects to Amphibians.

were reduced, liver and thyroid sizes were increased, and total glandular thyroxine concentrations in hatchlings were reduced after exposure to perchlorate. While both species were negatively affected by exposure, impacts on red-eared sliders were the most severe, suggesting the slider may be a more sensitive sentinel species for studying the effects of perchlorate exposure to turtles [79].

Reptilian Toxicity: Chronic

No data relevant to reptiles were found.

RECOMMENDED TOXICITY REFERENCE VALUES

Toxicity Reference Values for Mammals

TRVs for Ingestion Exposures for the Class Mammalia

The toxicological database for perchlorate is extensive and includes chronic as well as reproductive and developmental data. More than three species of two orders are represented, as are chronic NOAEL and LOAEL values.

Interpretation of these data in an ecological context is challenging; few adverse effects have been identified, yet the database is comprehensive. There is a general lack of adverse effects. Changes in thyroid structure and function (systemic hormone levels) have been found; however, the relevance of these endpoints to wildlife at the individual or population level of organization is questionable. Moreover, although changes in thyroid hormone levels and histology are consistent findings because of exposure, functional differences related to these changes have not been observed. No changes in reproduction because of exposure have been observed, yet thyroid hormone levels may be decreased along with proportionate thyroid colloid depletion. Mortality is particularly relevant but occurs only at relatively high exposures.

In the companion studies by York *et al.* [34,40], the authors stated the LOAELs for rats and rabbits, based on structural histologic changes to the thyroid, as 2.54 and 8.81 mg/kg/day, respectively; the species-comparable NOAELs were 0.254 and 0.762. York *et al.* [42] used doses that differed slightly from those of the prior studies, resulting in a reported LOAEL and NOAEL for rats of 2.54 and 0.847 mg/kg/day, respectively, based on similar histologic changes to the thyroid. Similar effects on the thyroid gland were also reported in studies using deer mice [48]. None of these histologic changes suggest organ failure, merely adaptation (e.g., changes in the size or number of follicular cells, changes in lumen shape, among others). No evidence of cellular necrosis has been reported, and no other adverse effects (other than changes in TSH, T3, and T4 levels) have been linked to these findings.

Available studies on rats indicate that a change in thyroid and pituitary hormone levels (T4, T3, and TSH) is a more sensitive endpoint than are observable histologic changes to the thyroid gland. (The lowest LOAEL reported for the rat is 0.0423 mg/kg/day, from a study in which Crofton [22] reported decreased T4 thyroid hormone levels.) The results of Crofton [22] and King [24] revealed a significant and rapid response in the thyroid hormone levels of rats following the animals' exposure to low concentrations of perchlorate. However, at higher doses with longer exposure durations or during a critical life stage, adverse whole-body responses (e.g., reproduction, fetal development, growth, and time to puberty) have not been observed [34,40,50]. Changes in ovarian follicle density and maturation was also found [45]; however, the biological significance of these findings is uncertain, particularly so because greater than five reproductive studies have failed to find a relationship between perchlorate exposure and reproduction [40,42–44,51]. The U.S. EPA [4] considers a perchlorate ion dose level of 25.4 mg/kg/day as a developmental LOAEL based on data reported in Argus 2000 [39] and as previously cited [4]. However, this finding was not considered biologically relevant by the study authors and, therefore, was not reported in the peer-reviewed publication.

The effects of pathologic changes in the thyroid gland to the health and ecology of the whole organism are poorly understood. No data showing any detrimental health effects linked to a decrement to systemic thyroid levels or to changes in thyroid histology as those mentioned in the studies discussed in the above could be found. Moreover, changes in iodine levels or cessation of perchlorate exposure may ameliorate these endpoints, suggesting they are adaptive in nature.

Due to the lack of definitive correlations between indirect endpoints (thyroid pathology and systemic thyroid hormone levels) and adverse health effects, TRVs could not be derived. Although several of the reviewed studies [37,40,42–44,46–48] met the minimum data set requirement of the Standard Practice, Section 2.2 [3], the data for the deer mice are equivocal and thus considered insufficient for TRV derivation.

The only marked adverse effect from perchlorate exposure found in the literature is death. Von Burg [16] reported the results of a 3-month rabbit study where nervous system effects were reported from exposures of 190 mg/kg/day; however, few details were provided, and no original data could be found. If the approximation method is used to extrapolate a TRV from the most sensitive LD_{50} data (rabbit; using an uncertainty factor (UF) of 100 for the NOAEL-based value and 20 for the LOAEL-based value), the result is a NOAEL-based TRV of 6.4 mg/kg/day and a LOAEL-based TRV of 32 mg/kg/day. These values are comparable, within an order of magnitude, to the NOAEL and LOAEL-based TRV values (i.e., 0.847 and 2.54 mg/kg/day, respectively) derived using the data from the most sensitive species (rat), as previously reported by York *et al.* [37,40–42] who used histological changes of the thyroid (i.e., thyroid hyperplasia) as an endpoint. Since these changes were reported as hyperplasia (i.e., increases in the number of normal cells), the biological significance of this finding is uncertain for ecological receptors and is not conducive for TRV derivation. These values are also protective of the only other nonconfirmed report of adverse effects (i.e., nervous system effects) in rabbits from exposures of 3 months in duration [16]. Together, these relationships provide a weight of evidence that shows the approximation approach is reasonable when applied to the acute mortality data. Thus, the TRVs for the Class Mammalia were derived from

Table 28.5 Summary of Relevant Amphibian Data for TRV Derivation

Test Type	Test Organism	Test Duration	Test Results: NOAEL(mg/L)	Test Results: LOAEL (mg/L)	Test Results: Effects Observed at the LOAEL	Study
Acute	*R. clamitans*	LC_{50}	5,100		Mortality (as sodium perchlorate)	[64]
	X. laevis	LC_{50}	2,780		Mortality (as sodium perchlorate)	[65]
	R. pipiens	96 hours	329		Mortality (as ammonium perchlorate)	[66]
Subacute	*X. laevis*	14 days	0.069	0.137	Inhibited time to metamorphosis	[72]
	X. laevis	14 days	Not determined	11.9	Inhibited metamorphosis	[63]
	X. laevis	5 days	$LC_{50} = 432$		Mortality	
	R. pipiens	7 days	$LC_{50} = 170$		Mortality	[66]
Chronic	*X. laevis*	5 months	Not determined	8,120	Decrease in red blood cells, reduced growth, immune effects	[6]
	X. laevis	70 days	Not determined	0.004	Delayed forelimb emergence	[63]
			0.015	0.125	Reduced hind limb length	
	X. laevis	70 days	0.06	11.9	Delayed forelimb emergence	[74]
			Not determined	0.06	Decreased male sex ratio	
	X. laevis	70 days	0.024	10.6	Decreased male sex ratio	[65]
			Not determined	0.024	Decreased hind limb length and NF stage	
	H. versicolor	70–100 days	EC50 3.63		Delays in metamorphosis; effects ameliorated when adding iodide	[67]

Table 28.6 Selected Ingestion TRVs for the Class Mammalia

TRV	Dose (mg/kg/day)	Confidence Level
NOAEL-based	6.4	Moderate
LOAEL-based	32	Moderate

Source: [16].

the LD_{50} for the rabbit (most sensitive for lethality) using an UF of 100 for the NOAEL-based value and a UF of 20 for the LOAEL-based value.

Table 28.6 presents the selected TRVs. A moderate level of confidence has been assigned to these TRVs given the numbers of species tested and the relative abundance of toxicity data for perchlorate. However, although many data are available, they generate considerable uncertainty regarding the relevance of structural and functional changes of the thyroid in wildlife. These changes may result in profound behavioral effects that could influence ecological interactions. More information is needed that focuses on neurodevelopmental endpoints that affect behavior important in maintaining populations and community structure. Presently, the methods by which these effects could be measured in controlled environments do not exist.

TRVs for Ingestion Exposures for Mammalian Foraging Guilds

TRVs specific to particular guild associations (e.g., small herbivorous mammals) have not yet been derived. However, the class-specific TRVs (Table 28.5) may be applied to herbivorous small mammals since both rats and rabbits are members of this guild. As with the class-specific TRVs, only two species are represented, and thus confidence in the TRVs is moderate. Data to derive TRVs for other guild associations (e.g., carnivorous mammals) is not available at this time.

TRVs for Inhalation Exposures for the Class Mammalia

Although one study [16] discusses the inhalation toxicity of perchlorate to mammals, no information on the dose required to produce the symptoms listed was provided in the study. Therefore, no relevant data are available for derivation of an inhalation TRV for mammals.

TRVs for Dermal Exposures for the Class Mammalia

No data are available at this time.

Toxicity Reference Values for Birds

TRVs for Ingestion Exposures for the Class Aves

Four avian studies were evaluated, with two [55,56] important in TRV derivation. McNabb *et al.* [55] evaluated the effects of perchlorate on thyroid hormone levels (2- and 8-week study) and growth parameters (8-week study) in 3- to 4-day post-hatch bobwhite quail. In the 2-week study, there was a decrease in thyroidal T4 hormone levels in the 0.0326 mg/kg/day perchlorate ion dose group, while T4 levels were unchanged in the 0.00326 mg/kg/day perchlorate ion dose group. In the 8-week study, all dose groups experienced impacts to thyroid hormone levels. A NOAEL and LOAEL were determined in the 8-week study using thyroid weight and tibia length as endpoints. Thyroid weight was significantly increased at dose levels of 65 mg/kg/day perchlorate ion or greater, and similar to controls at 33 mg/kg/day perchlorate ion or less. Tibia length was significantly decreased at 261 mg/kg/day, but similar to controls at 130 mg/kg/day. Changes in tibia length have been widely used as an indicator of growth rates in birds, though it is recognized that changes in thyroid weight are of uncertain significance.

In the second study [56], results from a series of experiments including those reported in McNabb *et al.* [55] are presented. It was found that reduced thyroidal T4 levels observed after a 2-week exposure to 0.5 to 5 mg/L ammonium perchlorate were restored to control levels by weeks 4 and 8 of exposure. Exposure to concentrations of 50 mg/L ammonium perchlorate or greater continued to result in a significant decrease in thyroidal T4 levels. Therefore, the 8-week NOAEL and LOAEL for thyroidal T4 levels are 0.326 (5 mg/L dose group) and 3.26 mg/kg/day (50 mg/L dose group) perchlorate ion, respectively. As presented in McNabb *et al.* [51], the NOAEL and LOAEL for thyroid weights are 33 mg/kg/day (500 mg/L dose group) and 65 mg/kg/day (1,000 mg/L dose group) perchlorate ion, respectively.

As with mammals, a significant and rapid response in thyroid hormone levels was observed at low doses of perchlorate. Additionally, the McNabb *et al.* [55,56] studies indicate that developing quail chicks appear to have limited ability to compensate for early thyroidal hormone effects resulting from exposure to low doses of ammonium perchlorate. Effects on growth parameters (thyroid weight and tibia length) were only observed at higher levels of exposure. Since birds, particularly nestlings, exhibit a relatively high rate of growth and are particularly vulnerable to predation, any reduction in growth has the potential to influence survival. Therefore, growth inhibition may be relevant to the health and ecology of the species within the class and thus constitutes data from which a TRV could be derived. Therefore, the data from McNabb *et al.* [51] were used to derive the avian TRVs where growth rates (indicated by changes in tibia length) were used.

Because the avian toxicity database for perchlorate lack in meeting the minimum data set requirements of the Standard Practice, Section 2.2 [3], TRVs based on an approximation of the NOAEL and LOAEL were developed for Class Aves using UFs. Since these growth changes were evaluated in a sensitive life stage in birds for growth parameters (chicks), these data are considered equivalent in value to chronic data (130 mg/kg/day and 261 mg/kg/day). Since only one species from a single taxonomic order are represented, an UF of 10 was applied to the NOAEL and LOAEL. Table 28.7 presents the selected TRVs. A low-moderate level of confidence has been given to these TRVs because these studies were of high quality, but lacking with regard to data from other species and lack of other developmental and reproductive data.

TRVs for Ingestion Exposures for Avian Foraging Guilds

Toxicity Reference Values specific to particular guild associations (e.g., herbivorous birds) have not yet been derived. The class-specific TRVs shown in Table 28.6 can be applied to herbivorous birds although the confidence in these TRVs is low. Data with which to derive TRVs for other guild associations (e.g., carnivorous birds) are not available at this time.

TRVs for Inhalation Exposures for the Class Aves

No data are available at this time.

TRVs for Dermal Exposures for the Class Aves

No data are available at this time.

Table 28.7 Selected Ingestion TRVs for the Class Aves

TRV	Dose (mg/kg/day)	Confidence Level
NOAEL-based	13	Low-Moderate
LOAEL-based	26	Low-Moderate

Sources: [55,56].

TRVs for the Class Aves

Although Sterner and Mattie [6] is potentially relevant for TRV derivation, the information provided in this study is extremely limited. Only one species was evaluated at a single test exposure concentration, and this concentration exceeded the LC_{50} and EC_{50} values reported in Dean *et al.* [64] and the LC_{50} values from Goleman *et al.* [63]. In contrast, the Goleman *et al.* studies provide multiple dose levels and an evaluation of multiple, biologically relevant endpoints (e.g., metamorphosis and growth). Goleman *et al.* [63], in particular, demonstrates significant dose-response relationships for three relevant endpoints, while Goleman *et al.* [74] indicates that effects observed for these endpoints are reversible if perchlorate exposure is removed. Since thyroid changes can occur following relatively brief exposures, and since the mechanism of delay metamorphosis and growth has not been fully elucidated in these species, a conservative approach is needed.

However, the data for delays in metamorphosis and in growth are inconsistent. Forelimb emergence was the most sensitive endpoint evaluated in Goleman *et al.* [63] at 0.004 mg/L perchlorate ion. However, Goleman *et al.* [74], using the same species, compound, and exposure duration, reported delayed forelimb emergence at much higher levels (11.9, but not at 0.06 mg/L). Goleman and Carr [65] reported that exposures of 10.6 mg/L perchlorate ion caused a depression in growth rates and in metamorphosis, but exposures of AP, not SP, at 23 μg/L (as perchlorate ion) caused a mild depression in growth rates and metamorphosis. Tietge *et al.* [72] reported delays in metamorphosis at 137 μg/L; however, the biological relevance of these values in this model, given the magnitude and variation, is questionable (57.5 ± 3.6 days relative to 54.1 ± 2.9 days in controls).

Goleman and Carr [65] investigated the relative influence of AP, SP, and SC in an effort to determine if the perchlorate ion was responsible for observed adverse effects. Although these acute data suggest the cationic portions of the molecule may help describe acute effects, depression in growth rates and changes in thyroid histology were relatively consistent between the two forms of perchlorate. Here, growth rate (and metamorphic stage) changes were reported from exposures to 10.6 mg/L perchlorate ion, and slight reductions were reported from exposures to only AP at 23 μg/L. Thyroid changes were scored at the 23 μg/L concentrations also, although the biological significance of these changes is uncertain. Greater ratios of females were found in the 10.6 mg/L exposures for both forms of perchlorate ion, also. Tietge *et al.* [72] found slightly differing results. Using the same species and compound as Goleman *et al.* [63,74] and Goleman and Carr [65], Tietge *et al.* [72] found no developmental changes but did identify a delay in metamorphic stage from

44-day exposures of perchlorate ion concentrations at 137 μg/L. Sparling and Harvey [66] have provided data to suggest toxicity is due primarily to the ammonium ion, not perchlorate, and, in doing so, have added another represented species. Sparling and Harvey, however, did not provide long-term data or data on the influence of sex ratio on perchlorate exposures. Sparling *et al.* [67] evaluated and found delays in metamorphosis in gray tree frogs (*Hyla versicolor*) by exposing tadpoles to 2.2 to 50 mg/L of potassium perchlorate. However, no differences were observed in treatments where iodide was added at environmentally relevant concentrations. These data suggest that the form of perchlorate as well as environmental iodide concentrations is important in understanding the potential for risk.

Since Goleman *et al.* [63,74], Tietge *et al.* [72], and Goleman and Carr [65] conducted exposures to animals from the embryonic stage, through the larval stage, to metamorphosis, exposures through all life stages have been examined. Tietge *et al.* [72] found histological changes to the thyroid at these levels, which was consistent with earlier findings [63,74]. Considering the preponderance of these data, a consistent NOAEL and LOAEL would be 0.023 and 0.06 mg/L for perchlorate ion, respectively. The latter value was based on changes in male sex ratios [74]. These data satisfy the requirements for a chronic study. Data from two species of two different families are included and are considered equivalent in terms of orders given the diversity in the Order *Anura*. Additionally, data from a field investigation of another species representing another family (*A. crepitans*) are used as an additional check on the value. Two chronic LOAELs and one chronic NOAEL are available. Therefore, these data meet the minimum data set requirements of the Standard Practice, Section 2.2 [3], and the NOAEL/LOAEL approach was used for TRV derivation. Using this corroborative approach to the data, the NOAEL-based TRV is 0.023 and the LOAEL-based TRV is 0.06 mg/L, respectively. A confidence level of "low" was assigned because of the inherent variability in the data set.

The field data for Theodorakis *et al.* [78] provide an additional line of evidence that even at low levels these values are protective of subtle responses of a native amphibian species in the field. It is important to note, however, that other compounds (e.g., nitrates), which may be found in aquatic environments, also affect the thyroid in a similar way and may enhance the probability for the manifestation of effects.

Toxicity Reference Values for Reptiles

No data are available at this time.

IMPORTANT RESEARCH NEEDS

Primarily due to the breadth of the toxicology studies provided to date, the mammalian TRVs derived for perchlorate have received a confidence level of Medium. However, the uncertainty regarding adaptive changes in the thyroid due to perchlorate exposure continues to be investigated, and clear levels of change that are instructive in predicting adverse effects have yet to be elucidated. The present mammalian value was derived, albeit with a relative large uncertainty factor and consistent with other endpoints, from mortality data. It may be true that long-term exposure to perchlorate may yield alterations in behavior or other effects that may have a profound influence on ecological interactions. Studies focused on development and behavior relevant to wildlife in a natural setting are needed, as are investigations using other species and taxonomic orders to provide a greater breadth of interspecific data. Moreover, the potential for adverse effects needs to be considered respective to environmental iodide levels (which may ameliorate toxicity) as well as the impact of the reserve capacity of the thyroid in understanding the likelihood of continuous exposure.

Toxicity Reference Value derivation for birds, amphibians, and reptiles was even more uncertain due to the paucity of toxicity data for birds and amphibians and the minimal toxicity data for reptiles. Additional avian data are limited by the availability of useful models. The amphibian data set would benefit from additional studies with other native species conducted under high-quality Good Laboratory Practice (GLP) conditions. Research studies should include experimental models of species that are genetically, biologically, and behaviorally similar to the wildlife that exhibit the greatest propensity for toxicant exposure. Experimental design should attempt to mimic both exposure type and duration and should include assessments of long-term effects.

DISCLAIMER

Use of trademarked name(s) does not imply endorsement by the U.S. Army but is intended only to assist in identification of a specific product.

REFERENCES

[1] Braverman LE. Clinical studies of exposure to perchlorate in the United States. Thyroid 2007;17:819–22.

[2] Bogazzi F, Tomisti L, Bartalena L, Aghini-Lombardi F, Martino E. Amiodarone and the thyroid: a 2012 update. J Endocrinol Invest 2012;35:340–8.

[3] USACHPPM. Standard practice for wildlife toxicity reference values. Aberdeen Proving Ground (MD): U.S. Army Center for Health Promotion and Preventive Medicine; 2000, Technical Guide 254.
[4] U.S. EPA. Perchlorate environmental contamination: toxicological review and risk characterization. External review (draft). U.S. Environmental Protection Agency, Office of Research and Development; 2002, NCEA-1-0503.
[5] Callahan C, Sprenger M, Long GC, Porter R. Perchlorate ecological risk studies – a report on literature reviews and studies conducted by the Ecological Impact/Transport and Transformation Subcommittee of the Interagency Perchlorate Steering Committee. Brooks Air Force Base (TX): US Air Force Institute for Environment, Safety, and Occupational Health Risk Analysis, Risk Analysis Directorate, Risk Assessment Division; 1998, IERA-RS-BR-TR-2001-0004.
[6] Sterner TR, Mattie DR. Perchlorate literature review and summary: developmental effects, metabolism, receptor kinetics and pharmacological uses. Wright-Patterson Air Force Base (OH): US Air Force Armstrong Laboratory; 1998, AFRL-HE-WP-TR-1998-0106.
[7] GeoSyntec Consultants. Alternative causes of wide-spread, low concentration perchlorate impacts to groundwater. Arlington (VA): Strategic Environmental Response and Development Program (SERDP); 2005, Final Report.
[8] Jackson WA, Karl Boehlke J, Gu B, Hatzinger PB, Sturchio NC. Isotopic composition and origin of indigenous natural perchlorate and co-occurring nitrate in the southwestern United States. Environ Sci Technol 2010;44:4869–76.
[9] Smith PN, Yu L, McMurry ST, Anderson TA. Perchlorate in water, soil, vegetation, and rodents collected from the Las Vegas wash, Nevada, USA. Environ Pollut 2004;132:121–7.
[10] Cheng Q, Perlmutter L, Smith PN, McMurry ST, Jackson WA, Anderson TA. A study on perchlorate exposure and absorption in beef cattle. J Agric Food Chem 2004;52:3456–61.
[11] Sanchez CA, Krieger RI, Khandaker N, Moore RC, Holts KC, Neidel LL. Accumulation and perchlorate exposure potential of lettuce produced in the Lower Colorado River region. J Agric Food Chem 2005;53:5479–86.
[12] Sanchez CA, Crump KS, Krieger RI, Khandaker NR, Gibbs JP. Perchlorate and nitrate in leafy vegetables of North America. Environ Sci Technol 2005;39:9391–7.
[13] Park J-W, Rinchard J, Anderson TA, Liu F, Theodorakis CW. Food chain transfer of perchlorate in largemouth bass, *Micropterus salmoides*. Bull Environ Contam Toxicol 2005;74:56–63.
[14] Kendall RJ. ER-1223: ecological risk assessment of ammonium perchlorate on fish, amphibians, and small mammals. Fort Belvoir (VA): Defense Technology Information Center; 2003, ADA 469870.
[15] Smith PN, Theodorakis D, Anderson T, Kendall R. Preliminary assessment of perchlorate in ecotoxicological receptors at the Longhorn Army Ammunition Plant (LHAAP), Karnak, TX. Ecotoxicology 2001;10:305–11.
[16] Von Burg R. Perchlorates. J Appl Toxicol 1995;15:237–41.
[17] Shigan SA. Substantiation of the maximum permissible concentration of ammonium perchlorate in water of reservoirs. Gig Sanit 1963;28:8–14.
[18] Environmental Resources Management (ERM), Inc. Extended literature review concerning NOAEL and LOAEL values for perchlorate. Exton (PA): Environmental Resources Management, Inc; 1995.
[19] Männistö PT, Ranta T, Leppäluoto J. Effects of methylmercaptoimidazole (MMI), propylthiouracil (PTU), potassium perchlorate (KClO4) and potassium iodide (KI) on the serum concentrations of thyrotrophin (TSH) and thyroid hormones in the rat. Acta Endocrinol 1979;91:271–81.

[20] Khan MA, Fenton SE, Swank AE, Hester SD, Williams A, Wolf DC. A mixture of ammonium perchlorate and sodium chlorate enhances alterations of the pituitary-thyroid axis caused by the individual chemicals in adult male F344 rats. Toxicol Pathol 2005;33:776–83.
[21] Keil D, Warren DA, Jenny M, EuDaly J, Dillard R. Effects of ammonium perchlorate on immunotoxicological, hematological, and thyroid parameters in B6C3F1 female mice. Charleston (SC): Medical University of South Carolina, Department of Medical Laboratory Sciences; 1999, Final Report No. DSWA01-97-0008.
[22] Crofton KM. Re-analysis of thyroid hormone data from the subchronic perchlorate study submitted by Springborn Laboratories. SLI Study No. 3455.1 [memorandum with attachment to Annie Jarabek]. Research Triangle Park (NC): U.S. Environmental Protection Agency, National Health and Environmental Effects Research Laboratory; 1998 July 21 [revised October 12 and November 18].
[23] Springborn Laboratories, Incorporated. A 90-day drinking water toxicity study in rats with ammonium perchlorate: amended final report [amended study completion date: June 3]. Spencerville (OH): Springborn Laboratories, Inc; 1998, Study No. 3455.1.
[24] King Jr JH. Effects of ammonium perchlorate on the thyroid hormone levels of the Sprague-Dawley rat. AFIT/GEE/ENV/95D-09 [thesis]. ADA 303621, Fort Belvoir (VA): Air Force Institute of Technology, Defense Technology Information Center; 1995, ADA 303621.
[25] Capen CC. Mechanistic data and risk assessment of selected toxic end points of the thyroid gland. Toxicol Pathol 1997;25:39–48.
[26] Isanhart JP, McNabb FMA, Smith PN. Effects of perchlorate on resting metabolism, peak metabolism, and thyroid function in the prairie vole (*Microtus ochrogaster*). Environ Toxicol Chem 2005;24:678–84.
[27] Zhang R, Wu FH, Zhou X, Peng KL. The effects of ammonium perchlorate on thyroid function and testis in male rats. Zhonghua Lao Dong Wei Sheng Zhi Ye Bing Za Zhi 2011;29:579–82 [Article in Chinese; Abstract at PMID: 22335154].
[28] Kessler FJ, Krüskemper HJ. Experimentelle Schilddrüsentumoren durch mehrjährige Zufuhr von Kaliumperchlorat [Experimental thyroid tumors caused by long-term potassium perchlorate administration]. Klin Wochenschr 1966;44:1154–6.
[29] Wu FH, Chen H, Zhou X, Zhang R, Ding M, Liu Q, et al. Pulmonary fibrosis effect of ammonium perchlorate exposure in rabbit. Arch Environ Occup Health 2013;68 (3):161–5.
[30] Lampé L, Módis L, Géhl Á. Effect of potassium perchlorate on the foetal rabbit thyroid. Acta Med Acad Sci Hung 1967;23:223–32.
[31] Brown-Grant K. Failure of orally administered perchlorate to affect deciduoma formation or pregnancy in the rat. J Reprod Fertil 1966;12:353–7.
[32] Brown-Grant K, Sherwood MR. Viability of the rat blastocyst following the oral administration of potassium perchlorate or potassium iodide to the mother. J Reprod Fertil 1971;27:265–7.
[33] Postel S. Placental transfer of perchlorate and triiodothyronine in the guinea pig. Endocrinology 1957;60:53–66.
[34] York RG, Brown WR, Girard MF, Dollarhide JS. Two-generation reproduction study of ammonium perchlorate in drinking water in rats evaluates thyroid toxicity. Int J Toxicol 2001;20:183–97.
[35] Greer MA, Goodman G, Pleus RC, Greer SE. Health effects assessment for environmental perchlorate contamination: the dose response for inhibition of thyroidal radioiodine uptake in humans. Environ Health Perspect 2002;110:927–37.
[36] National Research Council of the National Academies. Health implications of perchlorate ingestion. Washington, DC: National Academies Press; 2005, pp. 260.

[37] Letter to Acting Administrator–EPA, EPA Science Advisory Board, EPA-SAB-13–004, Subj: SAB Advice on Approaches to Derive a Maximum Contaminant Level Goal for Perchlorate, 29 May 2013.
[38] Argus Research Laboratories, Incorporated. Oral (drinking water) two-generation (one litter per generation) reproduction study of ammonium perchlorate in rats. Horsham (PA): Argus Research Laboratories, Inc; 1998, Protocol No. 1416-001.
[39] Argus Research Laboratories, Inc. Oral (drinking water) developmental toxicity study of ammonium perchlorate in rats. Horsham (PA): Argus Research Laboratories, Inc; 2000, Protocol No. 1416-003D (as cited in USEPA 2002).
[40] York RG, Brown WR, Girard MF, Dollarhide JS. Oral (drinking water) developmental toxicity study of ammonium perchlorate in New Zealand white rabbits. Int J Toxicol 2001;20:199–205.
[41] Argus Research Laboratories, Inc. Oral (drinking water) dosage-range developmental toxicity study of ammonium perchlorate in rabbits. Horsham (PA): Argus Research Laboratories, Inc; 1998, Protocol No. 1416-002P. Final Pilot Report.
[42] York RG, Barnett Jr J, Brown WR, Garman RH, Mattie DR, Dodd D. A rat neurodevelopmental evaluation of offspring, including evaluation of adult and neonatal thyroid, from mothers treated with ammonium perchlorate in drinking water. Int J Toxicol 2004;23:191–214.
[43] York RG, Lewis E, Brown WR, Girard MF, Mattie DR, Funk KA, et al. Refining the effects observed in a developmental neurobehavioral study of ammonium perchlorate administered orally in drinking water to rats. I. Thyroid and reproductive effects. Int J Toxicol 2005;24:403–18.
[44] York RG, Barnett Jr J, Girard MF, Mattie DR, Bekkedal MVK, Garman RH, et al. Refining the effects observed in a developmental neurobehavioral study of ammonium perchlorate administered orally in drinking water to rats. II. Behavioral and neurodevelopment effects. Int J Toxicol 2005;24:451–67.
[45] Baldridge MG, Stahl RL, Gerstenberger SL, Tripoli V, Hutz RJ. In utero and lactational exposure of Long-Evans rats to ammonium perchlorate (AP) disrupts ovarian follicle maturation. Reprod Toxicol 2004;19:155–61.
[46] Thuett KA, Roots EH, Gentles BA, Kendall RJ, Smith EE. Developmental toxicity of ammonium perchlorate administered orally in drinking water to deer mice (*Peromyscus maniculatus*). Toxicologist 2000;54(1):296.
[47] Thuett KA, Roots EH, Mitchell LP, Gentles BA, Anderson TA, Smith EE. *In utero* and lactational exposure to ammonium perchlorate in drinking water: effects on developing deer mice at postnatal day 21. J Toxicol Environ Health Part A 2002;65:1061–76.
[48] Thuett KA, Roots EH, Mitchell LP, Gentles BA, Anderson TA, Kendall RJ, et al. Effects of in utero and lactational ammonium perchlorate exposure on thyroid gland histology and thyroid and sex hormones in developing deer mice (*Peromyscus maniculatus*) through postnatal day 21. J Toxicol Environ Health Part A 2002;65 (24):2119–30.
[49] Roots EH, Thuett KA, Gentles BA, Kendall JR, Smith EE. Gestational through post-lactational exposure of ammonium perchlorate to deer mice (*Peromyscus maniculatus*). Toxicologist 2000;54(1):296.
[50] Stoker TE, Ferrell JM, Laws SC, Cooper RL, Buckalew A. Evaluation of ammonium perchlorate in the endocrine disruptor screening and testing program's male pubertal protocol: ability to detect effects on thyroid endpoints. Toxicology 2006;228:58–65.
[51] Smith PN, Severt SA, Jackson WA, Anderson TA. Thyroid function and reproductive success in rodents exposed to perchlorate via food and water. Environ Toxicol Chem 2006;25(4):1050–9.
[52] Bekkedal MVK, Carpenter T, Smith J, Ademujohn C, Maken D, Mattie DR. A neurodevelopmental study of the effects of oral ammonium perchlorate exposure on the

motor activity of pre-weanling rat pups. Wright-Patterson Air Force Base (OH): Naval Health Research Center Detachment, Neurobehavioral Effects Laboratory; 2000, Report No. TOXDET-00-03.

[53] Feinsilver L, MacNamee JK, McGrath FP, Oberst FW. Inhalation toxicity of combustion products of perchlorate – fuel propellants. U.S. Army Chemical Center (MD): Chemical Corps Medical Division; 1950, CMLEM-52.

[54] Wu FH, Guo HX, Lin MF, Chen ZZ, Zhou X, Peng KL. Biomarker expression in lung of rabbit with pulmonary fibrosis induced by ammonium perchlorate. Toxicol Ind Health 2011;27(3):235–41.

[55] McNabb FMA, Larsen CT, Pooler PS. Ammonium perchlorate effects on thyroid function and growth in bobwhite quail chicks. Environ Toxicol Chem 2004;23 (4):997–1003.

[56] McNabb FMA, Jang DA, Larsen CT. Does thyroid function in developing birds adapt to sustained ammonium perchlorate exposure? Toxicol Sci 2004;82:106–13.

[57] Kendall RT, Anderson E, Smith R, Patino S, McMurray J, Carr K, et al. Ecological risk assessment of explosive residues in rodents, reptiles, amphibians, and fish. Fort Belvoir (VA): Defense Technical Information Center; 2004, FY2002 Final Report. SERDP Project ER-1235 (ADA 573347).

[58] Rainwater TR, Wood MB, Millam JR, Hooper MJ. Effects of perchlorate on growth and behavior of a granivorous passerine, the zebra finch (*Taeniopygia guttata*). Arch Environ Contam Toxicol 2008;54(3):516–24.

[59] Gentles A, Surles J, Smith EE. Evaluation of adult quail and egg production following exposure to perchlorate-treated water. Environ Toxicol Chem 2005;24:1930–4.

[60] Coleman R, Everett PJ, Dodd JM. Ultrastructural observations on the droplets of experimentally induced goiters in *Xenopus laevis* with especial reference to the development of uyhlenhuth colloid cells. Z Zellforsch Mikrosk Anat 1968;84(4):490–6, Cited in: Callahan C, Sprenger M, Long GC, Porter R. Perchlorate ecological risk studies – a report on literature reviews and studies conducted by the Ecological Impact/Transport and Transformation Subcommittee of the Interagency Perchlorate Steering Committee. Brooks Air Force Base (TX): U.S. Air Force Institute for Environment, Safety, and Occupational Health Risk Analysis, Risk Analysis Directorate, Risk Assessment Division; 1998. IERA-RS-BR-TR-2001-0004. ADA 423410.

[61] Kinney KS, Felton SY, Cohen N. Sympathetic innervation of the amphibian spleen. Devel Comp Immunol 1996;20(1):51–9.

[62] Dumont JN, Bantle JA. FETAX analysis of ammonium perchlorate. (No further information.) Cited in: Callahan C, Sprenger M, Long GC, Porter R. Perchlorate ecological risk studies – a report on literature reviews and studies conducted by the Ecological Impact/Transport and Transformation Subcommittee of the Interagency Perchlorate Steering Committee. Brooks Air Force Base (TX): U.S. Air Force Institute for Environment, Safety, and Occupational Health Risk Analysis, Risk Analysis Directorate, Risk Assessment Division; 1998. IERA-RS-BR-TR-2001-0004. ADA 423410.

[63] Goleman WL, Urquidi LJ, Anderson TA, Smith EE, Kendall RJ, Carr JA. Environmentally relevant concentrations of ammonium perchlorate inhibit development and metamorphosis in *Xenopus laevis*. Environ Toxicol Chem 2002;21(2):424–30.

[64] Dean KE, Palacheck RM, Noel JM, Warbritton R, Aufderheide J, Wireman J. Development of freshwater water-quality criteria for perchlorate. Environ Toxicol Chem 2004;23(6):1441–51.

[65] Goleman WL, Carr JA. Contribution of ammonium ions to the lethality and antimetamorphic effects of ammonium perchlorate. Environ Toxicol Chem 2006;25 (4):1060–7.

[66] Sparling DW, Harvey G. Comparative toxicity of ammonium and perchlorate to amphibians. Bull Environ Contam Toxicol 2006;76:210–7.

[67] Sparling DW, Harvey G, Nzengung VA. Interaction between perchlorate and iodine in the metamorphosis of *Hyla versicolor*. In: Linder G, Krest S, Sparling DW, Little EE, editors. Multiple stressor effects in relation to declining amphibian populations. West Consohocken (PA): American Society for Testing and Materials (ASTM); 2003.
[68] Tietge JE, Butterworth BC, Haselman JT, Holcombe GW, Hornung MW, Korte JJ, et al. Early temporal effects of three thyroid hormone synthesis inhibitors in *Xenopus laevis*. Aquat Toxicol 2010;98(1):44–50.
[69] Brausch JM, Wages M, Shannahan RD, Perry G, Anderson TA, Maul JD, et al. Surface water mitigates the anti-metamorphic effects of perchlorate in New Mexico spadefoot toads (*Spea multiplicata*) and African clawed toads (*Xenopus laevis*). Chemosphere 2010;78(3):280–5.
[70] Weltje L, Simpson P, Gross M, Crane M, Wheeler JR. Comparative acute and chronic sensitivity of fish and amphibians: a critical review of data. Environ Toxicol Chem 2013;32(5):984–94.
[71] Opitz R, Kloas W. Developmental regulation of gene expression in the thyroid gland of *Xenopus laevis* tadpoles. Gen Comp Endocrinol 2010;168:199–208.
[72] Tietge JE, Holcombe GW, Flynn KM, Kosian PA, Korte JJ, Anderson LE, et al. Metamorphic inhibition of *Xenopus laevis* by sodium perchlorate: effects on development and thyroid histology. Environ Toxicol Chem 2005;24:926–33.
[73] Nieuwkoop PD, Faber J. Normal table of *Xenopus laevis* (Daudin). 2nd ed. Amsterdam: North-Holland Publishing Company; 1975.
[74] Goleman WL, Carr JA, Anderson TA. Environmentally relevant concentrations of ammonium perchlorate inhibit thyroid function and alter sex ratios in developing *Xenopus laevis*. Environ Toxicol Chem 2002;21(3):590–7.
[75] Dumont J. The effects of ammonium perchlorate on reproduction and development of amphibians, Fort Belvoir (VA): Defense Technology Information Center; 2008. SERDP Project ER-1236. ADA495519. Available from, http://www.dtic.mil/dtic/.
[76] Gosner KL. A simplified table for staging anuran embryos and larvae, Herpetologica 1960;16:183–90. Available from, http://froglet.us/Development/gosner_stages.html.
[77] Rollins-Smith LA, Davis AT, Blair PJ. Effects of thyroid hormone deprivation on immunity in post-metamorphic frogs. Dev Comp Immunol 1993;17(2):157–64.
[78] Theodorakis CW, Rinchard J, Carr JA, Park J-W, McDaniel L, Liu F, et al. Thyroid endocrine disruption in stonerollers and cricket frogs from perchlorate-contaminated streams in east-central Texas. Ecotoxicology 2006;15:31–50.
[79] Eisenreich KM, Dean KM, Ottinger MA, Rowe CL. Comparative effects of in ovo exposure to sodium perchlorate on development, growth, metabolism, and thyroid function in the common snapping turtle (*Chelydra serpentina*) and red-eared slider (*Trachemys scripta elegans*). Comp Biochem Physiol C Toxicol Pharmacol 2012;156:166–70.
[80] Dean JA. Lange's handbook of chemistry. 14th ed. New York: McGraw-Hill; 1992.
[81] Hazardous Substances Data Bank. Perchlorate [Internet]. Bethesda (MD): National Library of Medicine (US), National Institutes of, Health [cited October 2014].

CHAPTER 29

Wildlife Toxicity Assessment for Phenol

Christine Ann Arenal, Bradley E. Sample

Contents

INTRODUCTION

Phenol (C_6H_5OH), (CAS No. 108-95-2) a deliquescent crystalline solid found in petroleum products such as coal tar and creosote, is produced by natural degradation of organic wastes such as benzene. Phenol is produced for industrial/commercial use through distillation of petroleum, oxidation of cumene or toluene, and by hydrolysis of chlorobenzene. Phenol is predominantly used in the production of other compounds including bisphenol-A and other phenolic resins, caprolactam (nylon), aniline (polyurethane), alkylphenols (detergents and plasticizers), and xylenols (pesticides and antioxidants). Bisphenol-A is an intermediate in producing epoxy resins, and phenolic resins are used to make composite armor for military use.

Because phenol is toxic to bacteria, fungi, and viruses, it is also used in a number of medicinal and cleaning agents. While the compound is present in

Wildlife Toxicity Assessments for Chemicals of Military Concern
http://dx.doi.org/10.1016/B978-0-12-800020-5.00029-6

a number of over-the-counter personal care products, such as mouthwashes, ointments, and antiseptic and antipruritic lotions, the corrosive nature of the substance militates against its excessive application. Additionally, the cell-killing properties of the substance are not confined to microorganisms, suggesting that a narrow dosimetric window may exist between the substance's therapeutic and toxicological thresholds. As specified in the Hazardous Substances Data Bank (HSDB), phenol should never be used on pregnant women or on infants under 6 months of age (e.g., in products used for diaper rash) [1].

A substantial amount of information exists on the compound's potential threat to individuals who may become exposed to phenol in the workplace. Both the National Institute for Occupational Safety and Health and the Occupational Safety and Health Administration have established time-weighted average exposure limits of 5 parts per million (ppm) (19 milligrams per cubic meter [mg/m^3]), with a 15-minute ceiling value of 15.6 ppm (60 mg/m^3); and an immediately-dangerous-to-life-or-health value of 250 ppm (950 mg/m^3). The U.S. Environmental Protection Agency (U.S. EPA) has developed a reference dose (RfD) of 0.3 mg/kg/day from data on the developmental effects of the compound in laboratory rats, expressed as a decrease in maternal weight gain [2].

This Wildlife Toxicity Assessment (WTA) summarizes available information on the potential harmful effects of phenol to wildlife, emphasizing threshold doses for the onset of noncancer effects, as described in reports of experimental studies of the compound. Available dose-response data for phenol may support the establishment of Toxicity Reference Values (TRVs) that could serve as protective thresholds for wildlife exposed to phenol. This WTA is consistent with the protocols outlined in the U.S. Army Center for Health Promotion and Preventive Medicine (USACHPPM) Technical Guide (TG) 254, *Standard Practice for Wildlife Toxicity Reference Values* [3].

TOXICITY PROFILE

Environmental Fate and Transport

There is a substantial body of information on the dispersion of phenol in the environment; the HSDB [1] and the Agency for Toxic Substances and Disease Registry (ATSDR) [4] collectively provide summaries of the salient facts. The compound may be released to the air through storage tank vents, during loading and unloading operations, during wood burning, through emissions of waste incinerators and coal-fired power plants, and in automobile exhaust and cigarette smoke. Releases to streams and rivers may occur via the wastewater from chemical syntheses and manufacturing processes (e.g., resins, plastics, fibers, adhesives, rubber, etc.), from pulp mills and wood-treatment facilities, and during the use and discharge of phenol-containing commercial products (e.g., personal care items and disinfectants). Aquatic environments may also become contaminated with phenol through decomposition of organic wastes, and phenol may be released in effluent from municipal sewage treatment facilities. Soil may become contaminated with phenol from spills at manufacturing sites and by leaching from hazardous waste sites and landfills.

Transport and partitioning of phenol in environmental media are governed by its high solubility in water. Phenol released to the atmosphere may be washed out through rainfall and deposited in soil and surface water bodies. However, only limited amounts of airborne phenol will wash out in rainwater as the compound rapidly undergoes degradation by reaction with photochemically produced hydroxyl radicals (half-life 0.61 days). Owing to its high water solubility, phenol is also unlikely to sorb to sediments; it tends to leach to groundwater instead. However, the leaching process is mitigated by the relatively rapid biodegradation of phenol in soil by a number of aerobic and anaerobic bacteria (half-life $<$ 5–23 days) [4].

Phenol in sunlit surface waters will undergo photooxidation by hydroxyl and peroxyl radicals as well as biodegradation by some species of algae. Half-lives for biodegradation of phenol in water range from less than one day in fresh water to nine days in estuarine water. Phenol may also be taken up by higher plants; however, plants can readily metabolize it, limiting the potential for exposure via ingestion of plants grown in contaminated soils. Biodegradation of phenol may be hindered, and phenol may persist in the air, water, and soil for extended periods if releases are continual and concentrations become toxic to microorganisms capable of degrading phenol.

Notwithstanding the processes that would tend to remove phenol from the environment, the compound has been detected in all media. Phenol has been measured in air (0.03–44 parts per billion [ppb]), sediment ($>$10 ppb), groundwater (1.9 $>$ 10 ppb) and surface water (1.5 $>$ 100 ppb) [4]. Phenol has been measured in effluents at concentrations as high as 53 ppm. It has also been detected at "ppm" concentrations in groundwater near industrial facilities; for example, wood preservation plants and coal gasification installations.

The physical-chemical characteristics of phenol relevant to its environmental fate and transport are summarized in Table 29.1.

Table 29.1 Summary of the Physical-Chemical Properties of Phenol

CAS No.	108-95-2
Molecular weight	94.11
Color	Colorless to light pink
Physical state	Deliquescent crystals (liquefies by mixing with 8% water)
Melting point	40.9 °C
Boiling point	181.8 °C
Odor	Distinct aromatic, somewhat sickening sweet and acrid
Solubility in water	83 g/L, soluble in a number of organic solvents
Partition coefficients:	
Log K_{ow}	1.46
Log K_{oc}	1.21–1.96
Vapor pressure at 25 °C	0.35 mmHg
Henry's Law constant at 25 °C	3.0×10^{-7} m^3/mol
Conversion factors	1 ppm = 0.23 mg/m^3 1 mg/m^3 = 3.92 ppm

Sources: HSDB [1], ATSDR [4].

Summary of Mammalian Toxicity

Mammalian Oral Toxicity - Acute

Oral median lethal doses (LD_{50}) for phenol are listed in the HSDB [1], Lewis [5], and Berman *et al.* [6]. Primarily, the values are clustered in the range of 250–500 mg/kg although the value reported in cats (80 mg/kg) is considerably lower [1,5]. Lewis [6] specifies values of 317 mg/kg for rats, 270 mg/kg for mice, and 500 mg/kg for dogs. Berman *et al.* [6] used the simplified up/down procedure to derive an oral LD_{50} of 400 mg/kg in female F344 rats.

A number of groups have used a single dose of phenol to study toxicological endpoints other than lethality. For example, the Chemical Manufacturers' Association (CMA) sponsored a study in which five F344 rats/sex/group received a single gavage dose of 0, 1.5, 15, or 150 mg/kg ^{14}C-phenol [7]. The animals were kept in metabolic cages for 24 hours after dosing, and the amounts of radiolabel in urine, feces, and expired air were monitored. Two additional test groups consisting of five male rats each were given a single dose of either 150 mg/kg (high dose) or 1.5 mg/kg (low dose) ^{14}C-phenol. These male rats were cannulated in the jugular vein to permit monitoring of the area under the curve of the blood activity time-course. Greater than 90% of the radioactivity was recovered in the urine with few, if any counts expired in the breath. Of the few counts that remained in the carcass, most were centered in the liver and kidney, suggesting no tissue-selective retention. Blood levels of free phenol were very low and peaked 1 to 3 minutes after dosing. Additionally, muscle twitching was observed in the 150 mg/kg dose group immediately following dosing but disappeared within 45 minutes. Some radiolabeled glucuronides and sulfates of phenol and hydroquinone were identified, suggesting that the latter represent the primary conjugation products of orally administered phenol.

In one of a series of research reports wherein potential toxicants were tested using standard protocols, Moser *et al.* [8] used two dosing regimens to examine the neurotoxicological characteristics of phenol in female Fischer rats (8 rats/dose). In the first report, animals received a single oral dose of phenol at 0, 12, 40, 120, or 224 mg/kg. In a before-and-after experiment design that permitted each animal to serve as its own control, a battery of neurological function tests were used to [1] determine the animal's autonomic, sensorimotor, neuromuscular, and physiological (body weight, body temperature, piloerection) responses to phenol; and [2] follow the animal's level of activity and excitability, and document any related functional deficits. Two of the eight rats (25%) in the highest dose group died within 4 hours of dosing, and one of eight rats (13%) in the 120 mg/kg dose group died within 24 hours of dosing. Mild to severe whole-body tremors were observed in the 120 and 224 mg/kg dose groups within 1 to 2 minutes of dosing, but these rapidly disappeared. Decreased motor activity, crouched or flattened posture, decreased touch response, and body weight loss (7–8% within 24 hours) were also observed in the two highest dose groups. Neuromuscular impairment (e.g., tiptoe gait) was observed in the highest dose group. Based on these data, the 40 mg/kg dose represents an acute (single-dose) no observed adverse effect level (NOAEL), and the 120 mg/kg dose represents an acute lowest observed adverse effect level (LOAEL).

Mammalian Oral Toxicity: Subacute

The reports of Berman *et al.* [6] and Moser *et al.* [8] describe experiments in female F344 rats that delineate the systemic and neurological toxicity of phenol when administered subacutely. In each case, aqueous phenol was administered to F344 rats at 0, 4, 12, 40, or 120 mg/kg for 14 days. All the high-dose subjects died over the course of both studies. Because the deaths occurred at various times throughout the studies (e.g., one after the first dose and another after the eleventh dose in Moser et al. [8]), the authors indicate that death was not due to an accumulation of phenol. Moser *et al.* [8] also noted a 14%

decrease in body weight in the high-dose group. The principal systemic toxicity responses to phenol were necrosis of the renal tubules, formation of proteinaceous renal casts, and necrosis of the spleen and thymus [6]. A 12 mg/kg/day NOAEL was suggested for systemic effects of phenol. Moser *et al.* [8] noted neurological effects of phenol, including autonomic and activity deficits, with a subacute NOAEL of 12 mg/kg/day, which was similar to that reported for systemic effects, and a LOAEL of 40 mg/kg/day.

Mammalian Oral Toxicity: Subchronic

Only a single subchronic oral study was identified. The CMA sponsored a 13-week neurotoxicity study in which 15 Sprague Dawley rats/sex/group were exposed to 0, 200, 1,000, or 5,000 ppm phenol in drinking water [9]. According to the authors, the doses equivalent to these drinking water concentrations were 0, 18.1, 83.1, and 308.2 mg/kg/day in males and 0, 24.6, 107, and 359.8 mg/kg/day in females. Clinical signs and drinking water intake were recorded daily; body weight and food intake were monitored weekly. Animals were subjected to a functional observational battery of qualitative and quantitative neurological tests at 4, 8, 13, and 17 weeks (the latter 4 weeks after cessation of dosing). All animals were necropsied at term, with five animals/sex in the control and 5,000 ppm groups undergoing additional neuropathological evaluations. About 50% of the high-dose group became dehydrated, and the animals' food and water consumption was reduced throughout the treatment phase. The decreased water consumption and resulting dehydration were attributed to the aversion to the taste of phenol rather than to a direct toxicological effect. Few, if any, compound-related responses were observed in the functional battery at any dose level. Although a sporadic reduction in motor activity in high- and mid-dose animals was observed, the effects were considered secondary to the reduction in food and water consumption. No effects of phenol on central or peripheral nervous tissue were observed during necropsy or histopathology. Based on these findings, the authors considered the high-dose group (308 mg/kg/day in males and 360 mg/kg/day in females) to be the NOAEL for neurotoxicological effects of phenol in this experimental system.

Mammalian Oral Toxicity: Chronic

A single study documents the toxic effects of phenol in experimental animals under a chronic dosing regimen. As described in the U.S. EPA's IRIS record [2], the National Cancer Institute (NCI) conducted a chronic drinking water study in which 50 F344 rats and 50 B6C3F1 mice per sex per group were dosed with 0, 2,500, or 5,000 ppm phenol in drinking water for 103 weeks [10]. The drinking water concentrations correspond to estimated doses of 0, 260, 585 mg/kg/day for male rats; 0, 280, 630 mg/kg/day for female rats; and 0, 450, 660 mg/kg/day for male and female mice [2]. No toxicological effects could be unequivocally attributed to the compound. Although high-dose animals showed a reduction in body-weight gain, this response may have resulted from the marked reduction in water consumption in these groups, a response almost certainly related to taste aversion. In the absence of compound-related effects on mortality, behavior, morphology, or histopathology, this study identified LOAELs of 585 mg/kg/day and 660 mg/kg/day and NOAELs of 260 mg/kg/day and 450 mg/kg/day for rats and mice, respectively, based on body weight depression. As the reduction in body weight was attributed to taste aversion and not deemed a direct adverse effect of phenol, the LOAELs established based on this response may not be appropriate.

Mammalian Oral Toxicity: Other

Several studies evaluating reproductive effects of oral phenol exposure in mammals were identified. The Research Triangle Institute (RTI) carried out a teratological evaluation of phenol in CD rats on behalf of the National Institute of Environmental Health Sciences [11]. This study was used by the IRIS compilers as the principal study for developing a human health RfD for the compound [2]. Pregnant female CD rats (20–22 per group) were gavaged on gestation days (GDs) 6 to 15 with 0, 20, 60, or 120 mg phenol/kg/day in a volume of 5 ml/kg body weight. Dams were euthanized on GD 20 and their uteri examined for implantation losses, fetal number, and viability. Each live fetus was weighed and examined for external, visceral, or skeletal abnormalities. The doses used in the experiment were chosen because of a preliminary range-finding exercise that extended into the lethal range. In the range-finding study, seven of 10 rats given 125 mg/kg/day in a volume of 1 ml/kg died whereas only one of six rats given 160 mg/kg/day in a volume of 5 ml/kg died. The reduced toxicity observed at the higher dose was attributed to the larger dose volume, resulting in a decreased absorption rate. No maternal toxicity was evident in the sublethal phase of the study. The number of implantation sites and live fetuses did not differ among the groups. However, the proportion of litters with resorption sites was significantly increased in the low- and mid-dose groups, and was unchanged in the high-dose group. Fetal body weight demonstrated a dose-related decrease, with the high dose fetuses weighing significantly less than the controls. Based on this finding, the compound was determined to have developmental toxicity effects, with a NOAEL of 60 mg/kg/day and a LOAEL of 120 mg/kg/day.

Procter and Gamble, Incorporated (P&G) sponsored two reproductive/developmental studies in Sprague Dawley rats [12,13]. In the initial range-finding exercise, 10 pregnant female Sprague Dawley rats/group were gavaged on GDs 6 to 15 with 0, 60, 120, or 180 mg/kg/day [12]. Animals were euthanized on GD 15, and the uteri were examined for organ weight, fetal viability, early and late resorptions, number of corpora lutea, and total implants. The weights of major organs such as the liver, kidney, and stomach were recorded. Reduced food consumption and body-weight gain were evident in groups receiving phenol at the 120 and 180 mg/kg/day dose levels, but there were no other signs of maternal, reproductive, or developmental toxicity [12].

In the follow-up study, 25 pregnant female Sprague Dawley rats/group received 0, 20, 40, or 120 mg aqueous phenol/kg/day, three times/day by gavage (dose volume of 10 ml/kg) on GDs 6 to 15 [13]. This resulted in overall dose levels of 0, 60, 120, and 360 mg/kg/day. During the study, clinical signs and body weights were monitored, and food consumption was measured on GDs 0, 6, 9, 12, 16, and 20. All dams were euthanized on GD 20. A standard necropsy was supplemented by monitoring reproductive and developmental parameters such as the number of corpora lutea, the number and distribution of implantations, fetal viability, early and late resorptions, fetal weights, gross external fetal alterations, and soft tissue or skeletal alterations. In general, the effects of treatment were unremarkable. One high-dose dam died on GD 11, and several dams displayed clinical signs such as excess salivation and tachypnea. Maternal body-weight gain was reduced in dams receiving 120 mg/kg/day (11%) and 360 mg/kg/day (38%). Reduced maternal body-weight gain was associated with significant reductions in food consumption in the 120 mg/kg/day (11%) and 360 mg/kg/day (16%) groups. A NOAEL of 60 mg/kg/day and a LOAEL of 120 mg/kg/day were assigned for maternal effects. Fetal weights were reduced by 5 to 7% in the high-dose group. The effect on fetal weight, along with a reduction in the number of ossification sites for metatarsals, indicate a NOAEL of 120 mg/kg/day and a LOAEL of 360 mg/kg/day for developmental effects.

Narotsky and Kavlock [14] conducted a reproductive and developmental toxicity study of phenol in which approximately 20 pregnant female F344 rats/group were gavaged at doses of 0, 40, or 53.3 mg/kg/day on GDs 6 to 19. The dams were allowed to deliver their litters, and these litters were then evaluated for reproductive and developmental toxicity parameters. Nonsignificant reductions in body weight gain were observed in females, as was altered respiration in both dose groups. Additionally, one low-dose (7%) and two high-dose (13%) litters were fully resorbed, resulting in a significant reduction in litter size (number of live pups) and an increase in prenatal loss in the high-dose group. Of note is that these three females suffered severe respiratory signs. There were few other changes in treated groups compared to control although pups delivered by high-dose dams weighed significantly less than did those born to controls. The authors considered the difference, although significant, to have been unduly influenced by the progeny of a single dam, and therefore not necessarily indicative of a compound-related effect. The occurrence of kinked tails in progeny from the high-dose group could be used to assign a teratological NOAEL of 40 mg/kg/day. Based on the significant reduction in the number of live pups, 40 mg/kg/day could also be considered a reproductive NOAEL, with a corresponding LOAEL of 53.3 mg/kg/day.

Goodrich and CMA reported a two-generation reproductive study in which phenol was added to the drinking water of 30 pairs of Sprague Dawley rats [15]. The experiment was subsequently published in the open literature by Ryan *et al.* [16]. The concentrations employed were 0, 200, 1,000, and 5,000 ppm, which were equivalent to daily doses in the F_o generation of 0, 14.7, 70.9, and 301 mg/kg/day (males), and 0, 20, 93, and 320.5 mg/kg/day (females). Additionally, in the F_1 generation, the concentrations used were 0, 13.5, 69.8, and 319.1 mg/kg/day (males) and 0, 20.9, 93.8, and 379.5 mg/kg/day (females). Exposure occurred during an initial 10-week period prior to mating, through a 2-week mating period, and through gestation and lactation until weaning. Randomly chosen F_1 animals were then exposed for 11 weeks prior to mating, and then through mating, gestation, delivery, and weaning. A full slate of systemic, reproductive, and developmental toxicological parameters was evaluated in both generations, but the authors noted few obvious compound-related changes in any of the treated groups. Although significant reductions in water and food consumption, as well as corollary reductions in body weight and weight gain, were noted in the high-dose group, these effects were attributed to taste aversion to phenol in the drinking water. Mating performance, fertility, vaginal cytology, and male reproductive function were similar to controls in all dose groups. Prostate and uterus weights were reduced in the F_1 generation in all dose groups;

however, in the absence of underlying pathology or functional deficit in reproductive performance, these effects were not considered adverse. Litter survival was reduced in the 5,000 ppm group of both generations, suggesting a developmental NOAEL of 1,000 ppm, which was equivalent to approximately 70 mg/kg/day for males and 93 mg/kg/day for females.

A second RTI study evaluated teratological effects of phenol in CD-1 mice [17]. Pregnant female CD-1 mice (22–29 per group) were gavaged on GDs 6 to 15 with 0, 70, 140, or 280 mg phenol/kg/day in a volume of 10 ml/kg body weight. Dams were euthanized on GD 17 and their uteri weighed and examined for implantation sites and fetuses. Live fetuses were weighed and examined for external, visceral, or skeletal abnormalities. The high-dose group exhibited signs of maternal toxicity, including tremors, ataxia, reduced body weight, and weight gain, and increased mortality (11%). Prenatal viability (incidences of resorptions and dead fetuses) and fetal deformities were not significantly different among treatment groups. Uterine weight and fetal body weight demonstrated a dose-related decrease, with the high-dose groups weighing significantly less than the controls for both measures. Based on this finding, the compound was determined to have developmental toxicity effects with a NOAEL of 140 mg/kg/day and a LOAEL of 280 mg/kg/day.

Studies Relevant for Mammalian TRV Development for Ingestion Exposures

Several subacute, subchronic, chronic, and reproductive/ developmental studies were identified as appropriate for the development of TRVs for mammals. However, only one Order (Rodentia) and two families (Cricetidae and Muridae) are represented in the relevant studies.

Orally administered phenol is readily absorbed and is widely distributed throughout the body, with elevated tissue concentrations occurring in the liver and kidneys. The adverse effects observed following the ingestion of phenol are consistent with its generalized toxic effects for which there is no target organ specificity. This was evidenced in the widespread effects (e.g., necrosis of the kidney, spleen, and thymus and neurological deficits) observed in subacute (gavage) studies by Berman *et al.* [6] and Moser *et al.* [8]. Subchronic and chronic studies (all drinking water) did not demonstrate similar systemic toxicity effects [9,10]. Although decreased body weight was observed in these studies, this effect was attributed to decreased water consumption due to the poor palatability and aversion to phenol in drinking water.

The developmental and reproductive effects of exposure to phenol, as discussed earlier, were similar among studies; however, the magnitude of the effect and the dose at which the effects were observed varied among studies utilizing different dosing regimens. A contrast is evident in the toxic effects of phenol when administered as a bolus compared to similar doses administered throughout the day. For example, Ryan *et al.* [16] administered phenol in the drinking water and observed reproductive effects (e.g., reduced survival of F_1 pups) at the highest dose level (379.5 mg/kg), a dose that was approximately six times greater than the dose that produced similar effects in a gavage study [14]. The difference in effects between administration methods may be a result of the phenol serum levels not being linearly related to dose. At low doses, the rapid uptake, metabolism, and clearance of the compound results in almost all the phenol being excreted without entering the bloodstream [7]. Administration of the compound as a bolus may saturate these mechanisms, at least temporarily, thus allowing phenol to appear in the bloodstream and its toxic impacts to become apparent.

The systemic toxicity of phenol, as indicated by "phenol twitching behavior (PTW)," appears to be more closely related to peak blood levels than total dose. Namely, PTW coincided with peak blood concentrations following gavage dosing, but was not observed in drinking water exposure groups despite the total dose being higher than the high gavage dose [7]. Additionally, because of the more rapid absorption of the compound, the effects of phenol are increased when smaller dosing volumes are used [11,12,17].

In light of the effect of dosing regimen on phenol toxicity, and given that drinking water dosing may be more ecologically relevant than gavage dosing, development of a TRV based on drinking water studies may be appropriate. However, the effects of phenol observed in drinking water studies can largely be attributed to taste aversion and decreased water consumption rather than to toxicity. Additionally, precise determination of administered dose is difficult in drinking water studies because spillage can occur. As such, the P&G [13] study, which decreases the impact of gavage dosing and approximates the toxicokinetic profile of drinking water studies by dividing the dose into three administrations per day, may be the most appropriate study for TRV development. This is the only study conducted using a divided-dosing regimen, and thus support for a TRV based on this study may be drawn from other studies.

The NCI [11], Narotsky and Kavlock [14], and Ryan *et al.* [16] studies are particularly valuable as they include chronic, reproductive, and developmental studies that evaluate numerous relevant endpoints. The LOAELs and NOAELs for studies relevant to TRV derivation are listed in Table 29.2 and displayed in Figure 29.1.

Mammalian Inhalation Toxicity

One mammalian inhalation toxicity study was located. The CMA sponsored a 10-day inhalation toxicity experiment in which 20 F344 rats/sex/group were exposed 6 hours/day, 5 days/week for 2 weeks to phenol (nose only) at 0, 0.5, 5, or 25 ppm (18). Half of the animals at each dose were permitted a 2-week recovery period. All animals were monitored for clinical signs, food consumption, and body-weight changes. At term, hematological and clinical chemistry parameters were measured, and a full necropsy was performed to evaluate any morphological changes that might have occurred in response to treatment. Pieces of liver, kidney, spleen, and respiratory tract, as well as any gross lesions, were fixed and mounted for histopathological examination. The authors reported almost uniformly negative results for toxicological impacts of inhaled phenol in this experiment. Clinical signs such as chromodacryorrhea (i.e., red tears) and nasal discharge were evident in some of the animals in the study. However, these responses were reported in the published version of the study [19] as having been sporadic and not treatment-related whereas in the CMA [18] version of the study, these effects were reported to have been concentration-related in males during the second week of exposure to 5 and 25 ppm phenol. In the absence of abnormal nasal histopathology, these effects were not deemed adverse effects. Due to the equivocal nature of the responses, an appropriate NOAEL could not be determined.

Mammalian Dermal Toxicity

Two mammalian dermal toxicity studies were located. In one study, phenol was one of a number of compounds applied to the ears of 8 to 11 female ICR mice/group to measure swelling as an index of irritation [20]. Five dose groups (1.0, 1.25, 1.5, 1.75, and 2.0 mg) were evaluated. Because solvent alone (ethanol, in this case) was applied to one ear and phenol to the other ear, each animal was able to serve as its own control. Five microliters of the dose were applied to the dorsal and ventral surfaces of the ear. Phenol was highly effective in inducing ear swelling, as evidenced by striking dose-dependent increases in the incidence (13–100%) and degree of swelling.

In the other study, phenol was applied to a 3 cm^2 shaved area on the back of female B6C3F1 mice (6 per group) at dose levels of 0, 1.75, 3.5, and

Table 29.2 Summary of Relevant Mammalian Data for TRV Derivation

		Test Results			
Test Organism	**Test Duration**	**NOAEL (mg/kg/day)**	**LOAEL (mg/kg/day)**	**Effects Observed at the LOAEL**	**Study**
Rats (F344)	14 days	12	40	Necrosis of the kidney, spleen, and thymus	[6]
Rats (F344)	14 days	12	40	Neurological deficits	[8]
Rats (Sprague Dawley)	13 weeks	360	—	None	[9]
Rats (F344)	103 weeks	260	585	Body weight depression	[10]
Mice (B6C3F1)	103 weeks	450	660	Body weight depression	[10]
Rats (CD)	GD 6–15	60	120	Reduced fetal body weight	[11]
Mice (CD-1)	GD 6–15	140	280	Reduced fetal body weight and gravid uterine weight, maternal toxicity	[17]
Rats (Sprague Dawley)	GD 6–15	60	120	Reduced maternal body weight gain and food consumption	[12]
Rats (Sprague Dawley)	GD 6–15	60	120	Reduced maternal body weight gain and food consumption	[13]
		120	360	Reduced fetal body weights; reduced number of ossification sites for metatarsals	
Rats (F344)	GD 6–19	40	53.3	Reduced number of live pups (reproductive effects); kinked tails (teratogenic effect), reduced maternal body weight	[14]
Rats (Sprague Dawley)	10 or 11 weeks prior to mating through weaning	70–94	319–380	Reduced survival of F_1 pups	[16]

GD, gestational day; LOAEL, lowest observed adverse effect level; mg/kg/day, milligrams per kilogram per day; NOAEL no observed adverse effect level; TRV, toxicity reference value.

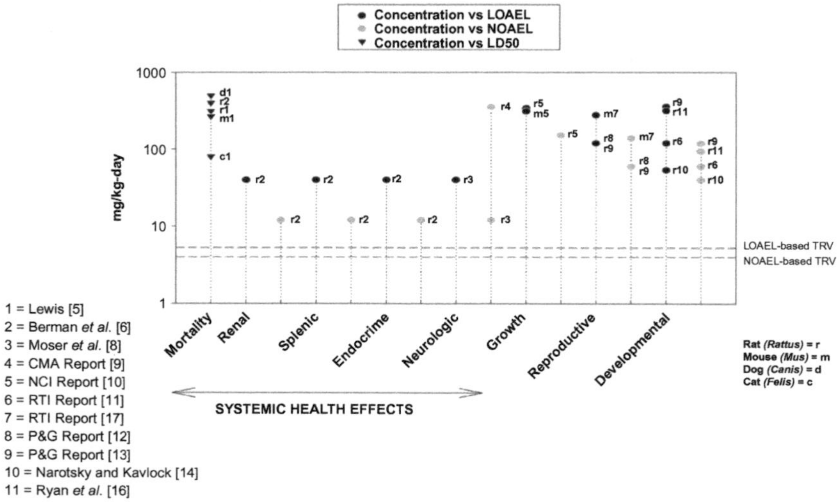

Figure 29.1 Phenol: Health Effects in Mammals. Letters by symbols represent test species, and numbers represent the study.

7.0 mmol/kg for 0, 5, 1, 2, 3, and 6 hours [21]. A dose-dependent increase in skin bifold thickness (a measure of skin inflammation) was observed, with all dose groups significantly increased over the control group. Additionally, oxidation of glutathione and protein thiols, decreased vitamin E, and total antioxidant reserves in the skin were observed in mice exposed to phenol.

Summary of Avian Toxicology

No studies that examined the toxic effects of phenol in birds were identified.

Summary of Amphibian Toxicology

Four studies on the effects of phenol on amphibian development were located; one used a short-term embryo exposure, and three used embryo and larval exposures.

Amphibian Toxicity: Other

Bernardini *et al.* [22] exposed *Xenopus* embryos (8 per dose level) to phenol at 0, 25, 50, 100, 150, 200, and 250 mg/L from 8 hours postfertilization (p.f.) through 120 hours p.f. Mortality was recorded over the course of the study. Embryos that had died and embryos that had been euthanized (i.e., embryos surviving to 120 hours) were evaluated for malformations, and head-to-tail length was measured. A concentration-dependent increase in both mortality and malformation rates was observed, with both responses increasing dramatically after 100 mg/L. There was also a significant concentration-dependent reduction in the mean length of embryos. Because significant growth retardation occurred at all exposure concentrations, 25 mg/L is considered the lowest observed adverse effect concentration (LOAEC) for this endpoint. Values for median lethal concentration (LC_{50}), median teratogenic concentration (TC_{50}), median teratogenic index (TI_{50}; equals LC_{50}/TC_{50}), 10% lethal concentration (LC_{10}), and 10% teratogenic concentration (TC_{10}) were developed in the study. The LC_{50}, TC_{50}, TI_{50}, LC_{10}, and TC_{10} values are 178, 141, 1.3, 32, and 42 mg/L, respectively.

Dumpert [23] examined the toxicity of phenol in the African clawed frog, *Xenopus laevis,* using concentrations of 0, 0.1, 1, 5, 10, and 50 ppm in a physiological salt medium. Exposure occurred during embryogenesis and larval development. Although the exact duration was not given, a 5-week period of exposure was indicated prior to the measurement of body length. All larvae exposed to the highest concentration of phenol (50 ppm) died within three weeks of exposure. Development of larvae exposed to the remaining exposure concentrations was similar to that of the controls, and no teratogenic effects of the compound were evident in this experiment. However, larvae exposed to 5 and 10 ppm phenol were slightly shorter than the controls. Although this effect was not statistically significant, it does indicate a possible growth retardation effect that is consistent with that reported by Bernardini *et al.* [22]. Therefore,1 ppm is considered the no observed adverse effects concentration (NOAEC), and 5 ppm the LOAEC.

Two companion studies conducted by the University of Kentucky Water Resources Research Institute [24,25] examined the toxicity of phenol in eight species of amphibians using a flow-through bioassay system. Birge *et al.* [24] exposed 50 to 130 embryos of *Rana pipiens*, *Rana catesbeiana*, *Rana palustris*, *Bufo fowleri*, and *Bufo americanus* to at least five test concentrations of phenol per species in doses ranging from 0.0007–21.8 mg/L from 0.5 to 6 hours p.f. through 4 days post-hatching. Using the same flow-through embryo-larval test system, Black *et al.* [25] exposed 50 to 125 embryos of *Ambystoma gracile*, *Rana temporaria*, and *Xenopus laevis* to phenol at concentrations of 0.002, 0.010, 0.12, 0.72, 1.45, 14.0, and 26.4 mg/L from 30 minutes p.f. through 4 days post-hatching.

Percent survival of normal organisms, expressed as the frequency in test organisms relative to controls, was determined at hatching and 4 days post-hatching. Percent egg hatchability included all embryos that completed the hatching process, regardless of teratogenic status. Teratogenesis was expressed as the percentage of survivors affected by gross, debilitating abnormalities at the time of hatching. Control-adjusted median lethal concentration (LC_{50}) was determined for each species, while LC_{10} and LC_1 values were determined for select species (*R. pipiens* and *R. catesbeiana*). These determinations were based on the combined frequencies for lethality and teratogenesis. The authors considered the LC_1 as the threshold for adverse effects and the LC_{10} concentration the level at which exposure begins to produce appreciable reproductive impairment [24].

Based on the LC_{50} values (in mg/L), the order of increasing tolerance for the 8 species tested was *R. pipiens* (0.04), *R. catesbeiana* (0.23), *R. temporaria* (0.27), *A. gracile* (0.38), *B. americanus* (>0.89), *B. fowleri* (2.45), *X. laevis* (7.68), and *R. palustris* (9.87). The LC_1 values for *R. catesbeiana* and *R. pipiens* were reported as 0.001 and 0.0011 mg/L, and the LC_{10} values for these species were 0.0085 and 0.0052 mg/L, respectively. Although no statistical tests were conducted, and NOAEC and LOAEC levels were not reported in these studies, teratogenesis reported in the control groups (typically <1%) occurred at concentrations that started at 0.0047 mg/L for *R. pipiens*; 0.12 mg/L in *R. temporaria*; 0.22 mg/L for *B. americanus*; 0.53 mg/L for *R. catesbeiana*; 1.86 mg/L for *R. palustris*; 10.2 mg/L for *B. fowleri*, and 14.0 mg/L in *X. laevis*. No teratogenic effects were observed in *A. gracile* at concentrations up to 26.4 mg/L.

Studies Relevant for Amphibian TRV Development for All Exposures

Each of the four studies identified for amphibians is relevant for TRV development. All the studies document phenol toxicity during the critical life stages of embryogenesis and larval development. The Bernardini *et al.* [22] study, however, was of limited duration and did not extend throughout embryogenesis. Dumpert [23] conducted a longer-term (>5 weeks) study of the effects of phenol on *Xenopus laevis*. Although the exact duration of the study was not reported, it occurred during a critical life stage (embryogenesis and larval development); as such, it may be considered a chronic study and is highly relevant for TRV development. The studies by Birge *et al.* [24] and Black *et al.* [25] were limited since they were of short duration, constituted only single replicates, and did not evaluate growth effects. However, because these two studies extend through embryogenesis and are the only studies that utilize amphibian species other than *Xenopus laevis*, they are important for TRV development. Eight species representing two Orders and four

Families of amphibians, including Anura: Ranidae, Bufonidae, Pipidae, and Urodela: Ambystomatidae, are represented by these studies.

The effects from exposure are consistent but vary in magnitude. Differences in the magnitude of effects observed in *X. laevis*, the only species utilized in multiple studies, may be attributed to differences in exposure protocols (static renewal versus flow-through) and duration of exposure (portion of embryo stage vs. embryo and larval stages). Two studies [22,23] identified LOAECs for growth retardation (25 mg/L and 5 mg/L, respectively) for *X. laevis*. Dumpert [23] identified a NOAEC for growth of 1 mg/L, but a NOAEC for growth could not be identified in the Bernardini *et al.* [22] study. The differences in growth effects in these studies may be attributable to the exposure duration and the measurement of growth during the larval phase in Dumpert [23]. Effects on amphibian growth are likely to be more substantial during the larval phase, during which rapid growth occurs.

Teratogenic effects of phenol, including intestinal and ocular malformations and generalized edema, were observed by Birge *et al.* [24], Black *et al.* [25], and Bernardini *et al.* [22]; however, LOAECs and NOAECs were not reported by any of these authors. The concentration expected to produce teratogenic effects in 10 or 50% of exposed *X. laevis* (TC_{10} or TC_{50}) were reported by Bernardini *et al.* [22] as 42 mg/L and 141 mg/L, respectively. The concentration of phenol that produced malformations ranged from 0.0047 mg/L in *R. pipiens* to 14.0 mg/L in *X. laevis* in the Birge *et al.* [24] and Black *et al.* [25] studies, respectively, although these changes may not be statistically different from those of the controls (82–92% effect rate in the controls). The study designs (i.e., single replicates) were not sufficient to allow for statistical tests.

Differences in the teratogenic effect concentration, as well as LC_{50} values, reported for *X. laevis* in these three studies are likely reflective of the differences in their exposure regimes. Bernardini *et al.* [22] used a static renewal system with daily test solution renewal whereas Birge *et al.* [24] and Black *et al.* [25] used a flow-through system with a retention time of 2.5 hours. Due to the relatively short half-life of phenol in fresh water (~1 day), exposure concentrations in Bernardini *et al.* [22] likely decreased substantially between renewals, resulting in exposure to lower-than-expected concentrations and potentially inflated effect concentrations. Both exposure scenarios may be ecologically relevant for amphibians. The data from studies relevant to TRV derivation are listed in Table 29.3 and displayed graphically in Figure 29.2.

Table 29.3 Summary of Relevant Amphibian Data for TRV Derivation

		Test Results				
Test Organism	**Test Duration**	**LC_{50} (mg/L)**	**NOAEC (mg/L)**	**LOAEC (mg/L)**	**Effects Observed at the LOAEC**	**Study**
Xenopus	120 hr p.f.	178	—	25	Reduced growth of larvae	[22]
Xenopus laevis	>5 week	—	1	5	Reduced growth of larvae	[23]
Rana pipiens	96 hr p.f.	0.04	—	0.0047	Teratogenesis	[24]
Rana catesbeiana	96 hr p.f.	0.23	0.062	0.53	Teratogenesis	
Rana palustris	96 hr p.f.	9.87	0.15	1.86	Teratogenesis	
Bufo fowleri	96 hr p.f.	2.45	0.53	10.2	Teratogenesis	
Bufo americanus	96 hr p.f.	>0.89	0.063	0.22	Teratogenesis	
Ambystoma gracile	96 hr p.f.	0.38	—	>26.4	Teratogenesis	[25]
Rana temporaria	96 hr p.f.	0.27	0.010	0.12	Teratogenesis	
Xenopus laevis	96 hr p.f.	7.68	1.45	140.0	Teratogenesis	

Key: hrs, hours; LC_{50}, median lethal concentration; LOAEC, lowest observed adverse effect concentration; mg/L, milligrams per liter; NOAEC, no observed adverse effect concentration; p.f., postfertilization; TRV, toxicity reference value.

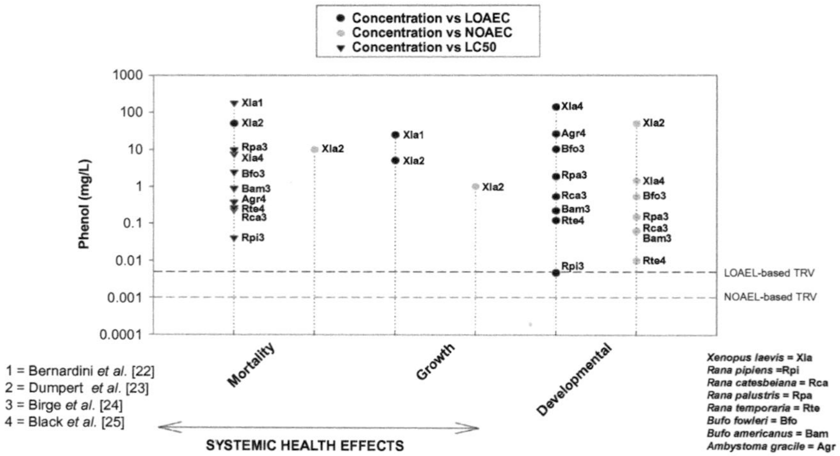

Figure 29.2 LOAEC-based, NOAEC-based, and LC50-based TRV. Letters by symbols represent test species, and numbers represent study.

Summary of Reptilian Toxicology

No data are available on the toxicity of phenol in reptiles.

RECOMMENDED TOXICITY REFERENCE VALUES

Toxicity Reference Values for Mammals

TRVs for Ingestion Exposures for the Class Mammalia

Studies relevant to the development of oral ingestion TRVs for mammals were previously discussed and are presented in Table 29.2 and Figure 29.1. Although subacute, subchronic, chronic, and reproductive/ developmental studies were available, only those of rats and mice are represented. Upon review of the data from the NCI study [10] that used similar methods and dosing regimens for both rats and mice, the animals appear to be equally sensitive to phenol, with LOAELs for body weight depression of 585 and 660 mg/kg/day, respectively. This is supported by the acute data in which LD_{50} values for rats were 317 and 400 mg/kg compared to 270 mg/kg for mice [5,6]. Additionally, the reproductive/developmental studies conducted by RTI indicate that developing rats and mice have similar sensitivity to phenol, with LOAELs for fetal weight of 120 and 280 mg/kg/day, respectively [11,17]. The LD_{50} data suggest that cats may be particularly sensitive to phenol (LD_{50} = 80 mg/kg); however, no long-term studies using cats were available.

Although organ necrosis (kidney, spleen, and thymus) was the most sensitive endpoint reported (Table 29.2, Figure 29.1), these effects were only observed in one of the subacute studies [6] in which the compound was administered via gavage. As noted earlier, toxic responses to phenol may be mitigated by administration route. That is, dosing by gavage may have overwhelmed the processes that metabolize and excrete phenol from the body, resulting in effects that are more adverse. In contrast, studies in which phenol was administered in the drinking water reported fewer toxic responses, particularly to body organs.

Similar reproductive and developmental effects were observed in studies that dosed via gavage and drinking water; however, the dose at which the effects were observed was lower in the studies that used gavage. For example, Narotsky and Kavlock [14] used gavage methods and observed fewer live pups at a dose of 53.3 mg/kg/day whereas Ryan *et al.* [16] administered the doses in drinking water and observed reduced survival in pups at 319 mg/kg/day. The LOAEL of 53.3 mg/kg/day reported by Narotsky and Kavlock [14] was the lowest among the five studies that reported reproductive [12,13] and developmental [11,13,14,16,17] effects. This LOAEL is also lower than the developmental NOAELs identified in the other studies (60–120 mg/kg/day), which may be a function of differing experimental designs (i.e., single dose gavage vs. divided dose gavage or drinking water) or dosing duration (i.e., during GDs 6–19 vs. GDs 6–15). Due to the differences in toxicity of phenol administered by gavage versus drinking water, the P&G [13] study was selected as an appropriate developmental study for TRV derivation. This study utilized a divided-dose gavage regime to decrease the impact of the bolus effect while avoiding the problems associated with drinking water studies (e.g., precise dosing, taste aversion). The LOAEL for developmental effects in this study [13] was 360 mg/kg/day, and the NOAEL was 120 mg/kg/day.

Reduced fetal or pup survival may adversely impact the ecology or health of the population, and reduced fetal body weights (an indicator of effects on growth and/or energy efficiency) may have the potential to adversely affect future fitness. Therefore, these endpoints are considered ecologically relevant. The reproductive and developmental studies are considered chronic in nature because they evaluate the effects of phenol during a critical life stage (i.e., during gestation) [3]. The NCI rat and mouse studies [10] are also considered chronic (duration of 103 weeks). However, the affected endpoint (depressed body weight in adults) appears to be less sensitive than developmental endpoints (Table 29.2, Figure 29.1) and may not be as ecologically relevant.

Since the minimum data set requirements as outlined in USACHPPM TG 254 [3] were not met (i.e., data representative of at least three species and two taxonomic orders were lacking), the approximation approach as prescribed by the guidance was used to develop oral ingestion TRVs for mammals. Using the approximation method, an uncertainty factor (UF) of 10 was applied to the NOAEL and LOAEL values reported by P&G [13] to account for potential interspecies differences, resulting in a NOAEL-based TRV of 12.0 mg/kg/day and a LOAEL-based TRV of 36.0 mg/kg/day. The class-specific NOAEL and LOAEL are presented in Table 29.4.

Table 29.4 Selected Ingestion TRVs for the Class Mammalia

TRV	Dose	Confidence Level
NOAEL-based	12.0 mg/kg/day	Low-Medium
LOAEL-based	36.0 mg/kg/day	Low-Medium

Source: [13].

Developmental effects were generally consistent across studies; however, data were available for only two species, and the data for mice were limited to a single study. Additionally, there is some uncertainty regarding the effect of dose administration methods on responses. Therefore, a confidence level of low-to-medium was assigned to the TRVs (Table 29.4).

TRVs for Ingestion Exposures for Mammalian Foraging Guilds

TRVs specific to particular guild associations (e.g., small herbivorous mammals) have not yet been derived. However, the class-specific TRVs shown in Table 29.3 may be considered to apply to herbivorous small mammals because rats are members of this guild. As with the class-specific TRVs, the confidence level of these TRVs is low-to-medium because only one species is represented and there is some uncertainty related to the effect of dose administration methods on observed responses.

TRVs for Inhalation Exposures for the Class Mammalia

Although one 10-day inhalation study was available [18], the only effects observed were sporadic red tears (an indication of stress in rats) and nasal discharge. Because these data were inconsistent across dose levels, and the relevance to the health and ecology of the whole organism is uncertain, this study was considered insufficient for TRV development.

TRVs for Dermal Exposures for the Class Mammalia

Two dermal exposure studies for mammals were available [20,21]. A clear dose-response relationship for ear swelling was observed in Patrick *et al.* [20]; however, no systemic endpoints were evaluated. Ear swelling is a localized response that shows improvement with time and has uncertain, if any, ecological relevance. Murray *et al.* [21] also showed localized swelling with a dose-dependent increase in skin bifold thickness on dorsal sections of mice. Although an accompanying oxidation of glutathione and protein thiols as well as decreased vitamin E and total antioxidant reserves in the skin were observed, the ecological relevance of these responses is also uncertain. Therefore, these data were not considered suitable for TRV development.

TRVs for Birds

No data with which to derive TRVs for birds were found.

TRVs for Amphibians

All four amphibian studies are relevant for TRV derivation (Table 29.3, Figure 29.2). Growth was the most sensitive endpoint in the studies using *X. laevis* and may be a suitable endpoint for TRV development. Growth in larval amphibians is an ecologically relevant endpoint. Reduced larval growth can adversely affect adult fitness through impacts on developmental rate, timing of metamorphosis, survival to and size at maturity, and recruitment to the breeding population. An unbounded LOAEC for growth depression of 25 mg/L is indicated by the results presented in Bernardini *et al.* [22], while a NOAEC of 1 mg/L and LOAEC of 5 mg/L were identified for growth depression in the Dumpert study [23]. Because that study extends throughout the embryonic period into the larval period and has a clearly identifiable NOAEC and LOAEC, it is more applicable to TRV development than the study by Bernardini *et al.* [22]. However, the studies by Birge *et al.* [24] and Black *et al.* [25] indicate that *X. laevis* is among the most tolerant of the eight species tested. The TRVs based even on the most sensitive endpoint in this species may not be protective of other species. The most sensitive species was *R. pipiens,* with an LC_{50} of 0.04 mg/L [24,25]. If LC_1 values are considered thresholds for toxic effects (similar to LED_{10}) [3], the authors suggest that the corresponding mean ED_{10} values could be considered equivalent to the LC_{10} value or the level at which appreciable reproductive impairment may occur; for example, *R. pipiens* – 0.0011, and 0.0052, and *R. catesbeiana* – 0.001, and 0.0085 mg phenol/L for LC_1 and LC_{10} values, respectively [24,25].

The relevant amphibian studies meet the minimum data set requirements of the Standard Practice [3]; therefore, the NOAEC/LOAEC approach for TRV development is appropriate. The LOAEC for the most sensitive species and endpoint was 0.0047 mg/L for teratogenic effects in *R. pipiens* [24]. However, given the lack of replicates involved, the dose at which a 1% response could be detected (LC_1) that integrates the threshold for teratogenic effects was used for the NOAEC-based value, and the 10% response level was used for the LOAEC-based value (LC_{10}). Table 29.5 presents the selected TRVs. A confidence level of low-to-medium has been assigned to these TRVs because although four studies, which examined a broad range of species, are available, the endpoints examined are limited, and in some cases, statistical comparisons are not possible. Additionally, the low level at which effects were reported for *R. pipiens* was not corroborated with other data. Moreover, the studies included only a small portion of the larval period

Table 29.5 Selected Aquatic TRVs for Amphibians

TRV	Dose	Confidence
NOAEC-based	0.001 mg/L	Low-Medium
LOAEC-based	0.005 mg/L	Low-Medium

Source: [24].

and did not extend through metamorphosis, a period of heightened sensitivity.

TRVs for Reptiles

No data with which to derive TRVs for reptiles were found.

IMPORTANT RESEARCH NEEDS

Although several studies were available and are consistent among reproductive and developmental effects, mammalian TRVs derived for phenol have been assigned a confidence level of low-to-medium. This is due to uncertainty related to the dose administration methods (i.e., gavage vs. drinking water). Therefore, additional studies that would clarify the impact of differences between these methods on the magnitude and frequency of endpoint responses are needed. Evaluation of additional species and taxonomic orders is also needed. In addition, toxicity studies that examine demographic factors such as birth, death, and recruitment would be of much greater ecological significance. The additional data would increase confidence in the mammalian TRVs and enable development of TRVs for specific foraging guilds. Inhalation and dermal studies on mammals were very limited for phenol and did not allow for TRV derivation. Amphibian toxicity data were available for eight species; however, sensitive endpoints were not included, and statistical power was low, resulting in low-to-medium confidence in the TRVs. Toxicity data for birds and reptiles were lacking. Before reliable avian, amphibian, and reptilian TRVs can be derived, phenol toxicity in these wildlife classes needs to be adequately characterized. Appropriate acute, subacute, subchronic, and especially chronic phenol toxicity data derived through biologically relevant exposure routes are needed. Research studies should include experimental models of species that are genetically, biologically, and behaviorally similar to the wildlife species that exhibit the greatest propensity for toxicant exposure.

REFERENCES

[1] HSDB. Phenol [Internet]. Bethesda (MD): National Library of Medicine (US), National Institute of Health [cited 2014]. Available at http://toxnet.nlm.nih.gov/.

[2] U.S. EPA. Integrated risk information system [Internet]. Cincinnati (OH): Office of Health and Environmental Assessment (US), National Center for Environmental Assessment [cited 2014]. Available at http://toxnet.nlm.nih.gov/ or http://www.epa.gov/IRIS.

[3] USACHPPM. Standard practice for wildlife toxicity reference values. Aberdeen Proving Ground (MD): U.S. Army Center for Health Promotion and Preventive Medicine; 2000, Technical Guide 254.

[4] ATSDR. Toxicological profile for phenol. Washington (DC): U.S. Department of Health and Human Services, Public Health Service; 2008.

[5] Lewis RJ, editor. Sax's dangerous properties of industrial chemicals. 8th ed. New York (NY): Van Nostrand Reinhold; 1992.

[6] Berman E, Schlicht M, Moser VC, MacPhail RC. A multidisciplinary approach to toxicity screening: I. Systemic toxicity. J Toxicol Environ Health 1995;45:127–43.

[7] CMA. Pharmacokinetics, metabolism, and distribution of ^{14}C-Phenol in fischer 344 Rats after gavage, drinking water, and inhalation exposure, with cover letter dated 07/13/94. Washington (DC): Chemical Manufacturers' Association; 1994, EPAOTS0557473.

[8] Moser VC, Cheek BM, MacPhail RC. A multidisciplinary approach to toxicological screening: III. Neurobehavioral toxicity. J Toxicol Environ Health 1995;45:173–210.

[9] CMA. Final report. A 13-week neurotoxicity study of phenol administered in the drinking water of the rat, with TSCA notice of receipt of test data (63 FR 67087) and cover letter dated 10/14/1998. Washington (DC): Chemical Manufacturers' Association; 1998, EPAOTS0559602.

[10] NCI. Bioassay of phenol for possible carcinogenicity. Washington (DC): National Institutes of Health; 1980, NCI-CG-TR-20/NTP No. 80-15/NIH Publ. No. 80-1759.

[11] RTI. Teratological evaluation of phenol (CAS No. 108-95-2) in CD rats. Research Triangle Park (NC): National Institute of Environmental Sciences; 1983, PB83247726.

[12] P&G. Range-finding maternal toxicity study with phenol in rats with cover letter 07/29/93. Cincinnati (OH): Procter & Gamble, Incorporated; 1993, EPAOTS0537777.

[13] P&G. Oral (Gavage) developmental toxicity study of phenol in rats with cover letter dated 07/29/97. Cincinnati (OH): Procter & Gamble, Incorporated; 1997, EPAOTS0573686.

[14] Narotsky MG, Kavlock RJ. A multidisciplinary approach to toxicological testing: II Developmental toxicity. J Toxicol Environ Health 1995;45:145–71.

[15] CMA. Final report: two-generation oral (Drinking Water) reproductive toxicity study of phenol in rats, with cover letter dated 07/21/1999. Washington (DC): Chemical Manufacturers' Association; 1999, FYI-OTS-0799-1356.

[16] Ryan BM, Selby R, Gingell R, Waechter JM, Butala JH, Dimond SS, et al. Two-generation reproduction study and immunotoxicity screen in rats dosed with phenol via the drinking water. Int J Toxicol 2001;20:121–42.

[17] RTI. Teratological evaluation of phenol (CAS No. 108-95-2) in CD-1 mice. Research Triangle Park (NC): National Institute of Environmental Sciences; 1983, PB85104461.

[18] CMA. Final report. Two-week (10-day) inhalation toxicity and recovery study of phenol vapor in the rat, with cover letter dated 1/16/1998. Washington (DC): Chemical Manufacturers' Association; 1998, EPAOTS0559328.

[19] Hoffman GM, Dunn BJ, Morris CR, Butala JH, Kimond SS, Gingell R, et al. Two-week (ten-day) inhalation toxicity and two-week recovery study of phenol vapor in the rat. Int J Toxicol 2001;20:45–52.

[20] Patrick E, Maibach HI, Burkhalter A. Mechanisms of chemically induced skin irritation. I. Studies of time course, dose response and components of inflammation in the laboratory mouse. Toxicol Appl Pharmacol 1985;81:476–90.
[21] Murray AR, Kisin E, Castranova V, Kommineni C, Gunther MR, Shvedova AA. Phenol-induced *in vivo* oxidative stress in skin: evidence for enhanced free radical generation, thiol oxidation, and antioxidant depletion. Chem Res Toxicol 2007;20:1769–77.
[22] Bernardini G, Spinelli O, Presutti C, Vismara C, Bolzacchini E, Orlandi M, et al. Evaluation of the developmental toxicity of the pesticide MCPA and its contaminants phenol and chlorocresol. Environ Toxicol Chem 1996;15:754–60.
[23] Dumpert K. Embryotoxic effects of environmental chemicals: tests with the South African clawed toad (*Xenopus laevis*). Ecotoxicol Environ Saf 1987;13:324–38.
[24] Birge WJ, Black JA, Kuehne RA. Effects of organic compounds on amphibian reproduction, vol. 121. Lexington: Water Resources Research Institute, University of Kentucky; 1980, 39. National Technical Information Service, Springfield, VA PB80-147523.
[25] Black JA, Birge WJ, McDonnell WE, Westerman AG, Ramey BA, Bruser DM. The aquatic toxicity of organic compounds to embryo-larval stages of fish and amphibians, vol. 133. Lexington: Water Resources Research Institute, University of Kentucky; 1982, 61. National Technical Information Service, Springfield, VA. PB82-224601.

CHAPTER 30

Wildlife Toxicity Assessment for Propylene

Lawrence R. Williams

Contents

INTRODUCTION

Propylene is a colorless gas and produced in vast quantities by the petrochemical industry throughout the developed world. As summarized by the International Agency for Research on Cancer (IARC), industrial output of propylene in the United States topped 10,248 thousand tons in 1992, and the demand is driven by its use as a chemical intermediate [1]. In 1994, propylene was ranked seventh among the top 50 domestically produced chemicals [2]. Propylene serves as a monomer for the production of polypropylene and as a substrate for important industrial chemicals including acrylonitrile, propylene oxide, propylene glycol, and isopropanol. Propylene is also used in the production of synthetic rubber and as a propellant or component in aerosols. Propylene is a compound produced as a by-product of petroleum refining and of ethylene production by steam-cracking of hydrocarbon feed stocks. The compound is released to the environment as a result of its widespread production and use. Propylene is also released in automobile exhausts

Wildlife Toxicity Assessments for Chemicals of Military Concern
http://dx.doi.org/10.1016/B978-0-12-800020-5.00030-2

and tobacco smoke. The compound is found in emissions from vegetation and combustion of organic matter.

This Wildlife Toxicity Assessment summarizes current knowledge of the toxicological impacts of propylene on wildlife. Evaluating the toxicity of the compound contributes to the derivation of Toxicity Reference Values (TRVs) that serve as screening-level benchmarks for wildlife inhabiting contaminated sites. The protocol for the performance of this assessment is documented in the U.S. Army Center for Health Promotion and Preventive Medicine (USACHPPM) Technical Guide (TG) 254, *Standard Practice for Wildlife Toxicity Reference Values* [3].

TOXICITY PROFILE

Environmental Fate and Transport

Propylene is an ever-present component of the atmosphere, with concentrations reaching 8.2 $\mu g/m^3$ at remote sites. However, concentrations can range from greater than 150 $\mu g/m^3$ in urban centers, largely because of emissions from automobile exhausts, chemical plants, and various other industrial facilities [4]. Propylene has been detected in marine and fresh waters in trace amounts: up to 9.3×10^{-7} ml/L in marine waters and 0.5 mg/L in fresh waters [4]. Since propylene is not lipophilic, bioaccumulation is unlikely. The IARC [5] reported that 92 companies worldwide, 24 of which are located in the United States, produce propylene. Volatilization of soil-borne propylene is likely, based on a high vapor pressure of 8.7×10^3 and a Henry's Law constant of 2.0×10^{-1} atm-m^3/mole at 25 °C) [4]. As listed in Table 30.1, these and other physical-chemical characteristics also favor volatilization of propylene from the surface of marine and freshwater bodies.

The Hazardous Substance Data Bank (HSDB) [4] discusses the terrestrial fate of propylene in the context of possible degradation by abiotic or biotic processes. For example, the 1,2-epoxide is a possible oxidative derivative of propylene although the extent of its abiotic formation in aqueous media is unclear. Aqueous concentrations of propylene are more likely to be degraded by micro-organisms.

A number of mechanisms have been suggested for how the chemical is degraded in the atmosphere. For example, the compound is expected to degrade rapidly in the atmosphere as it reacts with photochemically produced hydroxyl radicals. This process has a half-life of about 14.6 hours [4]. Nitrate radicals and ozone will also bring about propylene degradation, although at a slower rate.

Table 30.1 Summary of the Physical-Chemical Properties of Propylene

CAS No.	115-07-1
Molecular weight	42.08
Color	None
Physical state	Gas
Melting point	−185.2 °C
Boiling point	−47.4 °C
Odor	Slight
Solubility in water	446 mg/L at 20 − 25 °C: soluble in ethanol, diethyl ether
Partition coefficients:	
Log K_{ow}	1.77
Log K_{oc}	2.34 − 2.7
Vapor pressure at 25 °C	8.7×10^3 mm Hg
Henry's Law constant at 25 °C	2.0×10^{-1} atm.m^3/mole
Conversion factors	1 ppm = 1.72 mg/m^3 1 mg/m^3 = 0.58 ppm

Sources: IARC [1], C&EN [4], IARC [5].

Summary of Mammalian Toxicity

Mammalian Toxicity: Oral

Given the gaseous nature of propylene at ambient temperatures and pressures, there are no data on the toxicological effects of propylene when administered via the oral route.

Studies Relevant for Mammalian TRV Development for Ingestion Exposures

Not applicable.

Mammalian Inhalation Toxicity: Acute

A short-term (minutes) acute toxic effect to cats was reported in HSDB [4], where cats were exposed to propylene/air mixtures of up to 800,000 ppm (80%). No effects were observed at concentrations up to 31%, where there was an increase in the severity of effects as concentrations increased to 80%. The pharmacological effects that were observed included decreased blood pressure, rapid pulse, and the onset of an unusual ventricular ectopic beat. Equivalent concentrations induce deep anesthesia in rats. Due to the apparent tolerance to high concentrations of propylene in animals, a reliable compound-specific median lethal concentration (LC_{50}) is unavailable.

Although there are few if any toxicological consequences of exposure to propylene via inhalation, the ability of the compound or its metabolites to react with biological macromolecules has been demonstrated. For example, Eide *et al.* [6] exposed male Sprague Dawley rats to propylene (one of a range of alkenes under investigation) at 300 ppm for 12 hours/day on three separate days. Exposure took place in conically-shaped steel inhalation chambers with glass doors and walls. At termination, aliquots of blood and pieces of lung, brain, liver, kidney, and peripheral fat were measured for propylene, and the formation of deoxyribonucleic acid (DNA) adducts in liver and lymphocytes was monitored using the ^{32}P-postlabeling technique. The formation of alkene-hemoglobin adducts was detected using gas chromatography/mass spectrometry.

N-(2-hydroxyalkyl) valine adducts of hemoglobin and 7-alkylguanine adducts of DNA were detected consistently in these experiments, including *N*-(2-hydroxypropyl) valine and 7-propylguanine when propylene was used as the test compound. Earlier, Svensson and Osterman-Golkar [7] had investigated the formation of hydroxyalkyl derivatives of hemoglobin and DNA when male CBA mice were exposed acutely to either ^{14}C-labeled propylene or ^{14}C-labeled ethylene. ^{14}C-labeled propylene was dissolved in a buffered saline and animals were exposed in an unspecified chamber. Exposure to ^{14}C-labeled ethylene took place in an 11-liter glass chamber. They used the structural homology and the similarity in their reaction products of the two substrates to infer that the reactive intermediate of propylene was propylene oxide. Propylene oxide was subsequently detected in the blood of male F-344 rats exposed in groups of 8 to 6 or 600 ppm propylene for either 80 or 480 minutes [8]. Svensson *et al.* [9] in an acute toxicity study, administered ^{14}C-labeled propylene to male CBA mice, confirming that propylene oxide was the primary metabolic product of propylene. However, propylene oxide was six to 10 times less effective than ethylene oxide in forming hydroxyalkyl adducts.

Propylene was one of 19 hydrocarbon vapors that were administered to male F-344 N rats via inhalation (nose only) to quantify the rate of uptake [10]. Animals were exposed for 80 minutes on five separate days to an escalating concentration rate as follows: day 1, 1 ppm; day 2, 5 ppm; day 3, 20 ppm; day 4,000 ppm; and day 5,500 ppm. Uptake was determined by linking the vapor concentration and applied flow rate to the amounts of inhaled and exhaled propylene measured in real time, then normalized to the body weight of the animal, and the vapor concentration (i.e., the normalized uptake). Values between 1.3 and 1.9 nmoles/kg/min/ppm were obtained for propylene.

Mammalian Inhalation Toxicity: Subacute

The National Toxicology Program (NTP) sponsored a full-scale series of toxicological investigations on propylene in which a 14-day experiment served as a range finder for subsequent longer duration studies [11]. The subacute portion of the investigation involved exposing five F344 N rats and B6C3F1 mice/sex/group to 0, 625, 1,250, 2,500, 5,000, or 10,000 ppm propylene, 6 hours/day, and 5 days/week for 14 days. Animals were exposed to the compound in unspecified chambers. Animals were examined daily for mortality and clinical signs, and body weights were monitored at the start and end of the experiment. All animals were euthanized and necropsied, with examination of a full suite of organs and tissues. No compound-related differences were observed in survival, growth, behavior, histology, or morphology between controls and any of the treated groups.

Chronic exposure to propylene does not result in any increased incidence of tumors, yet does increase N7-hydroxypropylguanine (N7-HPGua) adducts in tissue DNA [12]. To investigate any potential for genotoxicity (mutagenicity or clastogenicity), male F344 rats were exposed via inhalation to up to 10,000 ppm propylene for 1, 3, or 20 days (6 hours/day, 5 days/week). The endpoints examined included gene (hypoxanthine-guanine phosphoribosyl transferase, or Hprt, splenocytes) and chromosomal (bone marrow micronucleus [MN]) mutations, hemoglobin (hydroxypropylvaline, HPVal) adducts in systemic blood, and DNA adducts (N7-HPGua) in several tissues. Similarly, exposed female and male F344 rats, implanted with bromodeoxyuridine (BrdU) minipumps, were evaluated for nasal effects (irritation via histopathology and cell proliferation via BrdU). Internal dose measures provided clear evidence for propylene exposure, with HPVal increased for all exposures; N7-HPGua was increased in all tissues from rats exposed for more than 1 day (except lymphocytes). Saturation of propylene conversion to propylene oxide was apparent from the adduct dose-response curves. There were no biologically significant genotoxic effects demonstrated at any exposure level, with no increase in Hprt mutant frequency or in bone marrow MN formation. In addition, no histopathological changes were noted in rodent nasal tissues or any induction of cell proliferation in nasal tissues. These results demonstrate that repeated exposure of rats to high concentrations of propylene (< or = 10,000 ppm) does not produce evidence of local nasal cavity toxicity or evidence of systemic genotoxicity to hematopoietic tissue, despite the formation of N7-HPGua adducts. In addition, these data indicate that formation of N7-HPGua does not correlate with any measure of genotoxic effect, neither mutagenic nor clastogenic [12].

Mammalian Inhalation Toxicity: Subchronic

The subchronic portion of the NTP's investigation of propylene featured the administration of 0, 625, 1,250, 2,500, 5,000, or 10,000 ppm propylene to 10 F344 N rats and B6C3F1 mice, 6 hours/day, 5 days/week for 14 weeks [11]. Animals were exposed to the compound in stainless steel and glass chambers. Animals were observed daily for mortality and clinical signs. Body weights were monitored at the start of the experiment and at weekly intervals. Necropsies were carried out on all animals at the end of the experiment, with samples of a full range of tissues and organs from animals that died before the end of the study, all controls, and all members of high dose groups examined histologically. Both test species showed no significant differences in survival, body weight, or behavior, and no compound-related effects in pathology or histopathology were observed.

Mammalian Inhalation Toxicity: Chronic

The chronic portion of the NTP study incorporated the exposure of 50 F344 N rats and 50 B6C3F1 mice/sex/group to 0, 5,000, or 10,000 ppm propylene for 24 months [11,13]. All subjects were monitored regularly for survival, clinical signs, and body weight, and a complete necropsy and histopathological examination were carried out on all subjects. In rats, there were no compound-related effects on survival, clinical signs, or body weights, no changes in gross morphology in any tissues or organs, and no obvious compound-related histopathological findings. There were some instances of inflammation of the lining of the nasal cavity. Males and females at 5,000 ppm showed increased incidence of squamous metaplasia although this was seen only in females at 1,000 ppm. Females also showed significantly higher epithelial hyperplasia at 10,000 ppm while males had significantly more nasal inflammation at 5,000 ppm. Together these results indicate that some effects on the nasal passage are probable but the lack of correlation to dose suggests a tenuous relationship.

Although not statistically significant, high-dose male mice displayed a slight but consistently smaller weight gain than did controls. Survival was near identical to controls if not better in some treatment groups. Treated mice displayed isolated and sporadic instances of some nonneoplastic lesions, including focal inflammation of the kidney. However, the incidence of these lesions, while greater in test animals than controls (0/50 in males, 1/50 in females), was essentially sporadic and not obviously related to exposure concentration (17/49 in low-dose males, 9/49 in high-dose males; 7/49 in low-dose females, 6/49 in high-dose females). The incidence of other nonneoplastic and neoplastic lesions among the groups showed similar low incidences and sporadic distributions with, in some cases, an apparent negative trend with increasing concentration. These data suggest an overall benign nature of the toxicological effects of propylene in the animals tested and do not allow the derivation of a plausible no observed adverse effect level (NOAEL) relevant to TRV development.

Ciliberti *et al.* [14] exposed 100 to 120 Sprague Dawley rats/sex/group and Swiss mice/sex/group to 0, 200, 1,000, or 5,000 ppm propylene, 7 hours/day, 5 days/week for 78 weeks, and published data that are consistent with the negative findings of the NTP study on propylene. Exposure took place in a stainless steel chamber with glass [14]. Clinical signs were examined three times a week, body weights were monitored every two weeks, and all animals were subjected to a gross examination at the time of weighing. All animals were maintained until they died naturally, at which point they were necropsied and samples of organs and tissues were examined histopathologically. The reported results of this investigation focused on the formation of neoplasms, the incidences of which appeared to be similar to controls at all exposure levels. No findings relating to the formation of non neoplastic lesions were reported.

Studies Relevant for Mammalian TRV Development for Inhalation Exposures

The capacity of propylene to induce anesthesia, with full recovery of faculties on cessation of exposure, is consistent with the negative toxicological results outlined earlier in the sections describing mammalian acute, subacute, subchronic, and chronic inhalation toxicity. Furthermore, the results suggest a relatively benign nature of propylene in biological systems and argue against the likelihood that a viable TRV for propylene can be derived from the available toxicological information. However, the formation of 2-hydroxypropyl derivatives of hemoglobin and guanine indicates that propylene has biochemical activity, most likely through its propylene oxide intermediate.

In contrast to ethylene, there were no published studies on the kinetics of propylene oxide formation from propylene. However, it is reasonable to speculate that, similar to ethylene oxide formation from ethylene [15], the rate of formation of propylene oxide from propylene may saturate below the toxicity threshold for the former substance. This would explain the benign effects of propylene when compared to the more severe effects of its metabolic product.

Mammalian Inhalation Toxicity: Other

Propylene has been included in two surveys on the genotoxicity of environmentally important compounds with negative findings. Ambiguous results were obtained from the compound in a modified version of the L5178Y mouse lymphoma forward mutation assay [16]. Additionally, the compound was negative for the induction of sex-linked recessive lethal mutations in *Drosophila melanogaster* [17].

Mammalian Dermal Toxicity

No data are available.

Summary of Avian Toxicology

No toxicological data for the effects of propylene on avian species were located. Ecotoxicological research on the effects of this compound on birds is recommended.

Amphibian Toxicology

No toxicological data for the effects of propylene on amphibian species were located. Ecotoxicological research on the effects of this compound on amphibians is recommended.

Reptilian Toxicology

No toxicological data for the effects of propylene on reptiles were located. Ecotoxicological research on the effects of this compound on reptiles is recommended.

RECOMMENDED TOXICITY REFERENCE VALUES FOR MAMMALS

TRVs for Ingestion Exposures for the Class Mammalia

At this time, it is not possible to derive a TRV for the oral route of exposure for propylene. Due to the gaseous nature of the compound, it cannot be administered orally.

TRVs for Inhalation Exposures for the Class Mammalia

The data that has been obtained from the inhalation studies presented in this document indicated that propylene is not toxic at the levels tested. It was reported in NTP [11] that even at concentrations of 10,000 ppm inhalation for 2 years, there was little toxicological consequence. Further need for studies characterizing the toxicity of propylene is evident given that propylene is present in the environment in large quantities.

IMPORTANT RESEARCH NEEDS

The limited availability of data on the toxicity of propylene to wildlife species precludes the development of a TRV. Hence, more studies of the compound and its derivatives are recommended. In particular, chronic toxicity studies on nonmammalian wildlife such as birds, reptiles, and amphibians are particularly warranted. The main challenge appears to lie in finding levels of propylene that result in toxicity to study organisms.

REFERENCES

[1] International Agency for Research on Cancer (IARC). IARC monographs on the evaluation of carcinogenic risks to humans. Vol. 60: some industrial chemicals. In: International Agency for Research on Cancer, editor. Propylene. Geneva, Switzerland: World Health Organization; 1994. p. 161–80.

[2] Chemical & Engineering News (C&EN). Production by the U.S. chemical industry [Heylin M, editor.]. Washington (DC): American Chemical Society; 1995, p. 38–44.

[3] U.S. Army Center for Health Promotion and Preventive Medicine (USACHPPM). Standard practice for wildlife toxicity reference values. Aberdeen Proving Ground, MD: U.S. Army Center for Health Promotion and Preventive Medicine; 2000, Technical Guide No. 254.

[4] Hazardous Substance Data Base (HSDB). Propylene. [Internet]. Bethesda (MD): National Library of Medicine (US), National Institute of Health: [cited 2013]. Available from: http://toxnet.njm.nih.gov/.

[5] International Agency for Research on Cancer (IARC). IARC monographs on the evaluation of carcinogenic risks to humans. In: International Agency for Research on Cancer. Ethylene and polyethylene. Geneva, Switzerland: World Health Organization; 1979. p. 157–86.

[6] Eide I, Hagemann R, Zahlsen K, Tareke E, Tornqvist M, Kumar R, et al. Uptake, distribution, and formation of hemoglobin and DNA adducts after inhalation of C2-C8 1-alkenes (olefins) in the rat. Carcinogenesis 1995;16(7):1603–9.

[7] Svensson K, Osterman-Golkar S. Kinetics of metabolism of propene and covalent binding to macromolecules in the mouse. Toxicol Appl Pharmacol 1984;73(3):363–72.

[8] Maples KR, Dahl AR. Blood levels of propylene oxide during propylene inhalation and effect on hepatic and nasal cytochrome P-450 concentrations. Drug Metab Dispos 1991;19(4):835–7.

[9] Svensson K, Olofsson K, Osterman-Golkar S. Alkylation of DNA and hemoglobin in the mouse following exposure to propene and propylene oxide. Chem Biol Interact 1991;78(1):55–66.

[10] Dahl AR, Damon EG, Mauderly JL, Rothenberg SJ, Seiler FA, McClellan RO. Uptake of 19 hydrocarbon vapors inhaled by F344 rats. Fundam Appl Toxicol 1988;10(2):262–9.

[11] National Toxicology Program (NTP). Toxicology and carcinogenesis studies of propylene in F344/N rats and B6C3F1 mice (inhalation studies). Technical Report No. 272. 1997/04/01 ed, National Toxicology Program; 1985.
[12] Pottenger LH, Malley LA, Bogdanffy MS, Donner EM, Upton PB, Li Y, et al. Evaluation of effects from repeated inhalation exposure of F344 rats to high concentrations of propylene. Toxicol Sci 2007;97(2):336–47.
[13] Quest JA, Tomaszewski JE, Haseman JK, Boorman GA, Douglas JF, Clarke WJ. Two-year inhalation toxicity study of propylene in F344/N rats and B6C3F1 mice. Toxicol Appl Pharmacol 1984;76(2):288–95.
[14] Ciliberti A, Maltoni C, Perino G. Long-term carcinogenicity bioassays on propylene administered by inhalation to Sprague-Dawley rats and Swiss mice. Ann N Y Acad Sci 1988;534:235–45.
[15] Bolt HM. The carcinogenic risk of ethene (ethylene). Toxicol Pathol 1998;26 (3):454–6.
[16] McGregor D, Brown AG, Cattanach P, Edwards I, McBride D, Riach C, et al. Responses of the L5178Y mouse lymphoma forward mutation assay: V.Gases and vapors. Environ Mol Mutagen 1991;17(2):122–9.
[17] Foureman P, Mason JM, Valencia R, Zimmering S. Chemical mutagenesis testing in Drosophila. X. Results of 70 coded chemicals tested for the National Toxicology Program. Environ Mol Mutagen 1994;23(3):208–27.

CHAPTER 31

Wildlife Toxicity Assessment for Phenanthrene

Sang Ho Lee, Wilfred C. McCain

INTRODUCTION

Phenanthrene (CAS No. 85-01-8) is one of at least 200 structurally related polycyclic aromatic hydrocarbons (PAHs) that are produced by incomplete combustion of organic matter [1]. The compound has been detected in tobacco smoke, automobile exhausts, barbecued meat, garbage incineration, and wood burning. Along with other PAHs, phenanthrene has also been detected in used motor oils, crude oils, and lubricating fluids. According to the International Agency for Research on Cancer (IARC) [2] and the Agency for Toxic Substances and Disease Registry (ATSDR) [3], phenanthrene is found in ambient air, surface and drinking water, and in some commercial food products. Wildlife could be exposed to phenanthrene in air and water if the concentration is sufficient. Phenanthrene is not commercially produced in the United States, but small amounts of it are imported to manufacture explosives, pesticides, drugs, certain dyes, phenanthrene-quinone, and for use in research [3].

This chapter summarizes the current knowledge of the toxicological impacts of phenanthrene on wildlife. Evaluating the toxicity of phenanthrene is intended to contribute to the derivation of toxicity reference values (TRVs) that could serve as screening-level benchmarks for wildlife near

Wildlife Toxicity Assessments for Chemicals of Military Concern
http://dx.doi.org/10.1016/B978-0-12-800020-5.00031-4

contaminated sites. The protocol for the development of this assessment is documented in the U.S. Army Center for Health Promotion and Preventive Medicine (USACHPPM) Technical Guide (TG) 254, *Standard Practice for Wildlife Toxicity Reference Values* [4].

Toxicity reference values (TRVs) for phenanthrene cannot be derived at this time. This is mainly due to a paucity of relevant data. This chapter should nevertheless contribute to the overall risk management process by helping to identify the most critical research needs.

TOXICITY PROFILE

Environmental Fate and Transport

Phenanthrene formed during burning of organic matter can pose an immediate inhalation and dermal exposure threats to wildlife. It can also result in delayed exposure once the emitted phenanthrene enters the food chain following soil and surface deposition. Natural sources of such emissions include forest fires and volcanoes. A more prevalent source of phenanthrene is anthropogenic and includes emissions from power stations and steel mills, the burning of garbage and wood, and the production of coal tar, coke, and asphalt [3]. Particulates and vapors from diesel emissions contain an abundance of three-ringed aromatic organic chemical compounds like phenanthrene. Overall, these chemical combustion by-products and other PAHs are widely distributed in environmental media, largely in the parts per billion (ppb)-concentration range.

Transport and partitioning of PAHs in the environment depend on the physical-chemical characteristics of each compound (Table 31.1). Phenanthrene is insoluble in water with an octanol-water partition coefficient (log K_{oc}) of 4.35, which suggests that most of the compound released to the environment will be bound to soil and organic sediments. With sediment aging, and increased contact time, phenanthrene becomes more tightly bound [5]. Other PAHs of relatively low molecular weight are expected to exhibit a similar distribution pattern in the environment with an octanol-water partition coefficient range of 3–4. Thus, the availability of phenanthrene for aquatic organism exposure will depend greatly on the equilibrium distribution of PAHs between sediment and water. Because of their low aqueous solubility, water-borne PAHs tend to adhere to particles that settle to the bottom or possibly remain suspended in the water column. Nonetheless, to some extent according to the data published by the ATSDR [3] lower molecular weight PAHs (e.g., phenanthrene) volatilize to the atmosphere at the surface of a contaminated body of water.

Table 31.1 Summary of the Physical-Chemical Properties of Phenanthrene

CAS No.	85-01-8
Molecular weight	178.2
Color	Colorless
Physical state	Plates, crystals, leaflets
Melting point	100 °C
Boiling point	340 °C
Odor	Faint aromatic
Solubility in water	1.2–1.3 mg/L at 25 °C: soluble in benzene, ethanol, ether, toluene, carbon tetrachloride, and carbon disulfide
Partition coefficients:	
Log K_{ow}	4.45–4.57
Log K_{oc}	4.15–4.36
Vapor pressure at 25 °C	6.8 10^{-4} mm Hg
Henry's Law constant at 25 °C	2.56 10^{-5} atm.m^3/mole
Conversion factors	1 ppm = 7.3 mg/m^3 1 mg/m^3 = 0.137 ppm

Sources: HSDB [1], ATSDR [3].

It is assumed that phenanthrene shares the capacity of other PAHs to bioaccumulate in terrestrial and aquatic food chains. This may result in greater exposure for organisms higher in the food chain. However, those organisms possessing the aryl hydrocarbon hydroxylase (AHH) enzyme systems should be able to metabolize the compound, thereby preventing bioaccumulation, thus species lacking this system might receive greater exposure and contribute to bioaccumulation. The compound may also undergo photolytic and oxidation reactions in the environment, with the reported formation of an arene oxide in the latter case [3]. By contrast, the major photoproduct of phenanthrene is 9, 10-phenanthrenequinone. Aerobic biodegradation of PAHs is likely to be a significant breakdown mechanism for compounds such as phenanthrene, since different species of bacteria, algae, and fungi possess the necessary molecular architecture (the AHH enzyme system) to catabolize the compound [3].

Summary of Mammalian Toxicity

Mammalian Toxicity: Oral

Secondary sources have pointed to the overall absence of toxicological information on phenanthrene [5,6]. For example, the U.S. Environmental Protection Agency's (U.S. EPA) Integrated Risk Information System (IRIS)

record for phenanthrene reports no data on which to derive a chronic oral reference dose (RfD) or inhalation reference concentration [6]. This record also reports inadequate data from which to derive a carcinogenic slope factor for the compound [6]. Similarly, the IARC concluded that the data were inadequate to permit an evaluation of the carcinogenicity of phenanthrene to experimental animals [2]. In general, the literature searches supporting an evaluation of the toxic potential of phenanthrene in wildlife have confirmed earlier conclusions of the lack of experimental evidence.

Mammalian Oral Toxicity: Acute

Simmon *et al.* [7] reported a median lethal dose (LD_{50}) for phenanthrene of 700 mg/kg in mice (strain unstated) when the compound was administered intraperitoneally. Yoshikawa *et al.* [8] found that single intraperitoneal injections of 150 mg/kg phenanthrene produced some slight hepatotoxicity and effects on blood chemistry in rats. Though not applicable to derivation of a TRV, these studies still provide toxicity information bearing on the potential effects to wildlife.

Mammalian Oral Toxicity: Subacute

No data are available.

Mammalian Oral Toxicity: Subchronic

No data are available.

Mammalian Oral Toxicity: Chronic

No data are available.

Mammalian Oral Toxicity: Other

No relevant studies are available.

Studies Relevant for Mammalian TRV Development for Ingestion Exposures

There are no studies available in the scientific literature that provides data directly applicable to TRV development for wildlife exposed to phenanthrene via ingestion. In fact, available information on the oral toxicity of phenanthrene was insufficient to enable the U.S. EPA to develop a human health RfD for the compound [6]. Therefore, a TRV for phenanthrene cannot be derived at this time.

Mammalian Inhalation Toxicity

No data are available.

Mammalian Dermal Toxicity

The capacity of phenanthrene to induce tumors in the mouse skin-painting assay was summarized in several reports including those by the ATSDR and the U.S. EPA's IRIS [3,5,6]. In general, the compound appears to be inactive in these systems. For example, dermal application of 5% phenanthrene in solvent, 3 times weekly for 1 year, did not induce skin tumors in mice [5,9]. Similarly, in two-stage skin initiation-promotion assays, the overwhelming weight of evidence indicates an inability of phenanthrene to cause papilloma formation, irrespective of the experimental protocol employed [5].

Mammalian Toxicity: Other, Acute

No data are available.

Mammalian Toxicity: Other, Subacute

In one of the few published experiments to explore the toxicological effects of phenanthrene, the compound was administered to male Holtzman and Charles River rats (12 animals/group) for 7 days via intraperitoneal injection (0.35 mg/rat/day) [10]. Rats were euthanized 3 days later and their body and liver weights were measured and recorded and compared to those of untreated controls. No statistically significant differences in body and liver weights were found between test versus control animals indicating that phenanthrene had little toxic impact on the subjects at the administered dose. There is no additional published data to derive the No Observed Adverse Effect Level (NOAEL) value for phenanthrene. Furthermore, the use of a single dose level precludes the delineation of the compound's toxicological threshold.

Studies Relevant for Mammalian TRV Development

In the single study that was identified as potentially relevant to TRV development, the absence of any toxicological consequences of phenanthrene administration in rats at the only dose tested did not allow a dependable low observed adverse effect level (LOAEL) or NOAEL to be identified [10]. Physiological experiments suggest that phenanthrene's inactivity is probably not related to failure to penetrate the intestinal mucosa [11] where, at least in male Sprague Dawley rats, the compound appeared to be more or less quantitatively absorbed [11].

The use of other protocols has pointed to the largely benign nature of phenanthrene in experimental systems. Thus, the compound exhibited no mutagenic potential in the Ames test, even in circumstances where the reactive metabolite, phenanthrene 9, 10 oxide, itself brings about gene reversion in *Salmonella typhimurium* strains TA98, TA100, TA1537, and TA1538 [12]. However, when Bucker *et al.* [12] included the epoxide hydratase inhibitor, 1,1,1-trichloropropene 2,3-oxide in the culture medium, they observed the formation of histidine-independent colonies, which suggests that phenanthrene is normally inactive in this system because its active metabolite normally cannot accumulate. Studies that are more recent have shown phenanthrene to be weakly positive in TA100, but ineffective in TA98 [13].

In the battery of experiments aimed at researching the ability of PAHs to induce or initiate skin tumors in mice, phenanthrene has been found to be inactive under most experimental conditions and does not behave like a complete carcinogen [3,5]. Results from these studies indicate that phenanthrene is relatively benign. Given the lack of noticeable toxic effects of the compound, a meaningful TRV cannot be derived. Phenanthrene may be regarded as a part of the overall issue of the toxic effects of multiple PAHs on wildlife.

Summary of Avian Toxicology

Toxicological data for the effects of phenanthrene on avian species was not located. Ecotoxicological research on the effects of this compound in birds is recommended.

Amphibian Toxicology

Toxicological data on the effects of phenanthrene on amphibian species were not located. Ecotoxicological research on the effects of this compound on amphibians is recommended.

Reptilian Toxicology

Toxicological data on the effects of phenanthrene on reptilian species were not located. Ecotoxicological research on the effects of this compound on reptiles is recommended.

RECOMMENDED TOXICITY REFERENCE VALUES

Toxicity Reference Values for Mammals

TRVs for Ingestion Exposures for the Class Mammalia

At this time, it is not possible to derive a TRV for the oral route of exposure for phenanthrene. No studies have shown toxicological effects of phenanthrene ingestion.

TRVs for Inhalation Exposures for the Class Mammalia

No data are available at this time.

TRVs for Dermal Exposures for the Class Mammalia

There is limited data suggesting that phenanthrene will not induce tumors when applied dermally [7]. Given these results and the lack of other suitable studies, it is not possible to derive a TRV for dermal exposure at this time.

Toxicity Reference Values for Amphibians

No data are available at this time.

Toxicity Reference Values for Reptiles

No data are available at this time.

IMPORTANT RESEARCH NEEDS

A recent study showed that exposure to environmental concentrations of phenanthrene in a Zebrafish model caused morphological changes and other developmental effects in the Zebrafish embryo [14]. This fact in addition to the limited amount of data available on the toxicity of phenanthrene reflects the need for more research on oral, dermal, and inhalation routes of exposure. In particular, chronic studies on mammals are necessary considering the limited data now available. Adequate characterization of the toxicity of phenanthrene is required before a reliable TRV can be derived. Wildlife toxicity assessments of compounds, having scant information on which to base a TRV, still form a part of the overall risk management process, not only by identifying research needs but also by helping to prioritize the allocation of finite resources in research.

REFERENCES

[1] Hazardous Substance Data Bank (HSDB). Propylene [Internet]. Bethesda (MD): National Library of Medicine (US), National Institutes of Health [cited 2001]. Available from: http://toxnet.njm.nih.gov/.

[2] IARC, IARC monographs on the evaluation of the carcinogenic risk of chemicals to humans: phenanthrene, vol. 32, World Health Organization, Geneva (Switzerland), 1983, p. 419–430.

[3] ATSDR, Toxicological profile for otto fuels II, U.S. Department of Health and Human Services, Atlanta (GA), 1995, Agency for Toxic Substances and Disease Registry (ATSDR). Toxicological profile for polycyclic aromatic hydrocarbons (Update). Atlanta (GA): U.S. Department of Health and Human Services, Public Health Service.

[4] USACHPPM, Standard practice for wildlife toxicity reference values, U.S. Army Center for Health Promotion and Preventive Medicine, Aberdeen Proving Ground (MD), 2000, Technical Guide No. 254.

[5] P.F. Landrum, B.J. Eadie, W.R. Faust, Variation in the bioavailability of polycyclic aromatic hydrocarbons to the amphipod Diporeia spp, Environ. Toxicol. Chem. 11 (1992) 1197–1208.

[6] U.S. EPA. Integrated risk information system. Phenanthrene [Internet]. Cincinnati (OH): USEPA Office of Health and Environmental Assessment (US), National Center for Environmental Assessment [cited 2001]. Available from: http://www.epa.gov/IRIS/.

[7] V.F. Simmon, H.S. Rosenkranz, E. Zeiger, L.A. Poirier, Mutagenic activity of chemical carcinogens and related compounds in the intraperitoneal host-mediated assay, J Natl Cancer Inst 62 (4) (1979) 911–918.

[8] T. Yoshikawa, L.P. Ruhr, W. Flory, D. Giamalva, D.F. Church, W.A. Pryor, Toxicity of polycyclic aromatic hydrocarbons. I. Effect of phenanthrene, pyrene, and their ozonized products on blood chemistry in rats, Toxicol Appl Pharmacol 79 (2) (1985) 218–226.

[9] F.J.C. Roe, G.A. Grant, Tests of pyrene and phenanthrene for incomplete carcinogenic and anticarcinogenic activity, Br Emp Cancer Campaign 41 (1964) 59–69 (as cited in U.S. EPA 2001).

[10] L.L. Gershbein, Liver regeneration as influenced by the structure of aromatic and heterocyclic compounds, Res Commun Chem Pathol Pharmacol 11 (3) (1975) 445–466.

[11] A. Rahman, J.A. Barrowman, A. Rahimtula, The influence of bile on the bioavailability of polynuclear aromatic hydrocarbons from the rat intestine, Can J Physiol Pharmacol 64 (9) (1986) 1214–1218.

[12] M. Bucker, H.R. Glatt, K.L. Platt, D. Avnir, Y. Ittah, J. Blum, et al., Mutagenicity of phenanthrene K-region derivatives, Mutat Res 66 (4) (1979) 337–348.

[13] R.P. Bos, J.L. Theuws, F.J. Jongeneelen, P.T. Henderson, Mutagenicity of bi-, tri- and tetra-cyclic aromatic hydrocarbons in the "taped-plate assay" and in the conventional salmonella mutagenicity assay, Mutat Res 204 (2) (1988) 203–206.

[14] L. Huang, C. Wang, Y. Zhang, M. Wu, Z. Zuo, Phenanthrene causes oculaar developmental toxicity in zebrafish embryos and the possible mechanisms involved, J. Hazard Mater. 261 (2013) 172–180.

CHAPTER 32

Wildlife Toxicity Assessment for Pyrene

Joseph P. Sullivan

Contents

INTRODUCTION

Pyrene (CAS No. 129-00-00) is a polycyclic aromatic hydrocarbon (PAH) consisting of four fused benzene rings. Pyrene is formed from incomplete combustion of organic chemicals. This Wildlife Toxicity Assessment summarizes current knowledge of the toxicological impacts of pyrene on wildlife. Evaluating the toxicity of pyrene will contribute to the derivation of Toxicity Reference Values (TRVs) for use as screening-level benchmarks for wildlife near contaminated sites. Pyrene is not produced commercially in the United States. The U.S. Army Center for Health Promotion and Preventive Medicine (now the U.S. Army Public Health Command) Technical Guide 254, *Standard Practice for Wildlife Toxicity Reference Values* [1] documents the protocol for the performance of this assessment.

Wildlife Toxicity Assessments for Chemicals of Military Concern
http://dx.doi.org/10.1016/B978-0-12-800020-5.00032-6

The production and processing of PAHs probably result in the direct release of only small amounts of PAHs. Pyrene is, however, an intermediate in the production of pigments. The largest emissions of PAHs result from incomplete combustion of organic materials during industrial processes and other human activities, including processing of coal, crude oil, and natural gas; coking, coal conversion, petroleum refining; production of carbon blacks, creosote, coal-tar, and bitumen; aluminum, iron, and steel production in plants and foundries; heating in power plants and residences; cooking; combustion of refuse; motor vehicle traffic; and environmental tobacco smoke. The emissions from wood stoves are 25 to 1,000 times higher as compared to those derived from charcoal-fired stoves. Additionally, in areas where wood burning predominates for domestic heating, the major portion of airborne PAHs derives from this source, especially in winter. Motor vehicle traffic is another source of pyrene. In countries with large areas of forest, fires can make an important contribution to PAH emissions. Combustion at coal-fired power plants and refuse incineration are additional sources [2]. The physical and chemical properties of pyrene are provided in Table 32.1 and Figure 32.1.

Table 32.1 Summary of Physical-Chemical Properties of Pyrene

CAS No.	129-00-00: $C_{16}H_{10}$
Molecular weight	202.26
Color	Colorless
Physical state	Solid
Melting point	151.2 °C
Boiling point	404 °C
Odor	None stated
Solubility in water	0.135 mg/L at 25 °C:
Solubility in other solvents	Soluble in carbon disulfide, toluene, alcohol, ether, benzene, and petroleum ether
Partition coefficients:	
Log K_{ow}	4.88
Log K_{oc}	4.8
Vapor pressure at 25 °C	8.92×10^{-5} mm Hg
Henry's Law constant at 25 °C	1.2×10^{-5} atm-m^3/mole
Vapor density	Not available
Conversion factors	Not available

Source: HSDB [3].

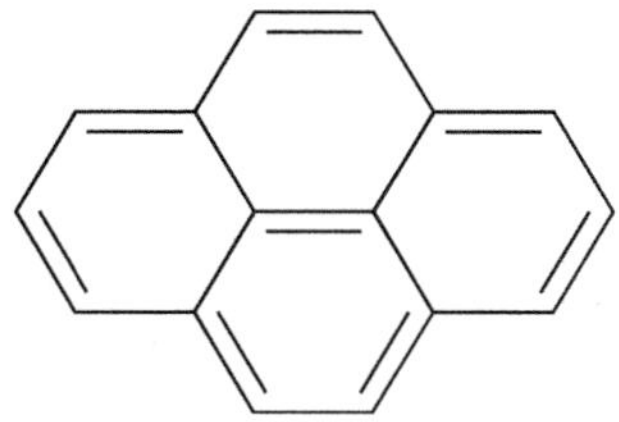

Figure 32.1 Chemical structure of Pyrene.

TOXICITY PROFILE

Environmental Fate and Transport

Pyrene is released into the environment as the result of incomplete combustion in motor vehicle exhaust; emissions from cigarette smoke; and smoke from coal-, oil-, and wood-burning stoves and furnaces. Both its production and its use as a starting material in the manufacture of dyes and optical brighteners may also account for its presence in the environment. Based upon a measured vapor pressure of 8.9×10^{-5} millimeters of Mercury (mm Hg) at 25 °C, pyrene will exist in both the vapor and particulate-phase in the ambient atmosphere [3]. Table 32.1 summarizes pyrene's additional physical-chemical properties.

Aerobic microbial action is the primary degradation route of pyrene in soil. The production of surfactants by *Pseudomonas fluorescens* 29L bacteria enhances the biodegradation of pyrene [4]. Yuan *et al.* [5] found that the degradation half-life in a sandy clay loam ranged from 2.9 to 9.9 days, which depended on the amount of compost added. Small amounts of compost were beneficial, but a 1:1 ratio slowed degradation. In sediments, the half-life ranged from 3.0 to 7.3 days when incubated at 30 °C [6]. The bacteria *Sphingomonas* sp. degraded 90% of the pyrene present in a mineral salt medium after 48 hours in the presence of water-extractable organic matter from manure but degraded only 40% of the pyrene after 48 hours in the absence of the water-extractable organic matter [7]. Anaerobic degradation is much slower, with half-lives of 77 to 321 days depending on the source of sediments tested [8].

When used at a dose of 200 micrograms per square centimeter (μg/cm^2), pyrene penetrated the skin of rats rapidly but was absorbed into the peripheral blood system more slowly. Excretion of pyrene continued after a dermal dose for a longer period of time than after an intravenous dose, suggesting that the pyrene remained in the skin and continued to be slowly absorbed from skin [9]. In another study that applied pyrene to the skin at doses that ranged from 2.0 to 15.0 milligrams per kilogram (mg/kg), the remaining pyrene on the skin was removed after 24 hours. Urinary excretion continued through day 3 in the low dose and through day 5 in the high dose. Excretion via the feces continued through day 4 in the low dose and through day 5 in the high dose. A lower proportion of the pyrene was excreted in urine at the high dose than at the low dose; conversely, a higher proportion of the pyrene was excreted via feces at the high dose. Between 70 and 80% of the applied dose was absorbed after 24 hours [10].

Following oral dosing, excretion was almost complete after 3 days. The total recovery of pyrene plus 1-hydroxypyrene found in urine and feces after oral administration of pyrene was 53.4% of the dose administered (7.2% pyrene and 46.2% 1-hydroxypyrene). After intraperitoneal injection, the total recovery was 3.1% of the administered dose (0.8% pyrene and 2.3% hydroxypyrene), which was significantly lower than after oral application. Finally, following intrapulmonary application, the total excretion of pyrene and hydroxypyrene in urine and feces was less than 0.004% of the amount administered, that is, 0.0009% as pyrene and 0.0024% as l-hydroxypyrene [11].

Rats given three daily oral doses of 20 mg/kg had blood concentrations reduced to approximately 1 microgram per milliliter (μg/mL) 6 hours after the final oral dose. On the fourth day, each group received an intravenous dose of 20 mg/kg. After intravenous administration, concentrations in blood, liver, lung, kidney, and brain had all diminished to the limits of detection by 3 or 4 hours post-administration. The biological half-life of pyrene in rats was 0.72 hours [12].

Laboratory rats exposed to contaminated soils containing 54 mg/kg pyrene showed pyrene in the lung (327 nanograms per gram (ng/g)) and the liver (82 ng/g) tissues [13]. The presence of pyrene in the lungs would not necessary indicate absorption, but its presence in the liver indicates that pyrene was absorbed and circulated throughout the rat's system. The route of exposure was not considered.

Pyrene was measured in blubber from stranded sperm whales (*Physeter macrocephalus*) in the southern North Sea during the winter of 1994–1995. Low but detectable levels of pyrene occurred in the whales, with up to 2 ng/g dry weight in the blubber [14].

Structure-Activity Relationship

The Log K_{ow} and PAH structure were related to toxic response to the immobilized recombinant bioluminescent *Escherichia coli* strain. Pyrene and benzo[a]pyrene are classified as pericondensed PAHs (PCPAHs), and naphthalene, anthracene, and phenanthrene are classified as catacondensed PAHs (CCPAHs). Only CCPAHs were found to cause toxicity and inhibit cellular metabolism, while PCPAHs did not affect any changes in bioluminescence responses. In addition, log K_{ow} values for CCPAHs were well correlated with EC_{50} values under the quantitative structure-activity relationship (QSAR) regression, and the PCPAHs with higher log K_{ow} values did not exhibit dose-related toxicity [15].

Summary of Mammalian Toxicity

Mammalian Oral Toxicity: Acute

Searches identified a single acute toxicity study using intraperitoneal injections in mice. Pyrene was not highly toxic, with a LD_{50} of 514 mg/kg [16]. Searches identified no oral dosing or short-term dietary studies.

Mammalian Oral Toxicity: Subchronic

In the only longer-term toxicity study, mice received between 75 and 250 mg/kg/day via oral gavage for 13 weeks. Mortality was not reported for any dose level. Multiple foci of renal tubular regeneration characterize nephropathy, a condition that is often accompanied by interstitial lymphocytic infiltrates and/or foci of interstitial fibrosis. At a dose of 250 mg/kg, pyrene increased the occurrence of these pathologies. Doses of 125 and 250 mg/kg reduced the relative and absolute kidney weights in mice. Based on nephropathy and decreased kidney weights, the low dose (i.e., 75 mg/kg body weight/day) was the no observed adverse effect level (NOAEL), and the 125 mg/kg body weight/day was the lowest observed adverse effect level (LOAEL) [17].

Reproductive Toxicity

Female mice aged 6 to 8 weeks received one intraperitoneal injection of 80 mg/kg pyrene. They experienced a slight but nonsignificant reduction in primordial oocytes [18].

Developmental Toxicity

Two 6-mg subcutaneous injections of pyrene to mice during pregnancy did not cause any tumors in the lungs or mammary glands of maternal mice or to offspring mice at 1 year post-administration [19]. In addition, two oral doses of pyrene at 50 mg/kg failed to induce benzo[a]pyrene hydroxylase in either the pregnant females or in the fetuses on days 17, 18, or 19 of pregnancy.

Benzo[a]pyrene hydroxylase is a mixture of mono-oxygenases that metabolize benzo[a]pyrene and similar PAH epoxides and/or hydroxylated products [20].

Mammalian Toxicity: Other

Effects on Enzymes

Following dermal exposure to 100 mg/kg, the activity of skin aryl hydrocarbon hydroxylase in neonatal rats increased by 139%, and liver aryl hydrocarbon hydroxylase increased by 42% over controls by 24 hours after treatment. Skin 7-ethoxycoumarin deethylase activity increased by 88%, and liver 7-ethoxycoumarin deethylase activity increased by 25% over controls 24 hours after treatment [21]. Intraperitoneal injections of up to 100 mg/kg in mice did not induce liver microsomal 7-ethoxyresorufin O-deethylation activity 72 hours following the injections [22].

Free-living Norway rats (*Rattus norvegicus*) were collected at a contaminated site near Lyon, France, where the soil pyrene concentration was 13.05 mg/kg. Additionally, a broad range of organics and metals, including zinc, copper, and lead contaminated the site. Hepatic monooxygenase activities were higher in both male contaminated rats (2-, 3- and 2-fold for ethoxy-, pentoxy-, and benzoxy-resorufin O-dealkylases, respectively) and female contaminated rats (3.5-, 8-, and 4.5-fold for ethoxy-, pentoxy-, and benzoxy-resorufin O-dealkylases, respectively) as compared to rats from an uncontaminated site. Superoxide dismutase was unaltered in contaminated rats. All organs of contaminated male and female animals (except male hepatic activity) showed a 1.1- to 2-fold increase in glutathione S-transferase with peroxidase activity [23]. Because many contaminants were present, it is not possible to determine the level of enzyme induction attributable to pyrene.

Mutagenicity

Dietary exposure to pyrene at 200 parts per million (ppm) for 8 weeks failed to produce any indications of mutagenicity in rats [24]. Dermal exposure of hairless mice to up to 5% (w/v) once daily for 3 days failed to produce an increase in the number of micronuclei in the skin at 24 hours after the last dose was applied [25]. He and Baker [26] also investigated micronuclei production in hairless mice following dermal exposure and found no micronuclei after 72 hours following a maximum dose of 2,500 μg/mouse. Pyrene at 10 or 20 μg/mL did not lead to the transformation of any Syrian hamster embryo cells.

Pyrene at 5, 10, or 20 μg/mL was ineffective at enhancing the transformation of Simian Adenovirus, SA7. Pyrene at 20 μg/mL did not induce deoxyribonucleic acid (DNA) repair in Syrian hamster embryo cells [27]. Pyrene was negative in both the hepatocyte Unscheduled DNA Synthesis and modified Ames tests [28]. Pyrene at 101.1 mg/L was negative for DNA repair and carcinogenicity in rat liver cells *in vitro*. At 20.2 mg/L, pyrene was negative in the Adult Rat Liver Epithelial Cell/Hypoxanthine-Guanine Phosphoribosyl Transferase Mutagenicity Assay, and at 202.3 mg/L, pyrene was negative in the Adult Rat Liver Epithelial Cell/Sister Chromatid Exchange Assay [29].

In the microbial Mutatox assay, pyrene was genotoxic at all concentrations (0.6 to 5 μg/tube) but showed no detectable activity in the Ames test [30]. In contrast, others who used the Ames test found that pyrene was genotoxic at a dose of 202.26 mg/L with metabolic activation [31]. Positive results occurred with *Salmonella typhimurium* strain TA97a, TA98, and TA100. Similarly, pyrene was mutagenic in *Salmonella* strains TA97, TA98, TA100, and TA1537 in the presence of S9 in the plate test at 0.4 to 5 μg/plate [32]. Pyrene produced a mutagenic response in *Salmonella* at 5 μg/mL without S-9 activation in the TA100 strain (indicating base pair alterations), but not in the TA98 strain (sensitive to frame-shift mutagens). Pyrene produced no mutagenic response in *Salmonella* at 5 μg/mL with S-9 activation in the TA100 strain or the TA 98 strain [33].

Planar PAHs, including pyrene, had a greater antagonistic effect on 1-nitropyrene-induced mutations than did linear PAHs (chrysene, anthracene, and naphthalene) [34]. Among PAHs, there is a tendency toward increasing activity with increasing numbers of fused rings [35].

Carcinogenicity

According to the World Health Organization (WHO), the carcinogenicity category of "questionable" was used to describe pyrene based on the results of studies in experimental animals suggesting pyrene is not carcinogenic. However, a few studies reviewed by the WHO suggested that the potential carcinogenicity of pyrene could not be ruled out [2]. Toxic equivalencies were assigned based on a review of the literature that focused mostly on carcinogenicity. Pyrene was assigned a toxicity equivalency factor (TEF) of 0.001 as compared to a TEF of 1.0 for benzo[a]pyrene since the former has little if any carcinogenic potency [36].

The chemical structural trait thought to influence the carcinogenicity of PAHs is the configuration of inherent centers of biochemical reactivity that determine the site of substitution of the methyl group in the bioalkylation reaction. It is predicted that naphthalene, pyrene, and perylene are carcinogenically inactive since these compounds possess four centers of chemical reactivity that are symmetrically equivalent [37].

The first study of pyrene carcinogenicity was conducted in 1937, wherein Walker carcinomas were transplanted into a rat model, and Rous fowl sarcomas were transplanted into chickens. Injections of 25 mg pyrene/rat on successive days did not impact the growth rate of the tumors in rats. The tumors in Chickens were larger after treatment with 30 mg pyrene/bird on three successive days, but this was an insignificant increase [38]. Dietary exposure to pyrene at 200 ppm for 8 weeks failed to produce any indications of carcinogenesis in rats [24].

The application of 25 or 50 μg pyrene to the skin of mice increased γ-glutamyl transpeptidase activity in mouse skin homogenate over controls, but the increase was insignificant. The transfer of glutamyl groups of peptides to other peptides and amino acids is catalyzed by γ-glutamyl transpeptidase and was proposed as a marker of neoplasia [39]. Failure to induce a significant increase in γ-glutamyl transpeptidase activity suggests low carcinogenic potential for pyrene. Dermal application to mice of 1 mg split into 10 sub-doses given every other day slightly increased the percentage of tumor-bearing mice, but the increase was insignificant [40].

Dermal exposure of 200 μg/mouse did not increase the number of melanocytes after 7 days [41]. Dermal application of pyrene twice weekly for 18 weeks with or without exposure to ultraviolet radiation four times weekly did not lead to an increase in tumor production [42]. Dermal application to mice of solutions containing 0.26% pyrene and 0.2% phenanthrene, in addition to 0.2% benzo[a]pyrene, did not increase the rate of tumor production, but the latency period decreased and a higher proportion of the tumors were malignant. Pyrene (0.1%) alone produced one benign tumor, and the addition of 0.001% benzo[a]pyrene resulted in no tumors [43].

Three intraperitoneal injections of up to 1,750 μg/animal in newborn mice over a 2-week period failed to increase the number of either lung adenomas or adenocarcinomas after 6 months [44]. In a related study, intraperitoneal injections of up to 200 mg/kg in mice increased the number of mice with lung adenomas and slightly increased the number of lung adenomas per mouse 8 months after dosing; however, the conclusions from the study state that pyrene was noncarcinogenic [45].

Immunotoxicity

Captive-reared male deer mice (*Peromyscus maniculatus*) received six intraperitoneal injections every other day ranging from 0.3 to 30 mg/kg. No clinical signs were noted, including body and liver weights. Macrophage pinocytosis was unaffected, but treatment with pyrene at 1, 3, and 30 mg/kg significantly stimulated Concanavalin A- and pokeweed mitogen-induced lymphocyte proliferation compared to controls. Doses of pyrene at 1, 3, 10, and 30 mg/kg suppressed plaque-forming cell (PFC) responses. Pyrene treatment did not affect CYP1A1 function as measured by ethoxy-resorufin O-dealkylase activity. For immunotoxicological responses, the NOAEL was 0.3 mg/kg, and the LOAEL was 1.0 mg/kg [46].

Blood Chemistry

Rats received intraperitoneal injections of 150 mg/kg of either pyrene or its ozonated derivatives, which produce more severe effects than pyrene itself. One rat treated with pyrene and one treated with ozonated pyrene died between 48 and 72 hours post-treatment. Treated animals exhibited hyperemia of the peritoneum and watery texture of the abdominal fat. Additionally, some animals had minimal congestion and swelling of the livers. Rats given ozonated pyrene had higher levels of peritonitis and small congested livers after 72 hours of treatment. Serum aspartate aminotransferase, alanine aminotransferase, and γ-glutamyl transpeptidase were not elevated in pyrene-treated rats or in ozonated pyrene-treated rats. Serum lactic dehydrogenase decreased at 24 hours but recovered by 72 hours following treatment with pyrene or ozonated pyrene. No effect occurred on serum glucose or creatinine, but serum bilirubin increased in ozonated pyrene-treated rats. Serological levels of urea nitrogen were unaffected. Generally, ozonated pyrene had slightly more pronounced effects on blood chemistry than pyrene [47]. In a related study, Yoshikawa *et al.* [48] found that intraperitoneal injection of 150 mg/kg pyrene did not affect the activity of serum aspartate aminotransferase, serum alanine aminotransferase, serum γ-glutamyl transpeptidase, serum lactic dehydrogenase, or serum sorbitol dehydrogenase at 24 or 72 hours after treatment. Blood urea nitrogen, serum creatinine, and glucose concentrations were unaffected. Gross liver observations revealed minor swelling in the livers of some animals.

Liver Toxicity

When rats were administered pyrene subcutaneously (total dose of 6,068 mg/kg over seven days) or via the diet (10,000 ppm for 10 days), it had no effect on whole body or liver weights. Following partial hepatectomy, these doses had no effect on liver regeneration [49].

Mammalian Toxicity: Inhalation

Male rats received head-only treatments that ranged from 200 to 800 mg/m^3 for 95 minutes. No deaths occurred. Blood concentrations indicated that absorption continued after exposure ceased, and the amount excreted via urine was similar to the amount excreted via feces [50].

Mammalian Toxicity: Dermal

Pyrene did not cause skin irritation following three daily doses of up to 5% (w/v) in hairless mice [25]. However, guinea pigs showed severe erythema following treatment with approximately 1 mg and exposure to ultraviolet radiation. Pyrene was highly phototoxic [51].

Summary of Avian Toxicology

Following oral intubation with a limited dose of 2,000 mg/kg pyrene, none of the treated Northern Bobwhite (*Colinus virginianus*) died, exhibited clinical signs of toxicity, or had gross pathological findings within 48 hours of dosing. Therefore, the 48-hour LD_{50} for Northern Bobwhite is greater than 2,000 mg/kg [52].

The aromatic fraction of Prudhoe Bay oil (as opposed to its aliphatic or its nitrogen-, oxygen-, and sulfur-containing heterocyclic fraction) caused embryotoxicity in chicken eggs when applied to the outside of the eggshell just below the air sac. Individual PAHs were untested in an attempt to identify the specific chemicals as the cause [53]. Brunström *et al.* [54] injected a mixture of PAHs (0.2 and 2.0 mg/kg) and pyrene alone (2.0 mg/kg) into chicken egg yolks. Several chick embryos that survived a dose of 2.0 mg/kg of the mixture exhibited abnormalities, including degenerative hepatic lesions (6/12), pericardial edema (5/12), microphthalmia (2/12), and subcutaneous edema (1/12). The prevalence of lesions was not investigated for pyrene alone. The mixture caused the deaths of 8 out of 20 embryos at the high dose, but pyrene alone did not cause any embryo mortality.

When a mixture of PAHs at a dose of 8.560 μg/egg was injected into chicken eggs, 0.550 μg (6.4%) remained after 18 days, with only 0.083 μg (1%) in the embryo. Similarly, when a mixture of PAHs totaling 17.80 μg was injected into eider eggs, 0.160 μg (0.9%) remained in the embryo after 24 days of incubation. When the process was performed with pyrene alone, the amount injected into the chicken egg was 0.540 μg/egg;

0.057 μg (10.6%) remained after 18 days, of which 0.016 μg (3%) was present in the embryo. For eider eggs, 1.10 μg pyrene was injected, and 0.019 μg (1.7%) remained in the embryo [55]. When laying hens were orally dosed with pyrene, residues appeared in the eggs 3 days following the first dose and were undetectable after 11 days. The majority of the residues in eggs were metabolites rather than parent-derived material [56]. Adult birds excrete pyrene in an altered form via the bile [57].

The level of contamination in buzzards (*Buteo buteo*) and tawny owls (*Strix aluco*) was determined for free-living individuals in three regions of Spain. Pyrene was the most commonly identified PAH in 10 buzzards and in two tawny owls. Pyrene occurred in different organs at various concentrations. Among different individuals, pyrene did not always occur in the same organ, and the relative concentrations among organs were not consistent [58].

Summary of Amphibian Toxicology

Exposure to pyrene-saturated swimming water produced significant decreases in liver and thigh mitochondrial O_2 consumption. Pyrene also significantly reduced liver and calf muscle mitochondrial membrane potential in exposed leopard frogs (*Rana pipiens*) [59]. Following exposure to 40.5 mg/L pyrene in tap water, leopard frogs were forced to swim to exhaustion. Pyrene-exposed frogs tired more quickly than unexposed frogs. There were no significant differences in the blood hematocrit or blood PO_2, and glycogen stores were similar. Mitochondrial O_2 consumption of the thigh muscle of pyrene-exposed frogs was significantly lower than that of control frogs. Significant decreases ($P \leq 0.05$) occurred in the mitochondrial membrane potential of pyrene-exposed frogs [60]. Exposure to pyrene-saturated water for 7 days reduced the swimming performance of leopard frogs, but they recovered to normal performance levels after 25 days in clean water. Mitochondrial O_2 consumption was $78 \pm 4\%$ of control immediately following exposure and recovered after 25 days in clean water [61]. In-water exposure to pyrene alters mitochondrial membrane potential of the frog liver and muscle, which ultimately limits adenosine triphosphate (ATP) synthesis [59]. Exposure to pyrene reduced O_2 consumption and ATP content. Reduced ATP explained the reduced muscle performance following exposure to pyrene [62].

Leopard frogs exposed to water saturated with pyrene experienced a reduction in the ionic conductance across their skin and a reduction in the conductance of sodium and chlorine ions. This effect was not seen in the bullfrog (*Rana catesbeiana*) [63]. Free-living green frogs (*Rana esculenta*) that were captured in The Netherlands and exposed to 0.0002 to 0.2 mg/L pyrene for 96 hours exhibited no effect for ethoxy-resorufin O-dealkylase activity in hepatocytes [64].

Pyrene tissue concentrations diminished in amphibian larvae after a peak, regardless of whether the animals continued to be exposed or were moved to clean water. Tissue concentrations in *Pleurodeles waltl* peaked after 24 hours and diminished with a half-life of 86 hours to a steady state concentration after 600 hours when the newts remained in contaminated water. After spending 24 hours in contaminated water, the test animals' depuration half-life was 172 hours following their removal to clean water [65].

Summary of Reptilian Toxicology

A single study reported pyrene exposure to reptiles, specifically, three freshwater turtle species: red-eared sliders (*Trachemys scripta elegans*), Chinese pond turtles (*Mauremys reevesii*), and Chinese softshell turtles (*Pelodiscus sinensis*). The study showed that sulfate conjugation is the dominant phase II metabolic pathway in these three species, but no rates of metabolism or adverse effects were provided [66].

RECOMMENDED TOXICITY REFERENCE VALUES

Toxicity Reference Values for Mammals

TRVs for Ingestion Exposures for the Class Mammalia

No acute or chronic toxicity studies for ingestion were available for pyrene. A single oral gavage subchronic (13-week) study with laboratory mice (CD-1 strain) reported no mortality but did report a NOAEL of 75 mg/kg/day and a LOAEL of 125 mg/kg/day for kidney effects [17]. The TRV Protocol [1] provides uncertainty factors of 20 for the NOAEL TRV and 4 for the LOAEL TRV when the TRVs are based on a subchronic LOAEL. Use of these uncertainty factors and the LOAEL of 125 mg/kg/day resulted in a NOAEL TRV of 6.25 mg/kg/day and a LOAEL TRV of 31.25 mg/kg/day (Table 32.2).

TRVs for Inhalation Exposures for the Class Mammalia

A single inhalation study exposed male rats to a range of doses up to 800 mg/m^3 for 95 minutes and reported no mortality [50]. Exposure duration for acute inhalation toxicity studies is usually between four and six hours [67]. Therefore,

Table 32.2 Selected Ingestion TRVs for the Class Mammalia

TRV	Dose	Confidence Level
NOAEL-based	6.25 mg/kg/day	Medium
LOAEL-based	31.25 mg/kg/day	Medium

Source: [8].

the exposure in this study is insufficient for use as a basis for a mammalian inhalation exposure TRV.

TRVs for Dermal Exposures for the Class Mammalia

Of the various studies that employed dermal applications of pyrene, only three reported a range of doses and exposed the animals for at least 24 hours [25,26,39]. These three studies did not report whether skin irritation or other signs of toxicity were present or whether any irritation occurred. Thus, these studies do not provide sufficient information with which to determine a TRV for mammalian dermal exposure.

TRVs for Birds

The single acute oral toxicity test failed to show mortality or other adverse effects at the highest dose tested. The only other avian toxicity studies reported effects following application to the eggshell or injection into the egg. None of the studies reported dose-ranging investigations or any effects caused by the treatment. Therefore, sufficient data with which to determine a TRV for birds are not available at this time.

TRVs for Amphibians

The studies that reported the impacts of pyrene on amphibians focused either on genotoxicity or on some physiological effect such as mitochondrial O_2 consumption. However, no mortality was reported for leopard frogs or bullfrogs following 7 days exposure to pyrene-saturated water. This suggests that, at a minimum, pyrene will not cause impacts on Anuran amphibians. However, a dose-ranging study that reports the effects of pyrene in amphibians is necessary to produce a TRV. Since no such study could be identified, a TRV for amphibians cannot be determined at this time.

TRVs for Reptiles

The TRVs for reptiles cannot be determined at this time.

IMPORTANT RESEARCH NEEDS

The lack of data on the toxicity of pyrene to wildlife species weakens the development of TRVs. Hence, further toxicological studies of the compound and its derivatives are recommended. The majority of studies reviewed in this WTA focused on genotoxic or carcinogenic effects. Thus, studies that focus on both acute and chronic toxicity of pyrene to mammals as well as nonmammalian wildlife (e.g., birds, reptiles, and amphibians) are warranted. Since pyrene is likely found with other PAHs, studies that examine multiple exposures might be useful.

REFERENCES

[1] USACHPPM. Standard practice for wildlife toxicity reference values. Aberdeen Proving Ground (MD): U.S. Army Center for Health Promotion and Preventive Medicine; 2000, Technical Guide 254.
[2] WHO. Environmental health criteria 202: selected non-heterocyclic polycyclic aromatic hydrocarbons. Geneva: United Nations Environment Programme, International Labour Organisation. International Programme on Chemical Safety; 1998.
[3] Hazardous Substances Databank (HSDB). Pyrene. National library of medicine [Internet]; 2002. [cited 2010 Feb 09]. Available from, http://toxnet.nlm.nih.gov/cgi-bin/sis/search2/f?./temp/~EfnoRl:1.
[4] Husain S. Effect of surfactants on pyrene degradation by *Pseudomonas fluorescens* 29 L. World J Microbiol Biotechnol 2008;24:2411–9.
[5] Yuan SY, Su LM, Chang BV. Biodegradation of phenanthrene and pyrene in compost-amended soil. J Environ Sci Health A Tox Hazard Subst Environ Eng 2009;44(7):648–53.
[6] Chang B-V, Chang IT, Yuan SY. Biodegradation of phenanthrene and pyrene from mangrove sediment in subtropical Taiwan. J Environ Sci Health A Tox Hazard Subst Environ Eng 2008;43(3):233–8.
[7] Kobayashi T, Murai Y, Tatsumi K, Iimura Y. Biodegradation of polycyclic aromatic hydrocarbons by *Sphingomonas* sp. enhanced by water-extractable organic matter from manure compost. Sci Total Environ 2009;407:5805–10.
[8] Chang B-V, Chang IT, Yuan SY. Anaerobic degradation of phenanthrene and pyrene in mangrove sediment. Bull Environ Contam Toxicol 2008;80:145–9.
[9] Payan JP, Lafontaine M, Simon P, Marquet F, Champmartin-Gendre C, Beydon D, et al. *In vivo* and *in vitro* percutaneous absorption of [^{14}C]pyrene in Sprague Dawley male rats: skin reservoir effect and consequence on urinary 1-OH pyrene excretion. Arch Toxicol 2008;82:739–47.
[10] Withey JR, Law FCP, Endrenyi L. Percutaneous uptake, distribution, and excretion of pyrene in rats. J Toxicol Environ Health 1993;40:601–12.
[11] Jacob J, Brune H, Gettbarn G, Grimme D, Heinrich U, Mohtashamipur E, et al. Urinary and faecal excretion of pyrene and hydroxypyrene by rats after oral, intraperitoneal, intratracheal or intrapulmonary application. Cancer Lett 1989;46:15–20.
[12] Lipniak M, Brandys J. Toxicokinetics of fluoranthene, pyrene and benz(a)anthracene in the rat. Polycyclic Aromat Compd 1993;3(2):111–9.
[13] Fouchecourt MO, Arnold M, Berny P, Videmann B, Rether B, Riviere JL. Assessment of the bioavailability of PAHs in rats exposed to a polluted soil by natural routes:

induction of EROD activity and DNA adducts and PAH burden in both liver and lung. Environ Res Sect A 1999;80:330–9.
[14] Holsbeek L, Joiris CR, Debacker V, Ali IB, Roose P, Nellissen JP, et al. Heavy metals, organochlorines and polycyclic aromatic hydrocarbons in sperm whales stranded in the southern North Sea during the 1994/1995 winter. Mar Pollut Bull 1999;38 (4):304–13.
[15] Lee HJ, Villaume J, Cullen DC, Kim BC, Gu MB. Monitoring and classification of PAH toxicity using an immobilized bioluminescent bacteria. Biosens Bioelectron 2003;18:571–7.
[16] Salamone MF. Toxicity of 41 carcinogens and noncarcinogenic analogs. In: De Serres FJ, Ashby J, editors. Evaluation of short-term tests for carcinogens. Report of the International Collaborative Programme. Amsterdam: Elsevier North-Holland; 1981.
[17] U.S. EPA. 13 week mouse oral subchronic toxicity of pyrene, TRL Study # 042-012. Study Conducted by Toxicity Research Laboratories, Muskegon, Washington DC: MI for the Office of Solid Waste; 1989. http://www.epa.gov/iris/subst/0445.htm.
[18] Mattison DR, Thorgeirsson SS. Ovarian aryl hydrocarbon hydroxylase activity and primordial oocyte toxicity of polycyclic aromatic hydrocarbons in mice. Cancer Res 1979;39:3471–5.
[19] Nikonova TV. Transplacental action of benzo(a)pyrene and pyrene. Translated from Byull Eksp Biol Med 1977;84(7):88–91.
[20] Neubert D, Tapken S. Prenatal induction of benzo(a)pyrene hydroxylases in mice. Arch Toxicol 1988;62:192–9.
[21] Mukhtar H, Link CM, Cherniack E, Kushner DM, Bickers DR. Effect of topical application of defined constituents of coal tar on skin and liver aryl hydrocarbon hydroxylase and 7-ethoxycoumarin deethylase activities. Toxicol Appl Pharmacol 1982;64:541–9.
[22] Shimada T, Sugie A, Yamada T, Kawazoe H, Hashimoto M, Azuma E, et al. Dose–response studies on the induction of liver cytochromes P4501A1 and 1B1 by polycyclic aromatic hydrocarbons in aryl hydrocarbon-responsive C57BL/6 J mice. Xenobiotica 2003;33(9):957–71.
[23] Fouchecourt MO, Riviere JL. Activities of cytochrome p450-dependent monooxygenases and antioxidant enzymes in different organs of Norway rats (*Rattus norvegicus*) inhabiting reference and contaminated sites. Chemosphere 1995;31(11/12):4375–86.
[24] Ito N, Tsuda H, Tatematsu M, Inoue T, Tagawa Y, Aoki T, et al. Enhancing effect of various hepatocarcinogens on induction of preneoplastic glutathione S-transferase placental form positive foci in rats—an approach for a new medium-term bioassay system. Carcinogenesis 1998;9(3):387–94.
[25] Nishikawa T, Nakamura T, Fukushima A, Takagi Y. Further evaluation of the skin micronucleus test: results obtained using 10 polycyclic aromatic hydrocarbons. Mutat Res 2005;588:58–63.
[26] He S, Baker R. Micronuclei in mouse skin cells following in vivo exposure to benzo[a] pyrene, 7,12-dimethylbenz[a]anthracene, chrysene, pyrene, and urethane. Environ Mol Mutagen 1991;17:163–8.
[27] Casto BC. Polycyclic hydrocarbons and Syrian hamster embryo cells: cell transformation, enhancement of viral transformation and analysis of DNA damage. In: Jones PW, Leber P, editors. Polynuclear aromatic hydrocarbons. Ann Arbor (MI): Ann Arbor Science Publishers, Inc.; 1979.
[28] Probst GS, McMahon RE, Hill LE, Thompson CZ, Epp JK, Neal SB. Chemically-induced unscheduled DNA synthesis in primary rat hepatocyte cultures: a comparison with bacterial mutagenicity using 218 compounds. Environ Mutagen 1981;3:11–32.

[29] Tong C, Brat SV, Telang S, Laspia MF, Fazio M, Reiss B, et al. Effects of genotoxic polycyclic aromatic hydrocarbons in rat liver culture systems. In: Cooke M, Dennis AJ, editors. Polynuclear aromatic hydrocarbons: formation; metabolism, and measurement. Columbus (OH): Battelle Press; 1983.
[30] Johnson BT. An evaluation of a genotoxicity assay with liver s9 for activation and luminescent bacteria for detection. Environ Toxicol Chem 1992;11:473–80.
[31] Pagnout C, Rast C, Veber AM, Poupin P, Férard JF. Ecotoxicological assessment of PAHs and their dead-end metabolites after degradation by *Mycobacterium* sp. Strain SNP11. Ecotoxicol Environ Saf 2006;65:151–8.
[32] Matijasevic Z, Zeiger E. Mutagenicity of pyrene in *Salmonella.* Mutat Res 1985;142:149–52.
[33] Bhatia AL, Tausch H, Stehlik G. Mutagenicity of chlorinated polycyclic aromatic compounds. Ecotoxicol Environ Saf 1987;14:48–55.
[34] Cherng SH, Lin ST, Lee H. Modulatory effects of polycyclic aromatic hydrocarbons on the mutagenicity of 1-nitropyrene: a structure-activity relationship study. Mutat Res 1996;367:177–85.
[35] Ho CH, Clark BR, Guerin MR, Barkenbus BD, Rao TK, Epler JL. Analytical and biological analyses of test materials from the synthetic fuel technologies IV. Studies of chemical structure—mutagenic activity relationships of aromatic nitrogen compounds relevant to synfuels. Mutat Res 1981;85:335–45.
[36] Nisbet ICT, LaGoy PK. Toxic equivalency factors (TEFs) for polycyclic aromatic hydrocarbons (PAHs). Regul Toxicol Pharmacol 1992;16:290–300.
[37] Flesher JW, Myers SR. Rules of molecular geometry for predicting carcinogenic activity of unsubstituted polynuclear aromatic hydrocarbons. Teratog Carcinog Mutagen 1991;11:41–54.
[38] Haddow A, Robinson AM. The influence of various polycyclic hydrocarbons on the growth rate of transplantable tumours. Proc R Soc Lond B Biol Sci 1937;122:442–76.
[39] Shukla Y, Kumar S, Mehrotra NK. Status of the γ-glutamyl transpeptidase enzyme activity in mouse skin exposed to polyaromatic hydrocarbons and tumor promotor TPA. Bull Environ Contam Toxicol 1989;43:454–8.
[40] El-Bayoumy K, Hecht SS, Hoffmann D. Comparative tumor initiating activity on mouse skin of 6-nitrobenzo[a]pyrene, 6-nitrochrysene, 3-nitroperylene, l-nitropyrene and their parent hydrocarbons. Cancer Lett 1982;16:333–7.
[41] Iwata K, Inui N, Takeuchi T. Induction of active melanocytes in mouse skin by carcinogens: a new method for detection of skin carcinogens. Carcinogenesis 1981;2 (7):589–93.
[42] Cavalier E, Calvin M. Photochemical coupling of benzo[a]pyrene with 1-methylcytosine; photoenhancement of carcinogenicity. Photochem Photobiol 1971;14:641–53.
[43] Warshawsky D, Barkley W, Bingham E. Factors affecting carcinogenic potential of mixtures. Fundam Appl Toxicol 1993;20:376–82.
[44] Busby Jr WF, Stevens EK, Martin CN, Chow FL, Garner RC. Comparative lung tumorigenicity of parent and mononitro-polynuclear aromatic hydrocarbons in the BLU:Ha newborn mouse assay. Toxicol Appl Pharmacol 1989;99:555–63.
[45] Nesnow S, Mass MJ, Ross JA, Galati AJ, Lambert GR, Gennings C, et al. Lung tumorigenic interactions in strain A/J mice of five environmental polycyclic aromatic hydrocarbons. Environ Health Perspect 1998;106(Suppl. 6):1337–46.
[46] Peden-Adams MM, Liu J, Knutson S, Dancik J, Bryant K, Bodine AB, et al. Alterations in immune function and CYP450 activity in adult male deer mice (*Peromyscus maniculatus*) following exposure to benzo[a]pyrene, pyrene, or chrysene. J Toxicol Environ Health A 2007;70(21):1783–91.

[47] Yoshikawa T, Ruhr LP, Flory W, Giamalva D, Church DF, Pryor WA. Toxicity of polycyclic aromatic hydrocarbons I. Effect of phenanthrene, pyrene, and their ozonized products on blood chemistry in rats. Toxicol Appl Pharmacol 1985;79:218–26.

[48] Yoshikawa T, Flory W, Ruhr LP, Giamalva D, Church DF, Pryor WA. Toxicity of polycyclic aromatic hydrocarbons II. Effects of NO_2-nitrated phenanthrene and pyrene on blood chemistry in rats. Vet Hum Toxicol 1987;29(1):25–30.

[49] Gershbein LL. Liver regeneration as influenced by the structure of aromatic and heterocyclic compounds. Res Commun Chem Pathol Pharmacol 1975;11(3):445–66.

[50] Withey JR, Burnett R, Law FCP, Abedini S, Endrenyi L. Pharmacokinetics of inhaled pyrene in rats. J Toxicol Environ Health 1994;43:103–16.

[51] Kochevar IE, Armstrong RB, Einbinder J, Walther RR, Harber LC. Coal tar phototoxicity: active compounds and action spectra. Photochem Photobiol 1982;36:65–9.

[52] Brausch JM, Blackwell BR, Beall BN, Caudillo C, Kolli V, Godard-Codding C, et al. Effects of polycyclic aromatic hydrocarbons in northern bobwhite quail (*Colinus virginianus*). J Environ Sci Health A Tox Hazard Subst Environ Eng 2010;73:540–51.

[53] Walters P, Khan S, O'Brien PJ, Payne JF, Rahimtula AD. Effectiveness of a Prudhoe Bay crude oil and its aliphatic, aromatic, and heterocyclic fractions in inducing mortality and aryl hydrocarbon hydroxylase in chick embryo *in ovo*. Arch Toxicol 1987;60:454–9.

[54] Brunström B, Broman D, Näf C. Embryotoxicity of polycyclic aromatic hydrocarbons (PAHs) in three domestic avian species, and of PAHs and coplanar polychlorinated biphenyls (PCBs) in the common eider. Environ Pollut 1990;67:133–43.

[55] Näf C, Broman D, Brunström B. Distribution and metabolism of polycyclic aromatic hydrocarbons (PAHs) injected into eggs of chicken (*Gallus domesticus*) and common eider duck (*Somateria mollissima*). Environ Toxicol Chem 1992;11:1653–60.

[56] Fournier A, Feidt C, Dziurla MA, Grandclaudon C, Jondreville C. Transfer kinetics to egg yolk and modeling residue recovered in yolk of readily metabolized molecules: polycyclic aromatic hydrocarbons orally administered to laying hens. Chemosphere 2010;78:1004–10.

[57] Chalmers JG, Peacock PR. The excretion of derivatives of certain carcinogenic and noncarcinogenic hydrocarbons in fowl bile. Biochem J 1941;35:1276–82.

[58] Amigo SG, Yusty MAL, Lozano JS. Determination of polycyclic aromatic hydrocarbons from buzzards (*Buteo buteo*) and tawny owl (*Strix aluco*) by liquid chromatography with fluorescence detection. J AOAC Int 2002;85(1):141–5.

[59] Bonner A, Warren S, Stabenau EK. Influence of pyrene on mitochondrial oxygen consumption and membrane potential in frogs (Abstract only). FASEB J 2006;20:492.5.

[60] Stabenau EK, Sasser A, Schulte C. The effects of pyrene exposure on exercise performance, muscle contraction, and mitochondrial O_2 consumption in the leopard frog (*Rana pipiens*). J Environ Sci Health A Tox Hazard Subst Environ Eng 2008;43(6):576–83.

[61] Bonner AF, Stabenau EK. Restoration of muscle function following pyrene exposure and depuration in *Rana pipiens* (Abstract only). FASEB J 2007;21:963.4.

[62] Reed LK, Carroll JR, McGirk ME, Stabenau EK. Mitochondrial oxygen consumption and ATP content in control and pyrene-exposed frogs (Abstract only). FASEB J 2009;23:739.10.

[63] Anderson KA, Stabenau EK. Effect of pyrene exposure on skin ionic conductance in *Rana pipiens* and *Rana catesbeiana* (Abstract only). FASEB J 2009;23:778.19.

[64] Rankouhi TR, Koomen B, Sanderson JT, Bosveld ATC, Seinen W, Van Den Berg M. Induction of ethoxy-resorufin-o-deethylase activity by halogenated aromatic hydrocarbons and polycyclic aromatic hydrocarbons in primary hepatocytes of the green frog (*Rana esculenta*). Environ Toxicol Chem 2005;24(6):1428–35.

[65] Garrigues P, Djomo JE, Ferrier V, Dauta A, Monkiedje A, Mvondo Ze A, et al. Toxicokinetics of polycyclic aromatic hydrocarbons from contaminated sediment by the amphibian larvae (*Pleurodeles waltl*). Polycyclic Aromat Compd 2004;24(3):207–19.
[66] Oroszlany B, Ikenaka Y, Saengtiengchan A, Oguri M, Nakayama SMM, Ishizuka M. Metabolism of pyrene, a polycyclic aromatic hydrocarbon in freshwater turtles. Jpn J Vet Res 2013;(61 Suppl.):S77–8.
[67] Pigott GH. Inhalation toxicity. In: 3rd ed. Ballantyne B, Marrs TC, Syversen T, editors. General and applied toxicology, vol. 2. West Sussex (UK): John Wiley & Sons, Ltd.; 2009.

CHAPTER 33

Wildlife Toxicity Assessment for Mustard Agents

Gunda Reddy, Sang Ho Lee

Contents

INTRODUCTION

Mustard gas is a colorless, viscous liquid that emits odor resembling the mustard plant, garlic, and horseradish from which the name was derived. Mustard gas refers to two classes of blister-causing chemical agents. The first class of mustard gas contains sulfur or sulfur mustard (CAS No. 505-60-2) represented by bis (2-chloroethyl) sulfide (HD), but it also includes other compounds such as 1,2-bis-(2-chloroethylthio)-ethane. A second class of mustard gas contains nitrogen or nitrogen mustard (CAS No. 51-75-2) that includes bis (2-chloroethyl) ethylamine (HN1), bis (2-chloroethyl) methylamine (HN2), and tris (2-chloroethyl) amine HN3. The cytotoxicity of mustard gas is due to its nonspecific alkylating properties. Alkylation of cellular materials by these agents result in impaired cellular metabolic functions, and the alkylation of guanidine moiety of deoxyribonucleic acid (DNA) strain prevent cell replication leading to cell death. The cytotoxicity is manifested clinically as formation of large clear fluid-filled blisters on skin and mucous membranes (i.e., eye and lung) that come in direct contact with the blister

Wildlife Toxicity Assessments for Chemicals of Military Concern
http://dx.doi.org/10.1016/B978-0-12-800020-5.00033-8

agents. HD also has mutagenic and carcinogenic properties, with a median lethal dose $(LD)_{50}$ value of 0.7 milligrams per kilogram (mg/kg) [1,2].

Mustard gas refers to a number of related, highly toxic alkylating chemical agents that are best known for their effects on the destruction ofbasal epidermal cells on contact, which results in large, clear, fluid-filled blisters on the exposed skin and the mucus membranes, including those of the lungs. Its nonselective alkylating property also results in attacks on DNA leading to systemic effects including the hematopoietic system, causing progressive leucopenia and lowering of the white blood cell level as well as internal and external bleeding [2].

Mustard gas is easily produced and readily deployed to wide areas cheaply to kill or incapacitate unsuspecting victims. Its combined effectiveness, low production cost, and ease of use make HD very appealing as an alternative to advanced weapon systems that are expensive and require advanced technical knowledge to manufacture and deploy. Indeed, chemical warfare agents as a whole, including HD, are termed the "poor man's atomic bomb." Unfortunately, they are non-discriminating in their selection of victims, and their deployment can result in massive collateral damage in terms of the number of unintended civilian victims. Nonetheless, many nations continued to manufacture and stockpile significant quantities of these agents.

Industrial-scale production of sulfur mustard gas was first realized in Germany in 1916 with subsequent deployment against Allied Forces during World War I. The Allied Forces subsequently succeeded in manufacturing sulfur mustard and began deploying it in 1917 [2]. Subsequently, a number of nations have used mustard agent on numerous occasions, with the latest confirmed use having been perpetrated by the Iraqi government against Iraqi Kurds in 1988 [3].

A number of accidental exposures to mustard gas have been reported in the United States and around the world since World War I. Most of these incidents resulted from exposure to improperly disposed canisters or during disposal of military stockpiles. Most of the European World War vintage stockpiles reportedly were disposed of by throwing the stockpiles into the open ocean waters, but the exact amount disposed of in this manner was unspecified. Historically, a method of sulfur mustard disposal was to dump it in coastal waters of the United States, in the mid-Atlantic and Mississippi basin areas and off the coast of California from the1940s to the 1960s. Over the past 50 years, similar sporadic incidents have occurred in various locations around the world, including the Baltic Sea and the Adriatic Sea.

Nitrogen mustard gas was developed in the 1920s and 1930s, much later than its predecessor, and many nations stockpiled it for possible use during

World War II [4]. Although these agents were never used as weapons, they have been used to treat certain types of lymphoma [5]. The U.S. government has stored chemical warfare agents in a number of storage sites on U.S. Army installations. Many of these sites have been identified as contaminated due to spillage of these agents. Due to the high cost of decontamination, a few of these former storage sites have been converted to wildlife refuge areas using federal government funds. Contaminated soil, buildings, and debris collected during cleanup operations are buried in leak-proof clay and plastic liners. The surrounding areas are converted to wildlife sanctuaries. Initially, the U.S. Environmental Protection Agency (U.S. EPA) monitors the contaminated sites. Once deemed safe by the U.S. EPA, the U.S. Fish and Wildlife Service assumes the monitoring functions. An example of this transformative process is the Rocky Mountain Arsenal cleanup project, in which a 27-square-mile area was converted into a national wildlife refuge at a cost of $2.2 billion. The area attracts more than 100 bald eagles each year. Information on monitoring mustard agent gas in wildlife animal populations was not located during the online web analyses undertaken for this chapter.

This Wildlife Toxicity Assessment (WTA) for nitrogen and sulfur mustard agents summarizes current knowledge of their fate and transport in the environment, their potential harmful impacts on wildlife, and reports their Toxicity Reference Values (TRVs), which may be important for the health of wildlife (e.g., mammals, birds, reptiles, and amphibians) exposed to these compounds. Bioconcentration of, or ground water contamination by sulfur mustard does not occur due to its high reactivity; however, its hydrolysis product thiodiglycol persists longer in the environment [6]. The end-points of toxicity relate to the likelihood of effects in individual organisms that may be relevant to a population of organisms in the wild but are not related to demographic rates or outcomes for any particular population. Unfortunately, other than laboratory studies that were conducted on rats and rabbits, published data could not be located on the toxicity of wildlife animals following exposure to mustard agents [6].

The protocol for this assessment is documented in the U. S. Army Center for Health Promotion and Preventive Medicine (USACHPPM) Technical Guide (TG) 254, the *Standard Practice for Wildlife Toxicity Reference Values* [7].

TOXICITY PROFILE

Environmental Fate and Transport

Both sulfur and nitrogen mustard gases are liquids of low volatility that are chemically stable under ambient conditions. Moreover, limited information

Table 33.1 Summary of the Physical-Chemical Properties of Sulfur Mustard and Nitrogen Mustard

	Sulfur Mustard	Nitrogen Mustard
CAS No.	505-60-2	51-75-2
Molecular weight	159.08	156.07
Color	Oily and colorless to light yellow	Colorless to yellow
Physical state	Volatile liquid	Oily liquid
Melting point	13–14 °C	–60 °C
Boiling point	215–217 °C	87 °C (at 2 mm Hg)
Odor	Sweet, agreeable or garlic-like	Aromatic "fish-like"
Solubility in water	684 mg/L at 25 °C: miscible with most organic solvents	Slight at 25 °C: miscible with organic solvents
Partition coefficients:		
Log K_{ow}	2.03, 2.41	0.91
Log K_{oc}	2.07, 2.43	1.87
Vapor pressure (mmHg)	0.11 at 25 °C, 0.09 at 30 °C	65 at 25 °C
Henry's Law constant at 25 °C	2.45×10^{-5} atm.m^3/mole	8.5×10^{-8} atm.m^3/mole
Conversion factors	1 ppm = 6.5 mg/m^3 1 mg/m^3 = 0.15 ppm	1 ppm = 6.37 mg/m^3 1 mg/m^3 = 0.16 ppm

Sources: ATSDR [2], HSDB [9].

exists on the environmental fate and transport of mustard agents. The physical-chemical properties of both sulfur mustard (HD) and nitrogen mustard (HN2) agents are listed in Table 33.1. Photolysis and hydrolysis of HD were described in the literature as possible environmental degradation pathways. However, the effect of simple dilution has not yet been addressed and may deserve consideration for future study.

Photolytic degradation of HD is expected to be one of the initial degradation pathways in cases of spills or airborne releases. This will be particularly true for spills onto solid surfaces. The extent of partition into the air will be limited since these agents are considerably heavier than air with low volatility, and most will linger on ground surface. Dispersion into air will be slow with vaporization occurring faster under high temperature and windy conditions. However, sufficient partitioning should exist in the immediate vicinity of liquid droplets that persist for prolonged periods [2].

A study conducted by the U.S. Army indicated that HD on soil surface should evaporate in 30 to 50 hours at 25 °C despite its high viscosity and low volatility of 0.11 millimeters of mercury (mmHg) at 25 °C [8]. Airborne concentrations of 3 to 5 parts per million (ppm) were reported in battlefields during World War I. The reported rate constant is 7.82×10^{-12} cm^3/molecule-second at 25 °C for the vapor phase reaction of HD with photochemically generated hydroxyl radicals. This, in conjunction with the fact that HD does not absorb ultraviolet radiation above 290 nanometers (nm), suggests that the photochemical hydrolysis of surface spills as consequence of partitioning into air is unlikely to be a significant mode of degradation. Rather, the hydrolysis will serve as a steady source of short-term reservoir for airborne dispersion of the area of contamination [9]. During World War I, mustard gas was reported to have been active in soil for periods extending over weeks or months [5].

Dispersion of sulfur mustard via air will be even slower below its melting point of 14 °C. Therefore, sulfur mustard will persist for prolonged periods, depending on its location, and will serve as a prolonged source of toxicant to the wildlife population. Sulfur mustard persistence in winter has been reported to be greater than 2 weeks and, if samples are covered in snow, longer than 4 weeks [10]. Other factors that affect liquid-air partitioning include pH, moisture content, porosity of the surface, and soil composition.

Hydrolysis is the main initial degradation pathway in cases of spills. Mustard gases are lipophilic and readily dissolve in organic medium [2] with moderate affinity for soil. Their solubility in water is poor, but they readily decompose via hydrolysis in water under ambient temperature to produce thiodiglycol (TDG) along with hydrogen chloride for sulfur mustard gas and nitrogen equivalent for nitrogen mustard gas as the main hydrolysis products. Estimates based upon the K_{ow} of each compound suggest that each has a comparatively low capacity for bioconcentration [9]. Despite their poor solubility in water, hydrolytic degradation of mustard compounds is important in such instances as buried canisters in moist soil or compounds inadvertently released into water. The specific gravity of HD is 1.27, which is significantly higher than sea water's specific gravity of 1.027 at the surface. Therefore, any HD released in water will sink to the bottom as a separate phase. It will slowly dissolve in water and degrade over time. HD is more stable in salt water than in fresh water by a factor of 2.5 due to the hydrolysis-inhibiting effect of the chlorine ion in sea water. Epstein *et al.* reported that no more than a few parts per million of HD will be detected in sea water in case of spillage [11]. However, HD is moderately persistent in the environment because of its low water solubility along with intermediate degradation products (i.e., thiodiglycols). Thus, old sulfur mustard canisters originally produced as chemical weapons and later disposed of in the ocean occasionally surface in fishing nets in Northern European and U.S. coastal waters.

TDG, along with other intermediates, is thought to coat the HD phase and thus retard the hydrolysis process since the hydrolysis is a surface phenomenon during which water and HD encounter each other. Micelle or droplet-like aggregates of liquid HD whose outer surface is coated with TDG is stable in the absence of water and soluble in water due to the soapy property of thioglycerol, similar to that of oxygen-equivalent triglycerol. The half-life of TDG is reported to be 6 weeks [12]. TDG is chemically relatively inert to chemical reaction and resists photoreaction, and these properties of TDG have been implicated in contributing to the relative stability of HD-TDG aggregates [8].

Thiodiglycol also shows lower toxicity and persists in the environment for months due to its very poor utilization by microorganisms. Medvedeva *et al.* [13] reported a concentration decrease of 50% after 1 year for TDG in a medically contaminated peaty clay soil characterized by a high rate of microbiological processing. In the case of highly contaminated peaty clay soil, the concentration was reduced by only 10% after 1 year due to the effect of high toxicity to soil organisms by the initial HD assault.

Using a Sequencing Batch Reactor laboratory method, Harvey *et al.* published a biodegradation study of the long-term impact of HD hydrolyzed products on the environment [1]. The purpose of the study was not only to examine the primary degradation but also the irreversible biodegradation of the initial hydrolyzed product, TDG. The study showed that 90.6% of the carbon incorporated into HD was removed as carbon dioxide and other biomass after 146 days of Sequencing Batch Reactor runs. Destruction of mustard gas by ozone via oxidation pathway has been reported in literature. This reaction occurs mainly by direct oxidation with intermediate radicals serving a secondary role [14].

HD TOXICITY SUMMARY

HD Mammalian Toxicity

Mammalian Oral Toxicity: Acute

Sulfur mustard is a strong, nondiscriminating alkylating agent and causes systemic toxicity in mammals in addition to respiratory assault and blister formation. Watson and Griffin [15], citing the U.S. Army, reported an HD oral LD_{50} of 17 mg/kg in rats. No other values were found for HD oral lethality in mammals.

Mammalian Oral Toxicity: Subacute

Hackett *et al.* [16] carried out a subacute toxicity study of HD in pregnant female Sprague Dawley rats and New Zealand white rabbits to determine dose ranges for subsequent teratological experiments. HD in sesame oil was administered by gavage on gestation days (GDs) 6 to 15 to rats at doses of 0, 0.2, 0.4, 0.8, 1.6, 2.0, and 2.5 mg/kg and in rabbits gavaged on GDs 6 to 19 at 0, 0.50, 1.0, 2.0, and 2.5 mg/kg. Body weights were monitored on GDs 0 and 6 to 16 in rats and on GDs 0 and 6 to 20 in rabbits. All surviving animals were euthanized: rats on GD 20 and rabbits on GD 30. Blood samples for hematological measurement were obtained from all survivors, and all animals were examined grossly for alterations to their major organs. Numbers of corpora lutea, implantation sites, resorptions, and live and dead fetuses were recorded, and all live fetuses were weighed and examined for gross external defects.

After GD 11, the body weights of rats receiving HD at 1.6 mg/kg/day and above tended to be lower than those of the controls. However, the weights of gravid uteri did not differ from those of the controls in any treatment group, and fluctuations in the number of corpora lutea/dam were probably unrelated to treatment. No consistent dose effect was found in hematological parameters. In rats, the increased incidence of inflammation of the mesenteric lymph nodes was considered by the authors as the most striking response to HD at a dose of 0.4 mg/kg and above. Esophageal lesions, probably representing initial site-of-impact effects of HD, were also seen at those dose levels. The findings of Hackett *et al.* [16] proposed a subacute lowest observed adverse effect level (LOAEL) of 0.4 mg/kg/day and a no observed adverse effect level (NOAEL) of 0.2 mg/kg/day for HD for rats.

Dose-related mortality in rabbits receiving HD was reported [16]. In this high-dose exposure rabbit study, only two of seven rabbits survived to GD 30. Hackett *et al.* [16] described a sequence of lesions marked with gastric hemorrhage by GDs 7 to 9, gastric ulceration by GD 10, onset of diarrhea in some animals by GD 11, and the onset of intestinal and respiratory infections by GDs 13 to 24. Enlarged mesenteric lymph nodes and inflamed or enlarged Peyer's patches were observed as gross abnormalities during scheduled gross necropsies. Because of the lack of either maternal or fetal effects, 0.5 mg/kg/day was considered the NOAEL for rabbits. Based upon lower fetus weights, 1.0 mg/kg day was considered the LOAEL for rabbits.

Mammalian Oral Toxicity: Subchronic

In a 13-week subchronic HD toxicity study, HD in sesame oil was administered by gavage to Sprague Dawley male and female rats at 0, 0.003, 0.01, 0.03, 0.1, and 0.3 mg/kg/day/5days (averaged for 7-day doses of 0, 0.0021, 0.0071, 0.021, 0.071, and 0.21 mg/kg/day, respectively) [17]. Clinical signs were monitored twice daily, body weights were recorded weekly, and all animals were subjected to an ophthalmological examination both at the start of the experiment and at term. To measure hematological parameters, blood samples were taken at the start and end of the experiment, and sera from the terminal blood samples were used to measure clinical chemistry parameters. At the end of study period, full necropsies were performed, and the weights of major organs were recorded. A full suite of organs and tissues from the high-dose and control members were examined histopathologically, as were potential HD target organs from the intermediate groups.

Few rats showed clinical signs attributable to HD exposure, and all but three survived to term. High-dose exposure rats showed reduction in body weight gain at 4 weeks and thereafter, but absolute and relative organ weights were the same as those of the controls. Additionally, there were no signs of HD-related changes in clinical chemistry or hematology. One of the few HD-related effects described by the authors was the observed onset of minimal forestomach epithelial hyperplasia in five animals of each sex administered 0.3 mg HD/kg day and in one male from the 0.1 HD mg/kg/day group. Hyperplasia appeared to be a site-of-impact phenomenon rather than a systematic effect, and it allowed derivation of a LOAEL of 0.21 mg/kg/day with an associated NOAEL of 0.071 mg/kg/day for oral exposure.

A 42-week, two-generation study of HD reproductive affects in Sprague Dawley rats subjected to subchronic exposure was reported by Sasser *et al.* [18]. Groups of rats (27 females and 20 males per group) were gavaged with 0, 0.03, 0.1, or 0.4 mg HD/kg/day (0.3 mg/kg for the first 10 weeks in the high-dose group) in sesame oil. This was followed by additional administration of HD for 5 days a week for 13 weeks prior to mating, 7 days a week throughout gestation, and 5 days a week through parturition and lactation. Within its dose group, the F_1 generation was exposed to HD in a similar premating (13-week), gestation, and parturition and lactation cycle. The authors observed no significant impacts of HD on reproduction or pregnancy at any dose level in either generation. The F_0 and F_1 rats of both sexes exposed to HD developed site-of-impact lesions that were characterized by thickening of the forestomach, squamous mucosa, and hyperkeratosis. The threshold dose for this effect of 0.03 mg/kg/day was an order of magnitude less than 0.3 mg/kg/day observed by the authors in their 13-week study [17]. This lower number suggests a time-weighted average HD LOAEL of 0.021 mg/kg/day.

A dual phase dominant lethal study of HD in Sprague Dawley rats was also reported by Sasser *et al.* [19]. In the first phase, 10 animals of each sex per group were exposed to HD that was dissolved in sesame oil by gavage at concentrations of 0, 0.08, 0.2, or 0.5 mg/kg/5 days a week (averaged for 7-day doses of 0, 0.06, 0.14, and 0.36 mg/kg/day) for 10 weeks. Male and female rats from each dose group were housed together. Reproductive and developmental parameters of the dams and fetuses were evaluated on GD 14. No observable effects from HD were present in any maternal, reproductive, or developmental parameter under investigation. In the second phase, 20 male Sprague Dawley rats per group were subjected to the same regimen as phase I prior to cohabitation with virgin females for 10 successive 5-day-per-week cycles. Females were evaluated for reproductive and developmental effects following this exposure.

Some observed reproductive and developmental parameter effects showed statistical significance to the elapse of time since paternal exposure to HD, with most significant effects seen in fetuses conceived during post-exposure weeks 2, 3, and 4. An increased number of fetal resorptions and a decrease in the percentage of live fetuses per litter with statistical significance were reported for high-dose group subjects. In addition, males that received high doses of HD demonstrated morphologically anomalous sperm production. Due to the sporadic nature of developmental effects observed that lacked a statistically significant correlation, the determination of LOAELs and NOAELs for the observed effects in the dominant lethal experiment were not clear-cut. However, the authors suggested the 7-day average of 0.36 mg/kg/day (or 0.5 mg/kg/dosage day) for the high dose as a reasonable LOAEL and the 7-day average of 0.14 mg/kg/day (or 0.2 mg/kg/dosage day) as a potential NOAEL for the effects in this study.

HD Mammalian Oral Toxicity: Chronic

No information was found in the literature that addressed chronic oral toxicity of mustard agents.

Studies Relevant for Mammalian Ingestion Exposure TRV Development

A comprehensive review by Dacre and Goldman [20] contains much of the information on the toxicology and pharmacology of mustard agents that is currently available. Systemic effects observed from the administration of HD either intravenously or intraperitoneally include a marked perturbation of the hematopoietic system with readily observable changes in bone marrow, spleen, and thymus. When HD is administered orally or via inhalation or dermal routes to animals, however, such systemic effects are obscured or delayed by site-of-impact effects such as lesions of the forestomach, necrosis of the bronchial passages and alveoli, and necrosis of the skin. Characteristic features of animal and human exposure to HD are the delayed onset of many of clinical symptoms caused by the exposure and the susceptibility of the victim to secondary infections due to initial damage to various organs.

A literature review by Gray [21] provided an overview of the effects of HD and HN2 on a microbiological level. The review emphasized the alkylating properties of the mustard agents which result in the inability of DNA and ribonucleic acid (RNA) strains to form proper inter- and intrastrand crosslinks among and between nucleic acids and proteins. A number of literature data provide extensive evidence supporting the mutagenic properties of mustard agents.

Four toxicological studies provide data relevant to the identification of LOAELs and NOAELs reflective of the oral toxicity of HD. In the first, Hackett *et al.* [16] conducted a preliminary range-finding study in pregnant Sprague Dawley rats and New Zealand white rabbits to set HD dose levels for a subsequent teratological study. In the rats, the most striking effects were onset mesenteric lymph node inflammation and some esophageal site-of-impact at doses of 0.4 mg/kg/day and above.

Forestomach necrotic lesions in Sprague Dawley rats were the primary effect of HD administered by oral gavage in both the subchronic and two-generational reproductive studies by Sasser *et al.* [17,18]. However, the doses at which these effects became apparent differed markedly between the two studies. The 13-week study by Sasser *et al.* [17], for example, obtained the effects at 0.3 mg/kg/day (averaged for a 7-day LOAEL of 0.21 mg/kg/day), the highest dose tested, but in the two-generation, 42-reproductive study conducted by the same researchers about the same time and using the same stocks, forestomach lesions were observed at 0.03 mg/kg/day (averaged for a 7-day LOAEL of 0.021 mg/kg/day), the lowest dose of HD tested. The reason for these apparent HD toxic effects of threshold disparity is unclear. However, Sasser *et al.* [19] also demonstrated toxicological effects from HD on sperm morphology and progeny development in the dominant lethal

Table 33.2 Summary of HD Oral Toxicity Mammalian Data Relevant to TRV Derivation

Test Organism	Test Duration	Test Results			Study
		NOAEL (mg/kg/day)	LOAEL (mg/kg/day)	Effects Observed at the LOAEL	
Rats (Sprague Dawley) and Rabbits (New Zealand white)	10 days	0.2	0.4	Esophageal lesions, inflammation of the mesenteric lymph nodes	[16]
	14 days	0.5	1.0	Reduced fetal weights	
Rats (Sprague Dawley	13 weeks	0.071	0.21	Epithelial hyperplasia of the fore stomach	[17]
Rats (Sprague Dawley	42 weeks	NA	0.021	Necrosis of the fore stomach mucosa	[18]
Rats (Sprague Dawley	10 weeks	0.14	0.36	Dominant lethal effects in males and changes in sperm morphology	[19]

NA = not applicable.

assay of Sprague Dawley rats. Table 33.2 and Figure 33.1 summarize data relevant to the derivation of mammalian wildlife TRVs.

HD Mammalian Oral Toxicity: Other

In the teratological part of the study by Hackett *et al.* [16], HD in sesame oil was administered by gavage at doses of 0, 0.5, 1.0, and 3.0 mg/kg/day to 20 to 26 pregnant female Sprague Dawley rats from GDs 6 to 15 and at doses of 0, 0.4, 0.6, and 0.8 mg/kg/day to 18 or 19 New Zealand white rabbits on GDs 6 to 19. Necropsies of rats were performed on GD 20 and of rabbits on GD 30. Major organs were examined for gross lesions. Among the reproductive parameters monitored were the numbers of corpora lutea, implantation sites, resorptions, and live and dead fetuses. All viable fetuses were weighed, and the incidence of visceral and skeletal anomalies was determined. Several of the monitored maternal reproductive/developmental responses in rats, such as the number of corpora lutea and implantation sites, intrauterine mortality, and the number of live fetuses per litter, were unaffected by HD. There were, however, statistically significant reductions in fetal body weights from dams receiving 1.0 and 2.0 mg HD/kg/day.

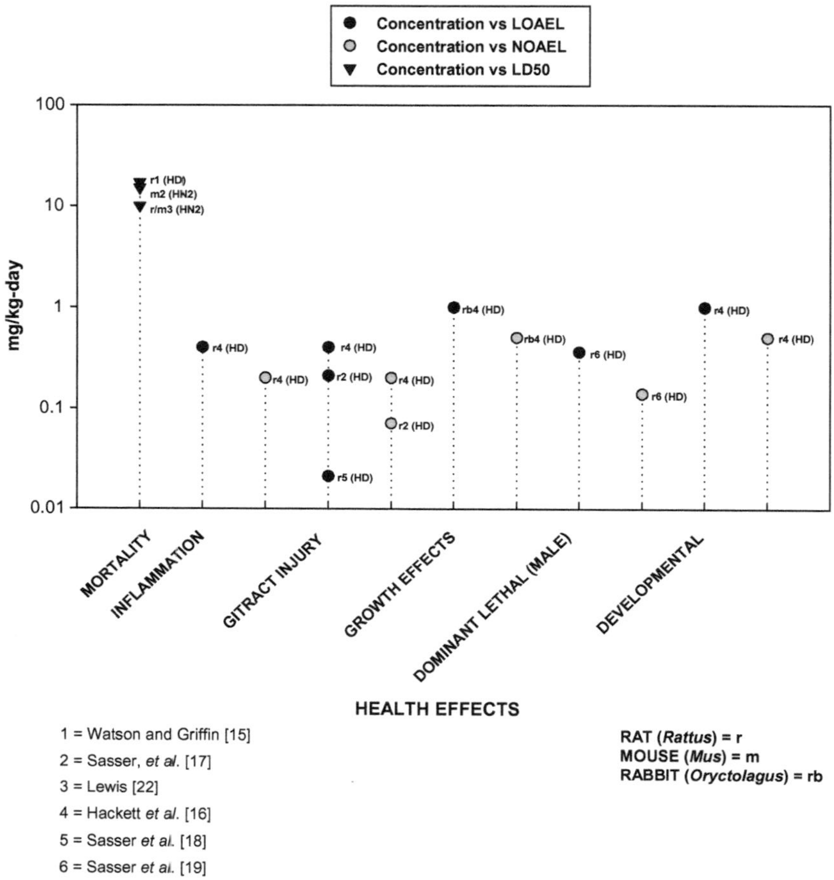

Figure 33.1 Mustard agents: Health effects in mammals. Letters by symbols represent test species, and numbers represent the study.

In addition, the number of minor anomalies, including misaligned sternebrae, was increased in fetuses from high-dose administered dams. In the teratological study in rabbits by Hackett *et al.* [16], a number of rabbits died due to injuries during gavage, and threshold doses could not be derived. There was some reduction in body weight among survivors, but HD exposure had no effect on reproductive measures (i.e., implantation losses and the number of live fetuses per litter). Furthermore, HD appeared not to have an effect on rabbits' fetal weight or length, and the placental weights were similar among the treatment groups. While there were sporadic instances of visceral and skeletal abnormalities among the fetuses, none was definitively related to treatment.

Mammalian HD Inhalation Toxicity: Acute

Following mustard agent inhalation, LC_{50} values were reported in the scientific literature [15,22] for HD: 13.3 to 25.2 milligrams per cubic meter per hour ($mg/m^3/hr$) in rats; 14.3 to 23 mg/m^3/hour in mice; 13.3 mg/m^3/hour in monkeys; 31.7 mg/m^3/hour in goats; 10 mg/m^3/hour in dogs; 11.7 mg/m^3/hour in cats; 17.1 mg/m^3/hour in rabbits; and 13.3 to 28.3 mg/m^3/hour in guinea pigs. While no experimental studies have examined the subchronic or chronic inhalation toxicity of HD, a number of single-dose inhalation and intertracheal instillation experiments have defined pathological changes that occur in an animal's airways when a laboratory animal is exposed to HD. For example, Calvet *et al.* [23] instilled 0.3 mg/kg HD in the trachea of five male Hartley strain guinea pigs/group and examined the trachea at 5 hours or 14 days after treatment. A number of observed histopathological lesions, including columnar cell shedding, formation of intraepithelial bullae, formation of atypical basal cells, and subepithelial edema, were not seen in control animals. Further evidence of the damage inflicted by HD on the respiratory airways was provided by Anderson *et al.* [24], who exposed male Crl:CD SD RR rats to 0.35 mg HD for 50 minutes, collected bronchial lavage after 1, 4, or 24 hours, and measured the concentration of total protein and gamma-glutamyl transferase and lactate dehydrogenase activity in the suspensions. While these parameters were increased posttreatment, elevation of enzyme activity as a percentage of the control was most marked 1 hour after treatment. The authors considered their data indicative of cellular damage occurring prior to the appearance of any overt pathological signs.

Langanberg *et al.* [25] measured the appearance of HD adducts in tissues from male hairless Crl: IAF (HA) guinea pigs following intravenous, cutaneous, and respiratory HD exposures. A nose-only apparatus was used to expose animals for a single 5-minute dose at the LC_{50}, and the formation of 7-HD-guanine was demonstrated in the DNA of respiratory tract tissues. Most of the 7-HD guanine was bound at the larynx and trachea 4 hours after the exposure.

Pant and Vijayaraghavan [26] examined the histomorphological and histochemical alterations induced by short-term exposure to HD via inhalation in female white mice (strain unstated). The mice received a single, 1-hour exposure to 84.6 mg/m^3 (which was twice the nominal LC_{50}), and were then observed over a 14-day period. Serial euthanizations occurred postexposure at 6, 24, and 48 hours, and at 7 days. Gross lesions of the lungs were observed, including hemorrhage, congestion, and atrophy, all of which led to necrosis and epithelial sloughing that became more severe with time. Histopathological damage to the liver, kidneys, and spleen was also observed.

More recently, Pons *et al.* [27] exposed male Sprague Dawley rats to HD via intratracheal instillation at concentrations of 0, 0.25, 0.05, or 0.1 mg/kg. Microsomal and cytosolic fractions of excised lung tissue were examined after 24 hours postexposure. While histopathological examination showed cellular architecture to be intact, increased expression of the 3A1 isomer of the cytochrome P450 group of enzymes was demonstrated by the Western blotting and sodium dodecyl sulfate polyacrylamide gel electrophoresis.

Mammalian HD Inhalation Toxicity

No information was found in literature on mammalian subacute, subchronic, chronic, or other mammalian HD inhalation toxicity studies.

Studies Relevant for Mammalian HD Inhalation Exposure TRV Development

No information was found in literature on mammalian HD longer-term inhalation exposure studies.

Mammalian HD Dermal Toxicity

A number of mustard agent dermal LD_{50} values have been reported in the scientific literature [15,21,28–30], including 5 to 18 mg/kg HD in rats, 20 to 155 mg/kg HD in mice, 40 to 100 mg/kg HD in rabbits, and 20 mg/kg HD in dogs.

Kadar *et al.* [31] characterized acute and delayed corneal lesions in rabbits resulting from mustard gas vapor exposure. In this study, rabbit eyes were exposed to sulfur mustard vapor at concentrations of 370 and 420 micrograms per liter (μg/L) for 2 minutes. Dose-dependent clinical signs peaked at 24 to 74 hours post-exposure and were characterized by eye closure, eyelid swelling, conjunctival hyperemia, corneal erosion, and inflammation. The onset of clinical symptoms began within 2 to 6 hours of exposure. Histopathologic changes were characterized by absence of epithelial layer, marked stromal edema, and cellular infiltration. Corneal recovery was observed following initial pathological change. However, delayed and more severe pathological changes followed this recovery with characteristic corneal edema, opacity, erosions, and neo-vascularization in 25 to 40% of cases.

Vijayaraghavan *et al.* [30] carried out a series of HD toxicology experiments on Swiss mice, one of which featured dermal application of HD in polyethylene glycol 300 at 0.25 and 0.5 times the dermal LD_{50} (calculated by the authors as 155 mg/kg in that animal model). Reduced glutathione and malondialdehyde was measured in samples of blood and excised liver, and fluctuations possibly related to an effect of HD on lipid peroxidation were observed. In another phase of the study, potential antidotes for mustard poisoning, such as vitamin E and the flavonoids gossypin and hydroxyethyl rutosides, were found to marginally increase survival rate.

The same research group examined the effect of dermally applied HD on the activities of antioxidant enzymes such as superoxide dismutase, catalase and glutathione peroxidase in blood cells (erythrocytes, leucocytes, and platelets) and body tissues (liver, kidney, spleen, and brain) [32]. The results from both of these experiments suggested that HD applied to the dermiscan be taken up systematically. Direct evidence of systemic take up of HD from the skin was provided by the radiolabeling experiments [33,34]. They applied HD to the shaved skin of rats (strain unstated) and detected such HD metabolites as thiodiglycol, thiodiglycol sulfoxide, 1,1-sulfonylbis-[2-(methylthio)ethane], 1,1-sulfonylbis[2-(methylsulfinyl)-ethane], and 1-methyl-2-[2-(methylthio)-ethylsulfonyl]-ethane in the urine. Husain *et al.* [32] concluded from these experiments that HD inhibited antioxidant activities in leucocytes and spleen and suggested a key biochemical step in HD-intoxication outcomes may be formation of reactive oxidant species. In a later series of experiments, these researchers applied HD dermally to the backs of male rats (strain unstated) and demonstrated that nifedipine improved survival of animals challenged with 1.5 times the dermal LD_{50} [35].

Zlotogorski *et al.* [36] proposed an experimental system for measuring the dermal effects of HD and the potential efficacy of antidotes by measuring the fluctuations over time of the degree of swelling in the outer ears of the rabbits. The swelling subsided slightly 12 hours after the application. However, histopathological examination of affected tissue biopsies revealed a number of necrotic changes, such as the formation of necrotic keratinocytes and subepidermal vesiculation.

HD Avian Toxicity Summary

No information was found in the literature on avian HD toxicity studies from which TRVs could be developed.

HD Amphibian Toxicity Summary

No information was found in the literature on amphibian HD toxicity studies from which TRVs could be developed.

HD Reptilian Toxicity Summary

No information was found in the literature on reptilian HD toxicity studies from which TRVs could be developed.

Nitrogen Mustard (HN2) Summary of Mammalian Toxicity

No relevant HN2 mammalian toxicity studies were identified in the literature despite its long use as a chemotherapy agent. There are, however, numerous study journal articles in the medical literature. These articles are generally about a half-century old, and their contents pertain to disease, patient, or treatment. Recent studies have used either animal tissues or cell lines *in vitro* and focused on cellular biochemical actions [21,36].

HN2 Mammalian Oral Toxicity

Although no animal studies on oral exposure of HN2 were identified in the literature, HN2 oral LD_{50} values in the 10 to 20 mg/kg range were reported in mice [17] with a corresponding acute HN2 oral LD_{50} of 10 mg/kg in both rats and mice and 5 mg/kg in rabbits.

HN2 Mammalian Inhalation Toxicity

A number of acute HN2 inhalation LC_{50} values have been reported in the scientific literature [15,22], including 20 mg/m^3/hour in rats, 750 mg/m^3/hour in mice, 333 mg/m^3/hour in dogs, and 83.3 mg/m^3/hour in rabbits. No experimental studies were identified that addressed either subchronic or chronic mammalian HN2 inhalation toxicity.

HN2 Mammalian Dermal Toxicity

A number of HN2 dermal LD_{50} values have been reported in the scientific literature [15,22,28–30], including 12 mg/kg in rats, 29 mg/kg in mice, and 12 mg/kg in rabbits.

HN2 Mammalian Toxicity: Other

Although there are no available studies addressing HN2 teratogenicity in mammals, HN2 has been shown to be a potent intraperitoneal and intravenous reproductive toxicant and teratogen. In one of many such studies, aqueous HN2 was injected in pregnant Swiss-Webster mice at doses of up to 3.2 mg/kg. Dose-dependent reductions in the number of litters, number of fetuses/litter, weight of fetuses; and increases in the number of soft tissue and skeletal anomalies were observed. However, since HN2 is only slightly soluble in water and readily hydrolyzes, it may be questionable to attribute those effects to the HN2 dose rather than to a by-product [37].

HN2 Avian Toxicity Summary

The toxicology of HN2 is primarily derived from developmental biology, embryology, and cellular and genetics research using chicken embryos. Developmental abnormalities were observed in a dose and developmental stage linked pattern when 0.025 to 0.045 mg/mL HN2 aqueous solutions were injected into fertilized eggs [38]. The extensive study demonstrated that the type of malformations depended on the dose, the stage of the embryo at the time of treatment, and the method of administration [39–41]. However, since HN2 is only slightly soluble in water and readily hydrolyzes, it may be questionable to attribute dose effects to HN2 rather than to hydrolysis by-product(s).

HN2 Amphibian Toxicity Summary

In experiments with newt (*Pleurodeles waltii*) larvae, they were exposed for 45 minutes in 1.25 μg/mL, 2.5 μg/mL, 5 μg/mL, and 10 μg/mL-HN2 aqueous solutions; rinsed in a series of distilled water baths; then allowed to develop in tap water. Dose-related survival effects on the larvae, along with range of limb development anomalies, were documented. However, since HN2 is only slightly soluble in water and readily hydrolyzes, it may be questionable to attribute dose effects to HN2 rather than to hydrolysis by-product(s) [42].

HN2 Reptilian Toxicology

No information was found on reptilian HN2 toxicity studies.

REFERENCES

[1] Harvey SP, Szafraniec LL, Beaudry WT, Earley JT, Irvine RL. Neutralization and biodegradation of sulfur mustard. Aberdeen Proving Ground (MD): Research, Development and Engineering Center; 1997 Feb, USARDEC Tech Report No. ERDEC-TR-388.

[2] Agency for Toxic Substances and Disease Registry (ATSDR). Toxicological profile for sulfur mustard. Atlanta (GA): U.S. Department of Health and Human Services, Public Health Service; 2003.

[3] Wils ER, Hulst AG, van Laar J. Analysis of thidiglycol in urine of victims of an alleged attack with mustard gas. Part II. J Anal Toxicol 1988 Jan-Feb;12(1):15–9.

[4] Fischer K. Georg wilhelm steinkopf [internet]. Available from: In: Schattkowsky M, editor. Sächsische Biografie [cited 28 December 2010]. Dresden (Germany): Institute für Sächsische Geschichte und Volkskunde; 2004.

[5] Centers for Disease Control and Prevention (CDC). Facts about Nitrogen Mustards. [Internet]. [cited 2013]. Available at: www.bt.cdc.gov/agent/nitrogenmustard/basics/facts.asp.

[6] Munro NB, Talmage SS, Griffin GD, Waters LC, Watson AP, King JF, et al. The source, fate, and toxicity of chemical warfare agent degradation products. Environ Health Perspect 1999 Dec;107(12):933–74.

[7] USACHPPM. Standard practice for wildlife toxicity reference values. Aberdeen Proving Ground (MD): U.S. Army Center for Health Promotion and Preventive Medicine; 2000, Technical Guide No. 254.

[8] Small MJ. Compounds formed from the chemical decontamination of HD, GB, and VX and their environmental fate. Ft. Detrick (MD): U.S. Army Medical Bioengineering Research and Development Laboratory; 1984, Tech Rep 8304; AD A149515.

[9] Hazardous Substances Data Bank (HSDB). Mustard Agents [Internet]. Bethesda (MD): National Library of Medicine (US), National Institutes of Health [cited 2001].

[10] Johnsen BA, Blanch JH. Analysis of snow samples contaminated with chemical warfare agents. Arch Belg 1984;(Suppl):22–30.

[11] Epstein J, Rosenblatt DH, Gallacao A, McTeague WF. Summary report on a data base for predicting consequence of chemical disposal operations. Edgewood Arsenal (MD): U.S Department of the Army; January 1973, Report Number: EASP 1200-12.

[12] Lee KP, Allen HE. Environmental transformation mechanisms of thidiglycol. Environ Toxicol Chem 1998;17(9):1720–6.

[13] Medvedeva N, Polyak Y, Zaytceva T. Microbiological destruction of mustard in soil Environ sci pollut control series (bioremediation of contaminated soils), 32; 2000, 151-76.

[14] Popiela S, Witkiewicza Z, Chrzanowski M. Sulfur mustard destruction using ozone, UV, hydrogen peroxide and their combination. J Hazard Mater 2008 May1;153 (1-2):37–43.

[15] Watson A, Griffin GD. Toxicity of vesicant agents scheduled for destruction by the chemical stockpile disposal program. Environ Health Perspect 1992 Nov;98:259–80.

[16] Hackett PL, Rommerein RL, Burton FG, Buschbom RL, Sasser LB. Teratology studies on lewisite and sulfur mustard agents: effects of sulfur mustard in rats and rabbits. Richland (WA): Pacific Northwest Laboratory; 1987, Stock Number: AD A187495. Prepared for U.S. Army Medical Research and Development Command, Ft. Detrick, Frederick, MD.

[17] Sasser LB, Miller RA, Kalkwarf DR, Buschbon RL, Cushing JA. Toxicological studies on lewisite and sulfur mustard agents: subchronic toxicity of sulfur mustard (HD) in rats. Richland (WA): Pacific Northwest Laboratory; 1989, Stock Number: AD A214555. Prepared for U.S. Army Medical Research and Development Command, Ft. Detrick, Frederick, MD.

[18] Sasser LB, Miller RA, Kalkwarf DR, Buschbon RL, Cushing JA. Toxicological studies on lewisite and sulfur mustard agents: two-generation reproductive study on sulfur mustard (HD) in rats. Richland (WA): Pacific Northwest Laboratory; 1989, Stock No.: AD A216423. Prepared for U.S. Army Medical Research & Development Command, Ft. Detrick, Frederick, MD.
[19] Sasser LB, Cushing JA, Kalkwarf DR, Buschbom RL. Toxicology studies on lewisite and sulfur mustard agents: modified dominant lethal study of sulfur mustard in rats. Pacific Northwest Laboratory; 1989, Stock Number: AD A214556. Prepared for U.S. Army Medical Research and Development Command, Ft. Detrick, Frederick, MD.
[20] Dacre JC, Goldman M. Toxicology and pharmacology of the chemical warfare agent sulfur mustard. Pharmacol Rev 1996 Jun;48(2):289–326.18.
[21] Gray PJ. A literature review on the mechanism of action of sulfur and nitrogen mustard. Ft. Belvoir, VA: Defense Technical Information Center; 1989 Jul, Number: AD-A 215070.
[22] Lewis RJ. Sax's dangerous properties of industrial chemicals. 8th Edition New York, NY: Van Norstrand Reinhold; 1992.
[23] Calvet JH, Jarreau PH, Levame M, D'Ortho MP, Lorino H, Harf A, et al. Acute and chronic respiratory effects of sulfur mustard intoxication in guinea pig. J Appl Physiol 1994 Feb;76(2):681–8(1985).
[24] Anderson DR, Byers SL, Clark CR, Schlehr JA. Biochemical alterations in rat lung lavage fluid following acute sulfur mustard inhalation. Inhal Toxicol 1997;9(1):43–51.
[25] Langanberg JP, van der Schans GP, Spruit HE, Kuijpers WC, Mars-Groenendijk RH, van Dijk-Knijnenburg HC, et al. Toxicokinetics of sulfur mustard and its DNA-adducts in the hairless guinea pig. Drug Chem Toxicol 1998;21(Suppl 1):131–47.
[26] Pant SC, Vijayaraghavan R. Histomorphological and histochemical alterations following short-term inhalation exposure to sulfur mustard on visceral organs of mice. Biomed Environ Sci 1999 Sept;12(3):201–13.
[27] Pons F, Calvet JH, Haag M, Raeppel V, Keravis T, Frossard N. Altered expression of lung cytochrome P450 3A1 in rat after exposure to sulfur mustard. Pharmacol Toxicol 2001 Jan;88(1):40–4.
[28] Vojvodic V, Milosavljevic Z, Boskovic B, Bojanic N. The protective effect of different drugs in rats poisoned by sulfur and nitrogen mustards. Fundam Appl Toxicol 1985 Dec;5(6Pt2):160–8.
[29] Anguelov A, Belchev L, Angelov G. Experimental sulfur mustard gas poisoning and protective effect of different medicines in rats and rabbits. Indian Vet J 1996;73(5):546–51.
[30] Vijayaraghavan R, Sugendran K, Pant SC, Husain K, Malhotra RC. Dermal intoxication of mice with bis(2-chloroethyl)sulphide and the protective effect of flavonoids. Toxicology 1991;69(1):35–42.
[31] Kadar T, Turetz J, Fishbine E, Sahar R, Chapman S, Amir A. Characterization of acute and delayed ocular lesions induced by sulfur mustard in rabbits. Curr Eye Res 2001 Jan;22(1):42–53.
[32] Husain K, Dube SN, Sugendran K, Singh R, Das Gupta S, Somani SM. Effect of topically applied sulfur mustard on antioxidant enzymes in blood cells and body tissues of rats. J Appl Toxicol 1996 May-Jun;16(3):245–8.
[33] Hambrook JL, Harrison JM, Howells DJ, Schock C. Biological fate of sulfur mustard (1,1 thiobis(2-chloroethane)): urinary and faecal excretion of 35S by rat after injection or cutaneous application of 35S-labelled sulfur mustard. Xenobiotica 1992 Jan;22 (1):65–75.
[34] Black RM, Hambrook JL, Howells DJ, Read RW. Biological fate of sulfur mustard, 1,1-thiobis(2-chloroethane). Urinary excretion profiles of hydrolysis products and beta-lyase metabolites of sulfur mustard after cutaneous application in rats. J Anal Toxicol 1992 Mar-Apr;16(2):79–84.

[35] Mazumder PK, Sugendran K, Vijayaraghavan R. Protective efficacy of calcium channel blockers in sulfur mustard poisoning. Biomed Environ Sci 1998 Dec;11(4):363–9.
[36] Zlotogorski A, Goldenhersh M, Shafran A. A model for quantitative measurement of sulfur mustard skin lesions in the rabbit ear. Toxicology 1997 Jun 27;120(2):105–10.
[37] Papirmeister B, Gross CL, Meier HL, Petreli JP, Johnson JB. Molecular basis for mustard-induced vesication. Fundam Appl Toxicol 1985 Dec;5(6Pt2):S134–49.
[38] Salzgeber B. Limb malformation production in chicken embryos treated with nitrogen mustard at 48 hours of incubation. Adv Exp Med Biol 1972;27:175–88.
[39] Blattner RJ, Williamson AP, Simonsen L, Robertson GG. Teratogenesis with cancer chemotherapeutic agents. J Pediatr 1960 Feb;56:285–93.
[40] Gibson JE, Becker BA. Teratogenicity of structural truncates of cyclophosphamide in mice. Teratology 1971 May;v4n2:141–50.
[41] Salzgeber B. Mechanisms of limb teratogenesis: malformations in chick embryo induced by nitrogen mustard. Adv Study Birth Defects 1979;V1:141–62.
[42] Lauthier M. Cell death and abnormalities in limb morphogenesis of pleurodeles waltii micah (Urodela, Amphibia) after nitrogen mustard treatment. Roux's Arch Dev Biol 1980;189(1):35–45.

CHAPTER 34

Wildlife Toxicity Assessment for Lewisite

Sang Ho Lee, Gunda Reddy

Contents

INTRODUCTION

The arsenical vesicant lewisite, or dichloro-(2-chlorovinyl) arsine (CAS No. 541-25-3), is a chemical warfare agent. Lewisite was sporadically produced and stockpiled in the United States since WWI, and other nations have produced, and possibly used, it as recently as the 1990s. Thus, there is potential for unplanned releases of lewisite to the environment. The majority of lewisite research was subject to wartime efficacies and availability constraints, with subsequent research influenced by substance characteristics. Lewisite is a

Wildlife Toxicity Assessments for Chemicals of Military Concern
http://dx.doi.org/10.1016/B978-0-12-800020-5.00034-X

structurally complex mixture of several compounds produced by the catalyzed reaction of arsenic trichloride and acetylene, with a molecular formula of $C_2H_2AsCl_3$. In the presence of moisture, it rapidly converts to the more stable but highly toxic lewisite oxide (2-chlorovinylarsenous acid). Synonyms include 2-chlorovinyldichloroarsine, 2-chlorovinyl arsonous dichloride, chlorovinylarsine dichloride, and beta-chlorovinyldichloroarsine [1–3].

This chapter summarizes current knowledge of the potential harmful impacts of lewisite on wildlife and also reports lewisite Toxicity Reference Values (TRVs) that may be of importance for the health of wildlife (mammals, birds, reptiles, and amphibians) potentially exposed to the substance. The toxicity endpoints relate to the likelihood of effects in individual organisms with possible relevance to a population of organisms in the wild, but they are unrelated to demographic rates or outcomes for any particular population.

The U.S. Army Center for Health Promotion and Preventive Medicine (USACHPPM) Technical Guide (TG) 254, *Standard Practice for Wildlife Toxicity Reference Values* [4] describes and documents the protocol for this assessment.

TOXICITY PROFILE

Environmental Fate and Transport

There is little information on the environmental fate and transport of lewisite; because of its physical chemical properties, lewisite is expected to degrade upon release to the environment (Table 34.1). Regardless of the method of degradation, the arsenic component will not be eliminated, and some degradation products will be toxic arsenicals, such as chlorovinyl arsenous oxide, whose toxicities have been well documented. When applied to the soil, lewisite quickly volatilizes or can convert to its oxide [5]. While lewisite is unlikely to bioaccumulate through food chains, its arsenic by-products may exhibit potential toxicity. No data on lewisite's fate in the atmosphere was found. However, an ultraviolet absorption spectrum of 200 to 350 nanometers suggests that some photodegradation might occur. Rapid hydrolysis may occur in the gas phase [6]. The compound may also react with ozone [7].

Lewisite is slightly soluble in water (0.5 grams per liter (g/L)) with rapid hydrolysis resulting in lewisite oxide and hydrogen chloride [2]. Upon standing in water, the toxic trivalent arsenic of lewisite oxide is converted to the less toxic pentavalent arsenic [8]. Given its ready ability to both volatilize and undergo hydrolysis to lewisite oxide in moist conditions, lewisite is not thought to persist in soil; however, it may have intermediate persistence in moist soil [9].

Table 34.1 Summary of Physical-Chemical Properties of Lewisite

CAS No.	541-25-32
Molecular weight	207.32
Color	Colorless when pure; "war gas" is amber to dark brown
Physical state	Volatile oily liquid
Freezing/Melting point	–18.2 to 0.1 °C (depending on purity)
Boiling point	190 °C
Odor	Very little when pure; Geranium-like
Solubility in: Water and dilute mineral acids	Insoluble (500 mg/L)
Solubility in: Organic solvents, oils, and alcohol	Soluble
Partition coefficients:	
Log K_{ow}	2.56
Log K_{oc}	2.1
	0.35 at 20 °C, 0.22 at 25 °C
Vapor density (Air = 1)	7.1
Liquid density (g/mL)	1.89 at 20 °C
Volatility (mg/m^3)	2,500 at 20 °C
Viscosity (Centipoise)	2.257 at 20 °C
Henry's Law constant at 25 °C	3.2×10^{-4} atm.m^3/mole
Conversion factors	1 ppm = 8.48 mg/m^3 1 mg/m^3 = 0.12 ppm

Sources: HSDB [7], MSDS [21].

Summary of Mammalian Toxicity

Because lewisite was designed for military use, it is generally accepted that it is a potent vesicant and system poison. The rapid onset of toxic effects from acute exposures to lewisite via vapor inhalation, direct dermal contact, or ingestion can be lethal [10]. Lewisite's vesicant property is characterized by immediate severe pain and possible corneal necrosis; percutaneous absorption toxicity is characterized by pulmonary edema, diarrhea, agitation, weakness, hypothermia, and hypotension [10]. Table 34.2 lists lewisite toxicity studies and identifies LOAELs and NOAELs of potential relevance to lewisite TRV development.

Mammalian Oral Toxicity: Acute

There are few data on acute oral lethality of lewisite, but oral median lethal dose (LD_{50}) values of 50 mg/kg and 5–15 mg/kg have been reported in rats and mice, respectively [2].

Table 34.2 Summary of Relevant Mammalian Data of Lewisite for TRV Derivation

Test Organism	Test Duration	Test Results: NOAEL (mg/kg/day)	Test Results: LOAEL (mg/kg/day)	Effects Observed at the LOAEL	Study
Rats (Sprague Dawley)	42 weeks	—	0.25	Two-generation reproductive study; reduction in body weights	[13] [14]
Rats (Sprague Dawley)	10 days	1.5	2.0	10% mortality; reduction in maternal and fetal body weights and number of viable fetuses	[12]
Rabbits (New Zealand white)	14 days	0.07	0.2	Calculated from mortality data	[12]
Rats (Sprague Dawley)	13 weeks	0.36	0.71	Inflammation of glandular stomach	[11]
Rats (Sprague Dawley)	13 weeks	0.71	1.43	Necrotic lesions of the forestomach	

Mammalian Oral Toxicity: Subacute

No specific subacute lewisite mammalian toxicity studies were identified (see Mammalian Oral Toxicity: Other).

Mammalian Oral Toxicity: Subchronic

In a 13-week subchronic lewisite toxicity study, 10 Sprague Dawley rats per sex per group received 0, 0.01, 0.1, 0.5, 1.0, or 2.0 mg/kg/day lewisite in sesame oil by gavage 5 days/week [11]. These values converted to 7 days/week average doses of 0, 0.007, 0.07, 0.36, 0.71, and 1.43 mg/kg/day, respectively. Clinical signs were monitored twice daily, body weights were recorded weekly, and all animals received an ophthalmological examination at the start and finish of the experiment. Hematological parameters of blood samples taken at the start and finish of the experiment were measured, and clinical chemistries were determined using blood serum. At euthanization, the weights of the major organs were recorded, and a complete necropsy was performed on each animal. Histopathology examinations were made on organs and tissues from some high-dose and control animals and on potential target organs from the intermediate groups.

The clinical chemistry results showed decreases in serum protein and creatinine concentrations and reductions in the activities of the enzymes glutamic oxaloacetic transaminase (SGOT) and glutamic pyruvic transaminase (SGPT) in high-dose males (2 mg/kg), while females at this high dose displayed elevated lymphocyte and platelet counts. These changes might be indicative of potentially important biochemical imbalances. The findings were inconsistent across dosages and genders, compounded by an insufficient number of surviving rats in some groups and an experiment that was designed neither to look for nor follow up on such effects. The gastrointestinal tracts of high-dose (2 mg/kg) rats of both sexes showed necrosis of the forestomach, evidenced by destruction of the stratified epithelium, infiltration of neutrophils, hemorrhage, edema, and proliferation of fibroblasts in the gastrointestinal tract. Inflammation of the glandular stomach was also observed in rats receiving 1.0 and 2.0 mg/kg/day. This observation suggested a 0.5 mg/kg/day dose (the average calculated for the 7-day dose of 0.36 equivalent) for oral toxicity NOAEL and a LOAEL of 1 mg/kg/day (equivalent dose for 7-day dose of 0.71 mg/kg/day) as appropriate levels to be considered.

Mammalian Oral Toxicity: Chronic

No studies addressing chronic lewisite oral toxicity were identified.

Mammalian Oral Toxicity: Other

This category of mammalian toxicity studies includes the developmental and reproductive toxicity studies described below. The developmental experiments were conducted with lewisite in pregnant female Sprague Dawley rats and New Zealand white rabbits [12]. The first experiment was conducted to select the dose levels for a subsequent main study. Lewisite 1 was administered to rats at doses of 0, 0.5, 1.0, 2.0, and 2.5 mg/kg via gavage in sesame oil on gestation days (GDs) 6 through 15. Pregnant rabbits were dosed orally with 0, 0.5, 1.0, 1.5, and 2.0 mg lewisite/kg on GDs 6 through 19. All surviving rats were euthanized on GD 20, and the surviving rabbits were euthanized on GD 30. Body weights were monitored on GDs 0, 6 to 16, and 20 in rats; and on GDs 0, 6 to 20, and 30 in rabbits. Blood samples were obtained from all survivors for hematological measurements, and all animals were examined for alterations in major organs, especially the gastrointestinal tracts. Numbers of corpora lutea, implantation sites, resorptions, and live and dead fetuses were recorded. All live fetuses were weighed and examined for gross external defects.

The authors reported no maternal or fetal effects of lewisite in rats at or below a dose of 1.0 mg/kg/day. However, there was 10% maternal mortality in rats receiving 2.0 mg/kg/day, with reduced maternal weight gain and fetal body weight in survivors. The number of viable fetuses in high-dose dams was lower than in controls. Because most of the animals died due to toxicity (in the higher-dose groups, 2 mg/kg) or dosing trauma (in the low-dose group), no clear-cut LOAEL or NOAEL was designated in rats.

In the main study, lewisite was administered via gavage in sesame oil to 25 pregnant female Sprague Dawley rats from GDs 6 to 15 at 0, 0.5, 1.0, and 1.5 mg/kg/day; and to 18 or 19 New Zealand white rabbits on GDs 6 to 19 at 0, 0.07, 0.2, and 0.6 mg/kg/day. Rats and rabbits were necropsied on GDs 20 and 30, respectively. The major organs were examined for gross lesions, and the following reproductive parameters were monitored and enumerated: corpora lutea, implantation sites, resorptions, and live and dead fetuses. All viable fetuses were weighed, and the incidences of visceral and skeletal anomalies were determined [12].

In rats, there were no effects of treatment on maternal body weights or evidence of other signs of maternal toxicity. Reproductive measures such as the number of corpora lutea, implantation sites, numbers of live and dead fetuses per litter, along with various teratological parameters, were also unaffected by treatment at all doses (0.5 to 1.5 mg/kg) tested.

In rabbits, significant dose-dependent mortality existed when mortality was corrected for deaths from other causes, and the authors reported the lewisite-related incidence of mortality to be 13, 46, and 69% in rabbits dosed at 0.07, 0.2, and 0.6 mg/kg/day, respectively. Other lewisite-induced effects included reductions in body weight gain in high-dose dams and decreases in hematocrit. The developmental or tautological effects of lewisite were seen in the high dose in rabbits, suggesting that the incidences of supernumerary ribs and reduced pelvic ossification might have been compound-related. This fetal effect suggests a LOAEL of 0.6 mg/kg/day with a corresponding NOAEL of 0.2 mg/kg/day although the high mortality precludes confidence in these effect levels in rabbits [12].

Two-generation reproductive toxicity studies were conducted in Sprague Dawley rats to determine the reproductive effects of lewisite. Rats of each sex per group were gavaged with 0, 0.1, 0.25, or 0.6 mg lewisite/

kg/day in sesame oil, 5 days/week, for 13 weeks prior to mating. The dosing schedule through the reproductive period was 7 days/week throughout gestation, and 5 days/week through parturition and lactation. The F_1 subjects were then exposed, within their dose group, to the same regime of premating (13-week), gestation, parturition, and lactation. A full suite of maternal and reproductive/development effects was monitored, but no significant lewisite-induced effects were observed on reproductive function or pregnancy in either generation. There was dose-dependent maternal and paternal mortality associated with lewisite exposure. In the F_0 generation, 11 females, and 4 males died at 0.60 mg/kg/day, and 4 females died at 0.25 mg/kg/day. Mortality rates were higher in the F_1 generation: 18 female and 6 male rats died at 0.60 mg/kg/day, 5 females and 2 males died at 0.25 mg/kg/day, and 2 females and 1 male died at 0.10 mg/kg/day. One female among the controls died, which occurred in the F_0 generation; there were no mortalities among the male control rats. There was also a significant reduction in body weight for F_0 females and F_1 males at 0.25 mg/kg/day. No effects level could be determined in rats as mortality was found in the F_0 generation [13,14].

STUDIES RELEVANT TO MAMMALIAN TOXICITY REFERENCE VALUE DEVELOPMENT: LEWISITE INGESTION EXPOSURES

Table 34.2 lists lewisite toxicity studies and identifies LOAELs and NOAELs of potential relevance to lewisite TRV development. Since Lewisite was designed for military use, it is generally accepted that lewisite is a potent vesicant and system poison, causing rapid onset of potentially lethal toxic effects from acute exposures via vapor inhalation, direct dermal contact, or ingestion [10]. Lewisite's vesicant property is characterized by immediate severe pain and possible corneal necrosis; percutaneous absorption toxicity is characterized by pulmonary edema, diarrhea, agitation, weakness, hypothermia, and hypotension [10]. Since lewisite in its pure form is volatile and undergoes rapid hydrolysis and degradation, it may not be ingested by wildlife. Several others have reviewed most of the available lewisite toxicology and pharmacological information [3,9,10,15,16].

Several toxicological studies [11–14] contain data relevant to the determination of lewisite toxicity reference values (TRVs). The subchronic study of lewisite in Sprague Dawley rats [11] identified the primary effect of lewisite to be necrotic lesions of the forestomach that occurred in animals receiving 1.0 mg/kg/day, equivalent to an average 7-day dose LOAEL of 0.71 mg/kg/day. The LOAEL may be consistent with that predicted to an average 7-day dose LOAEL of over 0.6 mg/kg/day from the 42-week, two-generation reproductive study [13,14] in which the high dose of 0.6 mg/kg/day failed to induce any overt toxic effects. In the subacute teratological study in Sprague Dawley rats and New Zealand white rabbits, rats showed reduction in fetal weights at the high-dose level of 2.0 mg/kg/day [12].

The teratology study of New Zealand white rabbits showed increased mortality [12]. Because mortality has clear implications for population sustainability, these results were used for mammalian TRV derivation. Further supporting the use of these data for TRV derivation is that rabbit mortality showed a clear dose-response and was statistically significant. However, Hackett *et al.* [12] did not statistically analyze the mortality data, so the raw data were obtained and survival analyses conducted to determine a NOAEL and LOAEL. Based on the statistical reanalysis, the NOAEL was 0.07 mg/kg, and the LOAEL was 0.2 mg/kg. The data also indicated that the rabbits were more sensitive to oral lewisite exposure than rats. Therefore, a TRV based on rabbit mortality is likely more protective of wildlife. The New Zealand white rabbit is a monogastric herbivore and potentially resembles numerous animals that could be exposed to lewisite. New Zealand white rabbits may also be genetically more similar to wild rabbits than the Sprague Dawley rat is to wild rats. Unlike the Sprague Dawley rats that were developed from similarly derived strains about 50 years ago for specific experimental purposes, the large, placid New Zealand white rabbit has been bred as a domestic meat animal for hundreds of years and appears to differ little in behavior from its wild relatives.

Mammalian Inhalation Toxicity: Acute

Although a number of lewisite inhalation median lethal dosages (LC_{50} values) were reported in the scientific literature [2,9,17], no studies deemed reasonably acceptable under current toxicology standards were identified. The U.S. Department of the Army study [18] reported mouse head-only, 10-minute, lewisite vapor exposure lethal concentrations (LCt_{50}) of 1,400–1,500 mg/min/m^3 (nominal concentration), 1,600 mg/min/m^3 (nominal concentration), and 1,500 mg/min/m^3 (analytical concentration), but experimental information was not provided. The same report [18] contained information on whole-body lewisite vapor exposure (Table 34.3).

Table 34.3 Acute Inhalation Toxicity Values of Lewisite in Mammals

Species	Time (min)	LCt_{50} (mg/min/m^3)	Concentration Determination
Mouse	10	900–1400	Nominal
	10	2,800	Nominal
	10	1,500	Analytical
	10	2,500–2,800	Nominal
	9–14	500	Analytical
Rat	9–25	1,500	Analytical
	60–180	580	Analytical
Guinea pig	9–14	1,000	Analytical
	60–180	470	Analytical
Rabbit	7.5–13	1,200	Analytical
	60–310	1,500	Analytical
Goat	100–255	1,250	Analytical
Dog	7.5–15	1,400	Nominal
Man	N/A	1,200–1,500	N/A

N/A = not applicable.
Sources: [2,9].

Acute effects of lewisite inhalation toxicity include pulmonary edema, diarrhea, restlessness, weakness, subnormal temperature, low blood pressure, a change in capillary permeability resulting in hemo-concentration (rise in blood cellular fraction), liver necrosis, and hemolytic anemia [18].

Mammalian Inhalation Toxicity: Other

No studies were identified on subacute, subchronic, chronic, or other inhalation toxicities in mammals.

STUDIES RELEVANT TO MAMMALIAN TOXICITY REFERENCE VALUE DEVELOPMENT: LEWISITE INHALATION EXPOSURES

Lewisite was designed as a highly reactive, toxic gas capable of eliciting a rapid response that precluded long-duration inhalation exposures. There are limited data on the inhalation toxicity of lewisite. Although a number of lewisite LCt_{50} values have been reported for a variety of mammalian species [2,18], the primary references could not be located. The values are relatively consistent over the range of mammalian species. The lowest LCt_{50} for exposures lasting about 10 minutes was 500 mg/min/m^3 in mice exposed for 9 to 14 minutes [18].

Mammalian Dermal Toxicity

Studies of dermal penetration and associated health effects in human volunteers exposed to very small doses of lewisite showed debilitating local, site-of-impact, and systemic effects, including elevations in plasma urea and reductions in the volume of urine [3]. A lewisite dermal exposure LD_{50} of 15 mg/kg was reported in rats, mice, and dogs [17]. Additionally, lewisite dermal exposure LD_{50} values of 20 and 24 mg/kg for rats and 38 mg/kg for dogs were reported [18]. Pulmonary edema is a likely mechanism of action [3]. Lethal dose low (LDLO) values of 4, 5, and 6 mg/kg for rabbits, 12 mg/kg for guinea pigs, 10 to 24 mg/kg for goats, and 38 mg/kg for humans have been reported [17,18]. Rice and Brown [19] exposed four 10 cm^2 areas of shaved skin on domestic white pigs to lewisite (0.3 mg/cm^2) and observed the resulting progression of epidermal lesions over a 21-day period. A sequence of necrotic lesions was seen, including epidermal degeneration, vacuolation of the basal keratinocytes, and inflammation of the dermis. A degree of recovery may have occurred at later time points.

STUDIES RELEVANT TO MAMMALIAN TOXICITY REFERENCE VALUE DEVELOPMENT: LEWISITE DERMAL EXPOSURES

Dermal exposure to lewisite can cause irritation, vesication, blistering, and a burning sensation at the site of application [3]. At high-exposure concentrations, lewisite causes pulmonary edema leading to death. Lewisite-induced mortality might provoke ecologically relevant effects by direct reduction of population size, which could be used as the basis for a dermal exposure TRV. While a number of lewisite dermal LD_{50} values for a variety of mammalian species were reported [2,18], the primary references could not be obtained. A dermal LD_{50} of 15 mg/kg in rats, mice, and dogs indicates that lewisite is toxic when exposure occurs via the dermal route; this data can serve as the basis for a dermal TRV.

SUMMARY OF AVIAN LEWISITE TOXICITY

No data were found on toxicity of lewisite in birds.

SUMMARY OF AMPHIBIAN TOXICITY

At a concentration of 0.5 parts per million (ppm), lewisite was lethal to tadpoles [20]. No other data were found on the toxicity of lewisite in amphibians.

SUMMARY OF REPTILIAN TOXICITY

No data were found on the toxicity of lewisite in reptiles.

RECOMMENDED TOXICITY REFERENCE VALUES

TRVs for Mammals

TRVs for Ingestion Exposures for the Class Mammalia

Results from the Hackett *et al.* subacute study showed that lewisite caused mortality in New Zealand white rabbits [12]. Statistical analysis of that data indicated that the NOAEL for mortality was 0.07 mg/kg, and the LOAEL was 0.2 mg/kg [12], and the study found dose-dependent mortality in rabbits given oral doses of lewisite. Unfortunately, there were no statistical analyses conducted on mortality in the original report; therefore, it was impossible to determine the level at which effects occurred, thus preventing the development of a TRV. To determine the statistical difference between lewisite treatment groups, the data from Hackett *et al.* [12] was reanalyzed using the Kaplan-Meir Survival analysis. Time-to-death (day) for each individual in the treatment groups was obtained from the original report. Animals surviving through the entire duration of exposure were censored. Between-treatment comparisons indicated that the time-to-death for rabbits exposed to 0.05 mg/kg/day was significantly shorter than that for control rabbits. The 0.05 mg/kg/day level represents the LOAEL. The 0.007 mg/kg/day treatment group did not show statistically significant differences in time-to-death compared to the control; therefore, this level represents the NOAEL.

Because of its clear ecological relevance, mortality was chosen as the endpoint for derivation of the TRV. Because the study was subacute, the approximation approach was used to derive the TRV. An uncertainty factor of 20 was used to derive the NOAEL-based TRV from a subacute NOAEL, and an uncertainty factor of 7 was used to derive the LOAEL-based TRV from a subacute LOAEL. These TRVs were assigned a medium confidence rating based on the available data [4]. Because there were no chronic studies, and fewer than three orders of mammals were represented (Table 34.4), a high confidence TRV could not be assigned.

TRVs for Inhalation Exposures for the Class Mammalia

Currently, there are few data available delineating the inhalation toxicity of lewisite other than those from a number of acute exposures. A Department of the Army technical report [18] summarized inhalation toxicity studies of

Table 34.4 Selected Ingestion TRVs of Lewisite for the Class Mammalia

TRV	Dose	Confidence Level
NOAEL-based	0.004 mg/kg/day	Medium
LOAEL-based	0.03 mg/kg/day	Medium

Source: [12].

Table 34.5 Selected Inhalation TRVs of Lewisite for the Class Mammalia

TRV	Dose	Confidence Level
NOAEL-based	5.0 mg/min/m^3	Low
LOAEL-based	25.0 mg/min/m^3	Low

Source: [18].

lewisite with derived LCt_{50} values that ranged from 470 to 2800 mg/min/m^3. The LCt_{50} for mice (500 mg/min/m^3) appears to be protective of most other species for durations of approximately 10 minutes. Since only acute data are available, the approximation approach was used to derive the inhalation TRVs. An uncertainty factor of 100 was used to obtain the NOAEL-based TRV from the acute LCt_{50} of mice (500 mg/min/m^3), and an uncertainty factor of 20 was used for the LOAEL-based TRV. The inhalation exposure TRVs for the NOAEL of 5.0 mg/min/m^3 and LOAEL of 25.0 mg/min/m^3 of dose are estimated (Table 34.5). Due to the lack of long-term studies and the difficulty in obtaining primary references, the inhalation TRV for lewisite was given a low confidence rating.

These inhalation-based TRVs are for general use for species within the Class Mammalia. To estimate the degree of hazard, exposure relative to the species of concern (i.e., breathing rate, body weight, tidal volume) must be accounted for.

TRVs for Dermal Exposures for the Class Mammalia

As is the case with inhalation toxicity data, there are few data describing the toxicity of lewisite via the dermal route, but a number of LD_{50} values are available. In rats, dermal exposure to 15 mg/kg causes mortality. Because only acute data were available, the TRV was derived using the approximation approach with uncertainty factors of 100 and 20 for the NOAEL- and LOAEL-based TRVs, respectively. The dermal exposure TRVs were derived on a dermal LD_{50} of 15 mg/kg; the NOAEL of 0.15 mg/kg and LOAEL of 0.75 mg/kg were estimated (Table 34.6). Due to the lack of

Table 34.6 Selected Dermal TRVs of Lewisite for the Class Mammalia

TRV	Dose	Confidence Level
NOAEL-based	0.15 mg/kg	Low
LOAEL-based	0.75 mg/kg	Low

Sources: [18,2].

toxicity data on longer-term durations, and the difficulty in obtaining primary references, the dermal TRV for lewisite was given a low confidence rating.

TRVs for Birds

No data for the determination of TRVs for birds were located.

TRVs for Amphibians

Insufficient data were available for the determination of TRVs for amphibians.

TRVs for Reptiles

Insufficient data were available for the determination of TRVs for reptiles.

IMPORTANT RESEARCH NEEDS

The lewisite toxicity in nonmammalian wildlife is limited. However, because lewisite undergoes rapid hydrolysis and degradation, there may be a need for TRVs pertaining to lewisite by-products. Research studies examining lewisite should include animal models of species that are genetically, biologically, and behaviorally similar to the wildlife most likely to be exposed. The experimental design should mimic both exposure type and duration and should include assessments of long-term effects.

REFERENCES

[1] Waters WA, Williams JH. Hydrolysis and derivative of some vesicant arsenicals. J Chem Soc 1950;18–22.

[2] Rosenblatt DH, Miller TA, Dacre JC, Muul I, Cogley DR. Problem definition studies on potential environmental pollutants. II. Physical, chemical, toxicological, and biological properties of 16 substances. Frederick (MD): U.S. Army Medical Bioengineering Research and Development Laboratory, Fort Detrick; 1975, Technical Report 7509, AD A030428.

[3] Goldman M, Dacre JC. Lewisite: its chemistry, toxicology, and biological effects. Rev Environ Contam Toxicol 1989;110:75–115.

[4] USACHPPM. Standard practice for wildlife toxicity reference values. Aberdeen Proving Ground (MD): U.S. Army Center for Health Promotion and Preventive Medicine; 2000, Technical Guide No. 254.
[5] Wardell EL. Lewisite (M-1): 1940 summary of physiologic and toxicological data. Edgewood Arsenal (MD): Chemical Warfare Service; 1941, Report Number EATR 285. ADB95953L.
[6] MacNaughton MG, Brewer J. Environmental chemistry and fate of chemical warfare agents. San Antonio (TX): South West Research Institute; 1994, SWRI Report 01-5864.
[7] Hazardous Substances Data Bank (HSDB). Lewisite. National library of medicine [Internet]; 2012. [cited September 2014]. Available from, http://toxnet.nlm.nih.gov/cgi-bin/sis/search2/f?./temp/~qfaQ7D:1.
[8] Epstein J. Nerve gas in public water. Public Health Rep 1956;71(10):955–62.
[9] Watson AP, Griffin GD. Toxicity of vesicant agents scheduled for destruction by the chemical stockpile disposal program. Environ Health Perspect 1992;98:259–80.
[10] Subcommittee on Chronic Reference Doses for Selected Chemical Warfare Agents, National Research Council, Division on Earth and Life Studies, Board on Environmental Studies and Toxicology. Review of the U.S. Army's health risk assessments for oral exposure to six chemical-warfare agents. Washington (DC): National Academy Press; 1999, Appendix F, Health Risk Assessment for Lewisite; p. 275–294.
[11] Sasser LB, Cushing JA, Mellick PW, Kalkwarf DR, Dacre JC. Subchronic toxicity evaluation of lewisite in rats. J Toxicol Environ Health 1996;47(4):321–34.
[12] Hackett PL, Sasser LB, Rommerein RL, Cushing JA, Buschbom RL, Kalkwarf DR. Teratology studies on lewisite and sulfur mustard agents: effects of lewisite in rats and rabbits. Frederick (MD): U.S. Army Medical Research and Development Command, Fort Detrick; 1987, Report No. PNL-6408, AD A198423.
[13] Sasser LB, Cushing JA, Kalkwarf DR, Mellick PW, Buschbom RL. Toxicological studies on lewisite and sulfur mustard agents: two-generation reproductive study of lewisite in rats. Frederick (MD): U.S. Army Medical Research and Development Command, Fort Detrick; 1989, Report No. AD A214311.
[14] Sasser LB, Cushing JA, Lindemeier CW, Mellick PW, Dacre JC. Two-generation reproductive study of lewisite in rats. J Appl Toxicol 1999;19(4):229–35.
[15] Dacre JC, Goldman M. Toxicology and pharmacology of the chemical warfare agent sulfur mustard. Pharmacol Rev 1996;48(2):289–326.
[16] Munro NB, Talmage SS, Griffin GD, Waters LC, Watson AP, King JF, et al. The sources, fate, and toxicity of chemical warfare agent degradation products. Environ Health Perspect 1999;107(12):933–74.
[17] Lewis Sr RJ. Sax's dangerous properties of industrial materials. 8th ed. New York (NY): John Wiley & Sons, Inc.; 1992.
[18] U.S. Department of the Army (DA). Chemical Agent Data Sheets, vol. 1. Aberdeen Proving Ground (MD): U.S. Department of the Army, Edgewood Arsenal; 1974, (Edgewood Arsenal Special Report). Technical Report Number EO-SR-74001.
[19] Rice P, Brown RF. The development of lewisite vapour-induced lesions in the domestic, white pig. Int J Exp Pathol 1999;80(1):59–67.
[20] Price CC, von Limbach V. Further data on the toxicity of various CW agents to fish. Washington (DC): National Defense Research Committee. Office of Scientific Research and Development; 1945, OSRD No. 5528, Division 9.
[21] DA Material safety data sheet (MSDS) for lewisite (L). Aberdeen Proving Ground (MD): Edgewood Research, Development, and Engineering Center, Chemical Biological Defense Command; 16 April 1988, rev. March 27, 1996.

CHAPTER 35

Wildlife Toxicity Assessment for Thiodiglycol

Christine Ann Arenal, Bradley E. Sample, Glenn J. Leach, Gunda Reddy

Contents

INTRODUCTION

Thiodiglycol (TDG), CAS No. 111-48-8, is an oily liquid that is used commercially as a solvent in antifreeze solutions, dyestuffs for printing, and as part of the process by which polyvinyl chloride is manufactured [1]. Within the terms of the Chemical Weapon Convention Treaty, TDG is considered a "Schedule 2" compound (i.e., a dual-use chemical with low to moderate commercial use and high-risk precursors) [2,3]. The compound is also formed when HD undergoes hydrolysis. Thus, TDG has been detected in animals and human beings exposed to HD, and in environmental media when HD is released to the environment. For example, TDG has been detected in the ground and surface waters near a chemical warfare agent disposal site [1]. Renewed interest in the environmental and human health impacts of TDG has arisen with the alleged use of HD by the military forces of Iraq against Iranians and their own Kurdish population during the 1980s [4,5].

Wildlife Toxicity Assessments for Chemicals of Military Concern
http://dx.doi.org/10.1016/B978-0-12-800020-5.00035-1

Toxicity information for HD indicates that it is highly toxic to animals and humans [6]. In contrast, toxicity data for TDG indicate the compound is much less toxic than HD, though information on the toxicity and environmental fate of TDG is limited [7]. There are no records for TDG in the U.S. Environmental Protection Agency's Integrated Risk Information System (U.S. EPA IRIS) database because of this limited information on the toxicity and environmental fate of TDG [7]. No occupational standards or guidelines have been set for this compound by the National Institute of Occupational Safety and Health (NIOSH), the Occupational Safety and Health Administration (OSHA), or the American Conference of Governmental Industrial Hygienists (ACGIH).

This Wildlife Toxicity Assessment (WTA) summarizes available information on the potential harmful effects of TDG to wildlife, emphasizing threshold doses for the onset of noncancer effects, as described in reports of experimental studies of the compound. Available dose-response data for TDG may support the establishment of Toxicity Reference Values (TRVs) that could serve as protective thresholds for wildlife exposed to TDG. This WTA is consistent with protocols outlined in the U.S. Army Center for Health Promotion and Preventive Medicine (USACHPPM) Technical Guide (TG) 254, *Standard Practice for Wildlife Toxicity Reference Values* [8].

TOXICITY PROFILE

Environmental Fate and Transport

Few data were found on the fate and transport of TDG in the environment, although some characteristics can be inferred from the compound's physical-chemical properties [9–12, and Table 35.1]. For example, the miscibility of TDG in water suggests that the compound will readily partition to or be dispersed in aqueous media. Additionally, the estimated K_{oc} of 11 (calculated using the log K_{ow} of –0.63) suggests that TDG would have very high mobility in soil [11]. The compound appears to be largely resistant to hydrolysis or photolysis [1]. In a recent study using photoactivated periodate to decompose total organic carbon (TOC) from hydrolysates of chemical warfare agents, TDG had the fastest rate of TOC loss at pH 3 [13]. Lee and Allen [14] studied the environmental fate of TDG using six soils that were considered to be characteristic of U.S. military installations. For five of these soils, 99 to 100% of the original mass of TDG remained in the aqueous phase after 24 hours. In this study, its sorption to soils was less than 10 milligrams per kilogram (mg/kg), while its degradation product,

Table 35.1 Summary of Physical-Chemical Properties of Thiodiglycol

CAS No.	111-48-8
Molecular weight	122.2
Color	Colorless
Physical state	Syrupy liquid
Melting point	−10.2 °C
Boiling point	282 °C
Odor	Similar to hydrogen sulfide
Solubility in water	Miscible with water at 20 °C, soluble in ether
Partition coefficients:	
Log K_{ow}	−0.63
Log K_{oc}	0.99
Vapor pressure at 25 °C	3.23×10^{-3} mmHg
Henry's Law constant at 25 °C	1.9×10^{-9} atm-m^3/mole
Conversion factors	1 ppm = 5 mg/m^3 1 mg/m^3 = 0.2 ppm

Sources: Munro *et al.* [1], Budavari *et al.* [9], Lewis [10], HSDB [11], BASF Corp. [12].

thiodiglycolic acid, showed sorption capacity from 19.9 to 427 mg/kg, depending on soil type. Lee and Allen [14] also found TDG and thiodiglycolic acid resistant to photolysis and hydrolysis. TDG is biologically converted to the latter with the formation of an intermediate [(2-hydroxyethyl) thio] acetic acid. TDG was slowly biodegraded under anaerobic conditions, reaching about 42% of applied dose after 185 days, with a lag period of 52 days [15]. In one study, the authors indicate that TDG has persisted for 60 years in soils contaminated with mustard gas [16]. The authors concluded that this persistence may be caused by the toxicity of mustard gas to microorganisms and soil chemistry constraints (e.g., pH and oxygen availability).

Microbial degradation of TDG has been demonstrated in the presence of strains of *Pseudomonas pickettii* and *Alcaligenes xylosoxidans*, both of which are capable of growing in media with TDG as the sole carbon source [17]. In Baltic Sea sediments, *Achromobacter* sp., *Pseudomonas* sp., and *Arthrobacter* sp. were isolated and found to be capable of using TDG as the sole source of carbon and energy [18]. In addition, bioreactor experiments have demonstrated degradation of the compound in the presence of sewage sludge [1]. These studies confirm that TDG in the environment can be degraded by microbial processes. However, Medvedeva *et al.* [19] reported that mustard gas hydrolysis products such as TDG significantly change the count and composition of soil microorganisms. Specifically, the total number of bacteria increased while the populations of other vital microbiota significantly decreased (as compared to controls, 66–98% decline for micromycetes and 47–91% decline for actinomycetes). They also found that the biological activity of the soil was decreased, which suggests TDG may adversely affect species composition and thus, biochemical processes in soil.

Summary of Mammalian Toxicity

Mammalian Oral Toxicity

Angerhofer *et al.* [20] studied the subchronic oral toxicity of TDG in Sprague Dawley rats. As part of this study, they investigated the approximate lethal dose (ALD), as well as conducted a range-finding, subacute 14-day trial. The results of the subchronic study were also reported in an abstract [21]. Additionally, these acute, subacute, and subchronic studies were summarized by Reddy *et al.* [7] as part of a toxicity assessment for TDG. Most recently, these results were published in the open literature by Angerhofer *et al.* [22] and the analysis was expanded to include calculation of a benchmark dose (BMD) using the subchronic results. Similarly, Houpt *et al.* [23] is the technical report for developmental studies conducted with Sprague Dawley rats. These results were also reported in an abstract [24] and in the open literature [25]. These and other available studies are described in the following sections.

Mammalian Oral Toxicity: Acute

There are few data on the acute toxicity of TDG in laboratory animals, although oral lethal dose $(LD)_{50}$ values of 3,960 mg/kg in guinea pigs of mixed sexes [11] and 6,610 mg/kg in adult male Wistar rats have been reported [26]. During study of the ALD, neat TDG administered by oral gavage at a dose of 9,900 mg/kg was not lethal to male or female Sprague Dawley rats [7,20,22]. In fact, no toxic effects other than slight lethargy for 1 hour after dosing were observed. A subcutaneous LD_{50} of 4,000 mg/kg for rats and mice and an intravenous LD_{50} of 3,000 mg/kg for rabbits were also reported [27]. However, these types of exposure are not applicable to oral pathways.

Mammalian Oral Toxicity: Subacute

A subacute (14-day) oral toxicity study was conducted with TDG in male and female Sprague Dawley rats to select a suitable dose for a subsequent 90-day study [7,20,22]. Rats were dosed orally with neat (single-chemical compound) TDG at dose levels 0 (control), 157, 313, 625, 1,250, 2,500, 5,000, and 9,999 mg/kg/day for 5 days per week for 2 weeks. During the 14-day study, food consumption, body weights, and any clinical signs were recorded. At the end of 14 days, blood samples were collected for hematology and clinical chemistry and gross necropsies were performed. In the high dose group (9,999 mg/kg/day), four of six males and five of six females died within the first 3 days of the 14-day study. TDG-dosed rats at 5,000 or 9,999 mg/kg /day showed decreased body weights and increased kidney weights. There were no treatment-related changes observed in hematological and clinical parameters. Based on these results, the lowest observed adverse effect level (LOAEL) was 5,000 mg/kg/day, and this dose was selected as the highest dose for the 90-day study.

Mammalian Toxicity: Subchronic

Subchronic studies relating to the effects of TDG on mammals were available. Angerhofer *et al.* [7, 20–22] describe an oral gavage study in which neat TDG was administered to 10 Sprague Dawley rats/sex/group at 0, 50, 500, or 5,000 mg/kg/day, 5 days per week, for 90 days. Food consumption and body weight changes were monitored weekly, blood and urine samples were collected at term, and samples of organs were excised for histopathological examination at necropsy. Few compound-related clinical signs of toxicity were observed. Deaths occurred in both the control and treatment groups and were not considered treatment related. Those effects that were observed included a reduction in body weight gain and lower absolute body weight in high-dose animals, though there were no significant differences in food consumption between treatment and control groups, except for the first 1 to 3 days in the 5,000 mg/kg/day group. There were no changes in the pathology or histology of any organ, although the relative weights of liver, testis, and brain were increased in high dose versus control rats. Finally, potential renal impacts in the high dose group were evidenced by the observed increase in kidney weight, with concomitant increase in the volume of urine and the urine specific gravity, and decrease in urine pH. Granular casts were also observed in the urine. Although urine pH was reduced and specific gravity increased in females from the 500 mg/kg/day dose group, the authors [7,20,21] considered this level to represent the no observed adverse effect level (NOAEL). A NOAEL of 357 mg/kg/day was derived by duration, adjusting the 500 mg/kg/day dose to account for the 5 days per week dosing regimen. The associated LOAEL would be 3,570 mg/kg/day.

Using the BMD approach, Angerhofer *et al.* [22] calculated a BMD and its 95% lower confidence limit for 10% effect levels (BMDL; benchmark dose level) using the body weight results for this subchronic study. In their BMD run, the female body weights did not meet the minimum requirements for a valid estimate, but the male body weights showed an adequate fit from several models. The most conservative BMDL estimate (372 mg/kg/day) from the Hill model with specified parameters was chosen. The BMD was 1,704 mg/kg/day.

Mammalian Toxicity: Chronic

No relevant studies were identified for chronic exposures.

Mammalian Toxicity: Other

In addition to the acute and subchronic studies just described, one developmental study and several *in vitro* and *in vivo* studies of TDG were available. Houpt *et al.* [23–25] evaluated the developmental toxicity of TDG in Sprague Dawley rats. Following an initial range-finding study, a gavage study was conducted on positively mated female rats given doses of 0, 430, 1,290, and 3,870 mg/kg/day on gestation days 5 through 9. Although the cause of death could not be determined, one female in the high dose group died prior to the end of the study. Additionally, maternal body weights and food consumption were negatively affected during part of gestation (days 16–20 for both parameters, as well as days 5–13 for food consumption) in the high dose group. There was an increased number of fetuses per dam in the high dose group; however, both fetal body weights by litter and by individual fetus were significantly lowered in this dose group compared to controls. TDG was not teratogenic at the dose levels tested, and did not affect other reproductive parameters measured (e.g., number of implantation sites, resorptions, number of live and dead fetuses, and sex ratio). The NOAEL for developmental oral toxicity in rats was 1,290 mg/kg/day, when administered during the major period of organogenesis. The corresponding LOAEL based on decreased fetal body growth was 3,870 mg/kg/day.

In toxicokinetic studies, sequential reports from the U.K.'s Chemical and Biological Defence Establishment at Porton Down, examined HD metabolism *in vivo* and pointed to the physiological role of TDG as a urinary constituent of human beings and animals exposed to HD. The studies build on work of Wils *et al.* [4,5] whose demonstration of the presence of TDG in the urine of hospitalized Iranian soldiers lent support to claims of HD use by Iraq in the Iran–Iraq war. While Wils *et al.* [4,5] derived reconstituted HD by the addition of concentrated hydrochloric acid to urine, the Porton group used a number of technical advances to isolate and detect hydrolysis products (such as TDG and its sulfoxide), and the products of combined glutathione conjugation and β-lyase activity (such as 1,1-sulphonylbis[2-(methylsulphinyl)ethane] and 1-methylsulphinyl-2-[2-(methylthio)ethylsulphonyl]ethane). For example, administration of double-labeled ^{14}C-labeled and ^{35}S-labeled HD to male Porton rats via the intraperitoneal route resulted in the appearance of a substantial number of hydrolysis and glutathione conjugation products of HD, with 60% of the load appearing in the urine within the first 24 hours of treatment [28]. Many of the same products were formed when HD was applied to the skin of Porton rats, although a much lower percentage was released to the urine, even for as long as 8 days after application [29]. To explain this finding, Hambrook *et al.* [30] studied the quantitative recovery of cutaneously applied ^{35}S-labeled HD, and showed that while most of the counts passed into the blood stream, a substantial proportion was retained in the skin, some released as a vapor, and only a small portion transported to the urine. Additionally, their *in vivo* and *in vitro* demonstration of covalent binding of ^{35}S-containing moieties to hemoglobin in red blood cells (RBCs) helped to explain the reduced release of HD metabolites to the urine [30].

The Porton group reported a number of studies in which they applied their technical advances in TDG detection to urine samples from human beings who had been exposed to HD. In all cases, hydrolysis products and the metabolites of glutathione conjugation and β-lyase activity appeared in the urine [31,32]. Of the former, the sulfoxide of TDG was considered to be a more important metabolite than TDG itself, while the latter, measured as a single 1,1-sulphonylbis-[2-(methylthio)ethane] derivative, were regarded as more discriminating markers of HD exposure than TDG or its derivatives, because of their complete absence from the urine of unexposed controls.

That the sulfoxide of TDG is the primary hydrolysis product of HD is supported by an experiment in which double-labeled TDG itself was administered intraperitoneally to Porton rats [33]. More than 90% of the load was released to the urine within 24 hours, and more than 90% of those counts appeared in a peak identified as TDG sulfoxide. Less than 1% was unchanged TDG.

Inferential evidence that TDG may have toxicological effects also can be drawn from a report by Brimfield *et al.* [34] who, in an *in vitro* experiment, demonstrated the compound's ability to inhibit the serine/threonine protein phosphatase activity of mouse liver cytosol. Vodela *et al.* [35] studied the effects of neat TDG on the glutathione antioxidant system in rats. TDG was given orally at doses of 1,250, 2,500, and 5,000 mg/kg/day for 14 days and at doses of 50, 500, and 5,000 mg/kg/day for 90 days to rats obtained from subacute and subchronic studies [20–22]. Glutathione reductase levels decreased in females but increased in males given 5,000 mg/kg/day for 14 days. No change in the glutathione antioxidant system occurred in the 90-day study. The rat glutathione antioxidant system is thus not a highly sensitive indicator for TDG subchronic exposure. In a related study, Vodela *et al.* [36] evaluated the effects of TDG on the hepatic mixed function oxidase (MFO) system and the cytosolic glutathione antioxidant system in male and female rats gavaged with TDG at 50, 500, and 5,000 mg/kg/day for 90 days. The authors reported an increase in pentoxyresorufin O-dealkylation (CYP2B1/B2) activity (5,000 mg/kg/day) and a significant decrease in cytochrome B5 (500 and 5,000 mg/kg/day), glutathione (500 and 5,000 mg/kg/day), glutathione S-transferases (all doses), and glutathione peroxidase in males (5,000 mg/kg/day). There were no significant differences in any of the parameters in female rats. These effects on the MFO and glutathione antioxidant systems generally occurred at a high dose level, indicating that these parameters are not highly sensitive to TDG subchronic exposure.

The genotoxic potential of TDG has been evaluated using limited *in vivo* and *in vitro* assays. Mutagenicity was studied by Stankowski [37] using four *Salmonella typhimurium* tester strains (TA98, TA100, TA1535, and TA 1537) and one *Escherichia coli* tester strain (WP2uvrA). Each strain was dosed with 33.3, 100, 333, 1,000, 3,330, and 5,000 μg per plate, with and without S9 activation and with concurrent vehicle and positive controls. Mutagenic effects were not observed at any dose level for any tester strain, with or without metabolic activation. In an *in vivo* study, five mice were gavaged with TDG in sterile water at 500, 1,000, and 2,000 mg/kg, and clastogenic activity, as well as disruption of mitotic activity in the micronuclei in mouse bone marrow, was evaluated [38]. No signs of clinical toxicity or cytotoxic effects in bone marrow were observed (i.e., TDG was not mutagenic in this assay). An *in vitro* study that evaluated the effects of TDG on chromosomal aberrations in Chinese Hamster Ovary (CHO) cells was conducted by Tice *et al.* [39]. Dose levels included 3, 4, and 5 milligrams per milliliter (mg/mL), and clastogenic activity was investigated. A significant increase in the percentage of metaphase cells containing chromosomal aberrations (i.e., chromatid and chromosomal breaks and chromatid type rearrangements) was observed with and without metabolic activation. A lowest effective dose of 5 mg/mL was identified in the absence of metabolic activation and 4 mg/mL was identified in the S9-metabolic activation system. Lastly, TDG was evaluated in a mammalian mutagenic assay using cultured mouse lymphoma L5178Ytk+/- cells and dose levels 0, 50, 158, 1,580, and 5,000 micrograms per milliliter (μg/mL) [40]. Mutant frequency was not significantly increased with or without metabolic activation. As these results indicate, TDG was negative in three assays (Ames test, mouse micronucleus, and mouse lymphoma) and positive in the *in vitro* assay (chromosomal effects in CHO cells). Reddy *et al.* [7] indicate that this positive response may not correspond to any change in the *in vivo* system.

Studies Relevant for Mammalian TRV Development for Ingestion Exposures

Although a degradation product of the extremely toxic chemical warfare agent HD, TDG has been shown to have a low toxicological impact on experimental animals when administered via the oral route. Thus, values for the oral LD_{50} range in excess of 3,900 mg/kg body weight, and the few compound-related effects reported in the available studies [20–25] were observed only at very high dose levels (i.e., in excess of 3,000 mg/kg/day). As indicated in Table 35.2 and Figure 35.1, the responses of oral administration of TDG were confined to reduced body weight gain in adults and fetuses, relative increases in some organ weights, and to functional deficits in urine production indicative of possible kidney impairment. These studies are of sufficient quality for TRV derivation. Additionally, the male body weight data in the subchronic (90-day) study show a dose-response relationship that was used to develop a subchronic BMD and BMDL [22].

Mammalian Toxicity: Inhalation

Acute inhalation toxicity tests were conducted on neutralized HD solution for the Department of Transportation (DOT) [41]. In these experiments, HD was neutralized by hot water (90 °C) hydrolysis to produce a less toxic solution, mostly TDG. Rats were exposed (nose only) to this (in an aerosol) at a concentration of 5.4 mg/L for 4 hours. No overt toxicity or deaths attributed to this hydrolysis product were observed during or after the post-exposure period of 14 days. Analysis of the solution showed only a trace amount of HD.

Mammalian Toxicity: Dermal

Several studies reported the toxicological effects of TDG in experimental animals via the dermal route. The first reported a dermal lethal dose $(LD)_{50}$ value of 20 mL/kg for rabbits [42], and in the second, mild skin irritation in rabbits exposed to 500 mg TDG was observed [43]. Union Carbide

Table 35.2 Summary of Relevant Mammalian Data for TRV Derivation

		Test Results			
Test Organism	**Test Duration**	**NOAEL (mg/kg/day)**	**LOAEL (mg/kg/day)**	**Effects Observed at the LOAEL**	**Study**
Rats (Sprague Dawley)	90 days	357	3,570	Reduced body weight gain; structural and functional deficits in the kidney	[20–22]; [7]
Rats (Sprague Dawley)	GD 5–9	1,290	3,870	Decreased maternal body weight and food consumption; decreased fetal body weight	[23–25]

GD, gestational day; mg/kg/day, milligram per kilogram per day.

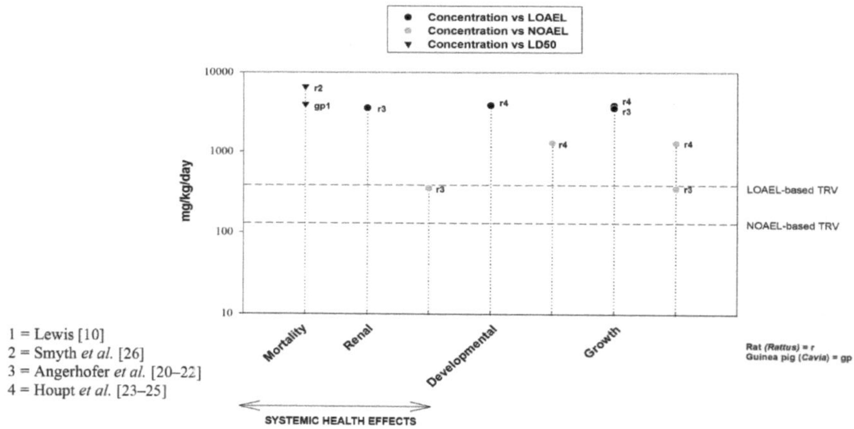

Figure 35.1 Thiodyglycol health effects to mammals.

[42] also reported moderate eye irritation to 500 mg TDG. Using a density of 1.18 g/mL for TDG [11], an LD_{50} of 20 mL/kg is equivalent to 23,600 mg/kg. Manthei *et al.* [44] studied the dermal effects of HD decontamination products, neat TDG, and HD in New Zealand white rabbits at a concentration of 1 mL/kg. Using a density of 1.18 g/mL [11], an equivalent dose for TDG is 1,180 mg/kg. No dermal irritation was observed for

HD decontamination products and mild dermal irritation that lasted for less than 48 hours was observed for neat TDG. In contrast, HD produced severe dermal irritation. Using a density of 1.18 g/mL for TDG as reported in Reddy *et al.* [7], a skin sensitizing test of TDG on guinea pigs was conducted following methods of Magnusson and Kligman [45]. TDG at 75% aqueous solution, when challenged to guinea pigs after 21 days of intradermal induction, showed no skin reactions in test and control animals. Therefore, TDG is not considered a skin sensitizer.

Other available studies focused on the dermal absorption of TDG. For example, Hambrook *et al.* [30] conducted a quantitative recovery study on the fate of ^{35}S-HD when applied to the skin of Porton rats. The results of this study demonstrated that, while urine is an important clearance route for the metabolic products of HD, significant portions of the load were retained in the blood through covalent binding of the labeled sulfur atoms to hemoglobin. Counts were detected in the blood for several weeks after an initial 6-hour exposure. The authors confirmed an earlier demonstration that cutaneous application of HD to Porton rats results in urinary formation of both hydrolysis products and glutathione conjugates/β-lyase metabolites of HD [29]. Recently Reifenrath *et al.* [46] studied dermal absorption of TDG and TDG-contaminated or spiked soils using freshly isolated pig skin in a flow-through cell system. They showed the percent absorption of TDG from Yolo soil (1.9% carbon) as $0.9 \pm 0.85\%$ and from Tinker soils (9.5% carbon) as $0.5 \pm 0.5\%$ as compared to about 20% absorption from acetone.

Summary of Avian Toxicity

No studies were identified that examined the toxicological effects of TDG in birds.

Summary of Amphibian Toxicity

No studies were identified that examined the toxicological effects of TDG in amphibians.

Summary of Reptilian Toxicity

No studies were identified that examined the toxicological effects of TDG in reptiles.

RECOMMENDED TOXICITY REFERENCE VALUES

Toxicity Reference Values for Mammals

TRVs for Ingestion Exposures for the Class Mammalia

Only two studies regarding the oral effects of TDG in mammals were available, though these results are presented or discussed within multiple reports and publications [7,20–25]. From these sparse data, it is difficult to determine the specific target organ for TDG, although Angerhofer *et al.* [20–22] reported adverse effects (e.g., increased kidney weight, increased urine output, and decreased urine pH) that may indicate functional deficits in the kidney. Based on these effects and a reduction in body weight gain, the high dose group in this study (3,570 mg/kg/day) was considered a sub-chronic LOAEL, and the next lower dose (357 mg/kg/day) was considered a sub-chronic NOAEL. Using the male body weight results in the subchronic (90-day) study, Angerhofer *et al.* [22] calculated a BMDL of 372 mg/kg/day and a BMD of 1,704 mg/kg/day. Houpt *et al.* [23–25] observed a decrease in maternal body weight and food consumption, as well as a decrease in fetal body weight in the high dose group (3,870 mg/kg/day), which they considered the LOAEL. The associated NOAEL for these growth and developmental effects was 1,290 mg/kg/day. This LOAEL is within the range of the LOAEL reported in the subchronic study.

The available toxicological studies include subchronic and gestational exposures; however, long-term studies were not available, and only one species was represented. Because Houpt *et al.* [23–25] evaluated the effects of TDG during a critical life stage (i.e., during gestation), this study is considered chronic in nature [8]. Additionally, a decrease in fetal body weight may result in reduced survival or fitness of the offspring. Reduced body weight may also reduce the survival or fitness of adults; therefore, the Angerhofer *et al.* [20–22] study is also relevant. However, the developmental study was chosen for TRV development because it is considered to represent chronic exposure and effects.

Houpt *et al.* [23–25] is a developmental toxicity study with relevant endpoints. Benchmark dose modeling of the fetal body weight data provided in Houpt *et al.* [23] was attempted; however, the BMD and BMDL were uninformative for TRV derivation purposes, in part due to the lack of any dose-dependent variability between low dose-range data and high dose-range data. In addition, the minimum data set requirements as outlined in Section 2.2 of USACHPPM TG 254 [8] were not met; namely, data were not available from at least three studies representing at least three species and two taxonomic orders. Therefore, the approximation approach as described in USACHPPM TG 254 [8] was used to develop oral ingestion TRVs for mammals. If an uncertainty factor (UF) of 10 is applied to the NOAEL and LOAEL from Houpt *et al.* [23–25] to account for potential interspecies differences, a NOAEL-based TRV of 129 mg/kg/day and a LOAEL-based TRV of 387 mg/kg/day can be derived.

Given the paucity of available data, it is useful to consider the mortality data in developing an appropriate TRV. If the approximation method is used to extrapolate a TRV from the most sensitive LD_{50} data (guinea pig; using an UF of 100 to estimate a NOAEL and a UF of 20 to estimate a LOAEL), it results in a NOAEL-based TRV of 39.6 mg/kg/day and a LOAEL-based TRV of 198 mg/kg/day. These are comparable to the NOAEL and LOAEL-based TRV values developed from Houpt *et al.* [23–25], and provide support for the appropriateness of the use of the Houpt *et al.* data. Moreover, TRVs that were developed from either the mortality or developmental data are protective of adverse effects (i.e., reduced growth and possible kidney impairment) in rats from subchronic exposures [20–22]. Together, these relationships provide a weight of evidence that shows the approximation approach is reasonable when applied to the developmental data. Therefore, the TRVs for the Class Mammalia were derived from the developmental NOAEL and LOAEL for the rat by applying an UF of 10.

Table 35.3 presents the selected TRVs. A low level of confidence has been given to these TRVs because the available data and representative species are severely limited.

TRVs for Ingestion Exposures for Mammalian Foraging Guilds

Toxicity Reference Values specific to particular guild associations (e.g., small herbivorous mammals) have not yet been derived. However, the class-specific TRVs shown in Table 35.3 may be considered to apply to herbivorous small mammals because rats are members of this guild. As with the class-specific TRVs, only one species is represented and toxicological data

Table 35.3 Selected Ingestion TRVs for the Class Mammalia

TRV	Dose	Confidence Level
NOAEL-based	129 mg/kg/day	Low
LOAEL-based	387 mg/kg/day	Low

Sources: [23–25].

are limited, so confidence in the TRVs is low. Data to derive TRVs for other guild associations (e.g., carnivorous mammals) were not identified.

TRVs for Inhalation Exposures for the Class Mammalia

Available mammalian inhalation data are limited to one acute study, in which no overt signs of toxicity or death were observed after 4 hours exposure to 5.4 mg/L TDG [41]. Although inhalation TRVs cannot be derived from these data, acute exposures of 5.4 mg/L TDG or less likely do not result in overt adverse effects.

TRVs for Dermal Exposures for the Class Mammalia

Of the studies evaluating the dermal toxicity of TDG to mammals, one showed mild dermal irritation that lasted less than 48 hours at a dose of 1,180 mg/kg [44]. In contrast, the LD_{50} for TDG was identified as 23,600 mg/kg [42]. Both of these are acute exposures. TDG was not found to be a skin sensitizer in guinea pigs [7]. Using the approximation method [8], the LOAEL-based TRV based on mild dermal irritation would be 1,180 mg/kg. A NOAEL-based TRV of 118 mg/kg can be estimated from this LOAEL by applying an UF of 10.

Given the paucity of available data, it is useful to consider the mortality data in developing an appropriate TRV. If the approximation method is used to extrapolate TRVs from the LD_{50} data (rabbits; using an UF of 100 to estimate a NOAEL and an UF of 20 to estimate a LOAEL), the resulting NOAEL- and LOAEL-based TRVs are 236 mg/kg and 1,180 mg/kg, respectively. These are comparable to the NOAEL- and LOAEL-based TRV values developed from Manthei *et al.* [44], and provide support for the selected TRVs.

These TRVs are presented in Table 35.4; however, confidence in these is low because the available data and representative species are severely limited.

Table 35.4 Selected Dermal TRVs for the Class Mammalia

TRV	Dose	Confidence Level
NOAEL-based	118 mg/kg	Low
LOAEL-based	1,180 mg/kg	Low

Source: [44].

Toxicity Reference Values for Birds

No data permitting derivation of avian TRVs for TDG were identified.

Toxicity Reference Values for Amphibians

No data permitting derivation of amphibian TRVs for TDG were identified.

Toxicity Reference Values for Reptiles

No data permitting derivation of reptilian TRVs for TDG were identified.

IMPORTANT RESEARCH NEEDS

Mammalian TRVs derived for TDG have low confidence because only one species is represented and toxicological data are limited. Therefore, additional species and taxonomic orders should be evaluated to provide a greater breadth of interspecific data. In addition, toxicity studies that examine demographic factors such as birth, death, and recruitment would have much greater ecological significance. The additional data would increase confidence in the mammalian TRVs and enable development of TRVs for specific foraging guilds. Inhalation and dermal studies on mammals were limited for TDG, and additional studies for these exposure routes are recommended. TRV derivation for birds, amphibians, and reptiles could not be performed due to a lack of toxicity data. Appropriate acute, sub-acute, sub-chronic, and especially chronic TDG toxicity data derived through biologically relevant exposure routes for these wildlife classes are needed. Research studies should include experimental models of species genetically, biologically, and behaviorally similar to wildlife exhibiting the greatest propensity for toxicant exposure. Experimental design should attempt to mimic both exposure type and duration and should include assessments of long-term effects.

REFERENCES

[1] Munro NB, Talmage SS, Griffin GD, Waters LC, Watson AP, King JF, et al. The sources, fate, and toxicity of chemical warfare agent degradation products. Environ Health Perspect 1999 Dec;107:933–74.

[2] Ember LR. Chemical arms treaty makes unprecedented demands of industry. Chem Eng News Rec 1993;7:7–18.

[3] Ember LR. Failure to ratify chemical arms pact would dampen U.S. Chemicals Trade. Chem Eng News 1996 Jan;19–22.

[4] Wils ER, Hulst AG, de Jong AL, Verweij A, Boter HL. Analysis of thiodiglycol in urine of victims of an alleged attack with mustard gas. J Anal Toxicol 1985;9:254–7.

[5] Wils ER, Hulst AG, Van Laar J. Analysis of thidiglycol in urine of victims of an alleged attack with mustard gas. Part II. J Anal Toxicol 1988;12:15–9.
[6] Dacre JC, Goldman M. Toxicology and pharmacology of the chemical warfare agent sulfur mustard. Pharmacol Rev 1996;48(2):289–326.
[7] Reddy G, Major MA, Leach GJ. Toxicity assessment of thiodiglycol. Inter J Toxicol 2005;24:435–42.
[8] USACHPPM. Standard practice for wildlife toxicity reference values. Aberdeen Proving Ground (MD): U.S. Army Center for Health Promotion and Preventive Medicine; 2000, Technical Guide No. 254.
[9] Budavari S, O'Neill MJ, Smith A, Heckelman PE. The Merck index: an encyclopedia of chemicals, drugs, and biologicals. 12th ed. Whitehouse Station (NJ): Merck and Co., Inc.; 1996.
[10] Lewis RJ. Sax's dangerous properties of industrial materials. 8th ed. New York (NY): Van Nostrand Reinhold; 1992.
[11] Hazardous Substances Data Bank (HSDB). Thiodiglycol [Internet]. Bethesda (MD): National Library of Medicine (US), National Institutes of Health [cited 2014]. Available from http://toxnet.nlm.nih.gov/.
[12] BASF Corporation. Global product strategy (GPS) safety summary: thiodiglycol [Internet]; 2011 Dec 1. Available from, http://www.safety-summaries.basf.com/group/corporate/safety-summaries/en/?indexterm=T.
[13] Tang X, Weavers LK. Using photoactivated periodate to decompose TOC from hydrolysates of chemical warfare agents. J Photochem Photobiol A Chem 2008;194:212–9.
[14] Lee KP, Allen HE. Environmental transformation mechanisms of thidiglycol. Environ Toxicol Chem 1998;17(9):1720–6.
[15] Sklyar V, Mosolowa TP, Kucheren IA, Deytyarova NN, Varfolomeyev SD, Kalyuzhmyi SV. Anaerobic toxicity and biodegradability of hydrolysis products of chemical warfare agents. Appl Biochem Biotech 1999;81:107–17.
[16] Li H, Muir R, McFarlane NR, Soileux RJ, Yu X, Thompson IP, et al. Soil biotransformation of thiodiglycol, the hydrolysis product of mustard gas: understanding the factors governing remediation of mustard gas contaminated soil. Biodegradation 2013;24:125–35.
[17] Ermakova IP, Starovoitov II, Tikhonova EB, Slepen'kin AV, Kashparov KI, Boronin AM. Thiodiglycol metabolism in *Alcaligenes xylosoxydans* subsp. *Denitrificans*. Microbiology 2002;71:519–24.
[18] Medvedeva N, Polyak Y, Kankaanpää, Zaytseva T. Microbial responses to mustard gas dumped in the Baltic Sea. Mar Environ Res 2009;68:71–81.
[19] Medvedeva N, Polyak Y, Kuzikova I, Orlova O, Zharikov G. The effect of mustard gas on the biological activity of soil. Environ Res 2007;106:289–95.
[20] Angerhofer RA, Michie MW, Leach GJ. Subchronic oral toxicity of thiodiglycol in rats. Aberdeen Proving Ground (MD): U.S. Army Center for Health Promotion and Preventive Medicine; 1998, Report No. 6415-38-97-05-01.
[21] Angerhofer RA, Michie MW, Leach GJ. Subchronic oral toxicity of thiodiglycol in rats. Toxicologist 1999;48(1-S):318.
[22] Angerhofer RA, Michie MW, Leach GJ, Johnson MS, Reddy G. Oral toxicity evaluation of thiodiglycol in Sprague-Dawley rats (Epub ahead of print August 27, 2014). Int J Toxicol 2014;33:393–402.
[23] Houpt JT, Crouse LCB, Angerhofer RA. Developmental toxicity of thiodiglycol in rats. Aberdeen Proving Ground (MD): U.S. Army Center for Health Promotion and Preventive Medicine; 2001, Report No. 7796-52-99-04-05.
[24] Houpt JT, Reddy G, Crouse LC. Developmental toxicity of thiodiglycol in rats. Toxicologist 2003;72(S-1):341.

[25] Houpt JT, Crouse LCB, Angerhofer RA, Leach GJ, Reddy G. Developmental toxicity of thiodiglycol in Sprague-Dawley rats. Int J Toxicol 2007;26:365–71.
[26] Smyth HF, Seaton J, Fischer I. The single dose toxicity of some glycols and derivatives. J Indust Hyg Toxicol 1941;23:259–68.
[27] Anslow LP, Karofsky DA, Jager BV, Smith HW. The intravenous, subcutaneous and cutaneous toxicity of bis (beta-chloroethyl0 sulfide (mustard gas) and of various derivatives. J Pharmacol Exp Ther 1948;93:1–9.
[28] Black RM, Brewster K, Clarke RJ, Hambrook JL, Harrison JM, Howells DJ. Biological fate of sulfur mustard, 1,1-thiobis(2-chloroethane): Isolation and identification of urinary metabolites following intraperitoneal administration to rat. Xenobiotica 1992;22:405–18.
[29] Black RM, Hambrook JL, Howells DJ, Read RW. Biological fate of sulfur mustard, 1,1-thiobis(2-chloroethane). Urinary excretion profiles of hydrolysis products and beta-lyase metabolites of sulfur mustard after cutaneous application in rats. J Anal Toxicol 1992;16((2):79–84.
[30] Hambrook JL, Howells DJ, Schock C. Biological fate of sulfur mustard 91,1-thiobis(2-chloroethane)): uptake, distribution and retention of ^{35}S in skin and in blood after cutaneous application of ^{35}S-sulfur mustard in rat and comparison with human blood in vitro. Xenobiotica 1993;23:537–61.
[31] Black RM, Read RW. Improved methodology for the detection and quantitation of urinary metabolites of sulfur mustard using gas-chromatography-tandem mass spectrometry. J Chromatogr B Biomed Appl 1995;665:97–105.
[32] Black RM, Read RW. Biological fate of sulfur mustard, 1,1'-thiobis (2-chloroethane): identification of β -lyase metabolites and hydrolysis products in human urine. 1995b. Xenobiotica 1995;25:167–73.
[33] Black RM, Brewster K, Clarke RJ, Hambrook JL, Harrison JM, Howells DJ. Metabolism of thiodiglycol (2,2-thiobis-ethanol): isolation and identification of urinary metabolites following intraperitoneal administration to rat. Xenobiotica 1993;23:473–81.
[34] Brimfield AA. Possible protein phosphatase inhibition by bis (hydroxyethyl) sulfide, a hydrolysis product of mustard gas. Toxicol Lett 1995;78:43–8.
[35] Vodela JK, Angerhofer RA, Michie MW, Leach GJ, Reddy G. Effect of thiodiglycol (2,2'-thiodiethanol) on the glutathione antioxidant system in rat erythrocytes. Environ Nutri Interac 1999;3:85–93.
[36] Vodela JK, Angerhofer RA, Michie MW, Leach GJ, Reddy G. Effects of subchronic oral exposure of thiodiglycol on hepatic mixed-function oxidase and cytosolic glutathione antioxidant system in rats. Environ Nutri Interac 1999;3:207–16.
[37] Stankowski Jr LF. *Salmonella-Escherichia coli*/mammalian-microsome reverse mutation assay with confirmatory assay with 2,2'-thidiethanol. Vienna (VA): Covance Laboratories, Inc; 2001, Contract No. DAAD05-99-P-1522. Covance Study No. 22283-0409OECD. Report.
[38] Erexson LG. In vivo mouse micronucleus assay with 2,2'-thiodiethanol. Aberdeen Proving Ground (MD): U.S. Army Center for Health Promotion and Preventive Medicine; 2001, Contract No. DAAD-05-99-P-1522. Covance Study No. 22283-0-455OECD. 2001. Report. Prepared by: Covance Laboratories Inc., Vienna, VA.
[39] Tice RR, Donner M, Udumundi A, Vasquez M. In vitro chromosomal aberration study in Chinese Hamster Ovary (CHO) cells. Aberdeen Proving Ground (MD): U.S. Army Center for Health Promotion and Preventive Medicine; 1997, Contract No. DAD05-91-C-00018. Prepared by Integrated Laboratory Systems, Durham, NC.
[40] Clark SL, Donner ME. Mouse lymphoma mammalian mutagenesis assay. Aberdeen Proving Ground (MD): U.S. Army Center for Health Promotion and Preventive

Medicine; 1998, Contract No. DAD05-91-C-00018. Prepared by Integrated Laboratory Systems, Durham, NC.
[41] Muse WT, Anthony JS, Thompson SA, Crouse CL, Crouse LC. Acute inhalation toxicity of chemically neutralized HD in rats. Final report. Aberdeen Proving Ground (MD): Edgewood Chemical Biological Center. U.S. Army Soldier and Biological Chemical Command; 1997, ADA384252.ECBCTR-067.
[42] NIOSH Registry of Toxic Effects of Chemical Substances (RTECS). Union carbide data sheet [Internet]. Atlanta (GA): Centers for Disease Control and Prevention (US), National Institute for Occupational Safety and Health [cited 1971]. Available from: http://www.cdc.gov/niosh/rtecs/.
[43] Carpenter CP, Smyth HF. Chemical burns of the rabbit cornea. Am J Ophthal 1946;29:1363–72.
[44] Manthei JH, Heitkamp DH, Way RA, Bone DM. Toxicological evaluation/verification of decontamination procedures/products from alternative technologies for chemical demilitarization: products of mustard gas (HD) neutralization and/or hydrolysis. Technical report. Aberdeen Proving Ground (MD): Edgewood Chemical Biological Center. U.S. Army Soldier and Biological Chemical Command; 1997, ERDEC-TR-419.
[45] Magnusson B, Kligmanj AM. The identification of contact allergens by anima: assay: the guinea pig maximization test. J Invest Dermatol 1969;52:268–76.
[46] Reifenrath WG, Kammen HO, Palmer WG, Major MM, Leach JG. Percutaneous absorption of explosives and related carbon compounds: an empirical model of bioavailabilty of organic nitro compounds from soil. Toxicol Appl Pharmacol 2002;182 (2):160–8.

INDEX

Note: Page numbers followed by *f* indicate figures and *t* indicate tables.

E

F

H

L

M

N

O

P